# Methods in Enzymology

Volume 176
NUCLEAR MAGNETIC RESONANCE
Part A
Spectral Techniques and Dynamics

# METHODS IN ENZYMOLOGY

EDITORS-IN-CHIEF

## John N. Abelson    Melvin I. Simon

DIVISION OF BIOLOGY
CALIFORNIA INSTITUTE OF TECHNOLOGY
PASADENA, CALIFORNIA

FOUNDING EDITORS

## Sidney P. Colowick and Nathan O. Kaplan

*Methods in Enzymology*

*Volume 176*

# Nuclear Magnetic Resonance

*Part A*

*Spectral Techniques and Dynamics*

EDITED BY

## Norman J. Oppenheimer

SCHOOL OF PHARMACY
DEPARTMENT OF PHARMACEUTICAL CHEMISTRY
UNIVERSITY OF CALIFORNIA, SAN FRANCISCO
SAN FRANCISCO, CALIFORNIA

## Thomas L. James

SCHOOL OF PHARMACY
DEPARTMENT OF PHARMACEUTICAL CHEMISTRY
UNIVERSITY OF CALIFORNIA, SAN FRANCISCO
SAN FRANCISCO, CALIFORNIA

ACADEMIC PRESS, INC.
Harcourt Brace Jovanovich, Publishers

San Diego   New York   Berkeley   Boston
London   Sydney   Tokyo   Toronto

ACADEMIC PRESS, INC.
San Diego, California 92101

*United Kingdom Edition published by*
ACADEMIC PRESS LIMITED
24-28 Oval Road, London NW1 7DX


LIBRARY OF CONGRESS CATALOG CARD NUMBER:    54-9110


ISBN    0-12-182077-7    (alk. paper)


PRINTED IN THE UNITED STATES OF AMERICA
89  90  91  92     9  8  7  6  5  4  3  2  1

# Table of Contents

## Section I. Basic Techniques

## Section II. Advanced Techniques

# Contributors to Volume 176

Article numbers are in parentheses following the names of contributors.
Affiliations listed are current.

FRITS ABILDGAARD (15), *Department of Chemistry, The H. C. Ørsted Institute, University of Copenhagen, DK-2100 Copenhagen, Denmark*

ROBERT S. BALABAN (16), *Laboratory of Cardiac Energetics, National Heart, Lung and Blood Institute, National Institutes of Health, Bethesda, Maryland 20892*

AD BAX (7, 8), *Laboratory of Chemical Physics, National Institute of Diabetes and Digestive and Kidney Diseases, National Institutes of Health, Bethesda, Maryland 20892*

BRUCE A. BERKOWITZ (16), *Laboratory of Cardiac Energetics, National Heart, Lung and Blood Institute, National Institutes of Health, Bethesda, Maryland 20892*

BRANDAN A. BORGIAS (9), *Department of Pharmaceutical Chemistry, University of California, San Francisco, San Francisco, California 94143*

L. R. BROWN (11), *Research School of Chemistry, The Australian National University, Canberra, A.C.T. 2601, Australia*

G. HERBERT CAINES (20), *Department of Chemistry, University of California, Santa Cruz, California 95064*

WALTER J. CHAZIN (6), *Department of Molecular Biology, Research Institute of Scripps Clinic, La Jolla, California 92037*

CLAUDIO DALVIT (6), *F. Hoffman-La Roche & Co., Department of Central Research Units, CH-4002 Basel, Switzerland*

B. T. FARMER II (11), *NMR Instrument Division, Varian Associates, Palo Alto, California 94303*

JAMES A. FERRETTI (1), *Laboratory of Chemistry, National Heart, Lung and Blood Institute, National Institutes of Health, Bethesda, Maryland 20892*

HENRIK GESMAR (15), *Department of Chemistry, The H. C. Ørsted Institute, University of Copenhagen, DK-2100 Copenhagen, Denmark*

RONALD L. HANER (21), *Department of Chemistry, University of California, Santa Cruz, California 95064*

JEFFREY C. HOCH (12), *Rowland Institute for Science, Cambridge, Massachusetts 02142*

P. J. HORE (3), *Physical Chemistry Laboratory, Oxford University, Oxford OX1 3QZ, England*

THOMAS L. JAMES (9), *School of Pharmacy, Department of Pharmaceutical Chemistry, University of California, San Francisco, San Francisco, California 94143*

MAX A. KENIRY (19), *Research School of Chemistry, The Australian National University, Canberra, A.C.T. 2601, Australia*

JENS J. LED (15), *Department of Chemistry, The H. C. Ørsted Institute, University of Copenhagen, DK-2100 Copenhagen, Denmark*

ROBERT E. LONDON (18), *Laboratory of Molecular Biophysics, National Institute of Environmental Health Sciences, National Institutes of Health, Research Triangle Park, North Carolina 27709*

JOHN L. MARKLEY (2), *Department of Biochemistry, College of Agricultural and Life Sciences, University of Wisconsin, Madison, Wisconsin 53706*

COURTNEY F. MORGAN (20), *Department of Chemistry, University of California, Santa Cruz, California 95064*

NERI NICCOLAI (10), *Dipartimento di Chimica, Università di Siena, 53100 Siena, Italy*

S. J. OPELLA (13), *Department of Chemistry, University of Pennsylvania, Philadelphia, Pennsylvania 19104*

NORMAN J. OPPENHEIMER (4), *School of Pharmacy, Department of Pharmaceutical Chemistry, University of California, San Francisco, San Francisco, California 94143*

MARK RANCE (6), *Department of Molecular Biology, Research Institute of Scripps Clinic, La Jolla, California 92037*

B. D. NAGESWARA RAO (14), *Department of Physics, Indiana University-Purdue University at Indianapolis, Indianapolis, Indiana 46205*

HEINRICH RODER (22), *Department of Biochemistry and Biophysics, University of Pennsylvania, Philadelphia, Pennsylvania 19104*

PAUL RÖSCH (17), *Department of Biophysics, Max-Planck-Institut for Medical Research, D-6900 Heidelberg, Federal Republic of Germany*

CLAUDIO ROSSI (10), *Dipartimento di Chimica, Universita di Siena, 53100 Siena, Italy*

THOMAS SCHLEICH (20, 21), *Department of Chemistry, University of California, Santa Cruz, California 95064*

STEVEN W. SPARKS (7), *Bone Research Branch, National Institute of Dental Research, National Institutes of Health, Bethesda, Maryland 20892*

P. L. STEWART (13), *Department of Chemistry, University of Pennsylvania, Philadelphia, Pennsylvania 19104*

DENNIS A. TORCHIA (7), *Bone Research Branch, National Institute of Dental Research, National Institutes of Health, Bethesda, Maryland 20892*

GERHARD WAGNER (5), *Biophysics Research Division, University of Michigan, Ann Arbor, Michigan 48109*

GEORGE H. WEISS (1), *Physical Sciences Laboratory, Division of Computer Research and Technology, National Institutes of Health, Bethesda, Maryland 20892*

PETER E. WRIGHT (6), *Department of Molecular Biology, Research Institute of Scripps Clinic, La Jolla, California 92037*

# Preface

NMR spectroscopy has undergone a remarkable transformation in the past decade which has few precedents in science. What was previously considered to be primarily an analytical tool for the study of small molecules has blossomed into one for investigating structure and dynamics, ranging from whole organisms to the atomic level. It is a field that is undergoing change at an astounding rate, with nearly continuous publication of powerful new techniques that further extend the range of experimental applicability. The applications of NMR spectroscopy to biochemistry, in general, and to enzymology, in particular, have also undergone parallel expansion. To date, however, there has not been an extended presentation in *Methods in Enzymology* of these applications.

In Volumes 176 and 177 we have attempted to serve two primary functions. The first is to rectify past omissions by providing a general background of modern NMR techniques, with a specific focus on NMR techniques that pertain to proteins and enzymology. The second is to provide a "snapshot" of the current state-of-the-art in NMR experimental techniques. Our overall goal is to provide information to enable the reader to understand a given technique, to evaluate its strengths and limitations, to decide which is the best approach, and, finally, to design an experiment using the chosen technique to solve a problem.

This volume covers basic and advanced NMR techniques, including two-dimensional NMR, and methods for studying protein dynamics, including rate constants and molecular motions. Volume 177 covers protein modifications for NMR, including isotope labeling, techniques for protein structure determination, enzyme mechanism methodology, and means for examining enzyme activity *in vivo*.

Although some techniques may be superseded by yet newer procedures and other techniques will certainly appear, it is our hope that the methods presented in these volumes will be of continuing value in the design and execution of NMR experiments for solving problems in protein and enzyme structure, function, and dynamics.

NORMAN J. OPPENHEIMER
THOMAS L. JAMES

# METHODS IN ENZYMOLOGY

VOLUME XIII. Citric Acid Cycle
*Edited by* J. M. LOWENSTEIN

VOLUME XIV. Lipids
*Edited by* J. M. LOWENSTEIN

VOLUME XV. Steroids and Terpenoids
*Edited by* RAYMOND B. CLAYTON

VOLUME XVI. Fast Reactions
*Edited by* KENNETH KUSTIN

VOLUME XVII. Metabolism of Amino Acids and Amines (Parts A and B)
*Edited by* HERBERT TABOR AND CELIA WHITE TABOR

VOLUME XVIII. Vitamins and Coenzymes (Parts A, B, and C)
*Edited by* DONALD B. MCCORMICK AND LEMUEL D. WRIGHT

VOLUME XIX. Proteolytic Enzymes
*Edited by* GERTRUDE E. PERLMANN AND LASZLO LORAND

VOLUME XX. Nucleic Acids and Protein Synthesis (Part C)
*Edited by* KIVIE MOLDAVE AND LAWRENCE GROSSMAN

VOLUME XXI. Nucleic Acids (Part D)
*Edited by* LAWRENCE GROSSMAN AND KIVIE MOLDAVE

VOLUME XXII. Enzyme Purification and Related Techniques
*Edited by* WILLIAM B. JAKOBY

VOLUME XXIII. Photosynthesis (Part A)
*Edited by* ANTHONY SAN PIETRO

VOLUME XXIV. Photosynthesis and Nitrogen Fixation (Part B)
*Edited by* ANTHONY SAN PIETRO

VOLUME XXV. Enzyme Structure (Part B)
*Edited by* C. H. W. HIRS AND SERGE N. TIMASHEFF

VOLUME XXVI. Enzyme Structure (Part C)
*Edited by* C. H. W. HIRS AND SERGE N. TIMASHEFF

# Section I

# Basic Techniques

## [1] One-Dimensional Nuclear Overhauser Effects and Peak Intensity Measurements

By JAMES A. FERRETTI and GEORGE H. WEISS

For the purposes of the present chapter the nuclear Overhauser effect (NOE) can be regarded as a phenomenon in NMR measurements allowing an experimenter to derive estimates of internuclear distances in molecules as well as information about rotational correlation times. The formal definition of the NOE is the change in the integrated intensity of an NMR signal from a nuclear spin when a neighboring spin is saturated. Measurements of the internuclear distances in solution, $r_{ij}$, are quite difficult by techniques other than the NOE, hence its widespread application to the study of configurations of biologically interesting macromolecules. The use of two-dimensional pulse techniques, high-speed minicomputers, and high-field NMR spectrometers greatly facilitate the determination of structure of proteins or polynucleotides exceeding 20,000 Da in solution.

The measured signal in an NMR experiment is known as a free induction decay (FID) which, in the absence of noise, can be expressed as

$$M(t) = M_0 \cos(\omega_0 t) \exp(-t/T_2^*) \tag{1}$$

where $M_0$ is the initial magnetization, $\omega_0$ is the carrier frequency, and $T_2^*$ is an effective spin–spin relaxation time.[1] In Fourier transform NMR (FTNMR) one considers, in place of $M(t)$, its Fourier transform

$$\hat{M}(\omega) = \int_0^\infty M(t) e^{-i\omega t} \, dt \tag{2}$$

On inserting Eq. (1) into the integral and taking the real part of the result one finds

$$\hat{M}(\omega) = M_0 T_2^* / [1 + (\omega - \omega_0)^2 (T_2^*)^2] \tag{3}$$

which is a Lorentzian curve centered at $\omega = \omega_0$. In our chapter we will discuss only the case of a single, isolated peak.

[1] R. R. Ernst, G. Bodenhausen, and A. Wokaun, "Principles of Nuclear Magnetic Resonance in One and Two Dimensions." Oxford Univ. Press (Clarendon), London and New York, 1967.

METHODS IN ENZYMOLOGY, VOL. 176

NOE Measurements

To define the notion of the NOE in the context of FTNMR we first note that the area under the Lorentzian peak defined in Eq. (3) is

$$\int_{-\infty}^{\infty} \hat{M}(\omega) \, d\omega = \pi M_0 \tag{4}$$

Let $\pi M_0$ be the area under the peak in the absence of a saturating radio-frequency (rf) field, and let $\pi M_0'$ be the peak area in the presence of a saturating rf field applied to the resonance of a nearby nucleus. The NOE is defined in terms of the relative difference between the two peak areas, $\eta$:

$$\eta = (M_0' - M_0)/M_0 \tag{5}$$

as

$$\text{NOE} = 1 + \eta \tag{6}$$

In turn, the value of the NOE factor is related to molecular parameters of interest by

$$\text{NOE} = f(\tau_c)/r_{ij}^6 \tag{7}$$

where $f(\tau_c)$ is a function of the rotational correlation time that depends on the model under consideration, and $r_{ij}$ is the distance between the nucleus whose peak area is being measured and the nucleus being strongly irradiated.[2] The NOE factor, which was originally determined by the saturation of a neighboring spin with an rf field, is here defined as a measure of change in the steady-state description of a nuclear spin system resulting from an external perturbation. There are other types of transient NOE measurements which we discuss later.

To understand the significance of $\eta$ or the NOE factor one needs to understand some basic aspects of the theory of the relaxation of spins in NMR spectroscopy. Details of the theory are discussed in a number of papers and monographs,[3,4] but we summarize here a few salient points. Consider a system consisting of two dipolar-coupled spin-$\frac{1}{2}$ nuclei, where the dipolar coupling implies that the atoms are close together in space. Let $\alpha$ and $\beta$ be two possible orientations of the magnetization vector of each spin in a magnetic field. The energy levels of such a two-spin system

[2] J. H. Noggle and R. G. Schirmer, "The Nuclear Overhauser Effect." Academic Press, New York, 1971.
[3] J. K. M. Sanders and J. D. Mersh, *Prog. NMR Spectrosc.* **15**, 353 (1982).
[4] A. E. Derome, "Modern NMR Techniques for Chemistry Research." Pergamon, Oxford, 1987.

are illustrated in Fig. 1.[2] In Fig. 1 the $W$ values are spin-lattice transition rates so that $W_0$ is the transition rate for transfer between the mixed spin states, $\alpha\beta$ and $\beta\alpha$, $W_1$ is the rate at which only one of the spins changes state, and $W_2$ is the rate of transitions $\alpha\alpha \rightarrow \beta\beta$. The parameter $\eta$ is found in terms of these rates by solving the differential equations for the relaxation of a weakly coupled two-spin system at equilibrium. This solution implies the relation,[2]

$$\eta = (W_2 - W_0)/(W_0 + 2W_1 + W_2) \tag{8}$$

where the numerator is often referred to as the cross-relaxation term. The denominator will be inversely proportional to the dipolar spin-lattice relaxation time, $T_1$, which establishes the relation between $T_1$ and $\eta$. It is evident from Eq. (8) that NOE will be greater or less than 1 depending on whether $W_2$ or $W_0$ dominates the cross-relaxation process. The rate, $W_2$, will dominate in the case of small molecules in a nonviscous medium and the NOE factor will be greater than 1. When the molecules are large, exemplified by proteins and polynucleotides which tumble slowly in solution, $W_0$ will be the dominant relaxation term and the NOE factor will be less than 1.

There are a number of alternative techniques available for the measurement of NOE factors. The classical one-dimensional technique uses the difference in peak areas as indicated in Eq. (4), in which the reference spectrum is measured in the absence of saturating irradiation, or is measured in a region in which there are no resonances. Consequently, two peak area determinations are required for an NOE measurement. An

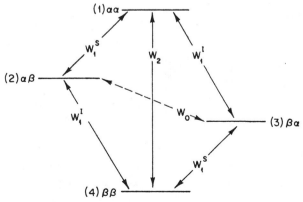

FIG. 1. Energy level diagram for two spins-$\frac{1}{2}$. The $W$ values are spin-lattice transition probabilities. The state of spin $I$ is listed first; e.g., $\alpha\beta$ means spin $I$ is $\alpha$ and spin $S$ is $\beta$. (From Noggle and Schirmer.[2])

alternative procedure is to subtract the time-dependent responses and then perform the Fourier transform. Empirically the latter procedure is often found to produce different spectra which are highly dispersive in character. This dispersive line shape is much more difficult to analyze and gives rise to unnecessarily large errors in the NOE estimate. Thus, we do not recommend this procedure, but rather prefer to make separate peak area measurements and then calculate the difference to obtain $\eta$. A time-dependent version of the experiment in which peak area differences are measured at various times after an irradiating field is applied produces a more direct estimate of the cross-relaxation terms.[5] An alternative version of this experiment uses measured peak areas after a steady state irradiating field is switched off. Both of these may be used to estimate values of $r_{ij}$,[6] but the data processing for the latter experiment is more complex.[6] More recently two-dimensional spectroscopy has been used[6] to measure the cross-relaxation behavior of a system involving a large number of peaks. We will not consider two-dimensional spectroscopy in this chapter, but rather restrict ourselves to a discussion of peak area estimation in the one-dimensional case. The spectra to be discussed may either be measured in the steady-state or transient experiments.

### Peak Intensity Measurements

The aim of any experiment is to measure parameters as accurately as possible within a reasonable amount of time. Both the technique for measuring peak areas and the choice of data analysis method are important for minimizing experimental uncertainties. From an experimental point of view, one must pay careful attention to the effects of the rf field on intensity measurements because it is possible to overload the spectrometer preamplifier and detection system, and because it is possible to partially saturate the resonances giving rise to the peaks. It is impossible to suggest an algorithm for minimizing the effects of systematic errors in peak intensity measurements. In addition to carrying out the necessary control experiments, one must pay close attention to such factors as the gain of the preamplifier and the noise figure of the receiver. Instrument manufacturers have tried to minimize systematic sources of error in the most recent generation of spectrometers.

Aside from potential systematic errors, at least four sources of error can be identified as having an effect on measurements of area. These

---

[5] S. Forsen and R. A. Hoffman, *J. Chem. Phys.* **39**, 2892 (1963).

[6] J. Jeener, B. H. Meier, P. Bachmann, and R. R. Ernst, *J. Chem. Phys.* **71**, 4646 (1979).

include (1) instrumental noise, which is generally modeled by adding a term to Eq. (1),

$$M(t) = M_0 \cos(\omega_0 t) \exp(-t/T_2^*) + n(t)$$

where $n(t)$ is random instrumental noise. It is generally assumed that this noise is unbiased, that is, the noise term averages to zero. An additional assumption is that the noise at any two times is uncorrelated, i.e., knowledge of the noise at one time tells one nothing about the value of the noise at any other time. The two assumptions about the nature of instrumental noise have not been substantiated by the results of experiments, but are usually made on the basis of theoretical convenience; (2) digitization error (that is, error due to the Fourier transform of the FID being available at uniformly spaced discrete frequencies only, rather than at all values of $\omega$); (3) truncation error [data are available only at a finite set of frequencies which cannot cover the entire range $(-\infty, \infty)$ in frequency]. There is still another type of truncation effect due to the necessity of collecting data for a finite, rather than an infinite amount of time; (4) unknown phase of a peak. This replaces the term $\cos(\omega_0 t)$ in Eq. (1) by $\cos(\omega_0 t + \phi)$, where $\phi$ is unknown. The phase error may manifest itself as a distorted baseline.

The resulting errors can be classified into two categories: (1) degradation of the precision of the estimate, and (2) bias, that is to say, error that occurs even in the absence of random noise. Bias in the estimate of the NOE factor is the error that results predominantly from truncation of the peak. Instrumental noise generally leads to a degradation in precision, whereas the last three types of errors listed above lead to a biased estimate of the area. However, by a proper choice of data-processing techniques, it is possible to partially overcome some effects of the errors just mentioned.

There are several investigations of the various types of error in NOE in the NMR literature. It has been shown, for example, that phase error can practically be eliminated by estimating the area in terms of the peak height multiplied by the half-width.[7] In a separate study, numerical integration and curve fitting were compared with respect to the accuracy and precision attainable in estimating peak areas.[8] In the case where the line shape is Lorentzian, curve fitting produces a more precise estimate of the peak area than does numerical integration. Curve fitting can be used only on presently available commercial instruments when the line shape is approximately Lorentzian. The technique of finding parameters by curve fitting implicitly takes into account the fact that the lineshape extends to

[7] G. H. Weiss, J. A. Ferretti, J. E. Kiefer, and L. Jacobson, *J. Magn. Reson.* **53,** 7 (1983).
[8] G. H. Weiss and J. A. Ferretti, *J. Magn. Reson.* **55,** 397 (1983).

infinity, thereby eliminating truncation error. In contrast, numerical integration requires that the measurement be truncated at finite values of frequency and implicitly assumes a fit by a polynomial. This procedure was shown to produce a greater bias or underestimation of the peak area than the method of curve fitting.

An important technique for minimizing error due to truncation of the signal in the time domain is through the use of apodization functions. These replace the measured signal, $M(t)$, by a new signal

$$M_1(t) = M(t)a(t) \qquad (9)$$

where $a(t)$ is known as the apodization function. Widely used apodization functions are the exponential $a(t) = \exp[-t/T]$ and the Gaussian $a(t) = \exp[-(t/T)^2]$, where $T$ is a parameter with the dimensions of time that is arbitrary but may be chosen to optimize some measure of accuracy. If $T$ is not too large, these apodization functions will effectively set the value of $M_1(t) = 0$ at a time at which $M(t)$ might still be measurable. This essentially eliminates truncation effects in the time domain. It is readily shown that, in the absence of bias and instrumental noise, the area under the peak generated by $M_1(t)$,

$$\hat{M}_1(\omega) = \int_{-\infty}^{\infty} M_1(t)e^{-i\omega t} \, dt \qquad (10)$$

is equal to the area under $\hat{M}(\omega)$ provided that $a(0) = 1$. However, when the frequency data are digitized and the resulting peak truncated, the estimate of area will no longer equal the true area. Thus, both bias and noise errors will generally occur in peak area measurements. It should also be noted that although $\hat{M}(\omega)$ is a Lorentzian, $\hat{M}_1(\omega)$ will not be Lorentzian unless the apodization function is a simple exponential.

Let us denote a true peak area by $A$ $(=\pi M_0)$ and the experimental peak area by $\hat{A}$. The estimate contains both bias and noise errors, which we later combine into a single parameter for the purpose of choosing an optimal apodization function. The measured area can be expressed as a sum of three components:

$$\hat{A} = A + B + \delta A \qquad (11)$$

where $A$ is the true area, $B$ is the bias, and $\delta A$ is the contribution from instrumental noise. Let the average value of $\hat{A}$ with respect to all possible noise be denoted by $\langle A \rangle$. Under the assumption that instrumental noise is unbiased, it follows that $\langle \delta A \rangle = 0$, and an expression for the bias can be written as

$$B = \langle \hat{A} \rangle - A \qquad (12)$$

As a measure of the noise, it is natural to choose the variance of $\delta A$, denoted by $\sigma^2(\delta A)$, which is given by

$$\sigma^2(\delta A) = \langle(\delta A)^2\rangle \tag{13}$$

Two investigations have appeared in the literature on the choice of apodization functions which take both bias and noise into account.[9,10] The first compares just two popular choices for the apodization function, the exponential and the Gaussian,

$$a_E(t) = \exp[-(t/T_E)] \tag{14a}$$
$$a_G(t) = \exp[-(t/T_G)^2] \tag{14b}$$

Let $\hat{M}_1(\omega)$ be available only at uniformly spaced points in $\omega$, the interval between adjacent points being $\Delta\omega$, and let the total number of measured points be $2m + 1$, $m$ being an integer (the total number of points was assumed to be odd for convenience in the analysis; no real restriction on the conclusions is implied by the assumption). The critical dimensionless parameter entering into the assessment of errors is

$$\varepsilon = [(2m + 1)T_2^*\Delta\omega/2]^{-1} \tag{15}$$

which is a measure of how well the peak is covered by the data points. This parameter is a simple consequence of the theory in Ref. 9. For the theory to be useful, it is important to work with values of $\varepsilon$ that are less than 0.1. Thus if $T_2^* = 1$ sec and $\Delta\omega = 0.5$ sec$^{-1}$, one would need 20 points to ensure the indicated value of $\varepsilon$. The condition of no truncation is equivalent to $m = \infty$ or $\varepsilon = 0$. It is intuitively evident that, in the presence of instrumental noise, there will be an optimal value of $\varepsilon$; if this parameter is too large, there will be too few points covering the peak while, if it is too small, much of the estimated area will be attributable to noise. Let $\beta_E$ be the dimensionless parameter $T_2^*/T_E$, and let $\beta_G$ be the analogous parameter for Gaussian apodization. These parameters can be set by the experimenter with some knowledge of $T_2^*$ for the particular molecule. The bias found in the case of exponential apodization is

$$B_E = 2M_0 \tan^{-1}[\varepsilon(1 + \beta_E)] \tag{16}$$

which is small whenever $\varepsilon$ is small. The expression for the bias in the case of Gaussian apodization is somewhat more complicated, but the important result is that when $\varepsilon$ is small the ratio of biases is

$$B_E/B_G \approx 1 + \beta_E \tag{17}$$

[9] G. H. Weiss, J. A. Ferretti, and R. A. Byrd, *J. Magn. Reson.* **71**, 97 (1987).
[10] G. H. Weiss and J. A. Ferretti, *Chemom. Intell. Lab. Syst.* **4**, 223 (1988).

This implies that the bias in the estimate of area will always be smaller when Gaussian apodization is used than it is for exponential apodization.

As we have mentioned, the bias is only one aspect of error that requires consideration when choosing an apodization function for use in the measurement of area. The second component of error is a reduction in the precision of the estimate. It is hardly likely that an experimenter will want to consider the two types of error separately, which suggests the use of a single criterion for the choice of experimental parameters, which combines both bias and precision. The choice of such a criterion is necessarily somewhat arbitrary. One such criterion that combines the bias and the variance of the estimate of area on an equal footing is based on a parameter $\Gamma$ defined by

$$\Gamma^2 = B^2 + \sigma^2(\delta A) \tag{18}$$

A comparison of exponential and Gaussian apodization functions has been carried out in terms of $\Gamma$ in Ref. 9. The results of this comparison are expressed in terms of the signal-to-noise ratio $S/N$, which is defined in terms of the root-mean-square average of the noise, $\sigma_n$, and the peak signal $M_0$, as $S/N = M_0/\sigma_n$. One finds, for the two types of apodization

$$\Gamma_E^2 = 4M_0^2(N/S)^{4/3}(1.924) \tag{19a}$$
$$\Gamma_G^2 = 4M_0^2(N/S)^{4/3}(0.750) \tag{19b}$$

where for exponential apodization we have set $\beta_E = 1$, which maximizes the signal-to-noise ratio of the peak. Thus, using this criterion that combines bias and precision, one finds that Gaussian apodization is to be preferred over exponential apodization for the estimation of area.

Some emphasis has been placed on exponential and Gaussian apodization functions because these options are included in software packages included with commercially available NMR spectrometers. It is possible to consider more general apodization functions of the form

$$a(t) = \exp[-(t/T)^\alpha] \tag{20}$$

which includes exponential and Gaussian apodization as the special cases $\alpha = 1$ and 2, respectively. The investigation of this more general apodization function was also based on the function $\Gamma$ whose definition is given in Eq. (18) of Ref. 10. The conclusions of the study can be summarized briefly: (1) The optimal value for $\varepsilon$ for all $\alpha$ is very close to

$$\varepsilon_m = 0.58(N/S)^{2/3} \tag{21}$$

(2) The value of $\Gamma$ can always be decreased by increasing the exponent $\alpha$. However, a prudent choice of that parameter will be between two and four. Lower values of $\alpha$ will increase the error in the estimate of area, and

larger values are quite sensitive to the experimenter's knowledge of $T_2^*$ (which is needed for the choice of the parameters $\varepsilon$ and $\beta$). (3) The parameter $\beta$ should be set equal to 1. Our discussion of the effects of apodization applies to the measurement of area under a single isolated peak in one-dimensional NMR measurements. A factor which might conceivably change some of these conclusions is baseline drift, but there is currently no analysis of errors taking this effect into account. Further work along present lines would be desirable for the case of overlapping peaks and two-dimensional NMR experiments.

There have been numerous reports in the literature using the NOE effect to determine internuclear distances. However, only one experimental study has been made of the variability of NOE factors and its effect on estimates of internuclear distances.[11] The results were reported in terms of the coefficient of variation of the estimate of $\eta$. This parameter, $C(\hat{\eta})$, is defined by

$$C(\hat{\eta}) = \sigma(\hat{\eta})/\hat{\eta} \tag{22}$$

where $\sigma(\hat{\eta})$ is the standard deviation of the estimate of $\eta$. The parameter $C(\hat{\eta})$ measures the fluctuations in the estimate relative to the estimate itself, and is therefore a dimensionless quantity. In measurements of the NOE factor made on vanillin, values of $C$ were found to be equal to 0.37 with a single pulse, 0.07 with 4 pulses, and 0.06 with 16 pulses. In general one expects that $C$ will vary with $n$, the number of pulses, as $C \approx n^{-1/2}$. One can convert the precision of the measured value of $\eta$ into an estimate for the precision of the measured value of internuclear distance by the relation[7]

$$\frac{\sigma(\hat{r})}{\hat{r}} = \frac{C(\hat{\eta})}{6(1 - 2\hat{\eta})} \tag{23}$$

For the measurements reported in Ref. 9 on vanillin, $\hat{\eta}$ was approximately equal to 0.2. Therefore, when $C = 0.07$, the relative standard deviation of the estimate of $r$ is approximately equal to 2%. These measurements indicate that the estimate of $r$ can be highly accurate provided that the signal-to-noise ratio is sufficiently large. In the reported experiments, that ratio was $\geq 100$. Similar results were obtained for measurements on 2,3,4-trichloroanisole.

[11] J. A. Ferretti, A. K. Weiss, and G. H. Weiss, *J. Magn. Reson.* **62**, 319 (1985).

## [2] Two-Dimensional Nuclear Magnetic Resonance Spectroscopy of Proteins: An Overview

*By* JOHN L. MARKLEY

### Introduction

NMR spectroscopy is the leading technique for obtaining structural and dynamic information at the atomic level about proteins in solution. It provides an efficient method for measuring dynamic, kinetic, and thermodynamic parameters (e.g., correlation times, order parameters, hydrogen exchange rates, and p$K_a$ values) in peptides and small proteins. NMR is the only serious competitor to single-crystal diffraction analysis for modeling three-dimensional structure of small proteins. NMR analysis is complementary to crystallography in providing a means of determining whether the protein structure is the same in solution as in the solid state.

X-ray diffraction and high-resolution NMR have different strengths and different experimental requirements. With X-ray crystallography, it is necessary first to obtain suitable protein crystals; the ultimate spatial resolution depends on properties of the crystal and the effort expended in refining the structure. An X-ray study provides few answers at initial stages. The major hurdles are crystallization, preparation of suitable heavy atom derivatives, and solution of the phasing problem. Once these have been overcome, the electron density map is amenable to analysis, except for regions of static or dynamic disorder. By contrast, NMR analysis can yield limited, but useful, molecular information from the first spectrum: a resolved coupling constant, for example, can indicate the existence of a particular chemical bond between two atoms, or a nuclear Overhauser effect (NOE)[1] can define the close proximity of two hydrogens in the molecule. The central problems in NMR spectroscopy are the resolution of signals from individual groups and their assignment in a sequence-specific and stereospecific manner (i.e., assignment to a particular atom or group of equivalent atoms such as methyl protons within a

---

[1] Abbreviations used are as follow: COSY, two-dimensional homonuclear correlated spectroscopy; $^{13}C\{^{13}C\}$DQC, two-dimensional $^{13}C$–$^{13}C$ double-quantum spectroscopy; SBC, single-bond correlated spectroscopy; HOHAHA, two-dimensional homonuclear Hartmann–Hahn spectroscopy; INADEQUATE, incredible natural abundance double-quantum transfer spectroscopy; MBC, multiple-bond correlated spectroscopy; NOE, nuclear Overhauser effect; NOESY, two-dimensional nuclear Overhauser effect spectroscopy; RELAY, two-dimensional relayed coherence spectroscopy; ROESY, two-dimensional rotating-frame (or transverse) NOE spectroscopy.

particular residue in the protein sequence). Relatively precise measurements of short ($<5$ Å) interproton distances can be made between any pair of hydrogens whose $^1$H NMR signals can be resolved and assigned. Dynamic and static disorder usually can be distinguished by NMR results, and NMR can provide rates of dynamic processes over a wide time scale (nanoseconds to seconds). In very large proteins, the sharpest and most easily resolved NMR signals come from less ordered regions; by contrast, mobile regions in crystals frequently do not yield interpretable electron densities. Other differences can result from differential solubility or differential crystallization: an NMR spectrum provides signals from all species present in solution with the signal strength proportional to their concentrations: an X-ray analysis shows the structures of those species that have crystallized from solution, and their relative abundance may differ from that of the original solution. As has been the case with several high-resolution X-ray structures, NMR analysis can provide evidence for multiple stable conformational forms of a protein. The effective molecular weight (hydrodynamic molecular weight at the protein concentration used) places limits on the degree of detail to be obtained from NMR analysis. At present, one can contemplate determining detailed solution structures of nonaggregating proteins of $M_r \leq 20,000$. Much more limited one-dimensional and two-dimensional investigations are being carried out with larger proteins ($M_r$ up to or in excess of 100,000); such studies will not be discussed here.

The great power of two-dimensional (2D) NMR methods[2,3] for protein spectroscopy stems from their ability to assist in solving the twin problems of resolution and assignment. Application of these methods[4] has led to nearly complete $^1$H resonance assignments in numerous small proteins (Table I), and extensive $^{13}$C[5,6] and $^{15}$N assignments[7,8] are proceeding rapidly. Once extensive assignments have been achieved, structural characterization based on homonuclear or heteronuclear 3-bond coupling con-

[2] R. R. Ernst, G. Bodenhausen, and A. Wokaun, "Principles of Nuclear Magnetic Resonance in One and Two Dimensions." Oxford Univ. Press (Clarendon), London and New York, 1987.

[3] H. Kessler, M. Gehrke, and C. Griesinger, *Angew. Chem. Int. Ed. Engl.* **27**, 490 (1988).

[4] K. Wüthrich, "NMR of Proteins and Nucleic Acids." Wiley (Interscience), New York, 1986.

[5] G. Wagner and D. Brühwiler, *Biochemistry* **25**, 5839 (1986).

[6] A. D. Robertson, G. I. Rhyu, W. M. Westler, G. I. Rhyu, and J. L. Markley, submitted for publication.

[7] G. Ortiz-Polo, R. Krishnamoorthi, J. L. Markley, D. H. Live, D. G. Davis, and D. Cowburn, *J. Magn. Reson.* **68**, 303 (1986).

[8] J. Glushka and D. Cowburn, *J. Am. Chem. Soc.* **109**, 7879 (1987).

TABLE I
LIST OF PROTEINS AND LARGE PEPTIDES FOR WHICH EXTENSIVE SEQUENCE-SPECIFIC PROTON NMR ASSIGNMENTS
OR NMR SOLUTION STRUCTURES HAVE BEEN DERIVED[a]

| Protein or peptide | Source | Extent of assignment[b] | References[c] | |
| --- | --- | --- | --- | --- |
| | | | Assignment | Solution structure |
| Acyl carrier protein | Escherichia coli | 77/77 | 1[d] | 79 |
| Acyl phosphatase | Rabbit | 58/98 | 94 | |
| Alamethicin | Trichoderma viride | 20/20 | 2 | |
| Anaphylatoxin C3a (des Arg-77) | Bovine | ~76/76 | 104 | 101,103 |
| Anaphylatoxin C5a | Bovine | 62/74 | 101 | |
| Antennapedia gene, homeodomain | Drosophila | 68/68 | 99 | |
| Anthopleurin-A | Anthopleura xanthogrammica | 33/49 | 3 | |
| Apamin | Apis mellifera | 18/18 | 4 | 86 |
| Atrial natriuretic factor (fragments 7–23) | Synthetic | 16/16 | 85 | 85 |
| Bacteriorhodopsin (fragments 34–65) | Synthetic | 32/32 | 111 | |
| Bombesin | Frog | 14/14 | 5 | |
| α-Bungarotoxin | Bungarus multicinctus | 74/74 | 6 | |
| Carboxypeptidase inhibitor | Potato | 38/39 | 7 | 7 |
| Cardiotoxin III | Formosan cobra | 60/60 | 8 | |
| Cardiotoxin | Naja mossambica mossambica | | | |
| CTXIIa | | 60/60 | 9 | |
| CTXIIb | | 60/60 | 9 | |
| Cecropin A | Hyalophora cecropia | 37/37 | 83 | 83 |
| Coat protein, micelle-bound | Pf1 bacteriophage | 11/46 | 10 | |
| Complement component C5a | Human | 63/74 | 76 | 76 |
| Conotoxin GI and analog | Conus geographus | 13/13 | 11 | |
| Crambin | Crambe abyssinica | 12/46 | 12 | |
| Cro repressor | Lambda phage | 65/66 | 13 | |

| Protein | Source | Assignment | Reference |
|---|---|---|---|
| Cytochrome b₅ (soluble fragment) | Bovine liver | | |
|  Oxidized | | 31/90 | 100 |
|  Reduced | | 31/90 | 100 |
| Cytochrome c | Horse heart | | |
|  Oxidized | | 60/104 | 75 |
|  Reduced | | 102/104 | 14 |
| | | 104/104 | |
| iso-1-Cytochrome c | Yeast | 45/108 | 96 |
| Dihydrofolate reductase | Lactobacillus casei | 32/162 | 15 |
| Eglin c | Leech | 70/70 | 16 |
| β-Endorphin | Human | 31/31 | 17 |
| Enterotoxin ST I (toxic domain) | Escherichia coli | 10/13 | 18 |
| Epidermal growth hormone | Murine | 53/53 | 19,20,84 |
| | Human | 48/48 | 21 |
| Ferredoxin I | | | 21 |
|  Oxidized | Anabaena 7120 vegetative form | 79/98 | 90 |
| Glucagon, micelle-bound | Human | 29/29 | 22 |
| Glycoprotein D-1, (antigenic domain) | Herpes simplex virus | 22/22 | 23 |
| Gramicidin A | Bacillus brevi | 15/15 | 24,25 |
| Growth hormone (fragments 96–133) | Bovine | 14/37 | 88 |
| Growth hormone releasing factor (fragment) | Human | 29/29 | 26 |
| Hemoglobin (α-chain) | Human | 17/141 | 27 |
| Hirudin | Hirudo medicinalis | 65/65 | 28, 29 |
| Histidine containing protein | Escherichia coli | 85/85 | 30, 110 |
| Histone H5 (globular domain) | Chicken | 79/79 | 31, 78 |
| Insectotoxin I₅A | Buthis eupeus | 35/35 | 32 |
| lac Repressor (headpiece) | Escherichia coli | 51/51 | 33, 34,35,108 |
| Leghemoglobin | | | |
|  I | Lupin | 95 | |
|  II | Lupin | 95 | |
|  II | Soybean | 95 | |

(continued)

TABLE I (continued)

| Protein or peptide | Source | Extent of assignment[b] | References[c] | |
| --- | --- | --- | --- | --- |
| | | | Assignment | Solution structure |
| Lysozyme | Chicken egg white | 121/129 | 36 | |
| | Human | 38/129 | 37 | |
| | T4 bacteriophage | 109 | | |
| α-Mating factor | Yeast | 13/13 | 38 | |
| Melittin, micelle-bound | Apis mellifera | 8/26 | 39 | |
| Metallothionein-2 | Rabbit liver | 62/62 | 40 | 41 |
| | Rat liver | 59/61 | 42 | 92 |
| Myelin basic protein, peptic peptide | Rabbit | 22/22 | 43 | |
| Myoglobin | Sperm whale | 30/153 | 44 | |
| α-Neurotoxin | Dendroapsis polylepsis polylepsis | 59/62 | 97 | |
| Neutrophil peptide 5 | Rabbit | 33/33 | 45 | |
| Ovomucoid third domain | Turkey egg white | | | |
| Virgin | | 56/56 | 46,47 | 77 |
| Reactive-site modified | | 56/56 | 47,48 | 77 |
| Pancreatic trypsin inhibitor | Bovine | | | |
| Native | | 58/58 | 48,50,51,52,107 | 52 |
| Circular | | 58/58 | 53 | |
| Parvalbumin, pI 5.0 | Esox lucius | 109/109 | 93 | 93 |
| Phoratoxin | Viscum album L. | 46/46 | 54 | 55 |
| Plastocyanin | French bean | >90/99 | 89 | |
| | Scenedesmus obliquus | 97/97 | 56 | 56,87 |
| | Spinach | 99/99 | 57 | |
| Proteinase inhibitor IIA | Bull seminal plasma | 57/57 | 58 | 59 |
| α1-Purothionin | Durum wheat | 45/45 | 60 | 61 |
| Ribonuclease A | Bovine pancreas | 121/124 | 62 | |

| | | | | |
|---|---|---|---|---|
| Ribonuclease T$_1$ | Aspergillus oryzae | 95/104 | 98 | |
| Ribosomal protein E-L30 | Escherichia coli | 56/58 | 63 | |
| Secretin | | 80 | 81 | |
| Serine proteinase inhibitor 2 | Barley | 82 | 82 | |
| Staphylococcal nuclease Ca$^{2+}$· pdTp ternary complex | Staphylococcus aureus (Foggi strain enzyme as "wild type") | | | |
| Recombinant wild type (with 7-residue N-extension) | | 129/156 | 105 | |
| Recombinant mutant H124L | | 121/149 | 106 | |
| Subtilisin inhibitor | Barley | 65/83 | 64 | 64 |
| Tendamistat | Streptomyces tendae | 74/74 | 65 | 66 |
| Thioredoxin | Escherichia coli | 101/108 | 67 | |
| Toxin ATX-Ia | Anemonia sulcata | 46/46 | 68 | |
| Toxin II | Radianthus paumotensis | 48/48 | 69 | |
| Transforming growth factor α | Human | 50/50 | 91 | |
| Trypsin inhibitor B-III | Peanut | 17/61 | 70 | |
| Trypsin inhibitor | Dendroapsis polylepsis polylepsis | | | |
| E | | 57/57 | 71 | |
| K | | 57/57 | 72 | |
| Ubiquitin | Human | 76/76 | 73,74 | |

[a] Adapted and updated from E. L. Ulrich, J. L. Markley, and Y. Kyogoku, *Protein Seq. Data Anal.* **2**, 23 (1989). This table is part of a database of NMR parameters and structural information being maintained at the National Magnetic Resonance Facility at Madison, Wisconsin. The list is updated as new protein data are published or brought to our attention.

[b] Approximate number of assigned amino acids (at least backbone assignments)/total number of amino acids in the molecule or fragment.

[c] Key to references: References to a solution structure indicate that a three-dimensional structure has been determined using NMR data and an algorithm that can provide a quantitative evaluation of the quality of the structure. Secondary and tertiary structure information is often reported in papers that list sequence specific assignments. (1) T. A. Holak and J. H. Prestegard, *Biochemistry* **25**, 5766 (1986); (2) G. Esposito, J. A. Carver, J. Boyd, and I. D. Campbell, *Biochemistry* **26**, 1043 (1987); (3) P. R. Gooley and R. S. Norton, *Eur. J. Biochem.* **153**, 529 (1985); (4) D. E. Wemmer and N. R. Kallenbach, *Biochemistry* **22**, 1901 (1983); (5) J. A. Carver, *Eur. J. Biochem.* **168**, 193 (1987); C. DiBello, L. Gozzini, M. Tonellato, M. G. Corradini, G. D'Auria, L. Paolillo, and E. Trivellone, *FEBS Lett.* **237**, 85 (1988); (6) V. Basus, M. Billeter, R. A. Love, R. M. Stroud, and I. D. Kuntz, *Biochemistry* **27**, 2763 (1988); (7) G. M. Clore, A. M. Gronenborn, M. Nilges, and

*(continued)*

Footnotes to TABLE I (*continued*)

C. A. Ryan, *Biochemistry* **26**, 8012 (1987); (8) C. Wu and T. H. Hseu, *28th Exp. NMR Conf.*, 1987; (9) G. Otting, W. E. Steinmetz, P. E. Bougis, H. Rochat, and K. Wüthrich, *Eur. J. Biochem.* **168**, 609 (1987); (10) R. A. Schiksnis, M. J. Bogusky, P. Tsang, and S. J. Opella, *Biochemistry* **26**, 1373 (1987); (11) Y. Kobayashi, T. Ohkubo, Y. Kyogoku, Y. Nishiuchi, S. Sakakibara, W. Braun, and N. Gō, *in* "Peptide Chemistry 1985" (Y. Kiso, ed.), Protein Research Foundation, Osaka, 1986, p. 259; (12) J. T. J. Lecomte, A. De Marco, and M. Llinas, *Biochim. Biophys. Acta* **703**, 223 (1982); (13) P. L. Weber, D. E. Wemmer, and B. R. Reid, *Biochemistry* **24**, 4553 (1985); (14) A. J. Wand, D. L. DiStefano, Y. Feng, H. Roder, and S. W. Englander, *Biochemistry* **28**, 186 (1989); (15) S. J. Hammond, B. Birdsall, M. S. Searle, G. C. K. Roberts, and J. Feeney, *J. Mol. Biol.* **188**, 81 (1986); (16) S. G. Hyberts, W. Marki, and G. Wagner, *Eur. J. Biochem.* **164**, 625 (1987); (17) O. Lichtarge, O. Jardetzky, and C. H. Li, *Biochemistry* **26**, 5916 (1987); (18) J. Gariepy, A. Lane, F. Frayman, D. Wilbur, W. Robien, G. K. Schoolnik, and O. Jardetzky, *Biochemistry* **25**, 7854 (1986); (19) G. T. Montelione, K. Wüthrich, H. E. C. Nice, A. W. Burgess, and H. A. Scheraga, *Proc. Natl. Acad. Sci. U.S.A.* **83**, 8594 (1986); G. T. Montelione, H. A. Scheraga, *Biochemistry* **27**, 2235 (1988); (20) K. H. Mayo and C. Burke, *Eur. J. Biochem.* **169**, 201 (1987); (21) R. M. Cooke, A. J. Wilkinson, M. Baron, A. Pastore, M. J. Tappin, I. D. Campbell, H. Gregory, and B. Sheard, *Nature (London)* **327**, 339 (1987); (22) G. Wider, K. H. Lee, and K. Wüthrich, *J. Mol. Biol.* **155**, 367 (1982); (23) M. P. Williamson, M. J. Hall, and B. K. Handa, *Eur. J. Biochem.* **158**, 527 (1986); (24) A. S. Arseniev, V. F. Bystrov, V. T. Ivanov, and Y. A. Ovchinnikov, *FEBS Lett.* **165**, 51 (1984); (25) G. E. Hawkes, L.-Y. Lian, E. W. Randall, K. D. Sales, and E. H. Curzon, *Eur. J. Biochem.* **166**, 437 (1987); (26) G. M. Clore, S. R. Martin, and A. M. Gronenborn, *J. Mol. Biol.* **191**, 553 (1986); (27) D. Dalvit and P. E. Wright, *J. Mol. Biol.* **194**, 329 (1987); (28) D. K. Sukumaran, G. M. Clore, A. Preuss, J. Zarbock, and A. M. Gronenborn, *Biochemistry* **26**, 333 (1987); (29) G. M. Clore, D. K. Sukumaran, M. Nilges, J. Zarbock, and A. M. Gronenborn, *EMBO J.* **6**, 529 (1987); (30) R. E. Klevit, G. P. Drobny, and E. B. Waygood, *Biochemistry* **25**, 7760 (1986); (31) J. Zarbock, G. M. Clore, and A. M. Gronenborn, *Proc. Natl. Acad. Sci. U.S.A.* **83**, 7628 (1986); (32) A. S. Arseniev, V. I. Kondakov, V. N. Maiorov, and V. F. Bystrov, *FEBS Lett.* **165**, 57 (1984); (33) E. R. P. Zuiderweg, R. Kaptein, and K. Wüthrich, *Eur. J. Biochem.* **137**, 279 (1983); (34) E. R. P. Zuiderweg, R. M. Scheek, R. Boelens, W. F. van Gunsteren, and R. Kaptein, *Biochimie* **67**, 707 (1985); (35) R. Kaptein, E. R. P. Zuiderweg, R. M. Scheek, R. Boelens, and W. F. van Gunsteren, *J. Mol. Biol.* **182**, 179 (1985); (36) C. Redfield and C. M. Dobson, *Biochemistry* **27**, 122 (1988); (37) J. Boyd, C. M. Dobson, and C. Redfield, *Eur. J. Biochem.* **153**, 383 (1985); (38) K. Wakamatsu, A. Okada, M. Suzuki, T. Higashijima, Y. Masui, S. Sakakibara, and T. Miyazawa, *Eur. J. Biochem.* **154**, 607 (1986); (39) L. R. Brown, W. Braun, Anil Kumar, and K. Wüthrich, *Biophys. J.* **37**, 19 (1982); (40) G. Wagner, D. Neuhaus, E. Wörgötter, M. Vašák, J. H. R. Kägi, and K. Wüthrich, *Eur. J. Biochem.* **157**, 275 (1986); (41) W. Braun, G. Wagner, E. Wörgötter, M. Vašák, J. H. R. Kägi, and K. Wüthrich, *J. Mol. Biol.* **187**, 125 (1986); A. Arseniev, P. Schultze, E. Wörgötter, W. Braun, G. Wagner, M. Vašák, J. H. R. Kägi, and K. Wüthrich, *J. Mol. Biol.* **201**, 637 (1988); (42) E. Wörgötter, G. Wagner, M. Vašák, J. H. R. Kägi, and K. Wüthrich, *Eur. J. Biochem.* **167**, 457 (1987); (43) E. Nygaard, G. L. Mendz, W. J. Moore, and R. E. Martenson, *Biochemistry* **23**, 4003 (1984); (44) C. Dalvit and P. E. Wright, *J. Mol. Biol.* **194**, 313 (1987); (45) A. C. Bach, M. E. Selsted, and A. Pardi, *Biochemistry* **26**, 4389 (1987); (46) A. D. Robertson, W. M. Westler, and J. L. Markley, *Biochemistry* **27**, 2519 (1988); (47) A. D. Robertson, G. I. Rhyu, W. M. Westler, and J. L. Markley, submitted for publication; (48) G. I. Rhyu and J. L. Markley, *Biochemistry* **27**, 2529 (1988); (49) G. Wagner and K. Wüthrich, *J. Mol. Biol.* **155**, 347 (1982); (50)

G. Wagner and D. Brühwiler, *Biochemistry* **25**, 5839 (1986); (51) J. Glushka and D. Cowburn, *J. Am. Chem. Soc.* **109**, 7879 (1987); (52) G. Wagner, W. Braun, T. F. Havel, T. Schaumann, N. Gō, and K. Wüthrich, *J. Mol. Biol.* **196**, 611 (1987); (53) W. J. Chazin, D. P. Godenberg, T. E. Creighton, and K. Wüthrich, *Eur. J. Biochem.* **152**, 429 (1985); (54) J. T. J. Lecomte, D. Kaplan, M. Llinas, E. Thunberg, and G. Samuelsson, *Biochemistry* **26**, 1187 (1987); (55) G. M. Clore, D. K. Sukumaran, M. Nilges, and A. M. Gronenborn, *Biochemistry* **26**, 1732 (1987); (56) J. Moore, W. J. Chazin, R. Powls, and P. E. Wright, *Biochemistry* **27**, 7806 (1988); (57) P. C. Driscoll, H. A. O. Hill, and C. Redfield, *Eur. J. Biochem.* **170**, 279 (1987); (58) P. Štrop, G. Wider, and K. Wüthrich, *J. Mol. Biol.* **166**, 641 (1983); (59) M. P. Williamson, T. F. Havel, and K. Wüthrich, *J. Mol. Biol.* **182**, 295 (1985); S. A. Sherman, A. M. Andrianov, and A. A. Akhrem, *J. Biomol. Struct. Dyn.* **5**, 785 (1988); (60) G. M. Clore, D. K. Sudumaran, A. M. Gronenborn, M. M. Teeter, M. Whitlow, and B. L. Jones, *J. Mol. Biol.* **193**, 571 (1987); (61) G. M. Clore, M. Nilges, D. K. Sukumaran, A. T. Brünger, M. Karplus, and A. M. Gronenborn, *EMBO J.* **5**, 2729 (1986); (62) A. D. Robertson, E. O. Purisima, M. A. Eastman, and H. A. Scheraga, *Biochemistry* **28**, in press (1989); (63) F. J. M. van de Ven and C. W. Hilbers, *J. Mol. Biol.* **192**, 389 (1986); (64) F. M. Poulsen, M. Kjaer, and M. Clore, *Exp. NMR Conf.*, 1987; (65) A. D. Kline and K. Wüthrich, *J. Mol. Biol.* **192**, 869 (1986); (66) A. D. Kline, W. Braun, and K. Wüthrich. *J. Mol. Biol.* **189**, 377 (1986); A. D. Kline, W. Braun, and K. Wüthrich, *J. Mol. Biol.* **204**, 675 (1988); (67) D. M. LeMaster and F. M. Richards, *Biochemistry* **27**, 142 (1988); (68) H. Widmer, G. Wagner, H. Schweitz, M. Lazdunski, and K. Wüthrich, *Eur. J. Biochem.* **171**, 177 (1988); (69) D. E. Wemmer, N. V. Kumar, R. M. Metrione, M. Lazdunski, G. Drobny, and N. R. Kallenbach, *Biochemistry* **25**, 6842 (1986); (70) S. Koyama, Y. Kobayashi, S. Norioka, Y. Kyogoku, and T. Ikenaka, *Biochemistry* **25**, 8076 (1986); (71) A. S. Arseniev, G. Wider, F. J. Joubert, and K. Wüthrich. *J. Mol. Biol.* **159**, 323 (1982); (72) R. M. Keller, R. Baumann, E.-H. Hunziker-Kwik, F. J. Joubert, and K. Wüthrich, *J. Mol. Biol.* **163**, 623 (1983); (73) D. L. DiStefano and A. J. Wand, *Biochemistry* **26**, 7272 (1987); (74) P. L. Weber, S. C. Brown, and L. Mueller, *Biochemistry* **26**, 7282 (1987); (75) Y. Feng, H. Roder, S. W. Englander, A. J. Wand, and D. L. DiStefano, *Biochemistry* **28**, 195 (1989); (76) E. R. P. Zuiderweg. D. G. Nettesheim, K. W. Mollison, and G. W. Carter, *Biochemistry* **28**, 172 (1989); (77) W. F. Walkenhorst, A. M. Krezel, P. Darba, A. D. Robertson, G. I. Rhyu, and J. L. Markley, "The Process of Protein Folding," AAAS Annual Meeting, 1989, Abstr. No. 25; (78) G. M. Clore, A. M. Gronenborn, M. Nilges, D. K. Sukumaran, and J. Zarbock, *EMBO J.* **6**, 1833 (1977); (79) T. A. Holak, M. Nilges, J. H. Prestegard, A. M. Gronenborn, and G. M. Clore, *Eur. J. Biochem.* **175**, 9 (1988); T. A. Holak, S. K. Kearsley, Y. Kim, and J. H. Prestegard, *Biochemistry* **27**, 6135 (1988); T. A. Holak, M. Nilges, and H. Oschkinat, *FEBS Lett.* **242**, 218 (1989); (80) A. M. Gronenborn, G. Boverman, and G. M. Clore, *FEBS Lett.* **215**, 88 (1987); (81) G. M. Clore, M. Nilges, A. T. Brünger, and A. M. Gronenborn, *Eur. J. Biochem.* **171**, 479 (1988); (82) G. M. Clore, A. M. Gronenborn, M. Kjaer, and F. M. Poulsen, *Protein Engineering* **1**, 305 (1987); (83) T. A. Holak, Å. Engström, B. J. Kraulis, G. Lindeberg, H. Bennich, T. A. Jones, A. M. Gronenborn, and G. M. Clore, *Biochemistry* **27**, 7620 (1988); (84) D. Kohda and F. Inagaki, *J. Biochem. (Tokyo)* **103**, 554 (1988); (85) E. T. Olejniczak, R. T. Gampe, Jr., T. W. Rockway, and S. W. Fesik, *Biochemistry* **27**, 7124 (1988); (86) J. H. B. Pease and D. E. Wemmer, *Biochemistry* **27**, 8491 (1988); (87) J. M. Moore, D. A. Case, W. H. Chazin, G. P. Gippert, T. F. Havel, R. Powls, and P. E. Wright, *Science* **240**, 314 (1988); (88) P. R. Gooley, S. A. Carter, P. E. Fagerness, and N. E. MacKenzie, *Proteins: Struct. Funct. Genet.* **4**, 48 (1988); (89) W. J. Chazin and P. E. Wright, *J. Mol. Biol.* **202**, 623 (1988); (90) B. H. Oh and J. L. Markley, *30th Exp. NMR Conf.*, 1989; (91) S. C. Brown, L. Mueller, and

*(continued)*

Footnotes to TABLE I (continued)

P. W. Jeffs, Biochemistry 28, 593 (1989); (92) P. Schultze, E. Wörgötter, W. Braun, G. Wagner, M. Vašák, J. H. R. Kägi, and K. Wüthrich, J. Mol. Biol. 203, 251 (1988); (93) A. Padilla, A. Cavé, and J. Parello. J. Mol. Biol. 204, 995 (1988); (94) V. Saudek, R. J. P. Williams, and G. Ramponi, J. Mol. Biol. 199, 229 (1988); (95) S. S. Narula, C. Dalvit, C. A. Appleby, and P. E. Wright. Eur. J. Biochem. 178, 419 (1988); (96) G. J. Pielak, J. Boyd, G. R. Moore, and R. J. P. Williams, Eur. J. Biochem. 177, 167 (1988); (97) A. L. Labhardt, E.-H. Hunziker-Kwik, and K. Wüthrich, Eur. J. Biochem. 177, 295 (1988); (98) E. Hoffmann and H. Rüterjans, Eur. J. Biochem. 177, 539 (1988); (99) G. Otting, Y. Qian, M. Müller, M. Affolter, W. Gehring, and K. Wüthrich, EMBO J. 7, 4305 (1988); (100) N. C. Veitch, D. W. Concar, R. J. P. Williams, and D. Whitford, FEBS Lett. 238, 49 (1988); (101) J. Zarbock, R. Gennaro, D. Romero, G. M. Clore, and A. M. Gronenborn, FEBS Lett. 238, 289 (1988); (102) A. E. Torda, B. C. Mabbutt, W. F. van Gunsteren, and R. S. Norton, FEBS Lett. 239, 266 (1988); (103) E. R. P. Zuiderweg, J. Senkin, K. W. Mollison, G. W. Carter, and J. Greer, Proteins: Struct. Funct. Genet. 3, 139 (1988); (104) D. G. Nettesheim, R. P. Edalji, K. W. Mollison, J. Greer, and E. R. P. Zuiderweg, Proc. Natl. Acad. Sci. U.S.A. 85, 5036 (1988); (105) D. Torchia, S. W. Sparks, and A. Bax, J. Am. Chem. Soc. 110, 2320 (1988); (106) J. Wang, D. M. LeMaster, and J. L. Markley. 30th Exp. NMR Conf., 1989; (107) E. Tuchsen and P. W. Hansen, Proteins: Struct. Funct. Genet. 3, 209 (1988); (108) J. DeVlieg, R. M. Scheek, W. F. van Gunsteren, H. J. C. Berendsen, and J. Thomason, Proteins: Struct. Funct. Genet. 5, 21 (1987); (109) L. P. McIntosh, F. W. Dahlquist, and A. G. Redfield, J. Biomol. Struct. Dynam. 5, 32 (Abstr. No. CA403) (1989); (110) R. E. Klevit and E. B. Waygood, Biochemistry 25, 7774 (1986); (111) A. S. Biochem. Suppl. 13A, 32 (Abstr. No. CA403) (1989); (110) R. E. Klevit and E. B. Waygood, Biochemistry 25, 7774 (1986); (111) A. S. Arseniev, I. V. Maslennikov, V. F. Bystrov, A. T. Kozhick, V. T. Ivanov, and Y. A. Ovchinnikov, FEBS Lett. 231, 81 (1988).

stants ($^3J$) and $^1H-^1H$ nuclear Overhauser enhancements (NOE's) follows fairly readily (Table I).

The wide selection of 2D NMR methods currently available for solution studies sort into three categories based on the physical mechanism mediating interactions between spins: cross-relaxation ($\sigma_{ij}$), chemical exchange, or spin–spin coupling ($^nJ_{ij}$) (Table II). The cross-relaxation experiment most important for protein studies is 2D $^1H\{^1H\}$ nuclear Overhauser enhancement spectroscopy, NOESY and ROESY discussed below; the same kinds of pulse sequences employed to determine NOE's are used to investigate chemical exchange phenomena. The most important nuclei involved in spin–spin couplings in proteins are $^1H$, $^{13}C$, and $^{15}N$, and the interactions may be homonuclear or heteronuclear in nature. To date, the vast majority of 2D NMR studies of proteins have involved $^1H\{^1H\}$ homonuclear interactions, because of the high sensitivity of the proton and its high natural abundance. Initial $^1H-^{13}C$ heteronuclear single-bond correlation 2D NMR studies of proteins employed $^{13}C$ detection ($^{13}C\{^1H\}SBC$) with $^{13}C$-enriched proteins[9-11] or at natural abundance.[12] Proton–$X$–nucleus heteronuclear correlation experiments are performed more efficiently, however, by detecting $^1H$. It is feasible to detect the less sensitive nuclei in proteins, $^{13}C$ and $^{15}N$, via their couplings to $^1H$, even at their natural abundance level (1.1% for $^{13}C$, 0.37% for $^{15}N$) (for a review, see Ref. 13). The selectivity and sensitivity of such experiments can be improved by $^{13}C$ or $^{15}N$ labeling. Isotope labeling also can enable 2D NMR approaches that take advantage of coupling between insensitive nuclei such as $^{13}C-^{15}N$[14-16] and $^{13}C-^{13}C$.[17-21]

This chapter presents a general overview of 2D NMR spectroscopy as it is practiced in our laboratory, the National Magnetic Resonance Facil-

[9] T.-M. Chan and J. L. Markley, *J. Am. Chem. Soc.* **104**, 4010 (1982).

[10] T.-M. Chan and J. L. Markley, *Biochemistry* **22**, 5996 (1983).

[11] C. L. Kojiro and J. L. Markley, *FEBS Lett.* **162**, 52 (1983).

[12] W. M. Westler, G. Ortiz-Polo, and J. L. Markley, *J. Magn. Reson.* **58**, 354 (1984).

[13] R. H. Griffey and A. G. Redfield, *Q. Rev. Biophys.* **19**, 51 (1987).

[14] M. Kainosho and T. Tsuji, *Biochemistry* **21**, 6273 (1982).

[15] W. M. Westler, B. J. Stockman, Y. Hosoya, Y. Miyake, and M. Kainosho, and J. L. Markley, *J. Am. Chem. Soc.* **110**, 6256 (1988).

[16] E. S. Mooberry, B.-H. Oh, and J. L. Markley, *J. Magn. Reson.*, in press (1989).

[17] B. H. Oh, W. M. Westler, P. Darba, and J. L. Markley, *Science* **240**, 908 (1988).

[18] B. J. Stockman, W. M. Westler, P. Darba, and J. L. Markley, *J. Am. Chem. Soc.* **110**, 4096 (1988).

[19] W. M. Westler, M. Kainosho, H. Nagao, N. Tomonaga, and J. L. Markley, *J. Am. Chem. Soc.* **110**, 4093 (1988).

[20] B. J. Stockman, M. D. Reily, W. M. Westler, E. L. Ulrich, and J. L. Markley, *Biochemistry* **28**, 230 (1989).

[21] B.-H. Oh, W. M. Westler, and J. L. Markley, *J. Am. Chem. Soc.* **111**, in press (1989).

ity at Madison, Wisconsin. The protein NMR field is in an exciting stage of rapid development; as a consequence, the methodology is in a state of flux. Many useful 2D NMR experiments are not covered here (newer variations on proton homonuclear correlated spectroscopy, isotope-filtered correlation spectroscopy, isotope-directed nuclear Overhauser effect spectroscopy, proton homonuclear multiple quantum spectroscopy, etc.). The reader is directed to other chapters in this volume for discussions of these techniques. The reader also is referred to monographs on 2D NMR and its applications to proteins.[2–4,22–27] The limited scope of this chapter does not permit a full discussion of numerous "tricks of the trade," such as optimization of the hardware configuration, uses of composite pulses, phase-cycling strategies, and details of processing and postprocessing operations. Since many of these are instrumentation dependent, the operating manuals supplied with the spectrometer or data workstation provide a good source of information.

*Instrumentation.* The major manufacturers of high-resolution NMR spectrometers (Bruker, General Electric, JEOL, and Varian) provide NMR hardware and software capable of carrying out routine 2D NMR experiments. Most modern spectrometers can perform standard variations of homonuclear COSY, RELAY, and NOESY experiments without modification. Rotating frame experiments (e.g., HOHAHA and ROESY) and heteronuclear 2D NMR experiments may require specialized electronic modules. Heteronuclear 2D experiments involving proton detection usually benefit from having a probe with the $^1H$ coil nearest to the sample (for optimal detection sensitivity) and the heteronuclear "$X$" coil outside. This geometry is the inverse of the normal heteronuclear probe architecture; hence these probes and the experiments in which they are used frequently are termed "inverse" or "reverse." The successful implementation of such experiments, especially at natural abundance, requires excellent electronics stability and clean transmitter frequency sources and signal amplification. $^{13}C$–$^{15}N$ 2D experiments require a probe tuned to four frequencies: $^{13}C$, $^{15}N$, $^1H$ (for decoupling or polarization transfer), and $^2H$ (for field-frequency lock).[15]

[22] A. Bax, "Two-Dimensional Nuclear Magnetic Resonance in Liquids." Reidel, London, 1982.

[23] O. W. Sørensen, Ph.D. Thesis, ETH Zürich (1984).

[24] N. Chandrakumar and S. Subramanian, "Modern Techniques in High Resolution FT NMR." Springer-Verlag, New York, 1987.

[25] A. Bax and L. Lerner, *Science* **232**, 960 (1986).

[26] D. E. Wemmer and B. R. Reid, *Annu. Rev. Phys. Chem.* **36**, 105 (1985).

[27] J. L. Markley, *in* "Protein Engineering" (D. L. Oxender and C. F. Fox, eds.), p. 15. Liss, New York, 1987.

*Background Information.* Two-dimensional NMR spectroscopy is an elaboration of the more familiar, one-dimensional (1D) pulse-Fourier transform experiment.[2] It is worth emphasizing that 1D methods have advantages over 2D methods for certain applications because of the higher speed of data acquisition. In 1D NMR, a short intense pulse at a single radio frequency (the transmitter pulse) serves to excite spins from a particular isotope (for example, all protons or all carbons). The response of the sample following the pulse, as manifested by the current generated in the receiver coil (the free induction decay, FID), is amplified, detected against the frequency of the transmitter, converted from analog to digital form, and then stored in computer memory or on a disk as amplitude vs time ($t_2$). The sequence can be repeated after a suitable delay which allows the spins to return toward equilibrium. Free induction decays, digitized during the *acquisition period,* are combined to improve the signal-to-noise ratio which increases as the square root of the number of transients averaged. Signal processing (baseline correction, zero filling, apodization, convolution, or a combination of these) of the averaged time-domain data can be carried out to increase resolution (at the expense of the signal-to-noise ratio) or to increase the signal-to-noise ratio (at the expense of resolution). The damped oscillatory components of the free induction decay are separated by Fourier transformation to give spectral peaks at their characteristic frequencies. Computational alternatives to Fourier transform analysis are discussed by Hoch [12], this volume.

A 2D NMR experiment differs from a 1D experiment by the addition to the pulse sequence of one or more transmitter pulses and one delay ($t_1$) that is incremented from one acquisition (or combined set of acquisitions) to the next. In general terms, the time axis divides into a *preparation period,* an *evolution period* (the time period that is incremented), a *mixing period,* and an *acquisition period.* As in 1D NMR, the data are digitized as a function of time, $t_2$. The 2D NMR raw data set consists initially of a series of these averaged FID's stored in different computer files, each having a different evolution time $t_1$. Hence the data can be viewed as a function of two time variables, $A(t_1, t_2)$. For each block with a given $t_1$, a Fourier transform is carried out with respect to the $t_2$ time axis. A second Fourier transform is carried out with respect to $t_1$. The result is spectral intensity as a function of two frequencies, $I(f_1, f_2)$. The additional frequency axis allows the correlation of magnetic properties of one nucleus with those of one or more other nuclei that interact with it as developed during the mixing time. Resolution in the $f_2$ dimension is determined by the digitization during the $t_2$ period; resolution in the $f_1$ dimension is limited by the number of incremented $t_1$ values.

Quadrature detection normally is used in acquiring data in the $t_2$ dimension to enhance sensitivity. It often is advantageous to obtain the equivalent of quadrature detection in the $t_1$ dimension as well. By obtaining pure-phase spectra, one can avoid the degraded resolution characteristic of peaks in mixed phase.

Strong solvent peaks (from water, for example) give rise to ridges that may mask peaks of interest. This complication is overcome by various schemes such as (1) selective excitation strategies that minimize the solvent signal, (2) selective saturation of solvent peaks, (3) improvement of probe characteristics, (4) postprocessing of the Fourier-transformed data by symmetrization or noise elimination strategies, or (5) suppression of single-quantum resonances by double-quantum filtering or relayed multiple-quantum approaches (see Fig. 1).

*Sensitivity.* Because 2D NMR data are acquired as a series of 1D spectra, which build up information about the time dimensions covered, signal averaging has to be distributed over both time dimensions. The time delays required for frequency labeling in the additional dimensions result in loss of signal intensity from $T_2$ (spin–spin or transverse relaxation) prior to data acquisition. As a consequence, nuclei yielding broad resonances do not show up well in 2D spectra. As a general principle, the length of the evolution period $t_1$ should be minimized to achieve the highest sensitivity. In 2D autocorrelated spectroscopy (COSY, for example), the 2D/1D sensitivity ratio is given by[28]

$$(S/N)_{2D}/(S/N)_{1D} = (2^{n-1})^{-1}(T_{2_{ab}}/2^{1/2}t_1)[1 - \exp(-t_1/T_{2_{ab}})]$$

where $T_{2_{ab}}$ is the spin–spin relaxation time of a coherence $ab$ between nuclei $a$ and $b$ during $t_1$, and $n$ is the number of coupled spins. As in 1D spectroscopy, sensitivity can be improved by a factor of $2^{1/2}$ by means of quadrature detection. Sensitivity can be maximized by optimizing sampling rates, acquisition times, and weighting functions applied to the time-domain data.

Improvements in probe design plus the higher sensitivity afforded by improved electronics and higher magnetic field strengths have enabled the collection of normal 2D NMR data sets for small proteins in time periods of 8 to 48 hr. Implementation of full-phase cycling routines places a lower limit on the number of acquisitions that must be averaged, and $T_1$ is the limiting factor in determining the repetition rate. Thus even with a sample at high concentration, the minimum experiment time may be 3 hr. Table II lists typical sample requirements and experiment times for various 2D NMR experiments. Higher concentrations than those listed in Table II are

[28] W. P. Aue, P. Bachmann, A. Wokaun, and R. R. Ernst, *J. Magn. Reson.* **29**, 523 (1978).

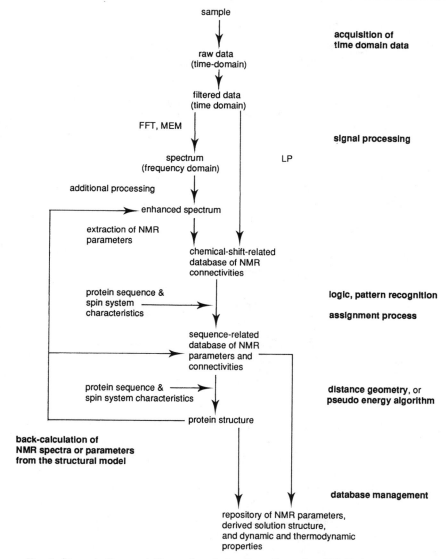

FIG. 1. Stages in the acquisition and analysis of two-dimensional NMR data. The kinds of analysis involved are signal processing, artificial intelligence (logic, pattern recognition, etc.), structure generation (distance geometry or molecular mechanics approaches), and database management. Abbreviations: FFT, fast Fourier transform; LP, linear prediction; MEM, maximum entropy method.

TABLE II

TYPICAL PRACTICAL SAMPLE REQUIREMENTS AND ACQUISITION TIMES FOR SELECTED
TWO-DIMENSIONAL NMR EXPERIMENTS AT 500 MHz[a]

| 2D NMR method[b] | Sample | | | Acquisition time (hr) |
|---|---|---|---|---|
| | Volume ($\mu$l) | Concentration (m$M$) | Amount ($\mu$mol) | |
| $^1$H correlation or 2D NOE methods | | | | |
| COSY, RELAY, HOHAHA, NOESY, ROESY) | 400 | 2 | 0.8 | 24 |
| $^1$H{$^{13}$C}SBC | | | | |
| natural abundance | 400 | 8 | 3.2 | 12 |
| 25% $^{13}$C enrichment | 400 | 1 | 0.4 | 6 |
| $^1$H{$^{13}$C}MBC | | | | |
| 25% $^{13}$C enrichment | 400 | 2 | 0.8 | 12 |
| $^1$H{$^{15}$N}SBC | | | | |
| natural abundance | 1200 | 10 | 12 | 12 |
| 98% $^{15}$N enrichment | 400 | 2 | 0.8 | 4 |
| $^{13}$C{$^{15}$N}SBC | | | | |
| 25% $^{13}$C, 98% $^{15}$N | 1200 | 4 | 4.8 | 72 |
| $^{13}$C{$^{13}$C}DQC | | | | |
| 25% $^{13}$C enrichment | 400 | 7 | 2.8 | 72 |

[a] Requirements will depend on the sensitivity of the instrument. In general, the higher the field, the higher the sensitivity.
[b] Pulse sequence nomenclature is defined in Table III.

desirable, provided that the protein is soluble and does not aggregate, since they can provide a greater signal-to-noise ratio or permit shorter experiment times. When the amount of protein is limiting and other factors are equal, our experience indicates that it is better to use a higher concentration in a 5-mm o.d. NMR tube (400-$\mu$l sample) than a lower concentration in a 10-mm o.d. NMR tube (2.5-ml sample). When a higher concentration can be tolerated, the signal-to-noise ratio can be increased by placing a more concentrated sample in a microcell insert that constrains the solution to the rf coil volume. For example, with a cylindrical microcell in a 10-mm o.d. NMR tube, we find that the required sample volume is typically 1.2 instead of 2.5 ml. Such characteristics depend on the probe construction and may differ from one probe design to another.

*Resolution.* Resolution in the $f_1$ dimension in a 2D spectrum is determined by the size of the spectral window observed and the number of $t_1$ increments for which data have been collected. The spectral width in the $f_1$ dimension is $1/2\Delta t_1$, where $\Delta t_1$ is the amount by which the evolution period is incremented. The spectral width in the $f_2$ dimension is 1/2DW, where DW is the dwell time or reciprocal of the digitization rate. For homonuclear 2D spectra, digital resolution in the $f_2$ dimension should be at least as great as that in the $f_1$ dimension and preferably four to eight times greater. Digital resolution in both dimensions can be increased by zero filling. Improvements in the measurements of frequencies, peak heights, coupling constants, and 2D cross-peak volumes can be achieved by zero filling by a factor of three to five times the original data size.[29] Owing to limitations in available computer memory and processing speed, practical realization of these theoretical improvements requires large mass storage, large main memory, and a fast processor.

## General Strategies in Two-Dimensional NMR of Proteins

Data acquisition, processing, and analysis can be divided into the following stages (Fig. 1): (1) The sample solution is subjected to a particular 2D NMR pulse sequence, and appropriate data are collected. (2) These raw time-domain data first are subjected to digital signal processing in the time domain and then are converted to the frequency domain via Fourier transformation or other suitable methods. (3) Additional signal processing or pattern recognition may be imposed at this point. (4) The spectral parameters associated with a signal at a particular position in the map (as defined by the two frequency axes, $f_1$ and $f_2$) are extracted. The data to be derived depend on the category of 2D experiment. They include strings of chemical shift values belonging to a particular spin system, and parameters associated with cross-peaks at particular chemical shifts: line widths ($w$), coupling matrices $^nJ_{ij}$, Overhauser effect connectivity matrices, and relaxation rates. Parameter extraction is basically a pattern recognition and database problem. Many operations now carried out by hand clearly can be facilitated by coupling interactive computer graphics to a database management system; such possibilities are discussed in the final section. (5) The chemical shift-related database next is converted to a sequence-related database with assignments of signals to particular atoms in particular amino acid residues. This stage requires logical analysis of the data with the imposition of knowledge of expected chemical shifts, NOEs, spin–spin couplings, etc.; it is an area where artificial intelligence methods should gain importance. (6) The sequence-related (assigned) data finally are catalogued and analyzed in terms of molecular structure, dynamics, kinetic properties such as hydrogen exchange rates or rates of dynamic equilibria, thermodynamic information about binding equilibria (e.g., $pK_a$ values) or conformational equilibria. General formats have

[29] A. G. Lindon and J. C. Ferrige, *Prog. NMR Spectrosc.* **14,** 27 (1980).

been designed for storing sequence-related files of NMR parameters and physical information derived from them.[30] Refinement stages compare back-calculated spectra or spectral parameters to original data, and adjustments are made to maximize the fit of the structural (or dynamic) model to the primary data.

### Acquisition of Time-Domain Data

Acquisition of high-quality NMR data is an essential prerequisite to further analysis. Characteristics (purity, solubility, molecular weight, aggregation, etc.) of the protein solution studied and experimental conditions (magnetic field homogeneity and stability, quality of radio frequency sources, noise figure of amplification stages, choice of pulse widths and delays, efficiency of solvent suppression, etc.) are important factors. Fundamental problems with the original data collected cannot be improved through computer manipulations and will limit extraction of spectral parameters (chemical shifts, intensities, line widths, coupling constants, etc.).

Of the large number (hundreds) of 2D NMR pulse sequences that have been described in the literature, only a limited number are applied routinely in protein spectroscopy. The 2D NMR pulse sequences that we have found to be most valuable (or most promising) in protein investigations are listed in Table III and described below. Most of the 2D NMR studies to date have concerned $^1H\{^1H\}$ investigations of proteins of $M_r \leq$ 15,000. Newer heteronuclear 2D NMR pulse sequences (Table III) that make use of $^{13}C$ labeling and/or $^{15}N$ labeling, possibly in conjunction with perdeuteration strategies, will extend the practical limit to $M_r$ 20,000 and above.

The following sections provide a brief overview of the various techniques available for the study of proteins. For more detailed discussions, see the appropriate chapters in this volume and Volume 177 of this series.

### J-Correlated 2D Spectroscopy

#### Hydrogen–Hydrogen Correlations

COSY. Two-dimensional homonuclear correlated spectroscopy (COSY)[31,32] generally is the first 2D experiment to be used in analyzing a protein. For those familiar with one-dimensional NMR spectroscopy, COSY provides the kind of information available from a single-frequency

---

[30] J. L. Markley, E. L. Ulrich, and Y. Kyogoku, *Protein Seq. Data Anal.* **2**, 23 (1989).

[31] J. Jeener, Ampère Summer School, Basko Polje, Yugoslavia (1971).

[32] K. Nagayama, A. Kumar, K. Wüthrich, and R. R. Ernst, *J. Magn. Reson.* **40**, 321 (1980).

TABLE III
SELECTED TWO-DIMENSIONAL NMR METHODS AND THEIR APPLICATION
IN PROTEIN SPECTROSCOPY

| Category | Systematic name[a] (popular name)[b] | Applications to protein spectroscopy | References[e] |
|---|---|---|---|
| J coupling | | Spin system analysis: efficient method of acquiring the kind of information provided by classical spin-decoupling analysis | |
| $^1H\{^1H\}$ Multiple bond | $^1H\{^1H\}$MBC (COSY) | Elucidation of direct coupling of $^1H^a$ to $^1H^b$ | 1 |
| | $^1H\{^1H\}$MBC-DQF (DQF-COSY) | Elucidation of direct couplings between protons having similar chemical shifts; resolution of coupling constants | 2 |
| Relayed | $^1H\{^1H\}$MBC-R (RELAY) | Elucidation of pairs of protons ($^1H^a$ and $^1H^c$) that have a mutual coupling partner $^1H^b$ ($^1H^a$ is coupled to $^1H^b$ and $^1H^b$ is coupled to $^1H^c$) | 3,4 |
| Total | $^1H\{^1H\}$TMT (TOCSY or HOHAHA) | Elucidation of all protons ($^1H^i$) belonging to a single spin system (within a given amino acid residue); assists in sorting $^1H$ resonances by residue and helps identify the amino acid type | 5,21 |
| $^1H\{^{13}C\}$ Direct | $^1H\{^{13}C\}$SBC[c] (HMQC) | Resolution of directly bonded $^1H$–$^{13}C$ pairs; cross assignment of their chemical shifts may assist in site-specific assignments within an amino acid residue | 6,7 |

(*continued*)

TABLE III (*continued*)

| Category | Systematic name[a] (popular name)[b] | Applications to protein spectroscopy | References[e] |
|---|---|---|---|
| $^1H\{^{13}C\}$ Multiple bond | $^1H\{^{13}C\}MBC$ (HMBC) | Correlation of a proton resonance with a carbon resonance 2–4 bonds distant; potential use for intra- or interresidue (sequential) assignments; determination of dihedral angles such as $\phi$ and $\chi_i$ | 8,11 |
| $^1H\{^{15}N\}$ Direct | $^1H\{^{15}N\}SBC^c$ (HMQC) | Resolution of directly bonded $^1H-^{15}N$ pairs; cross assignment of their chemical shifts may assist in site-specific assignments within an amino acid residue | 6,9,10 |
| $^1H\{^{15}N\}$ Multiple bond | $^1H\{^{15}N\}SBC$ (HMBC) | Assignment across the peptide bond; determination of the $\psi$ dihedral angle | 25,26 |
| $^{13}C\{^{15}N\}$ Direct | $^{13}C\{^{15}N\}SBC$ (HMQC) | Assignment across the peptide bond; discrimination of side-chain amide groups to identify Asn and Gln residues | 14–17 |
| $^{13}C\{^{13}C\}$ Direct | $^{13}C\{^{13}C\}DQC^d$ (INADEQUATE) | Elucidation of $^{13}C$ spin systems; sorts $^{13}C$ resonances by residues and assists in identifying the amino acid type | 11–13 |
| Cross relaxation $^1H\{^1H\}$ Laboratory frame | $^1H\{^1H\}NOE-LF$ (NOESY) | Identification of pairs of protons that are within 5 Å of one another; identification of multiple conformation states that interconvert on a ms time scale | 18,19 |

(*continued*)

TABLE III (*continued*)

| Category | Systematic name[a] (popular name)[b] | Applications to protein spectroscopy | References[e] |
|---|---|---|---|
| Rotating frame | $^1$H{$^1$H}NOE-RF (ROESY, CAMELSPIN) | Assists in distinguishing cross relaxation peaks from COSY-type peaks and in elucidating the influence of dynamics on the NOE | 20,23 |

[a] In order to keep up with the steady proliferation of 2D (and 3D) NMR experiments, a suitable systematic classification system of experiments seems a useful goal. Such a system would not be expected to displace the popular names of better known experiments (e.g., COSY and NOESY), but could provide guidance to the uninitiated. The approach used here is to specify the nuclei involved (observed nucleus outside the brackets; second nucleus inside the brackets), specify the kind connectivity, and append additional information about filtering order or relay step used. Abbreviations: DQF, double-quantum filtered; LF, laboratory frame; MBC, multiple-bond correlation; SBC, single-bond correlation; R, relay; RF, rotating frame; TMT, total magnetization transfer.

[b] COSY, Correlated spectroscopy; HMBC, heteronuclear multiple-bond correlation; HMQC, heteronuclear multiple-quantum correlation; HOHAHA, homonuclear Hartmann–Hahn; INADEQUATE, incredible natural abundance double-quantum transfer experiment; NOESY, NOE spectroscopy; ROESY, rotating-frame NOE spectroscopy; TOCSY, total correlated spectroscopy.

[c] We normally use pulse sequence "E" of Bax et al.[22] in this experiment. It is often useful to add a relay step to the heteronuclear correlation; this can be an additional coherence transfer step,[7] a HOHAHA relay,[26,27] or an NOE relay.[26,28]

[d] This experiment uses a variation on the 2D-INADEQUATE pulse sequence.[24]

[e] Key to references: (1) K. Nagayama, Anil Kumar, K. Wüthrich, and R. R. Ernst, *J. Magn. Reson.* **40**, 321 (1980); (2) M. Rance, O. W. Sørensen, G. Bodenhausen, G. Wagner, R. R. Ernst, and K. Wüthrich, *Biochem. Biophys. Res. Commun.* **117**, 479 (1983); (3) G. King and P. E. Wright, *J. Magn. Reson.* **54**, 328 (1983); (4) G. Wagner, *J. Magn. Reson.* **55**, 151 (1983); (5) A. Bax and D. G. Davis, *J. Magn. Reson.* **65**, 355 (1985); (6) A. Bax, R. H. Griffey, and B. L. Hawkins, *J. Magn. Reson.* **55**, 301 (1983); (7) G. Wagner and D. Brüwiler, *Biochemistry* **25**, 5839 (1986); (8) A. Bax and M. F. Summers, *J. Am. Chem. Soc.* **108**, 2093 (1986); (9) G. Ortiz-Polo, R. Krishnamoorthi, and J. L. Markley, *J. Magn. Reson.* **68**, 303 (1986); (10) J. Glushka and D. Cowburn, *J. Am. Chem. Soc.* **109**, 7879 (1987); (11) W. M. Westler, M. Kainosho, H. Nagao, N. Tomonaga, and J. L. Markley, *J. Am. Chem. Soc.* **110**, 4093 (1988); (12) B. J. Stockman, W. M. Westler, P. Darba, and J. L. Markley, *J. Am. Chem. Soc.* **110**, 4096 (1988); (13) B. H. Oh, W. M. Westler, P. Darba, and J. L. Markley, *Science* **240**, 908 (1988); (14) W. M. Westler, B. J. Stockman, Y. Hosoya, Y. Miyake, M. Kainosho, and J. L. Markley, *J. Am. Chem. Soc.* **110**, 6256 (1988); (15) B. J. Stockman, W. M. Westler, E. S. Mooberry, and J. L. Markley, *29th Exp. NMR Conf.,* 1988; Abstr. No. 122; (16) W. M. Westler, M. Kainosho, H. Nagao, N. Tomonaga, and J. L. Markley, *29th Exp. NMR Conf.,* 1988, Abstr. No. 137; (17) M. D. Reily, E. L. Ulrich, W. M. Westler, and

(*continued*)

Footnotes to TABLE III (*continued*)

J. L. Markley, *29th Exp. NMR Conf.*, 1988, Abstr. No. 135; (18) Anil Kumar, R. R. Ernst, and K. Wüthrich, *Biochem. Biophys. Res. Commun.* **95**, 1 (1980); (19) S. Macura, K. Wüthrich, and R. R. Ernst, *J. Magn. Reson.* **46**, 269 (1982); (20) A. Bax and D. G. Davis, *J. Magn. Reson.* **63**, 207 (1985); (21) L. Braunschwieler and R. R. Ernst, *J. Magn. Reson.* **53**, 521 (1983); (22) A. Bax, R. H. Griffey, and B. L. Hawkins, *J. Magn. Reson.* **55**, 301 (1983); (23) A. Bax, R. Freeman, and S. P. Kempsell, *J. Am. Chem. Soc.* **102**, 4849 (1980); (24) A. A. Bothner-By, R. L. Stephens, J. Lee, C. D. Warren, and R. W. Jeanloz, *J. Am. Chem. Soc.* **106**, 811 (1984); (25) G. M. Clore, A. Bax, P. Wingfield, and A. M. Gronenborn, *FEBS Lett.* **238**, 17 (1988); (26) A. M. Gronenborn, A. Bax, P. T. Wingfield, and G. M. Clore, *FEBS Lett.* **243**, 93 (1989); (27) B.-H. Oh, W. M. Westler, and J. L. Markley, *J. Am. Chem. Soc.* **111**, in press (1989); (28) K. Shon and S. J. Opella, *J. Magn. Reson.* **82**, 193 (1989).

decoupling experiment (Fig. 2), i.e., which spins are scaler coupled to one another. In a COSY plot, the 1D spectrum lies along the diagonal, and the off-diagonal elements are present at the intersection of chemical shifts of groups that are $J$ coupled. The "fingerprint" region (outlined in Fig. 3A)[32] contains ($^1H^N$, $^1H^\alpha$) cross-peaks from the peptide backbone. The degree of resolution of the "fingerprint" region of a COSY map obtained in $H_2O$ (Fig. 3B) is a good predictor of the success of sequence-specific assignments to be obtained without recourse to isotopic labeling. (For a detailed discussion of $^1H$ assignment strategies see Basus [7], Vol. 177, this series.)

We routinely use a standard ($d_r$–90°–$t_1$–90°–acq) pulse sequence, where $d_r$ is the relaxation delay and acq is the acquisition period ($t_2$), to minimize diagonal peaks while maximizing cross-peak intensities[35] and with phase cycling for quadrature detection and retention of pure-phase spectra.[36]

For higher resolution, better detection of cross-peaks near the diagonal, and suppression of the solvent signal (especially with samples dissolved in $H_2O$), DQF-COSY[37] obtained in the pure-absorption mode is the method of choice (cf. Rance *et al.* [6], this volume). We typically use a ($d_r$–90°–$t_1/2$–90°–$t_1/2$–90°–acq) pulse sequence. Pure-absorption spectra

[33] Z. Zolnai, S. Macura, and J. L. Markley, *Comput. Enhanced Spectrosc.* **3**, 141 (1986).

[34] A. D. Robertson, Ph.D. Thesis, University of Wisconsin, Madison (1988).

[35] A. Bax and R. Freeman, *J. Magn. Reson.* **44**, 542 (1981).

[36] G. Wider, S. Macura, A. Kumar, R. R. Ernst, and K. Wüthrich, *J. Magn. Reson.* **56**, 207 (1984).

[37] M. Rance, O. W. Sørensen, G. Bodenhausen, G. Wagner, R. R. Ernst, and K. Wüthrich, *Biochem. Biophys. Res. Commun.* **117**, 479 (1983).

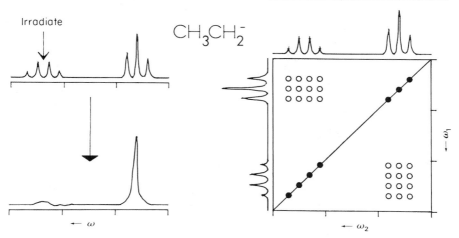

FIG. 2. Diagram with simulated data illustrating the similarity of information derived from a single-frequency 1D decoupling experiment (left) and 2D correlated spectroscopy (COSY) experiment (right).

can be obtained either by time-proportional phase incrementation (TPPI)[38] or by the States–Haberkorn–Ruben method.[39] On the Bruker spectrometers, we usually use TPPI with phase cycling as described by Rance *et al.*[37]

*RELAY.* The RELAY pulse sequence is used for identifying pairs of spins that are not coupled directly to one another but that share a mutual coupling partner. We obtain RELAY data[40] with the pulse sequence ($d_r$–90°–$t_1$–90°–$\tau$–180°–$\tau$–90°–acq), where $\tau$ is a delay which is adjusted according to the coupling constants of the spin system. Eight-step phase cycling[41] plus additional CYCLOPS cycling[42] completes the pulse program. Relay data can be obtained in absolute-value mode or pure-phase mode (relay peaks in pure phase, direct peaks in mixed phase), depending on the phase cycling used.

*HOHAHA.* Two-dimensional homonuclear Hartmann–Hahn magnetization transfer (HOHAHA) spectra[43] are acquired with the ($d_r$–90°–$t_1$–

[38] D. Marion and K. Wüthrich, *Biochem. Biophys. Res. Commun.* **113,** 967 (1983).
[39] D. J. States, R. A. Haberkorn, and D. J. Ruben, *J. Magn. Reson.* **48,** 286 (1982).
[40] G. Wagner, *J. Magn. Reson.* **55,** 151 (1983).
[41] A. Bax and G. Drobny, *J. Magn. Reson.* **61,** 306 (1985).
[42] D. I. Hoult and R. E. Richards, *Proc. R. Soc. London, Ser. A* **344,** 311 (1975).
[43] A. Bax and D. G. Davis, *J. Magn. Reson.* **65,** 355 (1985).

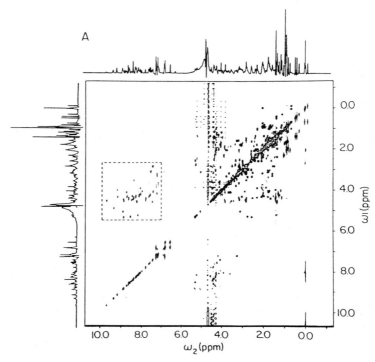

FIG. 3. Phase-sensitive double-quantum filtered COSY of turkey ovomucoid third domain (OMTKY3) in $H_2O$. The sample was 15 m$M$ OMTKY3, 0.2 $M$ KCl in 90% $H_2O$, pH 4.0. The sample temperature was 25°. The data, acquired at 500 MHz, consisted of 397 blocks of 64 summed transients, each containing 2K time-domain points. Both $t_2$ and $t_1$ data were multiplied by phase-shifted ($\pi/32$ in $t_2$, $\pi/6$ in $t_1$) sine-bell functions before zero filling to 4K × 2K data points ($t_2 \times t_1$) and double-Fourier transformation. (A) Full COSY plot. The 1D spectrum is shown on both the vertical and horizontal axes for comparison. The region enclosed by the box, known as the "fingerprint" region, contains cross-peaks due to coupling between backbone $^1H^\alpha$ and $^1H^N$. Much of the noise has been suppressed by using the MAKEUP program.[33] (B) Enlargement of the fingerprint region of the COSY plot showing assignments of the cross-peaks. (From Robertson.[34])

$SL_x$–MLEV17–$SL_x$–acq) pulse sequence in which $SL_x$ denotes a short (2 msec) spin-lock field applied along the $x$ axis, and MLEV17 is a composite pulse sequence (cf. Bax [8], this volume). HOHAHA spectra display both direct and relayed connectivities and are very useful for elucidating scalar-coupled networks. The intensities of peaks depend on the length of the spin-lock mixing time; we typically acquire HOHAHA spectra at mixing times of 55, 75, and 90 msec. This experiment is useful for

B

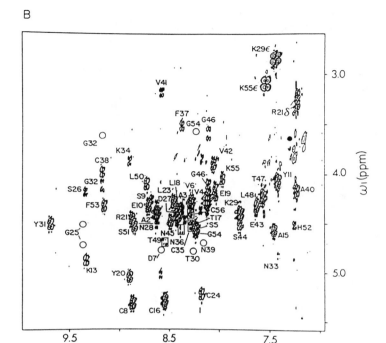

FIG. 3. (*continued*)

identifying resonances from the set of spins in a particular $^1$H spin system. Its advantages are that coherence transfer is efficient and data are largely pure phase. The results are complicated by the presence of rotating-frame NOE peaks; however, these can be distinguished since they are of opposite sign compared to the diagonal and coherence transfer peaks.

*Application of These Methods to $^1$H Spin-System Identification.* Figure 4[44] compares similar expanded regions of COSY, RELAY, and HOHAHA spectra of a small protein, turkey ovomucoid third domain (OMTKY3), dissolved in $^2$H$_2$O. Spin systems of three methyl-bearing amino acids (Leu, Thr, and Val) are identified. Leucine and Val both have two methyl groups coupled to a methine proton. Whereas the Val spin system has only an additional $^1H^\alpha$, identified as a pair of RELAY peaks,

[44] A. D. Robertson, W. M. Westler, and J. L. Markley, *Biochemistry* **27**, 2519 (1988).

the Leu spin system has $^1H^\alpha$ and $^1H^\beta$ protons identified as a string of cross-peaks in the HOHAHA spectrum. The Thr peaks are identified on the basis of their characteristic chemical shifts and coupling pattern as revealed in the RELAY spectrum.

Figure 5 illustrates how the spin systems can be extended to the $^1H^N$. The left-hand side shows the ($^1H^N$, $^1H^\beta$) region of the RELAY spectrum obtained with a sample dissolved, in $H_2O$; the right-hand side is the ($^1H^\alpha$, $^1H^\beta$) region of the COSY map. The portions of the alanine and valine spin systems identified by this comparison are indicated in Fig. 5.[44]

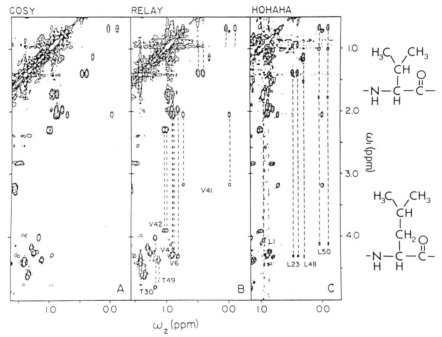

FIG. 4. Comparison of corresponding enlarged regions of COSY, RELAY, and HOHAHA maps. Methyl to methine cross-peaks are clearly present in the COSY plot (left). In the case of valine and threonine, the RELAY experiment produces methyl to $^1H^\alpha$ cross-peaks (center). Multiple relay steps are required to achieve methyl to $^1H^\alpha$ connectivities with the longer side chain of leucine; these are achieved by means of the HOHAHA experiment (right). The sample was 15 m$M$ turkey ovomucoid third domain (OMTKY3) in $^2H_2O$ containing 0.2 $M$ KCl. The pH* was 4.0, and the temperature was maintained at 25°. The COSY and RELAY spectra were acquired at 500 MHz; the mixing time in the RELAY experiment was 25 msec. The HOHAHA spectrum was acquired at 400 MHz with a mixing time of 101 msec. (Adapted from Robertson et al.[44])

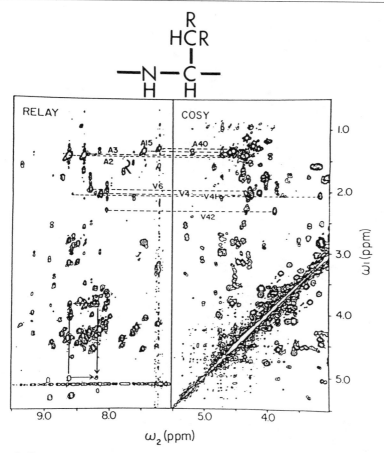

FIG. 5. Comparison of spectra illustrating how three-bond $^1H^\beta$–$C^\beta$–$C^\alpha$–$^1H^\alpha$ connectivities detected in a COSY experiment carried out in $^2H_2O$ (right) can be extended to the $^1H^N$ of the same residue by means of a RELAY experiment carried out in $^1H_2O$ (left). The sample was 15 m$M$ turkey ovomucoid third domain (OMTKY3) in 0.2 $M$ KCl, 25°. The relay mixing time was 40 msec. The arrows indicate RELAY connectivities from contaminating $N$-acetylglucosamine in the protein sample. (Adapted from Robertson et al.[44])

The HOHAHA pulse sequence is particularly valuable in elucidating the spin systems of long-chain amino acids. Figure 6 shows the identification of a $^1H^\alpha$ signal on the diagonal as that of a lysine through its total correlations with the $^1H^\beta$, $^1H^\gamma$, $^1H^\delta$ and $^1H^\varepsilon$ resonances of the side chain.[43]

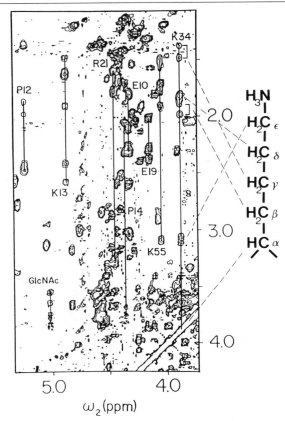

FIG. 6. Illustration of the use of the HOHAHA experiment to elucidate the proton spin systems of long-chain amino acids. The inset shows the side chain of Lys-34 with dashed lines connecting protons to their corresponding positions in the spectrum. Other proton spin systems identified in the figure by solid lines are those of proline (P), other lysines (K), glutamate (E), arginine (R), and contaminating $N$-acetylglucosamine (GlcNAc). Experimental conditions were as described for the HOHAHA spectrum in Fig. 4. (Adapted from Robertson et al.[44])

## Two-Dimensional Correlations Involving Less Sensitive Nuclei

### Proton Detection of Heteronuclei

$^1H\{^{13}C\}SBC$. Early $^1H$–$^{13}C$ heteronuclear correlation experiments on proteins were carried out by $^{13}C$ detection[45] for instrumental reasons. This

---

[45] J. Bodenhausen and R. Freeman, J. Am. Chem. Soc. **100**, 320 (1978).

method was practical for proteins enriched with $^{13}$C[10,11] but, at natural abundance $^{13}$C, required hundreds of milligrams of protein and $\geq$48-hr data experiment times.[12] By detecting the more sensitive nucleus,[46] $^1$H–$^{13}$C correlations can be obtained with the same amounts of protein required for a COSY experiment. $^1$H{$^{13}$C}SBC results can be obtained with small proteins at natural abundance[5,6]; $^{13}$C enrichment to a level of 15–30% is useful with larger proteins.[18–20] Higher levels of uniform $^{13}$C enrichment are to be avoided since long-range couplings complicate the spectrum.[19] $^1$H–$^{13}$C chemical shift correlations are useful for resolving overlaps in the $^1$H dimension and for identifying the origin of $^1$H signals (based on the chemical shift of the attached $^{13}$C). The $^1$H{$^{13}$C}SBC experiment can be performed without $^{13}$C decoupling during acquisition (cross-peaks appear as doublets split by $^1J_{^1H^{13}C}$),[5] but additional sensitivity can be obtained, and the spectrum can be simplified, if decoupling is used.[6] For decoupled spectra, an additional delay is inserted into the pulse sequence before detection to allow the antiphase magnetization to come into phase prior to acquisition. We normally use composite-pulse decoupling and acquire the spectrum in two sections [aliphatic ($^1$H, $^{13}$C) and aromatic ($^1$H, $^{13}$C)] so that the $^{13}$C frequency can be set at the middle of each section for optimal decoupling (about 5-W power for $^{13}$C).[6]

Since the $^1$H{$^{13}$C}SBC experiment is relatively easy to perform even at natural abundance $^{13}$C and the information is so valuable, it should be a routine part of the NMR analysis of any small protein. $^{13}$C assignments resulting from $^1$H–$^{13}$C correlations are interesting in their own right and are needed for the exploitation of $^{13}$C parameters such as chemical shifts and relaxation times. Figure 7 shows the $^1H^{\alpha}$–$^{13}C^{\alpha}$ region of the $^1$H{$^{13}$C}SBC spectrum of turkey ovomucoid third domain (natural abundance $^{13}$C).[6]

$^1H\{^{15}N\}SBC$. A similar approach can be used to obtain $^1$H–$^{15}$N correlations. The experiment can be carried out with small, soluble proteins at natural abundance,[7,8] but $^{15}$N enrichment facilitates the acquisition of data from larger proteins.[47,48] Figure 8 shows the $^1$H{$^{15}$N}SBC spectrum of flavodoxin from *Anabaena* 7120 ($M_r \sim 19,200$). The results have provided assignments of resonances from the flavin ring nitrogens of the prosthetic group.[48]

[46] A. Bax, R. H. Griffey, and B. L. Hawkins, *J. Magn. Reson.* **55,** 301 (1983).
[47] L. P. McIntosh, R. H. Griffey, D. C. Muchmore, C. P. Nielson, A. G. Redfield, and F. W. Dahlquist, *Proc. Natl. Acad. Sci. U.S.A.* **84,** 1244 (1987).
[48] B. J. Stockman, W. M. Westler, E. S. Mooberry, and J. L. Markley, *Biochemistry* **27,** 136 (1988).

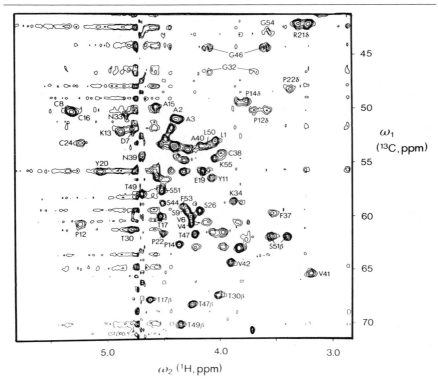

FIG. 7. A portion of the 2D $^1H\{^{13}C\}SBC$ spectrum of turkey ovomucoid third domain (OMTKY3). The region shown contains primarily ($^1H^\alpha$, $^{13}C^\alpha$) cross-peaks which are identified by residue type (one-letter amino acid code) and number. Other cross-peaks that fall in this region are assigned by position within the residue (Greek letter). Note the characteristic doublet pattern for glycine residues: two $^1H^\alpha$ coupled to a single $^{13}C_\alpha$. For clarity, some of the known assignments in this region are not indicated in the figure. The sample consisted of 15 m$M$ OMTKY3 (36 mg in 0.4 ml), 0.2 $M$ KCl, in $^2H_2O$. The sample pH* was 4.1 at 30° and 512 blocks of data (2048 time-domain points) were collected over 24 hr. Data in both dimensions were multiplied by Gaussian weighting functions, with maxima at 0.1 times the time-domain data size, prior to Fourier transformation. The line-broadening parameter was $-10$ Hz in the $t_2$ ($^1H$) dimension and $-40$ Hz in the $t_1$ ($^{13}C$) dimension. Data in the $t_1$ dimension were zero filled to 2048 data points before transformation. (From Robertson *et al.*[6])

$^1H\{X\}SBC$. Proteins may contain additional magnetically active nuclei of spin-$\frac{1}{2}$ (other than $^{13}C$ or $^{15}N$) that are coupled to protons. For example, the $^1H\{^{113}Cd\}SBC$ experiment has been applied to several proteins that bind $^{113}Cd$.[13,49]

[49] D. H. Live, C. L. Kojiro, D. Cowburn, and J. L. Markley, *J. Am. Chem. Soc.* **107**, 3043 (1985).

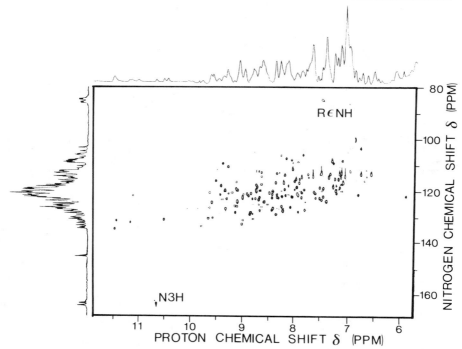

FIG. 8. Proton-detected $^{15}N$ ($^{1}H\{^{15}N\}SBC$) spectrum of oxidized [98% U-$^{15}N$]flavodoxin from *Anabaena* 7120 ($M_r$ 19,200).[54] Labeled cross-peaks are from the $^{15}N_3$–$^{1}H$ of the flavin ring of the prosthetic group and from the $^{15}N^{\varepsilon}$–$^{1}H^{\varepsilon}$ of arginine residues. The protein concentration was 4.4 m$M$ in 100 m$M$ phosphate buffer at pH 7.5. The $^{15}N$ frequency was 50.68 MHz, and the $^{1}H$ frequency was 500 MHz. The spectral width in each dimension was 4348 Hz with $^{15}N$ decoupled during acquisition. $^{1}H$ chemical shifts are referenced to (trimethylsilyl)propionic acid, and $^{15}N$ chemical shifts are referenced to external liquid ammonia. (B. J. Stockman, E. S. Mooberry, and J. L. Markley, unpublished data.)

## Carbon–Carbon Correlations

Carbon–carbon couplings in proteins contain a wealth of information. Although a prototype $^{13}C$–$^{13}C$ correlation experiment was carried out with an amino acid ([85% U-$^{13}C$]lysine) several years ago,[50] protein experiments were slow to follow. At natural abundance $^{13}C$ the abundance of $^{13}C$–$^{13}C$ pairs is exceedingly low (0.012%). We have found that random enrichment to a level of about 25% facilitates the detection of carbon–

[50] J. L. Markley, W. M. Westler, T.-M. Chan, C. L. Kojiro, and E. L. Ulrich, *Fed. Proc., Fed. Am. Soc. Exp. Biol.* **43**, 2648 (1984).

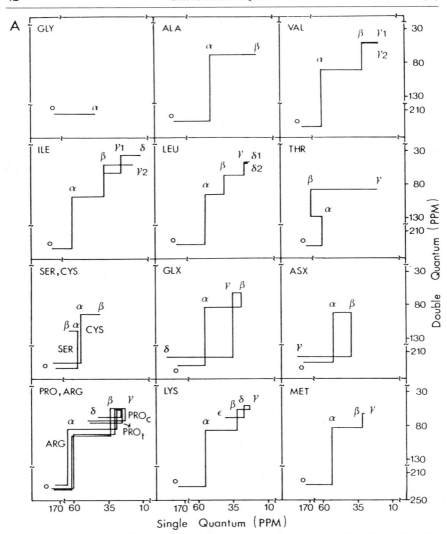

FIG. 9. Double-quantum carbon–carbon linkage patterns for the 20 amino acids. The horizontal axis is the normal $^{13}$C chemical shift scale, and the vertical axis is the double-quantum axis. Cross-peaks between directly bonded carbons $\alpha$ and $\beta$ appear at $\delta_A (\delta_A + \delta_B)$, and $\delta_B (\delta_A + \delta_B)$. (From Oh *et al.*[17] with correction of the spin system for tryptophan.[50a])

carbon connectivities; however, a higher random enrichment level of about 44% probably would be optimal for this experiment. Two separate 2D NMR methods for obtaining carbon–carbon correlations in proteins have been evaluated recently: carbon–carbon homocorrelated spectros-

[50a] B.-H. Oh and J. L. Markley, *Biopolymers*, in press (1989).

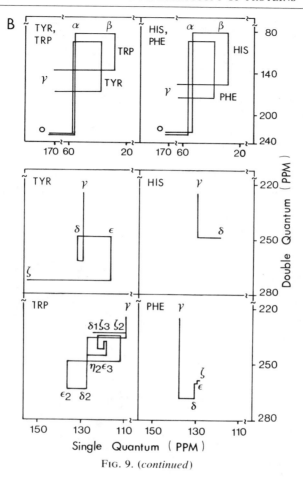

FIG. 9. (*continued*)

copy ($^{13}$C{$^{13}$C}HOMCOR) and carbon–carbon double-quantum correlation spectroscopy ($^{13}$C{$^{13}$C}DQC).[19] The latter method gives superior results in part because of the general correlation of chemical shifts of directly bonded carbon atoms. The sum of the chemical shifts (double-quantum axis) is plotted against the individual chemical shifts (single-quantum axis). Each carbon–carbon bond is represented by two cross-peaks, and the position of one can be predicted from the position of the other. This property lends itself to spectral symmetrization strategies.[51] Figure 9 shows the $^{13}$C{$^{13}$C} DQC linkage patterns expected for each of the 20 common amino acids in a random coil. The patterns are sufficiently

[51] P. Darba and J. L. Markley, unpublished.

distinctive so that 18 of the 20 amino acids can be identified unambiguously in a single 2D NMR experiment; only the Glu/Glu and Asp/Asn pairs may require additional data for their discrimination.

In the proteins studied so far by this method (cytochrome $c$-553, $M_r$ 10,000[20]; ferredoxin, $M_r$ 11,000[17,21]; flavodoxin, $M_r$ 19,200,[18,20] it has been possible in a single experiment to follow complete carbon–carbon connectivity chains with uniform enrichment (26% $^{13}$C) of all amino acid residues. Figure 10, which shows the $^{13}$C{$^{13}$C}DQC spectrum of flavodoxin from *Anabaena* 7120, illustrates the potential information content of this experiment. The carbon spin systems of one alanine and one tyrosine residue have been traced in Fig. 10, beginning at the carbonyl carbon and following through to the end of the side chain. One row of the spectrum,

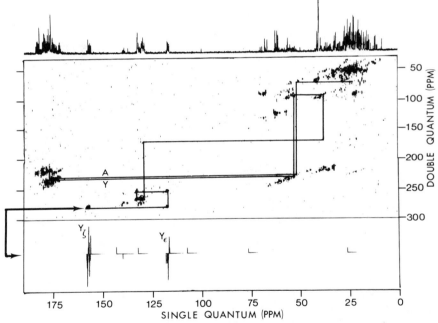

FIG. 10. Carbon–carbon correlations in oxidized [26% U-$^{13}$C]flavodoxin from *Anabaena* 7120 obtained by 2D $^{13}$C{$^{13}$C}DQC spectroscopy at 125.76 MHz. $^{13}$C chemical shifts are referenced to tetramethylsilane. Single- and double-quantum frequencies are plotted along the horizontal and vertical axes, respectively. The one-dimensional projection in the $\omega_2$ dimension is plotted at the top. Connectivities are outlined beginning at the carbonyl carbons for an analanine (A) and tyrosine (Y) residue. The row of the spectrum with a double-quantum chemical shift of 272.3 ppm is shown at the bottom of the figure. The antiphase doublets in this row that correspond to the $^{13}C^\zeta$–$^{13}C^\varepsilon$ correlation of the outlined tyrosine residue are labeled. (From Stockman *et al.*[19])

that containing information about $^{13}C^{\zeta}-^{13}C^{\varepsilon}$ correlations of the outlined tyrosine residue, is plotted at the bottom of the figure to illustrate the ease with which cross-peak signals can be distinguished. The carbon–carbon correlation results have been used to assign the carbon resonances of the noncovalently bound flavin mononucleotide cofactor of the flavodoxin. The $^{13}C\{^{13}C\}DQC$ method presents an alternative to selective isotopic enrichment for carbon assignments. For example, flavin carbon resonances have been assigned previously in other flavoproteins by incorporating $^{13}C$-labeled flavin into the nonlabeled protein.[52–54] Because of the presence of several closely spaced resonances, it was necessary to incorporate two or more selectively enriched flavins into the protein, either independently or as mixtures with varying levels of enrichment. By $^{13}C\{^{13}C\}DQC$ such assignments can be made directly with a single uniformly labeled sample without the necessity of organic syntheses and cofactor reconstitution. Carbon assignments can be extended readily to proton assignments by means of $^{1}H\{^{13}C\}SBC$ or $^{1}H\{^{13}C\}MBC$ experiments. The $^{13}C\{^{13}C\}DQC$ approach has been used recently to analyze and identify carbon spin systems of 75 of the 98 amino acid residues in *Anabaena* 7120 ferredoxin ($M_r$ 11,000) in its oxidized state.[17]

Aromatic amino acid spin systems frequently are difficult to identify on the basis of $^{1}H$ NMR data. Since coupling between the $^{1}H^{\beta}$s and ring protons is weak, NOESY data generally are needed to link the aliphatic AMX spin system with that of the aromatic spin system of the ring. Reliable spin system identifications may be difficult, particularly with larger proteins. A series of direct one-bond correlations along the $^{13}C^{\beta}-$ $^{13}C^{\gamma}-^{13}C^{\delta}$ chain circumvent this problem. Sequence-specific backbone proton assignments made in the conventional way[4] can be extended readily through quaternary carbons via $^{13}C-^{13}C$ correlations in the $^{13}C\{^{13}C\}DQC$ experiment. Aromatic ring proton–carbon correlations can then be used to assign the ring protons.[20] This through-bond technique circumvents ambiguities that may arise by using through-space (NOESY) connectivities between $^{1}H^{\beta}$s and aromatic ring side-chain protons to assign aromatic ring protons.[4] Additional quaternary carbon resonances, such as Tyr $^{13}C^{\zeta}$, Glu $^{13}C^{\delta}$, or Asp $^{13}C^{\gamma}$, which are useful reporters of pH titrations, also can be assigned directly in this manner. The positions of the proline [$^{13}C^{\gamma}$, ($^{13}C^{\gamma}$ + $^{13}C^{\delta}$)] and [$^{13}C^{\gamma}$, ($^{13}C^{\beta}$ + $^{13}C^{\gamma}$)] cross-peaks

[52] J. Verwoort, F. Müller, J. LeGall, A. Bacher, and H. Sedlmaier, *Eur. J. Biochem.* **151,** 49 (1985).

[53] W.-D. Beinert, H. Rüterjans, F. Müller, and A. Bacher, *Eur. J. Biochem.* **152,** 581 (1985).

[54] J. Verwoort, F. Müller, S. G. Mayhew, W. A. M. van den Berg, C. T. W. Moonen, and A. Bacher, *Biochemistry* **25,** 6789 (1986).

should provide a means of distinguishing whether the $X$–Pro peptide bond is cis or trans (Fig. 9).[17]

Sensitivity considerations limit the application of the $^{13}C\{^{13}C\}DQC$ experiment to proteins enriched with $^{13}C$. With current methods for incorporating stable isotopes into biotechnology-derived proteins,[27] this is becoming less of a problem. It should be noted that the amount of $^{13}C$-labeled protein required for this experiment is similar to that used routinely in 2D $^1H$ NMR experiments such as COSY and NOESY. Provided that a sample of ~25% U-$^{13}C$-labeled protein can be produced, this method appears to provide an expeditious approach to extensive spectral assignments (see Gerlt *et al.* [4], Vol. 177, this series, for strategies to incorporate $^{13}C$ into proteins).

## Carbon–Nitrogen Correlations

We recently have used a 2D NMR approach ($^{13}C\{^{15}N\}SBC$) to identify $^{13}C$–$^{15}N$ correlations in [26% U-$^{13}C$, 98% U-$^{15}N$]flavodoxin[15] (Fig. 11) and [26% U-$^{13}C$, 98% U-$^{15}N$]ferredoxin,[16] and *Streptomyces* [90% $^{13}C_0$–Met, 60% U-$^{15}N$]subtilisin inhibitor.[15] These results appear to provide a promising step toward the goal of sequential assignments based on one-bond couplings along the peptide backbone (see Dahlquist *et al.* [3], Vol. 177, this series, for a discussion of incorporation of $^{15}N$ into proteins).

## Cross-Relaxation 2D Spectroscopy

*NOESY.* Cross-relaxation is detected as a change in the intensity of one peak as a consequence of perturbing the population of spins involved in another transition. When the two resonances arise from the same nucleus residing in two conformational states of one molecule or two molecular species in dynamic equilibrium, the effect can provide information about the interconversion rates. When the two resonances arise from different nuclei and the cross-relaxation mechanism is via through-space dipole–dipole interaction, the results can provide information about the distance between the two nuclei. The 2D versions of these experiments are provided by the three-pulse 2D NMR sequence[55] ($d_r$–90°–$t_1$–90°–$\tau_m$–90°–acq), where the delay $t_1$ is incremented from one group of acquisitions to the next, $\tau_m$ is the "mixing time" which controls the time period during which cross-relaxation occurs, and acq is the spectral acquisition. Typical values for the mixing time are 30–300 msec. The length of $\tau_m$ is usually allowed to vary randomly by ±10% to suppress scalar-coupling

---

[55] Anil Kumar, R. R. Ernst, and K. Wüthrich, *Biochem. Biophys. Res. Commun.* **95,** 1 (1980).

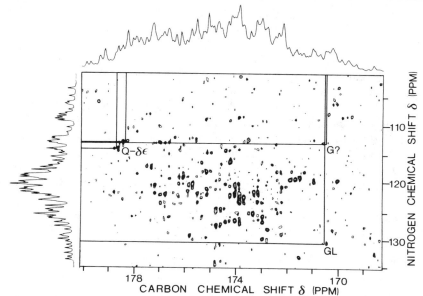

FIG. 11. Elucidation of one-bond carbon–nitrogen connectivities by means of a 2D $^{13}C\{^{15}N\}$SBC experiment. The sample was oxidized [26% U-$^{13}$C, 98% U-$^{15}$N]flavodoxin from *Anabaena* 7120 ($M_r$ 19,200). The data were acquired with a special 10-mm probe tuned to the following $^{1}$H, $^{2}$H, $^{13}$C, and $^{15}$N frequencies: $^{1}$H (400 MHz), $^{2}$H (61.40 MHz), $^{13}$C (100.58 MHz), and $^{15}$N (40.53 MHz). A sequential assignment is shown between a glycine $^{13}C_i'$ and a leucine $^{15}N_{i+1}$ (labeled "GL"). The position of a second glycine carbonyl determined by the $^{13}C\{^{13}C\}$DQC experiment is labeled "G?," but its coupling partner has not been identified. Two cross-peaks corresponding to glutamine side-chain amides ($Q - \delta\varepsilon$) are indicated in the figure and serve to distinguish the side-chain carbonyls of Asn and Gln from those of Asp and Glu. (From Westler *et al.*[15])

effects.[56] We generally use the phase cycling scheme described by Bodenhausen *et al.*[57]

Two-dimensional exchange spectroscopy can be used to correlate the chemical shifts of protons in two oxidation states of a protein in dynamic equilibrium[58,59] or two conformational forms, e.g., native and denatured states.[60] The cross-peak intensities on either side of the diagonal provide information about the rate of the reaction in each direction.

[56] S. Macura, K. Wüthrich, and R. R. Ernst, *J. Magn. Reson.* **46**, 269 (1982).
[57] G. Bodenhausen, H. Kogler, and R. R. Ernst, *J. Magn. Reson.* **58**, 370 (1984).
[58] J. Boyd, C. M. Dobson, and C. Redfield, *J. Magn. Reson.* **55**, 170 (1983).
[59] H. Santos, D. L. Turner, and A. V. Xavier, *J. Magn. Reson.* **58**, 344 (1984).
[60] C. M. Dobson, P. A. Evans, and K. L. Williamson, *FEBS Lett.* **168**, 1323 (1984).

In the absence of exchange, cross-peak intensities are determined by the nuclear Overhauser effect whose distance dependence falls off as $1/r^6$, where $r$ is the distance between the pair of protons that cross-relax each other.[61] The maximum distance sampled depends on the mixing time (e.g., a mixing time of 100 msec for a protein of $M_r$ 5000 yields cross-peaks from pairs of protons closer than about 3 Å). Identification of pairs of nuclei more distant than 5 Å becomes unreliable owing to competing relaxation mechanisms. The two-spin approximation is only valid for short mixing times.[62] Distances are obtained most accurately by measuring the ratios of cross-peak volumes to diagonal peak volumes as a function of the mixing time, and fitting the results[62–65] to a theoretical equation (however, see Borgias and James [9], this volume). Ideally, the correlation time for each dipole–dipole interaction should be known, but distances can be measured to within about $\pm 10\%$ accuracy by fitting the data to a quatratic approximation of the cross-relaxation build-up rates and by assuming that all correlation times are the same as that calculated from a known fixed distance.[66] NOESY data also are valuable in the analysis of amino acid spin systems (Fig. 12) and for elucidating secondary and tertiary structure (see below).

*ROESY.* The nuclear Overhauser effect that occurs under spin-lock conditions is known as the transverse or rotating-frame NOE (cf. Brown and Farmer [11], this volume). The experiment was proposed by Bothner-By and co-workers,[67] who named it "CAMELSPIN" from an ice-skating maneuver. More recently, 2D transverse NOE spectroscopy has been called "ROESY"[68] in analogy to "NOESY." The fundamental advantage of the experiment is that spin-locked NOE's are always positive and increase monotonically with increasing correlation time.[67] By contrast, the normal NOE is positive for short correlation times but becomes negative for longer correlation times. Since it goes through zero, there is a range of frequencies for which the NOE will not be detected. While NOESY analysis of the structures of moderate sized peptides is difficult or impractical, e.g., $M_r = 1000–2000$, ROESY offers a viable alternative.

[61] J. H. Noggle and R. E. Schirmer, "The Nuclear Overhauser Effect." Academic Press, New York, 1971.
[62] J. W. Keepers and T. L. James, *J. Magn. Reson.* **57,** 404 (1984).
[63] E. Suzuki, N. Pattabiraman, G. Zon, and T. L. James, *Biochemistry* **25,** 6854 (1986).
[64] P. A. Mirau and F. A. Bovey, *J. Am. Chem. Soc.* **108,** 5130 (1986).
[65] B. T. Farmer II, S. Macura, and L. R. Brown, *J. Magn. Reson.* **72,** 347 (1987).
[66] J. Fejzo, Z. Zolnai, S. Macura, and J. L. Markley, *J. Magn. Reson.,* in press (1989).
[67] A. A. Bothner-By, R. L. Stephens, J. Lee, C. D. Warren, and R. W. Jeanloz, *J. Am. Chem. Soc.* **106,** 811 (1984).
[68] A. Bax and D. G. Davis, *J. Magn. Reson.* **63,** 207 (1985).

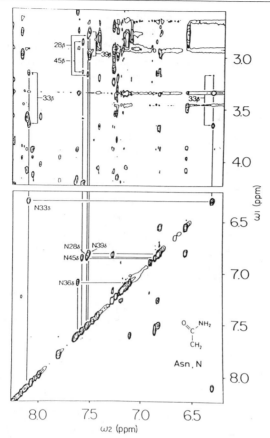

FIG. 12. Illustration of the use of NOESY data to correlate $^1H^\beta$ resonances to amide $^1H^\delta$ resonances from the side chains of asparagine residues. Four of the five asparagines of turkey ovomucoid third domain were assigned by backbone sequential assignments; the fifth (Asn-36) was assigned by difference. The pure-phase NOESY data were obtained at 500 MHz in H$_2$O at pH 4.0 and 25°. The protein concentration was 15 m$M$. (From Robertson.[34])

In addition, ROESY data provide a way of distinguishing cross-peaks that arise because of spin-diffusion rather than cross-relaxation because the two signals are 180° out of phase. This information is important for larger proteins, where spin diffusion is a problem. In our experience, ROESY data are quite helpful as a basis for deciding which NOESY cross-peaks arise from spin diffusion and thus should be disregarded for distance measurements.[66] ROESY may be used as well to distinguish between cross-relaxation and exchange effects in a protein.[68]

We use the ROESY pulse sequence ($d_r$–90°–$t_1$–$SL_x$–acq), where $SL_x$ is a spin-lock pulse along the $x$ axis and acq is the signal acquisition period $t_2$.[68] With the Bruker spectrometers, we use TPPI[38] for quadrature detection in $f_1$ and pure-absorption spectra, rather than the States–Haberkorn–Ruben method[39] as described by Bax and Davis.[68]

## Processing, Postprocessing, and Storage of Two-Dimensional NMR Data

Factors critical to data analysis, including accurate baseline or baseplane definition, resolution, and signal-to-noise ratio, are influenced by the available computer hardware and software. The amount of hard disk memory space limits the size of data files that can be handled, and data transfer rates to an array processor can limit the speed of these devices for large spectra. Array processors used to accelerate fast Fourier transform calculations ideally should handle floating-point data so as to provide the necessary precision. Software approaches can assist in dealing with the very large data sets that are required for high resolution and optimal analysis of NMR parameters. The programs described in the Appendix at the end of this volume have been developed in NMRFAM[69] for the Bruker Aspect (1,000, 2,000, or 3,000),[70–72] Silicon Graphics Iris 2400T, and 4D-series workstations.[17,51] Another software package for semiautomated spectral assignments in proteins is being developed on personal computers.[73]

## Methods for Sequential Assignment and Structural Analysis

*Sequential Assignment Strategies.* Methods for achieving sequential assignments in proteins (cf. Basus [7], Vol. 177, this series) can be divided into three categories based on the NMR parameter that provides transpeptide information (Table IV; Fig. 13). (1) Detection of short transpeptide contacts by 2D NOE spectroscopy: $d_{\alpha N}$, ($^1H_i^\alpha$, $^1H_{i+1}^N$); $d_{NN}$, ($^1H_i^N$, $^1H_{i+1}^N$); $d_{\beta N}$, ($^1H_i^\beta$, $^1H_{i+1}^N$) (Fig. 13), (2) detection of one-bond heteronuclear coupling: ($^{13}C_i'$–$^{15}N_{i+1}$), or (3) detection of multiple-bond heteronuclear coupling: $^2j_{C'H}N$ (two-bond $^{13}C_i'$–$N^\alpha$–$^1H_{i+1}^N$), $^3j_{C'H}\alpha$ (three-bond $^{13}C_i'$–$N^\alpha$–$C^\alpha$–$^1H_{i+1}^\alpha$), or $^3j_{H^\alpha N}$ ($^1H_i^\alpha$–$C^\alpha$–$C'$–$^{15}N_{i+1}$).

[69] Software requests should be addressed to Operations Assistant, National Magnetic Resonance Facility at Madison, Biochemistry Dept., University of Wisconsin–Madison, 420 Henry Mall, Madison, WI 53706.

[70] Z. Zolnai, S. Macura, and J. L. Markley, *J. Magn. Reson.* **80**, 60 (1988).

[71] Z. Zolnai, S. Macura, and J. L. Markley, *J. Magn. Reson.*, in press (1989).

[72] Z. Zolnai, S. Macura, and J. L. Markley, to be published.

[73] W. M. Westler, E. L. Ulrich, and J. L. Markley, *J. Cell. Biochem. Suppl.* **13A**, 30 (Abstr. CA315) (1989).

TABLE IV
TWO-DIMENSIONAL NMR STRATEGIES FOR AMINO ACID IDENTIFICATION AND
SEQUENCE-SPECIFIC ASSIGNMENT

| Method | Connectivities determined | NMR experiment used[a] | Protein labeling required | Requirement for $^1H_2O$ solvent[b] | Comments | References[d] |
|---|---|---|---|---|---|---|
| A | Distances[c] $d_{\alpha N}$, $d_{\beta N}$, $d_{NN}$ | NOESY | None | Yes | Requires elucidation of proton spin systems; NOE intensities will depend on the peptide backbone conformation; additional distance information is required for proline which lacks an $H^N$ | 1,2 |
| B | One-bond coupling $^1J_{C'N}$ | $^{13}C\{^{15}N\}SBC$ | $^{13}C$ and $^{15}N$ | No | Coupling independent of the backbone conformation | 3,4 |
| C | Multiple-bond coupling $^2J_{C'H^N}$ | $^1H\{^{13}C\}MBC$ | $^{13}C$ | Yes | Coupling will be independent of the peptide backbone conformation | 5–7 |
| | $^3J_{C'H^\alpha}$ | $^1H\{^{13}C\}MBC$ | $^{13}C$ | No | The three-bond coupling constant depends on the dihedral angle $\phi$; cross-peaks may be weak or missing for unfavorable backbone conformations | |
| | $^3J_{H^\alpha N}$ | $^1H\{^{15}N\}MBC$ | $^{15}N$ | Yes | The three-bond coupling constant depends on the dihedral angle $\psi$; cross-peaks may be weak or missing for unfavorable backbone conformations | |

[a] Pulse sequence nomenclature is defined in Table III.

(*continued*)

Footnotes to TABLE IV (*continued*)

  $^b$ If $^1H_2O$ is used as the solvent, solvent suppression by irradiation of the water peak may lead to saturation of backbone amide proton signals. This would lead to a break in the sequential assignment chain. This can be circumvented by changing the sample temperature to shift the water peak to a different location.

  $^c$ Distances are defined in Fig. 13.

  $^d$ Key to references: (1) K. Wüthrich, F. Wider, G. Wagner, and W. Braun, *J. Mol. Biol.* **155**, 311 (1982); (2) K. Wüthrich, "NMR of Proteins and Nucleic Acids." Wiley, New York, 1986; (3) M. Kainosho and T. Tsuji, *Biochemistry* **21**, 6273 (1982); (4) B. H. Oh, W. M. Westler, P. Darba, and J. L. Markley, *29th Exp. NMR Conf.*, 1988, Abstr. No. 138; B. J. Stockman, W. M. Westler, E. S. Mooberry, and J. L. Markley, *29th Exp. NMR Conf.*, 1988, Abstr. No. 122; W. M. Westler, B. J. Stockman, Y. Hosoya, Y. Miyake, M. Kainosho, and J. L. Markley, *J. Am. Chem. Soc.* **110**, 6256 (1988); (5) W. M. Westler, M. Kainosho, H. Nagao, N. Tomonaga, and J. L. Markley, *J. Am. Chem. Soc.* **110**, 4093 (1988); (6) G. M. Clore, A.'Bax, P. Wingfield, and A. M. Gronenborn, *FEBS Lett.* **238**, 17 (1988); (7) A. M. Gronenborn, A. Bax, P. T. Wingfield, and G. M. Clore, *FEBS Lett.* **243**, 93 (1989).

*Heteronuclear Approaches.* Sequential assignments based on heteronuclear one- to three-bond couplings along the peptide backbone of a protein require labeling the protein with $^{13}C$ and/or $^{15}N$.[29] This approach has been used successfully with larger proteins[14–16,19,74,75] that were not amenable to the $^1H\{^1H\}$ NOE assignment method. Advantage can be taken of the redundancy of possible transpeptide connectivities afforded by heteronuclear coupling (Fig. 13B and C) in order to increase the reliability of sequential assignments. The popularity of the heteronuclear approaches is expected to increase as it becomes easier and more economical to label proteins with $^{13}C$ and $^{15}N$.[27] Various strategies can be envisioned based on the protein-labeling pattern and the type of NMR spectroscopy to be used to determine the sequential assignment.

One-bond $^{13}C$–$^{15}N$ couplings in proteins can be observed in 1D $^{13}C$[14,74,75] or $^{15}N$ spectra or as passive coupling in $^1H\{^{13}C\}$ or $^1H\{^{15}N\}$ 2D spectra. If the carbonyl carbon of all residues $X$ is labeled with $^{13}C$ and the $\alpha$-nitrogen of all residues $Z$ is labeled with $^{15}N$, then the peptide bond between all dipeptides $X$–$Z$ will be identified by $^{13}C$–$^{15}N$ coupling. If the dipeptide is unique, then the residues are identified uniquely. Specific labeling of dipeptides $X$–$Z$ can be achieved conveniently by feeding the $[^{13}C']X$ and $[^{15}N]Z$ amino acids to microorganisms producing the protein of interest. Single-residue repeats are identified by feeding a single $^{13}C_0$-, $N_\alpha$-labeled amino acid. Care must be taken to defeat any metabolic pathways that lead to significant migration of the label into another amino acid type. The advantage of specific labeling is that the dipeptides can be

---

[74] M. Kainosho, N. Nagao, Y. Imamura, K. Uchida, N. Tomonaga, Y. Nakamura, and T. Tsujii, *J. Mol. Struct.* **126**, 549 (1985).

[75] M. Kainosho, H. Nagao, and T. Tsuji, *Biochemistry* **26**, 1068 (1987).

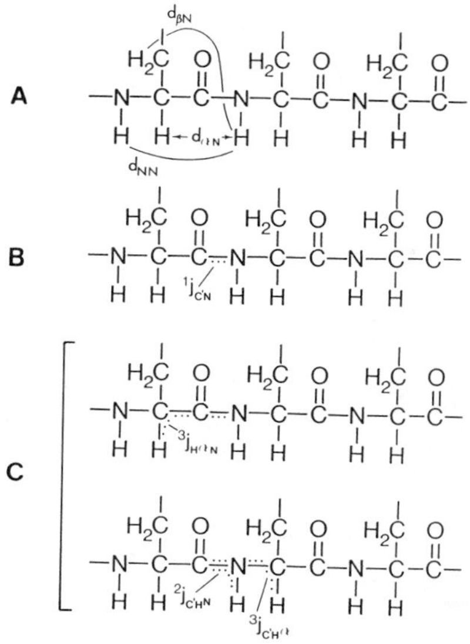

FIG. 13. Schematic peptide chain illustrating NMR connectivities used in various strate-gies for sequential assignments in proteins: (A) $^1$H–$^1$H NOE connectivities[4]; (B) one-bond $^{13}$C–$^{15}$N spin–spin coupling connectivity[14-16]; (C) multiple-bond $^1$H–$^{13}$C and $^1$H–$^{15}$N spin–spin coupling connectivities.[19,20] See Table IV for additional details.

identified unambiguously by a 1D experiment; this feature may be impor-tant in studies of protein conformational changes or ligand-binding stud-ies. The disadvantage of the approach is that a separate labeled protein is required for each dipeptide identification.

By making use of uniformly labeled proteins along with 2D NMR methods for spectral analysis, the number of labeled protein analogues can be reduced sharply. One approach would be to label uniformly with $^{13}$C so that the carbon spin systems can be identified by $^{13}$C{$^{13}$C}DQC spectroscopy and then to label selectively with a single $^{15}$N$^\alpha$ amino acid. The identity of residue $i + 1$ is specified by the [$^{15}$N]amino acid enriched, and the identity of residue $i$ is deduced from analysis of the carbon spin system.

Uniform $^{15}$N labeling of proteins produced by microorganisms is easy and relatively inexpensive. The proton–nitrogen multiple-bond correla-tion experiment provides a transpeptide connectivity based on 3-bond coupling ($^1H_i$–C$^\alpha$–C′–$^{15}N_{i+1}$). The magnitude of the coupling depends on the dihedral angle $\Psi$ and varies between $-6$ and $1$ Hz. Connectivities probably will not be observed for very small angles ($<3$ Hz).[75a] This

[75a] G. M. Clore, A. Bax, P. Wingfield, and A. M. Gronenborn, *FEBS Lett.* **238**, 17 (1988).

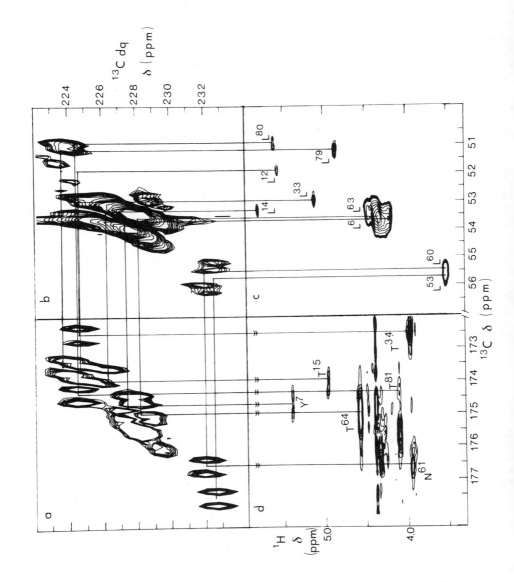

approach has been used to obtain several sequential assignments in the protein Mu *ner*.[75a,75b]

The proton–carbon multiple-bond correlation experiment[76] ($^1$H{$^{13}$C}MBC) allows transpeptide assignments to be achieved with proteins labeled with $^{13}$C at the backbone carbonyl positions. The protein could be labeled uniformly to assist in residue identification or only the carbonyl carbons could be labeled. Sequential assignments in the streptomyces subtilisin inhibitor (SSI) have been obtained by this method[19] (Fig. 14). Transpeptide connectivities are provided by 2-bond coupling, which is independent of protein conformation but in general will require data acquisition in $^1$H$_2$O ($^{13}C_i'-N-^1H_{i+1}^N$), or 3-bond coupling, which will be dependent on the dihedral angle $\phi$ ($^{13}C_i'-N-C^\alpha-^1H_{i+1}^\alpha$).

The most general approach may prove to be a combination of double $^{13}$C and $^{15}$N labeling with $^{13}$C{$^{15}$N}SBC spectroscopy (Figs. 11 and 13). The uniform $^{13}$C label will facilitate the residue identifications and spin system assignments and the transpeptide $^{13}C_i-^{15}N_{i+1}$ coupling will provide se-

[75b] A. M. Gronenborn, A. Bax, P. T. Wingfield, and G. M. Clore, *FEBS Lett.* **243**, 93 (1989).
[76] A. Bax and M. F. Summers, *J. Am. Chem. Soc.* **108**, 2093 (1986).

FIG. 14. NMR spectra of *Streptomyces* subtilisin inhibitor (SSI) containing [85% U-$^{13}$C]leucine. The sample was 100 mg of the labeled protein in 2 ml 0.05 *M* phosphate buffer, pH 7.3, in $^2$H$_2$O. The probe temperature was 61°. Dioxane in $^2$H$_2$O at 61° was used as an external reference and was assigned a chemical shift of 67.8 ppm with respect to (CH$_3$)$_4$Si. The data were collected on a Bruker AM-500 NMR spectrometer (500 MHz for $^1$H; 125.77 MHz for $^{13}$C). (a) $^{13}C'$, ($^{13}C' + ^{13}C^\alpha$) region of a resonance-offset-compensated $^{13}$C–$^{13}$C double-quantum-correlated ($^{13}$C{$^{13}$C}DQC) spectrum (see text). The total delay time for the double-quantum propagator was 9.08 msec. The recycle time was 1 sec, and 256 blocks of 4096 data points were collected. The probe size was 10 mm, and the $^{13}$C 90° pulse width was 14.5 μsec. (b) $^{13}C^\alpha$, ($^{13}C' + ^{13}C^\alpha$) region of the $^{13}$C{$^{13}$C}DQC spectrum. (c) $^1H^\circ$, $^{13}C^\alpha$ region of the one-bond proton–carbon chemical shift correlation ($^1$H{$^{13}$C)}SBC) spectrum obtained with 0.5 ml of the [85% U-$^{13}$C]leucine SSI solution in a 5-mm $^1$H NMR probe equipped with a broadband decoupling coil that was tuned to the $^{13}$C frequency. The recycle time was 1.5 sec; the delay for build-up of one-bond heteronuclear antiphase coherence was 3.57 msec; 512 blocks of 2048 data points were collected. Proton pulses were generated from the decoupler channel, and the 90° pulse width was 7 μsec. The decoupler and receiver reference frequencies were generated identically. The $^{13}$C pulses for excitation and decoupling were generated by first attenuating the low-power transmitter output of the Bruker AM-500 and then amplifying the signal with an external ENI 20-W amplifier. The power input to the $^{13}$C coil was about 5 W, giving a $^{13}$C pulse width of 105 μsec. (d) $^1H^\alpha$, $^{13}C'$ region of the multiple-bond proton–carbon chemical shift correlation spectrum of [85% U-$^{13}$C]leucine SSI. Conditions were as in (c), but 1024 blocks of 2048 data points were collected, and the $^{13}$C pulses were generated by the standard BSV-7 transmitter which gave a 90° pulse width of 11 μsec. The delay time for build-up of anti-phase multiple-bond coherence was 60 msec. (From Westler *et al.*[15])

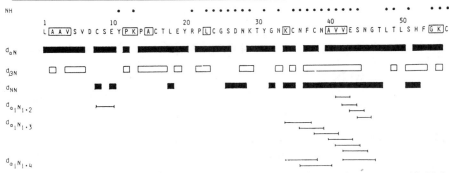

FIG. 15. Summary of short-range 2D NOE data obtained from turkey ovomucoid third domain (OMTKY3). The sequence of the protein is shown in one-letter code. Sequential assignment starting sites in the sequence are boxed. NOESY connectivities are indicated below the sequence. The dots above the sequence denote slowly exchanging $H^N$s as defined by observation of cross-peaks in the COSY map obtained with OMTKY3 freshly dissolved in $^2H_2O$ at pH 3.84 and 25°. (Adapted from Robertson *et al.*[44])

quential assignments. In principle, complete backbone assignments can be achieved by reference to two experiments: $^{13}C\{^{15}N\}SBC$ ($^{13}C_i'-^{15}N_{i+1}$ and $^{15}N_{i+1}-^{13}C^\alpha_{i+1}$) and $^{13}C\{^{13}C\}DQC$ ($^{13}C^\alpha_{i+1}-^{13}C_{i+1}'$). Recent experience with dual $^{13}C/^{15}N$ labeled proteins demonstrates feasibility of this approach.[15,16,73]

*Homonuclear Proton Approach.* The overwhelming majority of sequence-specific assignments obtained to date (Table I) have been achieved by strategy A in Table IV: detection of NOE connectivities between the $^1H^N$ of residue $i + 1$ and protons of the adjacent residue $i$.[4,77] In most cases, one or more of the interresidue distances ($d_{\alpha N}$, $d_{NN}$, or $d_{\beta N}$) (Fig. 13) can be inferred from NOESY cross-peaks. Information about the amino acid sequence and the spin system types to which these resonances belong (derived, for example, from COSY, RELAY, and HOHAHA data) is used to distinguish nearest neighbor NOE connectivities from longer range ones.

Starting points for sequential assignments are provided by unique single residues or unique di- and tripeptides whose identities can be deduced from spin-system analyses plus interresidue NOE connectivities. The residues enclosed in boxes (Fig. 15) in the sequence of turkey ovomucoid third domain (OMTKY3) are the ones that served as starting points for its sequential assignment.[44] Once starting points have been identified, they can be extended by combining NOESY and COSY data. Figure 16 shows two runs of backbone sequential assignments in OMTKY3: residues 34–39 and 40–46. Two-dimensional NOE cross-peaks supporting the full se-

[77] K. Wüthrich, F. Wider, G. Wagner, and W. Braun, *J. Mol. Biol.* **155**, 311 (1982).

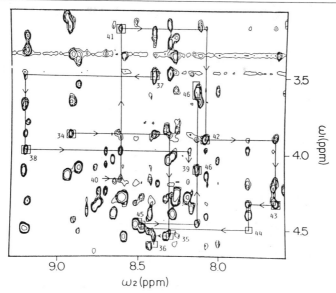

FIG. 16. An assignment spiral for turkey ovomucoid third domain based on interresidue $d_{\alpha N}$ connectivities and intraresidue COSY fingerprint region cross-peaks. The connectivities are represented according to the convention suggested by Wagner et al.[78] in which NOESY data alone are presented and the positions of COSY-type cross-peaks are indicated by boxes. The assignments are indicated by the numbers next to the boxes. Horizontal lines represent the NOESY $d_{\alpha N}$ connectivities used to link a resonance of residue $i$ to one of residue $i + 1$. The vertical line is then used to locate the COSY fingerprint ($^1H^\alpha$, $^1H^N$) cross-peak for residue $i + 1$. The assignments indicated in this figure are for residues 34–36, 37–39, and 40–46. (From Robertson et al.[44])

quential assignments are summarized in the $d_{\alpha N}$, $d_{NN}$, and $d_{\beta N}$ bars in Fig. 15.

*Determination of Secondary and Tertiary Structure.* Once extensive sequence-specific $^1H$ NMR peak assignments have been obtained, NOESY cross-peaks that represent longer range proton–proton connectivities can be catalogued (Figs. 17 and 18). Patterns of these NOE's are used in identifying turns and helical regions.[4] The $\alpha$ helix of OMTKY3 was identified on the basis of characteristic $^3J_{H^\alpha H^N}$ coupling and detection of 2D NOE cross-peaks corresponding to $d_{\alpha N(i, i+2)}$, $d_{\alpha N(i, i+3)}$, $d_{\alpha N(i, i+4)}$, and $d_{\alpha \beta (i, i+3)}$).[43] The triple-stranded antiparallel $\beta$ sheet was determined on the basis of medium and long-range NOE's (Fig. 19).[44]

Figure 20 compares the solution structures of OMTKY3 and OMTKY3* as determined from 2D NMR data. (OMTKY3* is the reac-

[78] G. Wagner, D. Neuhaus, E. Wörgötter, M. Vasak, J. H. R. Kägi, and K. Wüthrich, *Eur. J. Biochem.* **157,** 275 (1986).

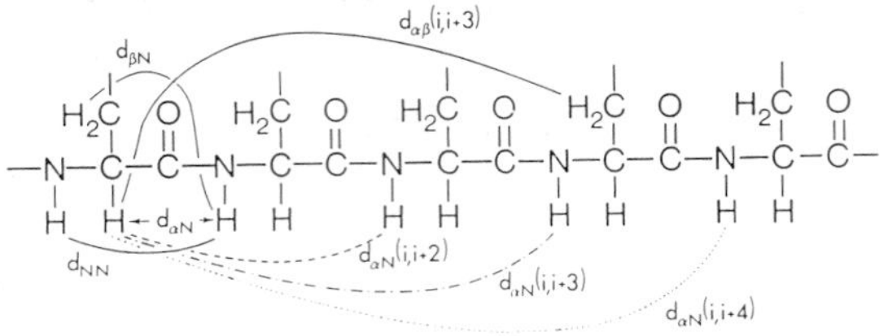

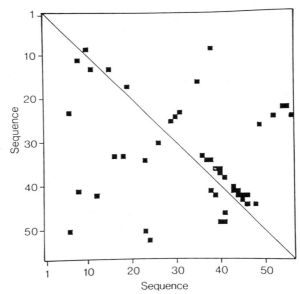

FIG. 17. Illustration of short-range and longer-range sequential distances in a protein identifiable from 2D NOE spectra. Secondary structural elements (helices and turns) can be deduced from patterns of these connectivities in conjunction with information on $^3J_{H\alpha_HN}$ coupling constants and hydrogen exchange rates.[4] From Robertson.[34]

tive-site modified form of OMTKY3 in which the Leu$^{18}$-Glu$^{19}$ peptide bond has been cleaved.) These structures currently are at relatively low resolution; they will be refined by back-calculation of 2D NOE spectra which will provide additional assignments as well as the means of checking previous assignments and distance measurements.

FIG. 18. Map of $^1$H–$^1$H contacts in turkey ovomucoid third domain derived from 2D NOE data. Only medium- and long-range contacts are presented. The region above the diagonal contains main-chain contacts while that below the diagonal represents contacts involving side-chain protons. Only a single contact between any two residues is indicated. (From Robertson.[34])

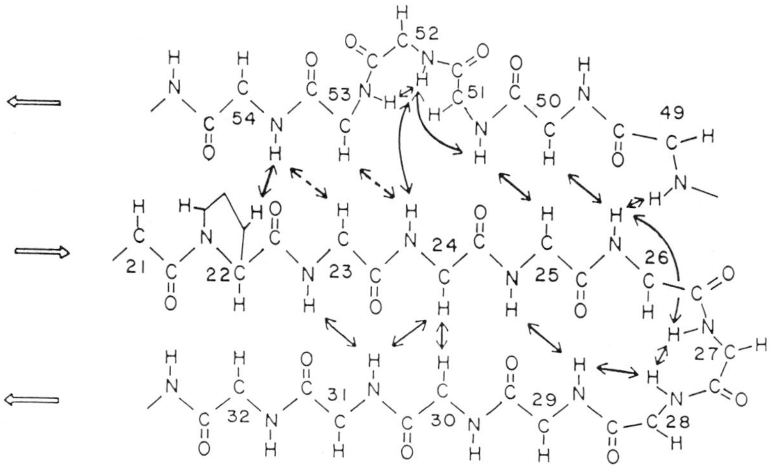

FIG. 19. Three-stranded antiparallel β-pleated sheet in turkey ovomucoid third domain (OMTKY3) deduced from NMR analysis. The solid arrows represent NOE's used to determine the structure. The dashed lines indicate possible NOE connectivities that were obscured by overlapping cross-peaks. (From Robertson et al.[44])

## Future Prospects

*Higher-Dimensional NMR.* Three-dimensional or 4D NMR methods offer attractive possibilities for protein studies, either as combinations of ¹H homonuclear 2D methods[81] or as a marriage between heteronuclear and ¹H homonuclear methods.[82,83] Promising applications of 3D NMR to protein spectroscopy have been reported.[82–85] The initial experience has been that low digital resolution in the third dimension is sufficient to achieve the expected additional resolution of cross-peaks. The long data accumulation times required for 3D experiments (3 days or more) place a

---

[79] J. Thomasson, M. Day, and I. D. Kuntz, "A Vectorized Distance Geometry Program" (to be published).

[80] W. F. Walkenhorst, A. M. Krezel, P. Darba, A. D. Robertson, G. I. Rhyu, and J. L. Markley, "The Process of Protein Folding," AAAS Annual Meeting, 1989, Abstr. 25.

[81] C. Griesinger, O. W. Sørensen, and R. R. Ernst, *J. Magn. Reson.* **73**, 574 (1987).

[82] G. W. Vuister, R. Boelens, and R. Kaptein, *J. Magn. Reson.* **80**, 176 (1988).

[83] S. W. Fesik and E. R. P. Zuiderweg, *J. Magn. Reson.* **78**, 588 (1988).

[84] H. Oschkinat, C. Griesinger, P. J. Kraulis, O. W. Sørensen, R. R. Ernst, A. M. Gronenborn, and G. M. Clore, *Nature (London)* **332**, 374 (1988).

[85] H. Oschkinat, C. Cieslar, A. M. Gronenborn, and G. M. Clore, *J. Magn. Reson.* **81**, 212 (1989).

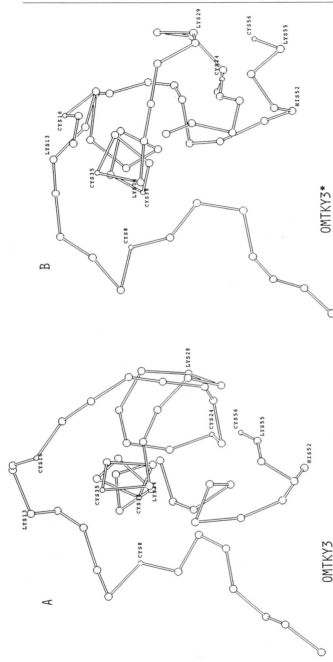

**OMTKY3**

**OMTKY3\***

FIG. 20. Comparison of solution structures for (A) virgin (OMTKY3) and (B) reactive-site modified (OMTKY3\*) turkey ovomucoid third domain. The peptide bond between residues 18 and 19 is cleaved in the modification reaction. The structure of OMTKY3 was determined from 311 interproton distances (119 sequential backbone, 16 sequential side-chain, 144 medium- and long-range, and 31 intraresidue constraints) derived from NOE spectra collected in water with mixing times of 100 and 180 ms. The structure of OMTKY3\* is based on 355 interproton distances (98 sequential backbone, 31 sequential side-chain, 145 medium- and long-range, and 77 intraresidue constraints). The NOE spectra were base-plane corrected, and the number of levels were counted on contour plots. Distances were sorted into four groups (2.5, 3.0, 3.5, and 4.0 Å based on the cross-peak intensities). These distances were used as upper-bound limits with the lower bound limits set to the van der Waals contact distances. The dg900 program[79] available on the Cray X-MP at the San Diego Supercomputing Center was used for computations. This program employs the metric matrix distance geometry algorithm for structure calculation. Ten structures were generated for each protein; the structures shown are those which had the fewest number of input distance violations. From Walkenhorst et al.[80]

high requirement on protein stability in solution. Several groups are developing software for processing and graphic display of 3D NMR data.

*Alternatives to the Fourier Transform.* The fast Fourier transform (FFT) always has been the mainstay for processing pulsed NMR data.[86] The advantages of FFT are the high computational speed, ease of implementation, and linearity of the transform. However, the FFT can cause large distortions in the spectrum if the time domain signal is truncated or if some of the data points are not collected. In 2D NMR, the $f_1$ dimension often is truncated because of limited accumulation time, or the initial part of the data is missing owing to delays incorporated into the pulse sequence. Both of these factors lead to distortions in FFT spectra that can be suppressed by digital filtering, but at the expense of decreased resolution or signal-to-noise ratio.

The maximum entropy method (MEM)[87,88] and various linear prediction (LP) methods (LPSQV, LPZ, etc.)[89–93] have been explored recently as alternatives to the FFT (Fig. 1) (cf. Hoch [12], this volume). Claims have been made for a higher signal-to-noise ratio and for decreased spectral distortions. The linear prediction method offers the further advantage of direct extraction of spectral parameters. Current disadvantages of these approaches are lengthy computation times and possible loss of spectral information under certain conditions.

*Symmetry and Pattern Recognition.* Symmetry characteristics of 2D NMR spectra have been used for some time to eliminate noise in spectra that have mirror symmetry across the diagonal ridge.[94] This "diagonalization" or "triangular multiplication" method is an example of more general procedures that use the mathematics of group theory. Recently, several investigators[95–99] have suggested the use of symmetry point groups and the tenets of group theory for pattern recognition in various forms of

[86] R. R. Ernst and W. A. Anderson, *Rev. Sci. Instrum.* **37**, 93 (1966).
[87] E. D. Laue, J. Skilling, and J. Staunton, *J. Magn. Reson.* **63**, 418 (1985).
[88] S. Sibisi, J. Skilling, R. G. Brereton, E. D. Laue, and J. Staunton, *Nature (London)* **311**, 446 (1984).
[89] H. Barkhuijsen, J. de Beer, W. M. M. J. Bovée, and D. van Ormondt, *J. Magn. Reson.* **61**, 465 (1985).
[90] H. Gesmar and J. J. Led, *J. Magn. Reson.* **76**, 183 (1988).
[91] J. Tang, C. P. Lin, M. K. Bowman, and J. R. Norris, *J. Magn. Reson.* **62**, 167 (1985).
[92] J. Tang and J. R. Norris, *J. Magn. Reson.* **69**, 180 (1986).
[93] A. E. Schussheim and D. Cowburn, *J. Magn. Reson.* **71**, 371 (1987).
[94] R. G. Baumann, G. Wider, R. R. Ernst, and K. Wüthrich, *J. Magn. Reson.* **44**, 402 (1981).
[95] P. Pfändler, G. Bodenhausen, B. U. Meier, and R. R. Ernst, *Anal. Chem.* **57**, 2510 (1985).
[96] P. Pfändler and G. Bodenhausen, *J. Magn. Reson.* **70**, 71 (1986).
[96a] P. H. Bolton, *J. Magn. Reson.* **68**, 180 (1986).
[97] P. Pfändler, M. Nović, H. Oschkinat, S. Wimperis, G. Jaccard, U. Eggenberger, D. Limat, and G. Bodenhausen, *28th Exp. NMR Conf., 1987*, Abstr. Tuesday AM.

2D NMR spectroscopy. As an example, the antiphase doublets that arise in most experiments involving coherence transfer can be classified by their symmetry to the $C_{2h}$ point group. Based on the techniques of projection operators, antiphase doublets can be projected from a 2D NMR spectrum. Other types of peaks in 2D NMR spectra can be classified to other point groups, e.g., $D_{2d}$ for an $AX$ cross-peak. These techniques should be useful for automated "peak picking" in both homonuclear and heteronuclear correlated spectroscopy. Additional approaches, such as cluster analysis or principal component analysis,[100,101] also show promise for peak-pattern recognition.

*Semiautomated Analysis.* In the process of assigning resonances, the spectroscopist often makes use of information from several different NMR spectra plus auxiliary information such as the sequence of the protein and any other structural information available. Graphics-based computer programs that access the database and can implement automatic or semiautomatic user-assisted strategies should speed up this analysis considerably.[102] Several groups have been developing pattern recognition techniques to automate peak picking,[95-99,103] while others have produced algorithms for automating sequence-specific peak assignments,[104-107] or recognition of secondary structural elements.[108]

*Database Management.* Analysis of the structure of a biological macromolecule by NMR spectroscopy requires a tremendous amount of information. Chemical exchange, dynamics, and other studies involve extensive series of spectra, especially if measurements are made as a function of additional variables such as pH and temperature. Each spectrum collected may contain up to several hundred peaks to be cataloged. Current methods of data analysis usually involve manual calculations and data tabulation. An obvious improvement would be the use of a data-

[98] J. C. Hoch, F. M. Poulson, S. Hengyi, M. Kjaer, and S. Ludvigsen, *28th Exp. NMR Conf., 1987*, Abstr. MK 31.
[99] S. Glaser and H. R. Kalbitzer, *J. Magn. Reson.* **74**, 450 (1987); K. P. Neidig and H. R. Kalbitzer, *Magn. Reson. Chem.* **26**, 848 (1988).
[100] Z. L. Mádi, B. U. Meier, and R. R. Ernst, *J. Magn. Reson.* **74**, 565 (1987).
[101] B. U. Meier, A. L. Mádi, and R. R. Ernst, *J. Magn. Reson.* **74**, 565 (1987).
[102] H. Grahn, F. Delaglio, M. A. Delsuc, and G. C. Levy, *28th Exp. NMR Conf. 1987*, Abstr. WK 30.
[103] B. U. Meier, G. Bodenhausen, and R. R. Ernst, *J. Magn. Reson.* **60**, 161 (1984).
[104] S. W. Englander and A. J. Wand, *Biochemistry* **26**, 5953 (1987).
[105] E. L. Ulrich and J. L. Markley, *13th Int. Conf. Magn. Reson. Biol. Syst.*, 1988, Abstr. S8-6.
[106] M. Billeter, V. J. Basus, and I. D. Kuntz, *J. Magn. Reson.* **76**, 400 (1988).
[107] C. Cieslar; G. M. Clore, and A. M. Gronenborn, *J. Magn. Reson.* **80**, 119 (1988).
[108] J. Brugge, B. G. Buchanen, and O. Jardetzky, *J. Comput. Chem.* **9**, 662 (1988).

base management program and a "scientific spread sheet" software package.[105]

*NMR Data Bases.* The Protein Data Bank of X-ray crystallographic data for proteins and nucleotides has proved very useful to the scientific community. Methods for predicting protein structures, analysis of protein domain structures, studies of amino acid hydrophobicities, investigations of protein structural homology, and other studies have all benefited from a centralized and accessible repository of structural data. With the advent of sequential assignment techniques and methods to determine three-dimensional structures from NMR NOE data, an explosion in the amount and quality of NMR structural information about biopolymers is occurring. An accessible repository of these data should prove to be a valuable research tool.[30] Correlations of NMR parameters (chemical shifts, line widths, coupling constants, NOE's, etc.) and derived values (interatomic distances, $\tau_c$, $S^2$, $k_{ex}$, $pK_a$, etc.) with structural features of biopolymers could lead to new insights about structure–function relationships. Data bank holdings are useful in theoretical studies of the effects of ring currents and other structural features on chemical shifts[109]; they also provide a convenient means for determining which proteins have been investigated and to what extent.

## Conclusion

The protein NMR field appears to be entering a new phase. Available and newly emerging 2D NMR methods make it possible to resolve and assign signals from a large number of $^1H$, $^{13}C$, and $^{15}N$ nuclei in proteins of $M_r \leq 20,000$. The size barrier appears to be advancing into the 20,000–30,000-Da range and may be expected to rise higher still. However, it seems clear that $T_2$ relaxation will prevent a dramatic increase in this limit. Future challenges lie in the areas of extracting useful information from 1D and 2D NMR data more quickly and more reliably. The modeling of 3D structures from NMR data is undergoing rapid development.[4,110–114] Advances in stereospecific assignment of prochiral protons[115–117] are lead-

[109] K.-H. Gross and H. R. Kalbitzer, *J. Magn. Reson.* **76**, 87–99 (1988).

[110] G. M. Clore and A. M. Gronenborn, *Protein Eng.* **1**, 275 (1987).

[111] K. Wüthrich, *Science* **243**, 45 (1989).

[112] K. Wüthrich, *Acc. Chem. Res.* **22**, 36 (1989).

[113] R. Kaptein, R. Bolens, R. M. Scheek, and W. F. van Gunsteren, *Biochemistry* **27**, 5389 (1988).

[114] O. Lichtarge, C. W. Cornelius, B. G. Buchanen, and O. Jardetzky, *Proteins: Struct. Funct. Genet.* **2**, 340 (1988).

[115] D. M. LeMaster, *FEBS Lett.* **223**, 191 (1987).

[116] P. L. Weber, R. Morrison, and D. Hare, *J. Mol. Biol.* **204**, 483 (1988).

[117] P. C. Driscoll, A. M. Gronenborn, and G. M. Clore, *FEBS Lett.* **243**, 223 (1989).

ing to structures of higher resolution. Much more structural and dynamic information derivable from NMR experiments remains to be exploited, however. It will be fascinating to see how these powerful tools develop further and how they can be applied to the solution of important problems in protein chemistry.

### Acknowledgments

It is a pleasure to thank my collaborators and co-workers who have contributed greatly to what is discussed here, in particular, Denise A. Benway, T.-M. Chan, Prashanth Darba, Jasna Fejzo, Masatsune Kainosho, Christopher L. Kojiro, Andrej Krezel, Slobodan Macura, Douglas C. McCain, Ed S. Mooberry, Byung-Ha Oh, Gilberto Ortiz-Polo, Michael D. Reily, Gyung Ihm Rhyu, Andrew D. Robertson, James F. Shurts, Brian J. Stockman, M. Albert Thomas, Eldon L. Ulrich, William F. Walkenhorst, Jinfeng Wang, W. Milo Westler, Michael H. Zehfus, and Zsolt Zolnai. This research was supported by grants from the National Institutes of Health (GM35976), National Science Foundation (DMB-84-10222), and U.S. Department of Agriculture (USDA/SEA 88-37262-3406). NMR data were collected at the National Magnetic Resonance Facility at Madison, Wisconsin [supported by NIH Grant RR02301, with equipment purchased with funds from the University of Wisconsin, NSF Biological Instrumentation Program (Grant DMB-8415048), NIH Biomedical Research Technology Program (RR02301), NIH Shared Instrumentation Program (Grant RR02781), and the U.S. Department of Agriculture].

# [3] Solvent Suppression

## By P. J. HORE

### Introduction

Fourier transform (FT) proton NMR ($^1$H NMR) of dilute aqueous solutions is fraught with difficulties. The potentially enormous signal from the water protons ($\sim 100\ M$) can cause severe dynamic range problems in the receiver, the analog-to-digital converter (ADC), and the computer. Unless great care (and often ingenuity) is exercised the weak solute resonances are submerged beneath an immense, broad water peak with its attendant spinning side bands, spurious harmonics, and digitization noise. The numerous strategies developed to deal with these problems (which are of course not restricted to water) are known by the term *solvent suppression*.

Unfortunately no single strategy is ideal in all circumstances. The "best" approach depends on a number of factors, including the sample, the type of information desired, the spectrometer, the skill of the operator, the time available for setting up the experiment, the degree of suppression required, the NMR experiment to be performed, and so on. In the following pages I attempt to summarize the practical advantages and disadvantages of the major methods of solvent suppression, without going into the theory behind them. The emphasis is on techniques developed during the last 5 years: earlier methods have been extensively reviewed elsewhere.[1-6] Although I shall concentrate on water suppression, most of the methods are applicable to other solvents and other nuclei.

### Hardware and Software

Dynamic range problems can be alleviated by making instrumental modifications and by altering the way in which digitized signals are accumulated and treated.[4,7] At the ADC stage, the most commonly perceived problem—that the solute signals may be smaller than one-half of the least significant bit (LSB) in the ADC and thus undetectable—may not be a problem at all. Thermal noise or random variability in the solvent signal from one transient to the next ensures that the weak signals of interest are in fact digitized.[4] Quantization harmonics of the water peak, arising from the $\pm\frac{1}{2}$LSB uncertainty in the digitized signals, can be minimized by setting the carrier to coincide with the solvent and by signal averaging.[4] There thus seems to be little reason to increase the ADC resolution beyond 16 bits unless the thermal noise and signal variabilities are smaller than one-half the LSB or extensive signal averaging is not required. Double-precision accumulation of the digitized signals avoids the need for scaling and block averaging[8] (to prevent memory overflow) and a floating point Fourier transform reduces the rounding and scaling errors during the FT operation.[9,10] Other methods for avoiding memory overflow (data shift accumulation and alternate delay accumulation[11]) are useful only in

[1] A. G. Redfield, *NMR: Basic Princ.* **13**, 137 (1976).
[2] A. G. Redfield, this series, Vol. 49, p. 253.
[3] A. G. Redfield and S. D. Kunz, *in* "NMR and Biochemistry" (S. J. Opella and P. Lu, eds.), p. 225. Dekker, New York, 1979.
[4] J. C. Lindon and A. G. Ferrige, *Prog. NMR Spectrosc.* **14**, 27 (1980).
[5] P. J. Hore, *J. Magn. Reson.* **55**, 283 (1983).
[6] C. J. Turner, *Prog. NMR Spectrosc.* **16**, 311 (1984).
[7] S. R. Maple and A. Allerhand, *J. Magn. Reson.* **72**, 203 (1987).
[8] J. W. Cooper, *in* "Topics in $^{13}$C NMR Spectroscopy" (G. C. Levy, ed.), Vol. 2, p. 391. Wiley, New York, 1976.
[9] J. W. Cooper, *J. Magn. Reson.* **22**, 345 (1976).
[10] J. W. Cooper, I. S. Mackay, and G. B. Pawle, *J. Magn. Reson.* **28**, 405 (1977).
[11] K. Roth, B. J. Kimber, and J. Feeney, *J. Magn. Reson.* **41**, 302 (1980).

the absence of double-precision arithmetic. Differential sampling,[12] over-sampling,[13] and notch filters[14] may also reduce problems at the ADC stage. In the receiver there appears to be ample scope for improving the dynamic range and linearity of the preamplifiers and mixers supplied with commercial spectrometers.

Even if all the dynamic range problems can be eliminated, one is still left with an enormous, broad $H_2O$ resonance of unpredictable shape, possibly with spinning side bands, obscuring many of the solute resonances. Cosmetic improvements may be afforded by data massaging[4,15,16] but it seems unlikely that the offending peak can be removed completely and routinely especially if nearby solute signals are to be retained without distortion. Moreover, any small instrumental instabilities that modulate the amplitude of the water peak from transient to transient will prove disastrous in two-dimensional (2D) NMR experiments. There is thus a need for some attenuation of the $H_2O$ signal, ideally before it reaches the receiver. Most of the methods to be discussed depend critically on good shimming: probe modifications that improve the spatial homogeneity of the static field ($B_0$) are well worth considering.[17]

### Deuteration

The most obvious method of water suppression is to use deuterium oxide ($D_2O$) in place of $H_2O$.[2] The major problem is that labile NH and OH protons exchange for deuterons and so disappear from the spectrum. The protons lost in this way contribute to the residual water signal (HDO and $H_2O$) which may cause dynamic range problems itself. Methods for the preparation and handling of samples in $D_2O$ are described elsewhere in this volume.[18] If the protons of interest exchange in a time shorter than it takes to record a spectrum, it is essential to work in $H_2O$ (usually with 10% $D_2O$ to provide a lock signal).

### Presaturation

Continuous weak (and therefore frequency-selective) radiofrequency irradiation prior to excitation and acquisition to eliminate the water mag-

[12] S. Davies, C. Bauer, P. Barker, and R. Freeman, *J. Magn. Reson.* **64,** 155 (1985).
[13] M. A. Delsuc and J. Y. Lallemand, *J. Magn. Reson.* **69,** 504 (1986).
[14] A. G. Marshall, T. Marcus, and J. Sallos, *J. Magn. Reson.* **35,** 227 (1979).
[15] P. Plateau, C. Dumas, and M. Guéron, *J. Magn. Reson.* **54,** 46 (1983).
[16] P. T. Callaghan, A. L. MacKay, K. P. Pauls, O. Soderman, and M. Bloom, *J. Magn. Reson.* **56,** 101 (1984).
[17] R. W. Dijkstra, *J. Magn. Reson.* **72,** 162 (1987).
[18] N. J. Oppenheimer, this volume [4].

netization is undoubtedly the simplest and most effective suppression method.[19] The following points must be borne in mind. Irradiation (usually from the decoupler) should ideally be gated off during data acquisition to avoid Bloch–Siegert shifts despite the reduction in the level of suppression this entails. The choice of decoupler power and irradiation time is usually a compromise: high decoupler power will give efficient elimination of the solvent resonance but will also saturate protons with nearby chemical shifts; long preirradiation times give good saturation but reduce the efficiency of signal averaging if the delay between transients has to be increased. Inhomogeneously broadened lines are difficult to saturate: careful shimming is therefore mandatory. Spatial inhomogeneity of the decoupler field is beneficial in that it dephases the precessing solvent magnetization.[19] However, regions of the sample that experience both a very inhomogeneous $B_0$ field and a weak decoupler field are difficult to shim and to saturate and give rise to broad "humps" that may extend over several hundred hertz from the water.[17] These undesirable resonances may be attenuated by shielding parts of the probe,[17] especially the leads to the radiofrequency coil. Other advantages of preirradiation are that it can easily be used in conjunction with most existing NMR experiments and that time-shared irradiation allows more than one peak to be saturated, although generally with some loss of efficiency. Improvements in the degree of suppression can be obtained by cycling the phase of the decoupler relative to the transmitter,[19,20] by making the preirradiation coherent with the carrier,[21] by including a homospoil pulse between the preirradiation and excitation,[2] and by shifting the irradiation frequency by small amounts from one transient to the next.

The principal drawback is that saturation transfer may reduce the intensity of exchanging protons.[22,23] Appreciable loss of signal is to be expected unless the exchange rate is at least five times slower than the spin-lattice relaxation of the protons in question. Cross-relaxation from water to solute protons can have a similar effect.[22,23] Cutting down the length of irradiation gives less time for saturation transfer and cross-relaxation but requires a stronger and thus less selective decoupler field to achieve the same level of suppression.[22] Conditions can be chosen to minimize the rates of exchange and hence the loss of signal (e.g., acidic pH for proteins, basic pH for nucleic acids[24]) but this may be undesirable

[19] D. I. Hoult, *J. Magn. Reson.* **21**, 337 (1976).
[20] J. D. Cutnell, J. Dallas, G. Matson, G. N. La Mar, H. Rink, and G. Rist, *J. Magn. Reson.* **41**, 213 (1980).
[21] E. R. P. Zuiderweg, K. Hallenga, and E. T. Olejniczak, *J. Magn. Reson.* **70**, 336 (1986).
[22] I. D. Campbell, C. M. Dobson, and R. G. Ratcliffe, *J. Magn. Reson.* **27**, 455 (1977).
[23] J. D. Stoesz, A. G. Redfield, and D. Malinowski, *FEBS Lett.* **91**, 320 (1978).
[24] K. Wüthrich, "NMR of Proteins and Nucleic Acids." Wiley, New York, 1986.

or even impossible depending on the sample. Inevitably some protons lie directly under the water or are sufficiently close to it in frequency that they do not escape saturation. Any protons bleached in this way are invisible in both one- and two-dimensional experiments, but can sometimes be revealed by exploiting the temperature dependence of the water chemical shift. Spin diffusion from these saturated protons to the remainder of the molecule may lead to a general reduction in sensitivity, especially in large molecules.

Relaxation

Methods that exploit inherent or induced differences in $T_1$ or $T_2$ are unique among the major suppression techniques in permitting undistorted detection of resonances very close to the water (because there is no frequency selection). However, like presaturation, they are subject to the problems of saturation transfer and cross-relaxation.

The basic $T_1$ experiment (WEFT, water-eliminated Fourier transform[25]) involves the inversion recovery pulse sequence 180°–$\tau$–90°–acquire. Both pulses are nonselective and the delay $\tau$ is chosen (by trial and error) so that the relaxing solvent magnetization is passing through a null at the moment the 90° pulse is applied. Unless the solute protons relax much more rapidly (or slowly) than the water, there is a loss of signal and a distortion in the relative intensities of protons relaxing at different rates. WEFT is less successful for $H_2O$ than for the HDO signal in $D_2O$ because of the shorter $T_1$ of $H_2O$ and the more pronounced radiation damping[26] during $\tau$. As originally described,[25] WEFT requires complete relaxation of the solvent magnetization between transients. This, together with the $\tau$ delay, makes the experiment very slow, especially for HDO. In practice, both the repetition time and $\tau$ can be reduced to increase the efficiency of signal averaging. Excellent spatial homogeneity of the radiofrequency field ($B_1$) is needed to get an accurate 180° pulse (composite pulses[27] do not seem to have been used in this context), otherwise different regions of the sample go through a null at different times. Improved suppression can be obtained by introducing a homospoil pulse during $\tau$ to eliminate any transverse magnetization.[28] An important variant of WEFT uses a selec-

[25] S. L. Patt and B. D. Sykes, *J. Chem. Phys.* **56**, 3182 (1972).
[26] R. Freeman, "A Handbook of Nuclear Magnetic Resonance," p. 177. Longman, New York, 1987.
[27] M. H. Levitt, *Prog. NMR Spectrosc.* **18**, 61 (1986).
[28] F. W. Benz, J. Feeney, and G. C. K. Roberts, *J. Magn. Reson.* **8**, 114 (1972).

tive 180° pulse so that only the water is inverted.[29–31] This removes the need for differential spin-lattice relaxation and avoids intensity distortions for the majority of the solute protons, which experience only the nonselective 90° pulse.

When the solvent $T_1$ is much smaller than that of the solute, fast pulsing to achieve a steady state in which the water is partially saturated can be advantageous.[28,32]

When the solvent $T_2$ is much shorter than the solute $T_2$ values, spin-echo methods (e.g., 90°–$\tau$–180°–$\tau$–acquire) may be used for solvent suppression[33] (WATR, water attenuation by $T_2$ relaxation). The delay $\tau$ is chosen so that the solvent $xy$ magnetization has relaxed completely before the start of acquisition (an echo is needed to avoid problems with phase shifts associated with delayed acquisition). In practice the Carr–Purcell–Meiboom–Gill sequence with a high-pulse repetition rate should be used to minimize echo phase modulation by scalar couplings. If the solvent $T_2$ is not intrinsically short, it may be made so by the addition of reagents that are paramagnetic[34,35] (e.g., $MnCl_2$) or have exchangeable protons[36–38] (e.g., $NH_4Cl$). The latter have two drawbacks: each reagent is only effective over a restricted pH range[37] and rather high concentrations are required (typically 0.5 $M$), which may cause difficulties if the solute undergoes a conformational change at high ionic strength. Careful choice of reagent and its concentration is needed lest solute protons are also relaxed. Both approaches are capable of impressive water suppression. Neither is satisfactory for solutes (e.g., macromolecules) with short $T_2$ values.

## Selective Excitation

An alternative to preirradiation is the use of pulse sequences that deliver zero net excitation at the frequency of the water resonance. The idea is to leave the water magnetization accurately along the $+z$ axis while exciting spins at different chemical shifts into the $xy$ plane. The

[29] R. K. Gupta, *J. Magn. Reson.* **24**, 461 (1976).
[30] G. A. Morris and R. Freeman, *J. Magn. Reson.* **29**, 433 (1978).
[31] C. A. G. Haasnoot, *J. Magn. Reson.* **52**, 153 (1983).
[32] J. Lauterwein and I. P. Gerothanassis, *J. Magn. Reson.* **51**, 153 (1983).
[33] D. L. Rabenstein and A. A. Isab, *J. Magn. Reson.* **36**, 281 (1979).
[34] R. G. Bryant and T. M. Eads, *J. Magn. Reson.* **64**, 312 (1985).
[35] T. M. Eads, S. D. Kennedy, and R. G. Bryant, *Anal. Chem.* **58**, 1752 (1986).
[36] D. L. Rabenstein, S. Fan, and T. T. Nakashima, *J. Magn. Reson.* **64**, 541 (1985).
[37] D. L. Rabenstein and S. Fan, *Anal. Chem.* **58**, 3178 (1986).
[38] S. Connor, J. Everett, and J. K. Nicholson, *Magn. Reson. Med.* **4**, 461 (1987).

water magnetization only deviates from equilibrium for the duration of the pulse sequence (typically a few milliseconds) so minimizing saturation transfer and cross-relaxation effects. More than 20 selective excitation methods have been proposed,[3,5,15,30,39–59] differing in their ease of implementation, sensitivity to instrumental imperfections, the offset dependence of the excitation, and the phase corrections needed in the spectrum. Some sequences require painstaking optimization of experimental parameters (pulse and delay lengths, phase shifts, etc.) to achieve acceptable results, others do not. The following general points can be made. (1) Sequences that require the carrier to be on resonance with the water [e.g., $90°-\tau-\overline{90}°$, also called jump and return[43] (JR) or $1-\overline{1}$, where the overbar indicates a 180° phase-shifted pulse] are usually superior to those that must have the carrier close to the resonances of interest (e.g., $90°-\tau-90°$[46] and Redfield's 2–1–4 method[41]). The suppression produced by the former is less sensitive to $B_0$ and $B_1$ inhomogeneity.[5] The latter require either twice the spectral width or a separate spectrum to be recorded with the carrier in a different position, if resonances on both sides of the water are to be observed. However, sequences with only one phase ($90°-\tau-90°$, etc.) are not prey to errors in phase shifts. (2) Of the sequences that deliver on-resonance suppression, those involving soft pulses (e.g., the

[39] S. Alexander, *Rev. Sci. Instrum.* **32**, 1066 (1961).

[40] A. G. Redfield and R. K. Gupta, *J. Chem. Phys.* **54**, 1418 (1971).

[41] A. G. Redfield, S. D. Kunz, and E. K. Ralph, *J. Magn. Reson.* **19**, 114 (1975).

[42] J. M. Wright, J. Feigon, W. Denny, W. Leupin, and D. R. Kearns, *J. Magn. Reson.* **45**, 514 (1981).

[43] P. Plateau and M. Guéron, *J. Am. Chem. Soc.* **104**, 7310 (1982).

[44] V. Sklenář and Z. Starčuk, *J. Magn. Reson.* **50**, 495 (1982).

[45] D. L. Turner, *J. Magn. Reson.* **54**, 146 (1983).

[46] G. M. Clore, B. J. Kimber, and A. M. Gronenborn, *J. Magn. Reson.* **54**, 170 (1983).

[47] P. J. Hore, *J. Magn. Reson.* **54**, 539 (1983).

[48] H. Bleich and J. Wilde, *J. Magn. Reson.* **56**, 154 (1984).

[49] J. H. Gutow, M. McCoy, F. Spano, and W. S. Warren, *Phys. Rev. Lett.* **55**, 1090 (1985).

[50] Z. Starčuk and V. Sklenář, *J. Magn. Reson.* **61**, 567 (1985).

[51] Z. Starčuk and V. Sklenář, *J. Magn. Reson.* **66**, 391 (1986).

[52] C. Yao, Y. Simplaceanu, A. K.-L. C. Lin, and C. Ho, *J. Magn. Reson.* **66**, 43 (1986).

[53] G. A. Morris, K. I. Smith, and J. C. Waterton, *J. Magn. Reson.* **68**, 526 (1986).

[54] M. P. Hall and P. J. Hore, *J. Magn. Reson.* **70**, 350 (1986).

[55] M. H. Levitt and M. F. Roberts, *J. Magn. Reson.* **71**, 576 (1987).

[56] M. H. Levitt, J. L. Sudmeier, and W. W. Bachovchin, *J. Am. Chem. Soc.* **109**, 6540 (1987).

[57] M. H. Levitt, *J. Chem. Phys.* **88**, 3481 (1988).

[58] C. Wang and A. Pardi, *J. Magn. Reson.* **71**, 154 (1987).

[59] A. T. Hsu, W. W. Hunter, P. Schmalbrock, and A. G. Marshall, *J. Magn. Reson.* **72**, 75 (1987).

NERO methods[55-57]) require more careful setting of $B_1$ but suffer less from switching transients. (3) Sequences with even numbers of pulses (e.g., 1–$3$–$3$–$\bar{1}$ [5,45]) are less prone to problems with nonrectangular pulse shapes than those with odd numbers (e.g., OBTUSE: $\bar{1}$–$3$–$8$–$\bar{3}$–$\bar{1}$ [53]). (4) Some sequences (e.g., $90°$–$\tau$–$\overline{90}°$) have no internal compensation for the effects of relaxation and radiation damping and so may give poorer suppression than expected. (5) Sequences involving small flip angle pulses ($\leq 5°$) often give problems with switching transients and require attenuation of the transmitter to enable pulse lengths to be set with reasonable accuracy.

The degree of suppression obtained for a given sequence is crucially dependent on the nature and magnitude of the spectrometer imperfections and instabilities: a technique that works splendidly on one machine may fail miserably on another. In this respect the "best" sequence is a matter of personal preference. On a more objective note, however, the various approaches can be distinguished by two important features: the frequency dependence of the excitation they deliver and the degree of phase correction needed in the resulting spectrum. The excitation null at the water frequency should be broad and flat to cope with $B_0$ inhomogeneity and spinning side bands but not so broad as to suppress resonances of interest close to the water. Large frequency-dependent phase corrections cause rolling baselines especially in the presence of broad or overlapping solute resonances or a strong residual water peak[15,57]: these are particularly intrusive in two-dimensional spectra.

To allow the reader to judge the properties of some selected sequences, calculated values of three quantities, defined as follows (see Fig. 1):

$$\psi = \Delta\nu_{band}/\Delta\nu; \qquad \Gamma = \Delta\nu_{null}/\Delta\nu; \qquad \Pi = |\Delta\phi|/\psi$$

are set out in Table I. $\Delta\nu_{band}$ is the half-width of the excitation band at 90%

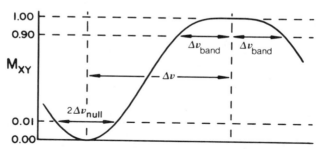

FIG. 1. Schematic excitation spectrum of a pulse sequence used for solvent suppression. The absolute value transverse magnetization, $M_{xy} = (M_x^2 + M_y^2)^{1/2}$ is drawn as a function of frequency, $\nu$. $M_{xy} = 1$ usually corresponds to a $90°$ pulse. Note the nonlinear scale.

TABLE I

PROPERTIES OF PULSE SEQUENCES FOR SOLVENT SUPPRESSION

| Sequence[a,b] | Abbreviated form | $\psi$ | $\Gamma$ | $\Pi^c$ | Notes |
|---|---|---|---|---|---|
| $90°-\frac{1}{2}\tau-\overline{90}°$ | JR | 0.29 | 0.0064 | 0 | d |
| $\alpha-\tau-\overline{\alpha}$ ($2\alpha = 90°$) | | 0.46 | 0.0045 | 220 | d |
| $\alpha-\tau-\overline{2}\alpha-\tau-\alpha$ ($4\alpha = 90°$) | $1-\overline{2}-1$ | 0.33 | 0.054 | 420 | e |
| $\alpha-\tau-\overline{3}\alpha-\tau-3\alpha-\tau-\overline{\alpha}$ ($8\alpha = 90°$) | $1-\overline{3}-3-\overline{1}$ | 0.27 | 0.13 | 620 | f, g |
| $9.0°-2\tau-48.6°-\tau-\overline{48.6°}-2\tau-\overline{9.0°}$ | | 0.54 | 0.071 | 960 | h |
| $5.4°-\tau-5.4°-\tau-45°-\tau-\overline{45°}-\tau-\overline{5.4°}-\tau-\overline{5.4°}$ | | 0.46 | 0.088 | 960 | h |
| $7.1°-4\tau-10.9°-\tau-\overline{55.1°}-\tau-55.1°-\tau-10.9°-4\tau-\overline{7.1°}$ | | 0.76 | 0.038 | 2020 | i |
| $6.8°-\tau-4.5°-\tau-7.4°-\tau-\overline{53.3°}-\tau-53.3°-\tau-7.4°-\tau-4.5°-\tau-\overline{6.8°}$ | | 0.65 | 0.060 | 1320 | i |
| $\alpha-\tau-3\alpha-\tau-\overline{3}\alpha-\tau-\overline{\alpha}$ ($4\alpha = 90°$) | $1-3-\overline{3}-\overline{1}$ | 0.35 | 0.0018 | 510 | j |
| $3\alpha-\tau-5\alpha-\tau-15\alpha-\tau-\overline{15}\alpha-\tau-\overline{5}\alpha-\tau-\overline{3}\alpha$ ($26\alpha = 90°$) | $3-5-15-\overline{15}-\overline{5}-\overline{3}$ | 0.83 | 0.0014 | 930 | j |
| $\overline{\alpha}-\tau-\overline{3}\alpha-\frac{1}{2}\tau-8\alpha-\frac{1}{2}\tau-\overline{3}\alpha-\tau-\overline{\alpha}$ ($8\alpha = 90°$) | OBTUSE | 0.44 | 0.041 | 540 | k |
| $\overline{\alpha}-\tau-\overline{5}\alpha-\tau-10\alpha-\frac{1}{2}\tau-32\alpha-\frac{1}{2}\tau-\overline{10}\alpha-\tau-\overline{5}\alpha-\tau-\overline{\alpha}$ ($32\alpha = 90°$) | OBTUSE | 0.55 | 0.033 | 930 | k |
| $90°$ ($h = 1/\sqrt{15}$) | | 0.41 | 0.0069 | 410 | l |
| $2\alpha-\overline{4}\alpha-4\alpha-\overline{\alpha}-2\alpha$ ($6\alpha = 44°; h = 0.205$) | $2-1-4$ | 0.25 | 0.014 | 380 | m, n |
| $\alpha-\overline{2}\alpha-\alpha$ ($\alpha = 34.2°; h = 0.337$) | Soft $1-\overline{2}-1$ | 0.33 | 0.054 | 480 | o |
| $\alpha-\overline{2}\alpha-\alpha$ ($\alpha = 108°; h = 1.24$) | Soft $1-\overline{2}-1$ | 0.32 | 0.055 | 120 | o |
| $81\tau'-120°-50\tau'-115°-225\tau'-115°-154\tau'-$ $115°-225\tau'-115°-50\tau'-\overline{120}°-81\tau'$ ($h = 10.0$) | NERO-1 | 0.54 | 0.079 | 80 | p |
| $14\tau'-80°-165\tau'-\overline{200}°-50\tau'-85°-300\tau'-$ $\overline{85}°-50\tau'-\overline{200}°-165\tau'-\overline{80}°-14\tau'$ ($h = 2.0$) | NERO-1' | 0.52 | 0.074 | 60 | q |
| $48\tau'-\overline{110}°-115\tau'-\overline{130}°-240°-360\tau'-$ $240°-\overline{130}°-115\tau'-\overline{110}°-48\tau'$ ($h = 2.0$) | NERO-2 | 0.62 | 0.029 | 80 | r |

| | | | | | |
|---|---|---|---|---|---|
| $66\tau'-\overline{125}°-165\tau'-\overline{50}°-105\tau'-175°-175°-510\tau'-$ <br> $175°-105\tau'-\overline{50}°-165\tau'-\overline{125}°-66\tau'$ $(h = 5.0)$ | NERO-2' | 0.65 | 0.027 | 50 | $q$ |
| $74\tau'-\overline{45}°-100\tau'-\overline{155}°-25\tau'-200°-210\tau'-$ <br> $200°-25\tau'-\overline{155}°-100\tau'-\overline{45}°-74\tau'$ $(h = 2.0)$ | NERO-3 | 0.58 | 0.033 | 20 | $q$ |
| $90°-\tfrac{1}{2}\tau-\overline{90}°-(90°-\tau-\overline{90}°)_{ex}$ | 1–1 echo | 0.17 | 0.14 | 0 | $s, t, u$ |

$^a$ $\tau = (2\Delta\nu)^{-1}$; $\tau' = (360\Delta\nu)^{-1}$; $h = \Delta\nu_{exc}/\Delta\nu$; $\Delta\nu_{exc} = \gamma B_1/2\pi$.

$^b$ Except where otherwise stated, the pulses are infinitely hard ($h = \infty$).

$^c$ $\Pi$ is quoted in degrees.

$^d$ P. Plateau and M. Guéron, J. Am. Chem. Soc. 104, 7310 (1982).

$^e$ V. Sklenář and Z. Starčuk, J. Magn. Reson. 50, 495 (1982).

$^f$ D. L. Turner, J. Magn. Reson. 54, 146 (1983).

$^g$ P. J. Hore, J. Magn. Reson. 55, 283 (1983).

$^h$ Z. Starčuk and V. Sklenář, J. Magn. Reson. 66, 391 (1986).

$^i$ M. P. Hall and P. J. Hore, J. Magn. Reson. 70, 350 (1986).

$^j$ C. Wang and A. Pardi, J. Magn. Reson. 71, 154 (1987).

$^k$ G. A. Morris, K. I. Smith, and J. C. Waterton, J. Magn. Reson. 68, 526 (1986).

$^l$ S. Alexander, Rev. Sci. Instrum. 32, 1066 (1961).

$^m$ A. G. Redfield, S. D. Kunz, and E. K. Ralph, J. Magn. Reson. 19, 114 (1975).

$^n$ At the center of the excitation band the spins experience a 44° pulse. All other sequences deliver a 90° rotation.

$^o$ C. Yao, V. Simplaceanu, A. K.-L. C. Lin, and C. Ho, J. Magn. Reson. 66, 43 (1986).

$^p$ M. H. Levitt and M. F. Roberts, J. Magn. Reson. 71, 576 (1987).

$^q$ M. H. Levitt, J. Chem. Phys. 88, 3481 (1988).

$^r$ M. H. Levitt, J. L. Sudmeier, and W. W. Bachovchin, J. Am. Chem. Soc. 109, 6540 (1987).

$^s$ V. Sklenář and A. Bax, J. Magn. Reson. 74, 469 (1987).

$^t$ M. von Kienlin, M. Decorps, J. P. Albrand, M. F. Foray, and P. Blondet, J. Magn. Reson. 76, 169 (1988).

$^u$ "ex" signifies EXORCYCLE. See text.

of the maximum excitation; $\Delta\nu_{null}$ is the half-width of the null at 1% of the maximum excitation; $\Delta\nu$ is the offset from the null of the center of the excitation band and $\Delta\phi$ is the maximum variation in the phase of transverse magnetization excited within the 90% band.

The following general points can be made. Sequences with small $\Gamma$ may give poor suppression if the water resonance is inhomogeneously broadened. The closer $\psi$ is to unity, the wider is the range of offsets over which there is approximately uniform excitation. Smaller values of $\Pi$ are to be preferred unless baseline correction is feasible or absolute value spectra can be tolerated.

The hard pulse sequences in Table I are quoted for infinite $B_1$ with delays in terms of $\Delta\nu$: the shape of the excitation profile, and hence the values of $\psi$, $\Gamma$, and $\Pi$, will differ slightly for smaller $B_1$ especially if the delays are taken from the end of one pulse to the beginning of the next[52] rather than between the pulse centers. For the soft pulse sequences, the value of $\Delta\nu_{exc}$ ($= \gamma B_1/2\pi$) relative to $\Delta\nu$ must be set reasonably accurately. The overall flip angle produced by the hard pulse sequences may be altered, without too much effect on $\Delta\nu$ or on the shape of the excitation spectrum, by scaling the durations of all the constituent pulses. The soft pulse sequences cannot, in general, be modified in this way: changes in the pulse durations and/or $B_1$ usually alter $\Delta\nu$. The NERO sequences[55–57] have much better phase properties than most of the earlier sequences but the phase gradient is not linear in frequency; if phase correction is deemed necessary, the software supplied with commercial spectrometers will not suffice.

Other sequences are available that need smaller frequency-dependent phase corrections and/or have narrower nulls and thus wider bands of excitation. Linear phase gradients may be refocused by adding a 180° pulse at the end of the pulse sequence[60] (and a $\overline{180°}$ pulse in front to minimize radiation damping during acquisition), e.g., $\overline{180°}$–1–$\overline{3}$–3–$\overline{1}$–180°–$\tau_R$–acquire. Very accurate 180° pulses are required to obtain good suppression. A more promising approach, proposed independently by Bax[61] and Decorps,[62,63] is the "1–1 echo": $90°$–$\frac{1}{2}\tau$–$\overline{90°}$–$(90°$–$\tau$–$\overline{90°})_{ex}$–acquire, i.e., two JR pulses, the first with one-half the delay of the second, with phase cycling of the second and the receiver according to the EXORCYCLE scheme. Although the suppression on each transient is poor (of

[60] G. J. Galloway, L. J. Haseler, M. F. Marshman, D. H. Williams, and D. M. Doddrell, *J. Magn. Reson.* **74,** 184 (1987).

[61] V. Sklenář and A. Bax, *J. Magn. Reson.* **74,** 469 (1987).

[62] M. von Kienlin, M. Decorps, J. P. Albrand, M. F. Foray, and P. Blondet, *J. Magn. Reson.* **76,** 169 (1988).

[63] P. Blondet, M. Decorps, S. Confort, and J. P. Albrand, *J. Magn. Reson.* **75,** 434 (1987).

the same order as a single JR), that after the four-step phase cycle is excellent. The 90% excitation band is rather narrow but there are no frequency-dependent phase shifts (Table I). Several related sequences have been proposed,[62] e.g., $90°-\frac{1}{2}\tau-\overline{90}°-(45°-\tau-\overline{45}°)_{ex}$–acquire.

An arbitrarily narrow null may be achieved with combinations of hard and soft pulses—a soft pulse to excite only the solvent and a hard pulse to bring it back to the $+z$ axis while at the same time exciting the solute protons. The width of the null is determined by the properties of the soft pulse. Various schemes exist, including (soft pulses are italicized) $\overline{90}°-90°$ [30,64]; $90°_{-y}-90°_x$–spin lock$_y$[65]; $\overline{45}°-90°-\overline{45}°$,[64] and shaped pulses consisting of a narrow strong pulse centered on a weak, broad pulse 180° out of phase.[49] The last two produce strong frequency-dependent phase errors because of the delay between the hard pulse and acquisition. Amplitude-modulated soft pulses (e.g., Gaussian) give better results than rectangular pulses.[64] These methods, though promising, make severe demands on instrumental stability, amplifier linearity, etc.,[64] and pulse-shaping devices are yet to become widely available on commercial spectrometers.

### Two-Dimensional NMR

The majority of the above techniques can be used for 2D NMR. The most popular and satisfactory approach is presaturation[24,66,67] (gated off during $t_1$ and $t_2$ to avoid Bloch–Siegert shifts): alternative methods are only worth considering if resonances of interest are attenuated by saturation transfer, cross-relaxation, or direct saturation.

Of the various methods that exploit differences in relaxation behavior,[68–70] selective WEFT appears to be the most versatile[68]: a soft 180° pulse can be used to null the water prior to the first pulse of any 2D experiment. The water magnetization that recovers during $t_1$ is converted into transverse magnetization by subsequent pulses but may be canceled to some extent by phase cycling. WATR has been used for homonuclear 2D-$J$ spectroscopy[71] and shift correlation[36] but is less generally applicable because of the need for a spin echo element in the pulse sequence.

[64] V. Sklenář, R. Tschudin, and A. Bax, *J. Magn. Reson.* **75**, 352 (1987).

[65] V. Sklenář and A. Bax, *J. Magn. Reson.* **75**, 378 (1987).

[66] G. Wider, R. V. Hosur, and K. Wüthrich, *J. Magn. Reson.* **52**, 130 (1983).

[67] G. Wider, S. Macura, A. Kumar, R. R. Ernst, and K. Wüthrich, *J. Magn. Reson.* **56**, 207 (1984).

[68] J. P. K. Tong and G. Kotovych, *J. Magn. Reson.* **69**, 511 (1986).

[69] L.-F. Kao and V. J. Hruby, *J. Magn. Reson.* **70**, 394 (1986).

[70] P. A. Mirau, *J. Magn. Reson.* **73**, 123 (1987).

[71] D. L. Rabenstein, G. S. Srivatsa, and R. W. K. Lee, *J. Magn. Reson.* **71**, 175 (1987).

Selective excitation is currently the most attractive of the alternatives to preirradiation. NOESY $(90°-t_1-90°-\tau-90°-t_2)$ is easily modified by replacing the final read pulse with a selective excitation sequence.[61,64,72-76] This only works satisfactorily when transverse water magnetization is removed prior to the read sequence with a homospoil pulse during the evolution period, $\tau$. Suppression methods that do not generate phase gradients are preferable as these allow pure absorption phase spectra to be recorded. Cross-peaks are only lost if neither of the coupled spins is significantly excited by the read sequence. The rotating frame NOE experiment (CAMELSPIN, also known as ROESY) can also be modified in this way.[77]

Water suppression in COSY $(90°-t_1-90°-t_2)$ experiments is less straightforward. The simplest strategy is to substitute selective sequences for both 90° pulses.[64,69,78,79] Absolute value presentation is essential if either or both of the sequences produce phase gradients in the excited magnetization. The minimum requirement for the appearance of a cross-peak between two scalar-coupled spins is that the first pulse excites one spin of the pair and the second pulse excites the other. There are thus two perpendicular trenches running parallel to the $f_1$ and $f_2$ axes, crossing at the water diagonal peak, in which cross-peaks are suppressed. The sequence $90°-t_1-\overline{90°}-t_2$ (cf. jump and return) generates a hole (rather than trenches) at the position of the water diagonal peak[80] but the need to keep the two pulses 180° apart in phase precludes quadrature detection in $t_1$. A similar modification of CAMELSPIN has been suggested.[81]

In more complex 2D experiments, it is often crucial that spins at different offsets experience pulses of the same phase: in these circumstances selective excitation sequences that produce phase gradients cannot be used.[57] An attractive strategy is to restrict the selective pulse element to the final read pulse of the 2D sequence and ensure that the water $xy$ magnetization is minimized prior to the read pulse. HOHAHA,[61,77] ZZCOSY,[82] and both homonuclear[83] and heteronu-

[72] J. D. Cutnell, *J. Am. Chem. Soc.* **104**, 362 (1982).

[73] A. L. Schwartz and J. D. Cutnell, *J. Magn. Reson.* **53**, 398 (1983).

[74] C. A. G. Haasnoot and C. W. Hilbers, *Biopolymers* **22**, 1259 (1983).

[75] H. Santos, D. L. Turner, and A. V. Xavier, *J. Magn. Reson.* **58**, 344 (1984).

[76] V. Sklenář, B. R. Brooks, G. Zon, and A. Bax, *FEBS Lett.* **216**, 249 (1987).

[77] A. Bax, V. Sklenář, G. M. Clore, and A. M. Gronenborn, *J. Am. Chem. Soc.* **109**, 6511 (1987).

[78] P. J. Hore, *J. Magn. Reson.* **56**, 535 (1984).

[79] M. McCoy and W. S. Warren, *Chem. Phys. Lett.* **133**, 165 (1987).

[80] E. Guittet, M. A. Delsuc, and J. Y. Lallemand, *J. Am. Chem. Soc.* **106**, 4278 (1984).

[81] D. Piveteau, M. A. Delsuc, E. Guittet, and J. Y. Lallemand, *J. Magn. Reson.* **71**, 347 (1987).

[82] E. R. P. Zuiderweg, *J. Magn. Reson.* **71**, 283 (1987).

[83] J. H. Prestegard and J. N. Scarsdale, *J. Magn. Reson.* **62**, 136 (1985).

clear[61,84] shift correlation experiments have been successfully modified in this way.

## Other Methods

Various, less general methods have been suggested. Inverse polarization transfer suppresses the resonances from protons not coupled to $^{13}C$ and is of use in metabolic studies with $^{13}C$-labeled substrates.[85,86] Less satisfactory are homo- and heteronuclear echo phase modulation[87–89] and multiple quantum filtration[90,91]: these rely on difference methods to cancel the solvent after several scans and so are prone to dynamic range distortions in the receiver and ADC unless care is taken to avoid exciting the solvent.[92]

Two older techniques are tailored excitation[93] and rapid scan correlation.[94–96] Neither is straightforward to implement and neither has been widely used for solvent suppression.

The problems associated with double resonance (for selective decoupling and NOE measurements) in conjunction with solvent suppression[20,97] can be circumvented to a large degree by two-dimensional methods (COSY and NOESY).

Solvent suppression for NMR experiments *in vivo* is beyond the scope of this chapter. Many of the methods discussed here can be adapted but new problems associated with inhomogeneous magnetic fields must be faced.

## Acknowledgments

I am indebted to A. Bax, M. Decorps, D. M. Doddrell, and M. H. Levitt, who kindly provided preprints of their work prior to publication and to Jonathan Boyd, Robert Cooke, Simon Davies, Ray Freeman, Jan Friedrich, James Keeler, and Gareth Morris for valuable advice and criticism.

[84] A. G. Redfield, *Chem. Phys. Lett.* **96**, 537 (1983).

[85] W. M. Brooks, M. G. Irving, S. J. Simpson, and D. M. Doddrell, *J. Magn. Reson.* **56**, 521 (1984).

[86] J. M. Bulsing and D. M. Doddrell, *J. Magn. Reson.* **68**, 52 (1986).

[87] J. Bornais, S. Brownstein, and S. Bywater, *J. Magn. Reson.* **52**, 120 (1983).

[88] C. L. Dumoulin and E. A. Williams, *J. Magn. Reson.* **66**, 86 (1986).

[89] C. J. Hardy and C. L. Dumoulin, *Magn. Reson. Med.* **5**, 58 (1987).

[90] D. G. Davis an D. Cowburn, *J. Magn. Reson.* **62**, 128 (1985).

[91] C. L. Dumoulin, *J. Magn. Reson.* **64**, 38 (1985).

[92] W. M. Brooks, M. G. Irving, and D. M. Doddrell, *J. Magn. Reson.* **79**, 348 (1988).

[93] B. L. Tomlinson and H. D. W. Hill, *J. Chem. Phys.* **59**, 1775 (1973).

[94] J. Dadok and R. F. Sprecher, *J. Magn. Reson.* **13**, 243 (1974).

[95] R. K. Gupta, J. A. Ferretti, and E. D. Becker, *J. Magn. Reson.* **13**, 275 (1974).

[96] Y. Arata and H. Ozawa, *J. Magn. Reson.* **21**, 67 (1976).

[97] K. Hallenga and W. E. Hull, *J. Magn. Reson.* **47**, 174 (1982).

# [4] Sample Preparation

By NORMAN J. OPPENHEIMER

This chapter is intended to provide short descriptions of important aspects of sample preparation. While it is by no means comprehensive, hopefully the procedures will prove useful and will spark development of new methods. The topics discussed in this chapter are focused on handling protein samples and include: suppressing bacterial growth, avoiding paramagnetic impurities, and methodology of $D_2O$ exchange in samples.

## Suppression of Bacterial Growth

The typical protein or nucleic acid NMR sample can provide a rich growth medium for bacteria or fungi, especially when incubated at 37°. Note that it is the excretion of proteases and nucleases that is the primary cause of sample damage, thus microbial contamination should be considered for samples that show long-term instability. The most effect antimicrobial agent typically used in NMR spectroscopy is $D_2O$. Deuterium is an effective antimicrobial agent at concentrations >80%. Thus for $^1H$ NMR spectroscopy of biomolecules conducted in $D_2O$, there should be few, if any, problems from bacterial or fungal growth. The same cannot be said for corresponding experiments conducted in $H_2O$ solutions. If problems arising from microbial contamination are suspected, the following strategies can be used to suppress growth.

Extremes of pH or temperature are generally effective in blocking growth of most organisms. There are, however, bacteria that can grow at extremes of temperature or pH, thus maintaining samples below pH 4 or at temperatures above 45° is no guarantee that microbial growth can be ignored. Metal chelators, e.g., EDTA (ethylenediaminetetraacetic acid) or EGTA [ethylene glycol bis($\beta$-aminoethyl ether)tetraacetic acid], combined with rigorous exclusion of divalent metal ions including $Ca^{2+}$, $Mg^{2+}$, and transition metals, will strongly suppress microbial growth. Since many proteases and hydrolases also require divalent metal ions for activity, exclusion or sequestering of divalent metals can further protect against damage.

For experiments where divalent metals cannot be excluded, a number of chemical antimicrobials can be used. Sodium azide is very effective against both prokaryotes and eukaryotes at concentrations <50 $\mu M$. Fluoride at similar concentrations is also an effective antimicrobial; however, it cannot be used in conjunction with divalent metal ions because it forms precipitates, e.g., calcium fluoride. Note that both hydrazoic acid and hydrogen fluoride are weak acids and their acid forms are volatile, hence degassing or vigorous lyophilization of solutions with a pH below 7 can lower their concentrations. An alternative to these ions is the addition of 10 $\mu$l of deuterated toluene to suppress growth. Toluene, on the other hand, has a number of properties, including volatility and hydrophobicity, that decreases its utility with most protein samples.

Antibiotics can be used as a last resort. Chloramphenicol is probably the best antimicrobial for this purpose. It is reasonably inert chemically, compatible with divalent metal ions and has a very broad spectrum of antimicrobial activity at concentrations of 10–50 $\mu M$. This low concentration should not pose a problem for $^{13}$C NMR spectroscopy, where sample concentrations above 1 m$M$ are typical. It has nonexchangeable protons that can present a problem for $^1$H NMR spectroscopy, especially at low sample concentrations.

### Paramagnetic Impurities

The vigor with which elimination of paramagnetic impurities is pursued is directly proportional to the importance of such contamination to the particular experiment being conducted. There are three ways to deal with paramagnetic impurities: avoidance, sequestering, and removal.

*Avoidance.* A few simple precautions should eliminate most common sources of paramagnetic ions. Chromic acid-washed glassware should never be used in the preparation of any solutions that will eventually go into an NMR tube. Never clean NMR tubes in chromic acid. The best way to clean glassware is with a product like Nochromix, a sulfuric acid-based oxidative cleaning solution that does not contain transition metals. Deposits of precipitated protein in an NMR tube can be removed using strong detergents and long pipe cleaners (25–30 cm) with the end folded over to avoid scratching the glass (scratches on either the inside or outside of the NMR tube can degrade spectral resolution). Pipe cleaners, however, usually contain either a steel or copper wire. If these metals pose a problem, then pipe cleaners should be used only as a last resort (adding 0.1 $M$ EDTA to soap solution can help cut down on contamination).

Aqueous solutions and reagents used for NMR experiments should not come in contact with stainless steel. High-performance liquid chromatography (HPLC) systems containing stainless steel tubing and fittings can be a significant source of paramagnetic contamination, particularly when large volumes of dilute solution must be concentrated in order to prepare a sample. Aqueous solutions containing chloride or formic acid are notorious for leaching paramagnetic ions from stainless steel. Formic acid is particularly incompatible with most common stainless steels such as surgical stainless steel, e.g., hypodermic needles. HPLC systems constructed of glass and plastic avoid these problems. Other items to use include platinum syringe needles, Teflon-coated spatulas, etc.

*Removal of Paramagnetic Ions.* Chelex is the best means for the routine removal of paramagnetic ions. It contains the same metal-binding moiety as EDTA. It is especially useful for processing the large volumes of solutions produced by column chromatography. Like EDTA, Chelex is most effective at neutral pH and in the absence of competing, nonparamagnetic divalent metal ions like $Mg^{2+}$ or $Ca^{2+}$, and Chelex loses its effectiveness below pH 4 as the carboxylates become protonated. Chelex has relatively low nonspecific adsorptivity; thus high recoveries of most proteins are expected.

Passing a typical NMR sample through a small Chelex column has all the problems inherent in handling small volumes. Dilution of the sample is unavoidable, since it is necessary to wash the column in order to recover all the sample. However, an entire column is not necessary. The concentration of metal ion contaminants in an NMR sample is usually so small that only a few particles of Chelex are needed to sequester the metals. Using these particles as an insoluble form of EDTA, a sample can be incubated in a small vial. Slow and continuous swirling for about 30 min will assure thorough contact of the solution with the beads before the solution is pipetted off. In this way a 0.5-ml sample can be processed with a minimal loss due to adsorption and with no significant dilution.

*Sequestering Paramagnetic Ions.* EDTA is the most widely used agent for sequestering paramagnetic ions in aqueous solution. It also has the fewest nonexchangeable proton resonances. EGTA is marginally more effective at complexing transition metal ions, but it also has more nonexchangeable protons. The concentration of chelator needed depends on the extent of metal contamination. A concentration of between 5 and 10 $\mu M$ is usually sufficient, in most cases, to sequester any adventitious metals. If paramagnetic broadening is still observed in the presence of 50 $\mu M$ EDTA, then treatment with Chelex should be considered.

Although EDTA has proton resonances, if there are trace amounts of paramagnetic ions, these will tend to broaden out the EDTA resonances, hence suppressing them. Most paramagnetic metal ions are in fast exchange with EDTA, hence observation of sharp resonances from the EDTA indicates the absence of significant metal ion contamination, thus the concentration of EDTA can be decreased in future samples. Unfortunately, deuterated EDTA is not yet available commercially. The protons of the $\alpha$-methylene, however, can be exchanged by boiling the $Mg^{2+}$ salt of EDTA in $D_2O$.[1]

## Exchanging Protons with Deuterons

The effectiveness of techniques for solvent suppression (see [3], this volume) have made conducting experiments in $H_2O$ nearly as simple as in $D_2O$. However, for experiments that need to observe resonances in the vicinity of the residual HDO resonance, there is no replacement for efficient exchange of the protons for deuterons. This section discusses methodologies used in the preparation of samples for NMR in $D_2O$.

*Care in Drying of NMR Tubes.* Precision-bore NMR tubes should not be dried at high temperatures (never above 100°) nor for prolonged periods above 50°. These conditions promote creep of the glass resulting in distortions that can increase side bands and interfere with proper spinning of the NMR tubes. Tubes can be dried (inverted) in a drying oven set at 37°. If rapid drying is needed, either drying in a vacuum or blowing filtered nitrogen through the tube is preferable to heating. Volatile solvents such as acetone are only marginally useful for rapid drying because of potential contamination from any residue left on the NMR tube. Air or nitrogen used to dry NMR tubes should be filtered to avoid contamination with dust particles. Note that short of flame drying the tube (which will invariably warp the tube), an NMR tube will always retain a layer of adsorbed water on its surface. Removal of this layer is impractical because any methods that will remove it will probably also ruin the NMR tube; however, the layer can be easily replaced by $D_2O$.

*Exchanging with $D_2O$.* The surface layer of water is a significant source of proton contamination for both NMR tubes and any other item that will come in contact with the NMR solvents, e.g., syringes and pipetter tips. This source of protons can be dealt with by a combination of preconditioning of items prior to use and exchanging solution after they have been prepared.

Soaking items in $D_2O$, drying them, then storing them in a $D_2O$-saturated atmosphere can be ungainly and even ineffective. The desired goal is to have an NMR tube or other items that have had the protons replaced with deuterons rather than to have the item rigorously dry, i.e., anhydrous. Storage and handling preexchanged items at a high level of deuteration can be difficult since proton sources abound. For example, $P_2O_5$ contaminated with protons will exchange the isotope, even though it is a

[1] R. P. Houghton, "Metal Complexes in Organic Chemistry," p. 117. Cambridge Univ. Press, London and New York, 1979.

powerful drying agent. For most cases, prerinsing of syringes or pipetter tips just prior to use, combined with direct lyophilization and exchange of buffers in the NMR tube, is usually sufficient to provide a high level of deuterium exchange. Such procedures also eliminate the need for maintaining a proton-free storage environment by making the exchange of deuterons for protons the last step of sample preparation.

*Lyophilization.* A few general comments regarding lyophilization as it relates to handling protein-containing NMR samples are in order. Lyophilization is a widely used method in biochemical research for the concentration of samples and the exchange of $H_2O$ molecules with $D_2O$. However, lyophilization is not necessarily a benign process. Lyophilization is a three-stage process: freezing, removal of bulk ice, and finally removal of bound water/ice. Problems can arise at each of these stages that result in loss of activity and increased turbidity (protein aggregation).

When an aqueous solution freezes, pure crystals of ice form, leaving a solution increasingly enriched in the various solutes until either a eutectic mixture is formed that freezes or the solutes begin to crystallize themselves. For example, what may start as a 15 mg/ml protein solution in 0.1 $M$ salt can become a protein solution of nearly 500 mg/ml in 3 $M$ salt. When the sample freezes the protein does not freeze itself but remains surrounded by its bound water in a concentrated salt solution. Furthermore, depending on the buffer, there can also be accompanying pH changes (either increases or decreases) of up to 3 pH units associated with freezing.[2] Any or all of these factors can contribute to protein instability.

Removal of the bulk water occurs with little disruption of protein structure. The tightly bound water molecules, however, come off much more slowly. Their total removal requires vigorous conditions of elevated temperature and can cause conformational changes leading to denaturation or aggregation. Even relatively small and thermally stable proteins such as R-60 plasmid dihydrofolate reductase or bovine pancreatic trypsin inhibitor can show observable turbidity following lyophilization. Therefore the use of lyophilization to exchange protons with deuterons must be accompanied by the appropriate controls to demonstrate that the protein remains fully functional.

Lyophilization can also influence the pH of a solution where there are potentially volatile species present. The typical NMR sample of a protein contains numerous weakly acidic or basic functional groups on both the protein and potentially also in the buffer. The presence of volatile counterions can cause unanticipated changes in pH upon lyophilization of solutions. The term "volatile ion" is used here to mean an ion whose

[2] S. S. Larsen, *Arch. Pharm. Chem., Sci. Ed.* **1**, 41 (1973).

nonionized form is volatile. Common volatile ions include ammonium, triethylammonium, pyridinium, bicarbonate, acetate, and formate. In solution these ions exist in equilibrium with their neutral, volatile form. By removing the volatile form, lyophilization can alter the position of the acid/base equilibria, hence change the pH of the solution. For example, while starting at pH 8, repeated lyophilization of an unbuffered solution of the ammonium salt of AMP will eventually yield a solution with a pH <4. The bicarbonate/carbonate system is the most common offender. These species can readily form in solutions of base (NaOH, KOH, etc.) by reaction with atmospheric carbon dioxide if the solutions are not carefully prepared or stored. The presence of significant concentrations of bicarbonate will cause an increase in the pH upon repeated lyophilization due to conversion of bicarbonate to carbon dioxide and the more basic carbonate ion. The less buffered the system, the more caution that should be exercised when using lyophilization in sample preparation, if maintaining a specific pH is critical.

*In-Tube Lyophilization.* A practical method for both conditioning NMR tubes and exchanging the buffers or sample with deuterium at the same time involves the direct lyophilization of the sample in the NMR tube. Subject to the caveats regarding lyophilization discussed above, the method has a number of attractive features. These include simultaneous exchange of both the sample and the surface of the tube, ability to concentrate dilute samples in the NMR tube by repeated cycles of addition and lyophilization, ability to prepare samples in advance that can be stored dry until needed, and, after obtaining spectra, ability to dry samples in the tube for long-term storage. Note that this technique requires some degree of manual dexterity and practice is strongly encouraged before trying it on an actual sample. The resulting speed and efficiency of proton removal should be worth the effort it takes to master the technique.

Since water expands when it freezes, simply freezing an aqueous solution at the bottom of a thin-walled glass tube carries with it a major risk of shattering the tube. The following procedure for freezing the solution as a thin film on the surface of the NMR tube, when properly executed, will permit up to 0.5 ml of an aqueous solution to be frozen in an NMR tube with a minimal risk of cracking the tube.

The procedure is outlined in Fig. 1 and summarized briefly here. Note that when exchanging a sample only 0.2–0.3 ml of $D_2O$ are needed per exchange. The first step is to flow the solution over the entire interior surface of the NMR tube in order to exchange the layer of surface water molecules. The solution is frozen as a thin film on the surface of the bottom half of the tube and then lyophilized. Drying times will vary depending on the quality of the vacuum, but at 50 $\mu$m, 0.2 ml of solution can

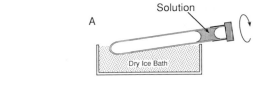

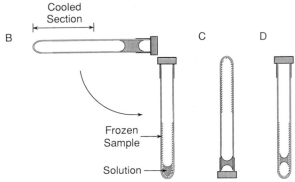

FIG. 1. Procedure for freezing a solution as a thin film on the surface of an NMR tube. (A) After flowing the solution over the entire surface of the NMR tube, the sample is brought to the top of the tube. Then holding the tube nearly horizontal begin cooling the lower half of the NMR tube in a dry ice/ethanol bath. Keep rotating the tube to assure uniform cooling of the surface. The tube should be ready in only a few seconds, before the solution can flow down to the cooled section of the tube. (Note the cap on the NMR tube; alternatively a septum or Parafilm can be used to seal the tube.) (B) The NMR tube is then taken out of the cold bath and rapidly swung so that the solution flows over the cold surface of the NMR. A film of ice will form as the solution flows down the cold surface of the NMR tube. (C) For sample volumes greater than 0.15 ml there will usually be some unfrozen solution at the bottom of the tube. If there is unfrozen solution, then immediately invert the tube (holding onto the warm upper half) and drain the solution to the top of the tube (tapping the tube with your fingers or swinging the tube helps). With the remaining solution now at the top of the tube, begin cooling the lower half of the tube as before and repeat the cycle until all the solution is frozen. When the sample is fully frozen, immerse it in the dry ice bath to cool it down to the bath temperature. Finally remove it from the bath, wipe it dry, then begin lyophilizing. (Note that crackling sounds will be heard coming from the NMR tube when a properly frozen sample is cooled down to the final temperature of the dry ice bath. Do not be alarmed; this is caused by cracking of the ice film as it contracts from the cooling. No strain is put on the glass wall of the tube.) Use common sense in handling a cold sample, e.g., exposure to atmospheric moisture will quickly lay down a layer of $H_2O$. Also it is not critical to exclude every trace of water for the first exchange lyophilization, whereas great care should be exercised for the final lyophilization. (D) It is extremely important to avoid forming a "plug" of solution in the cold part of the NMR tube. If a plug forms, as shown in the figure, and the solution will not drain to the top of the tube, immediately warm the tube (your hand or warm water will do) and start over again.

usually be dried in less than 30 min. An ethanol/dry ice bath (bath temperatures ca. −77°) is ideal for the freezing procedure. Baths maintained at
−40° by refrigeration coils do not cool the glass surface sufficiently to
allow ready formation of thin films of frozen solution (there is a distress-
ing tendency for the solution to supercool which can lead to rapid freezing
of the entire solution). Liquid nitrogen is not recommended because the
solution freezes too fast, promoting build-up of a "plug" of ice that can
crack the tube (see Fig. 1).

Lyophilization can be conducted in either a large vacuum container or
using a special fitting made to hold NMR tubes for degassing. For all but
the most critical experiments, capping the NMR tube with Parafilm is
sufficient for in-tube lyophilization. The Parafilm membrane holds the
solution in the tube while it is being manipulated and is then punctured
just prior to application of vacuum. The Parafilm is then replaced with a
conventional cap after the final lyophilization. Alternatively the NMR
tube can be sealed with a septum. The vacuum can be applied through an
18-gauge platinum needle using the filter system shown in Fig. 2. The

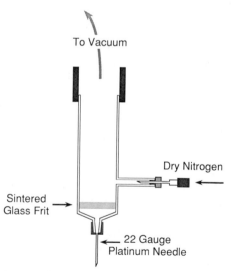

To Vacuum

Dry Nitrogen

Sintered
Glass Frit →

← 22 Gauge
Platinum Needle

FIG. 2. This type of sample filter (available from a number of suppliers of NMR tubes) is
ideal for providing filtered vacuum, air, or nitrogen to a sample. The nitrogen gas should be
regulated through a Firestone valve, since it allows purging of systems with alternating
application of vacuum and nitrogen or other gases at atmospheric pressure, without complex
regulators. A 22-gauge platinum needle is shown for use with freeze/thaw degassing of
samples. If used with direct lyophilization, a larger platinum needle of at least 18-gauge is
needed.

sample should be kept under constant observation while the vacuum is first applied since this is the time when the sample is most likely to melt if it is going to do so (at the first sign of liquid formation, begin bleeding in nitrogen in order to prevent bumping of the sample). Melting of a sample under high vacuum usually means the presence of a eutectic mixture that has too low a freezing point to be sustained by the cooling from subliming of the ice. It usually signals the need for an alternative approach to removal of the water.

Note that some form of shielding should be used to protect against implosion any time a vacuum is applied directly to an NMR tube (as opposed to putting the NMR tube in a large lyophilization flask where there is no direct pressure against the tube). Tubes are typically most susceptible to damage when inserting a needle through a septum because of the strain on the rim. In our experience, we have never had a tube implode. However, shielding the tube is a precaution well worth taking, both for safety and for recovery of sample. A convenient shield is a plastic drinking straw of ca. 7-mm (1/4-in.) bore with the bottom folded over and taped. It can be slipped over the NMR tube and held in place with either Parafilm or removable tape. (Translucent straws also make excellent sheaths for transporting and storing 5-mm NMR tubes. The typical cap for NMR tubes will fit snugly into the straw, thus holding the tube away from the plastic surface. Tubes stored in this way have survived drops of over 1 m onto concrete surfaces.)

When finished lyophilizing, the tubes should be vented with filtered dry nitrogen (or nitrogen pretreated with $D_2O$) to avoid contamination with atmospheric water. A typical sequence for exchanging a sample would involve three lyophilizations as follows. Start with the sample in 99% $D_2O$ (first lyophilization). After it is dry, then redissolve it in 99% $D_2O$ and lyophilize for the second time. Finally dissolve the dry sample in "100%" $D_2O$, and conduct the third lyophilization. It is important to flow the sample over the entire surface of the NMR tube before each lyophilization step in order to exchange the surface water. At this point the residual HDO peak of the sample should be limited by the proton content of the $D_2O$ being used (typically between 5 and 20 m$M$).

*Alternative Methods for Exchanging Protons and Concentrating Samples.* Not all proteins can be successfully lyophilized, especially those with surface sulfhydryls. Therefore alternative methods for exchanging or concentrating the samples must be used. The two best methods are membrane filtration (which can accomplish both exchanging and concentrating the sample) and size exclusion/molecular sieve chromatography (exchange only).

*Membrane filtration:* Centrifugal filtration systems like the Amicon Centricon microconcentrators, or their functional equivalent, are excellent for exchanging, equilibrating, or concentrating larger proteins ($M_r >$ 15 K). This type of filter has many advantages over pressure-driven concentrators, including speed of operation, the fact that they will not go to dryness, and ease of conducting multiple equilibrations. Their physical size is ideally suited for handling small NMR samples and minimizes losses due to surface adsorption. Their utility is diminished for lower molecular weight proteins because of increased losses through the membrane. Recently, however, an $M_r$ 3000 cut-off Centricon filter has been introduced which should allow handling of proteins with molecular weights down to 5000–6000 with minimal loss. As with all such operations, the system must be rinsed with the appropriate buffer before applying the sample to avoid any possibility of contamination. Note that while this procedure will exchange the protein, the NMR tube itself, as well as the associated buffers and cofactors, can be exchanged by in-tube lyophilization.

*Size-exclusion gel filtration for exchanging samples:* Although in principle gel filtration can be used for proteins of all sizes, it is most appropriate for smaller proteins (<15 kDa). The concept is simple, but execution requires great attention to detail. A short column containing a low-molecular-weight exclusion gel (typically $M_r$ 1000–2000) is prepared in a narrow column. A 1- to 2-ml plastic syringe capped with a septum and fitted with a syringe filter (typically 0.45 $\mu$m) and a Luerlock platinum needle is sufficient, as shown in Fig. 3. A bed volume of 1–2 ml is adequate to process samples for a 5-mm NMR tube. The column is equilibrated with the quality of $D_2O$ that will be used in the NMR experiment. The solution is layered on top using a 22-gauge platinum needle (total volume applied to the column should be $\leq$ 10% of the column volume for best results). The protein solution is forced into the column using dry nitrogen from a syringe, then more $D_2O$ is carefully layered onto the top of the column (if needed, a second needle pushed through the septum can be used to vent the column). The sample is forced through the column at a steady rate, and after a predetermined volume has been eluted the appropriate fraction is collected. Typically dilution by a factor of two is expected because of the tradeoff between maximizing concentration of the protein versus maximizing total recovery. Because of its small size and rapid diffusion, $H_2O$ quickly equilibrates with the full volume of the gel, thus the sample can be eluted quickly. Residence times of 30–60 sec on the column are usually sufficient to remove most of the protons. Note that for higher molecular weight proteins, exchange of protons will be hampered by the large num-

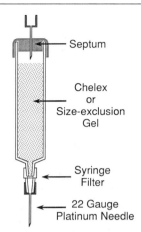

FIG. 3. Schematic of a typical small column for handling NMR samples. The syringe filter is used to hold back the column packing as well as provide a final filtration of the sample. Small disposable columns in the 1- to 2-ml size are now readily available from microbiological suppliers.

bers of relatively slowly exchanging protons (dissociation rates of greater than 5 sec). Such proteins are best handled by centrifugal filtration.

In summary, for proteins that can survive and for lower molecular weight proteins, lyophilization is best. Centrifugal filtration systems, with the availability of the lower molecular weight cut-off filters, are ideal for both exchanging and equilibrating proteins of molecular weight above 5000. Gel filtration represents the most technically demanding and least flexible approach to exchanging protons.

Other Techniques

*Freeze-Thaw Degassing of Samples.* The freezing technique outlined for exchanging protons can also be used in freeze-thaw cycles for degassing NMR samples (see Fig. 1). A septum or other gas-tight closure is fitted over the top of the NMR tube and the sample is then frozen as described. The large surface generated by freezing the solution as a thin film promotes rapid and thorough degassing of samples.

Vacuum can be applied through a platinum needle connected to a sintered glass filter available from NMR suppliers (see Fig. 2), and dry nitrogen gas can be provided through the side arm using a Firestone valve. The typical NMR tube septum will hold a vacuum for a few hours, thus the tube can be pulled off the needle, manipulated, e.g., refrozen or thawed, and then reattached. A 22-gauge platinum needle is best for degassing since it reduces tearing of the septum.

*Removal of Other Solvents by Lyophilization.* Freezing in the NMR tube followed by lyophilization can be employed with other solvents that have freezing points near 0°, e.g., dimethyl sulfoxide (DMSO), dioxane, or benzene. One advantage with these solvents is that they are unlikely to crack NMR tubes because their volume decreases upon freezing. DMSO is usually considered a difficult solvent to remove but by freezing a DMSO solution into a thin film on the inside of the NMR tube, it can be readily removed by lyophilization. A 0.5-ml sample can usually be brought to dryness in under 3 hr. Mixed solvent systems are most likely to form low freezing point eutectics and will probably need to be dried on a centrifugal vacuum evaporating system like a Speed Vac.

*Filtration of Samples.* Samples should be filtered just prior to addition to the NMR tube to remove any particulates. The most convenient and indeed best way to filter a sample is to use a 0.45-$\mu$m syringe filter. Such filters are widely used for filtration of HPLC samples. Their compact size, which is no bigger than the top of a typical disposable syringe needle, and their negligible retention volume make them ideal for use with small NMR samples.

# Section II

# Advanced Techniques

## [5] Heteronuclear Nuclear Magnetic Resonance Experiments for Studies of Protein Conformation

*By* GERHARD WAGNER

Nuclear magnetic resonance (NMR) has become a powerful method to characterize protein structure and dynamics,[1] features which are essential for protein function. NMR is a probe technique which relies on the assignment of many resonances which can then be employed to measure structural parameters, such as distances, torsion angles, exchange rates, or relaxation times. At present, mostly proton resonances are used to probe structural aspects of proteins. They seem to be more powerful probes than other nuclei to study protein structure, and thus proton assignments are of higher priority. With homonuclear experiments almost complete proton assignments can be achieved for proteins, and the information obtainable from proton probes is well exploited. Nowadays other nuclei have attracted new interest in the demand to increase the number of assigned probes for structural studies. Structural information which can be obtained from these other nuclei is complementary to that obtainable from proton NMR. It includes relaxation time measurements and measurements of heteronuclear long-range coupling constants. In addition, heteronuclear experiments may resolve ambiguities of proton resonance assignments.

A heteronuclear experiment was among the first two-dimensional (2D) NMR experiments proposed.[2] Nevertheless, heteronuclear experiments found little interest for studies of proteins, possibly because of the low inherent sensitivity, the low natural abundance of most spin-$\frac{1}{2}$ heteronuclei, and because of instrumentation problems. It has been realized early that proton-detected heteronuclear experiments are more sensitive than experiments with detection of the nucleus with lower gyromagnetic ratio.[3,4] Indeed, the first heteronuclear experiment proposed was designed for proton detection.[2] However, commercial NMR spectrometers were not equipped for inverse detection experiments until recently. Therefore, early heteronuclear 2D NMR experiments have used detection of low γ nuclei. To improve the sensitivity of such experiments, isotope enrich-

[1] K. Wüthrich, "NMR of Proteins and Nucleic Acids." Wiley, New York, 1986.
[2] W. P. Aue, E. Bartholdi, and R. R. Ernst, *J. Chem. Phys.* **64**, 2229 (1976).
[3] A. A. Maudsley and R. R. Ernst, *Chem. Phys. Lett.* **50**, 368 (1977).
[4] A. A. Maudsley, L. Müller, and R. R. Ernst, *J. Magn. Reson.* **28**, 463 (1977).

ment was employed, or very high sample volumes were used.[5–10] The first heteronuclear experiment with proton detection on a protein was performed with a $^{199}$Hg adduct of ribonuclease.[11] Most extensively, heteronuclear NMR was applied for studies of metallothionein-2 where such experiments were crucial for characterization of the solution conformation.[12,13]

Nowadays heteronuclear 2D NMR experiments are becoming popular for studies of proteins for various reasons. (1) Commercially available spectrometers are being routinely equipped for performing proton-detected heteronuclear experiments. (2) The potential of homonuclear 2D NMR experiments is highly developed so that the $^1$H spectra of small proteins can be assigned almost completely, and there is a desire to extend the assignments to the complete proton spectrum. This is important to avoid erroneous assignments of cross-peaks, in particular in two-dimensional nuclear Overhauser effect spectroscopy (NOESY) spectra. Remaining ambiguities for resonance assignments are usually due to resonance overlap, in particular when approaching larger proteins. It seems that these problems can be overcome in part by using the resolution in a third dimension, employing heteronuclear experiments. (3) The development of modern genetic techniques makes possible uniform or semiselective isotope enrichment in proteins. Thus the problem of the low natural abundance can be overcome. This chapter shall provide a survey of heteronuclear 2D NMR experiments which are useful for studies of proteins. The chapter will deal only with heteronuclear experiments of nuclei at natural abundance. No isotope enrichment studies will be discussed; they are covered in Volume 177 of this series. A treatment of $^{113}$Cd NMR is also omitted. Only 2D NMR techniques will be discussed because only such techniques provide the resolution required for analysis of protein spectra. Throughout the chapter the abundant nucleus with high gyromagnetic ratio $\gamma$ will be denoted $I$ or proton, the rare nucleus with low $\gamma$ will be denoted $S$, carbon, or nitrogen. When explaining pulse sequences, use

[5] T. M. Chan and J. L. Markley, *J. Am. Chem. Soc.* **104**, 4010 (1982).

[6] C. Kojiro and J. L. Markley, *FEBS Lett.* **162**, 52 (1983).

[7] K. Tokura and T. Miazawa, *Eur. J. Biochem.* **123**, 127 (1982).

[8] H. Santos and D. L. Turner, *FEBS Lett.* **184**, 240 (1985).

[9] H. Santos and D. L. Turner, *FEBS Lett.* **194**, 73 (1986).

[10] W. M. Westler, G. Ortiz-Polo, and J. L. Markley, *J. Magn. Reson.* **58**, 354 (1984).

[11] D. A. Vindusek, M. F. Roberts, and G. Bodenhausen, *J. Am. Chem. Soc.* **104**, 5452 (1982).

[12] M. Frey, G. Wagner, M. Vasak, O. W. Sørensen, D. Neuhaus, E. Wörgötter, J. H. R. Kägi, R. R. Ernst, and K. Wüthrich, *J. Am. Chem. Soc.* **107**, 6847 (1985).

[13] J. D. Otvos, H. R. Engeseth, and S. Wehrli, *J. Magn. Reson.* **61**, 579 (1985).

will be made of the operator product formalism as described by Sørensen et al.[14]

## Methods

Numerous heteronuclear experiments have been described in the literature.[15,16] In this chapter a selection of experiments is presented which appear to have high impact on protein NMR. Central aspects of heteronuclear experiments are to overcome the problem of low sensitivity and to employ the most efficient experimental schemes for heteronuclear correlation.

### Sensitivity

The sensitivity of heteronuclear experiments is determined by the polarization of the nuclei which are excited during the preparation period, by the magnetic moment of the detected nuclei, and by the number of molecules, $N_{IS}$, which have both NMR-active $I$ spins and $S$ spins ($N_{IS}$ is essentially the natural abundance of $S$ spins).[15] The polarization of the spins to be excited is given by the difference of the two diagonal elements of the equilibrium density operator $\sigma_0$. In the high-temperature limit we have

$$\sigma_0 = [1/\text{Tr}(\mathbb{1})]\{1 + [(\hbar\gamma_{exc}B_0/kT)I_z] + \ldots +\}$$

This means that the polarization is proportional to $\hbar\gamma_{exc}B_0/kT$. The precessing transverse magnetization to be detected is

$$\langle M_{x,y}(t)\rangle = \gamma_{det}\hbar\text{Tr}[I_{x,y}\sigma(t)]$$

The signal observed is proportional to the voltage induced by the precessing transverse magnetization in the receiver coil. This induced voltage $V_{ind}(t)$ is proportional to the time derivative of the precessing transverse magnetization. Considering the $x$ component, $M_x(t)$, we have

$$V_{ind}(t) = dM_x(t)/dt = d(M_0 \cos \Omega t)/dt$$
$$= -\Omega M_0 \sin \Omega t = \gamma_{det}B_0 M_0 \sin \Omega t$$

The noise of the experiment is proportional to the square root of the

[14] O. W. Sørensen, G. W. Eich, M. H. Levitt, G. Bodenhausen, and R. R. Ernst, *Prog. NMR Spectrosc.* **16**, 163 (1983).
[15] R. R. Ernst, G. Bodenhausen, and A. Wokaun, "Principles of Nuclear Magnetic Resonance in One and Two Dimensions." Oxford Univ. Press (Clarendon), London and New York, 1987.
[16] R. H. Griffey and A. G. Redfield, *Q. Rev. Biophys.* **19**, 51 (1987).

gyromagnetic ratio of the detected nucleus, $\gamma_{det}^{1/2}$. In total, the overall signal-to-noise ratio is proportional to $N_{IS}\gamma_{exc}\gamma_{det}^{3/2}B_0^2$. Considering this, it is obvious that heteronuclear experiments starting with $I$-spin polarization and ending with $I$-spin detection have the highest signal-to-noise ratio and their sensitivity compared to $^1H-^1H$ 2D experiments is lowered only by the lower natural abundance of the $S$ spins. In addition, it has to be considered that the repetition rate of the pulse sequences has to be adjusted to the relaxation of the excited nucleus. For small molecules carbon and nitrogen nuclei have longer relaxation times than protons. Thus, experiments which start with proton excitation can be run with a higher repetition rate. The price of the high sensitivity in proton-detected heteronuclear experiments is that the majority of proton signals which are not coupled to an $S$ spin have to be suppressed and, furthermore, one has to accept the lower resolution, which is usually obtained along the $\omega_1$ axis, for the carbon or nitrogen spectrum.

### Heteronuclear Correlated Spectra

Heteronuclear correlated experiments provide the most useful information among heteronuclear techniques with respect to analysis of crowded protein spectra. They can yield assignments of carbon or nitrogen resonances. These assignments are, however, of little interest per se unless they can be utilized for extracting structural information. Only a few heteronuclear correlation techniques have been applied to proteins so far; they will be discussed below together with other techniques which may become important in the future.

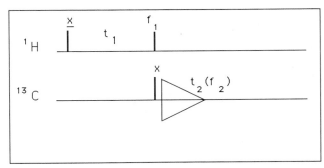

Fig. 1. Heteronuclear COSY sequence with carbon detection. The narrow vertical bars symbolize 90° pulses. The simplest phase cycle is $f_1$: $+x,-x; f_2$: $+,-$. The phase program underlined in this and the following figures should be incremented in steps of 90° together with $t_1$ in order to achieve discrimination of positive and negative frequencies along $\omega_1$ (time proportional phase incrementation), or a second data set should be recorded with the phase of this pulse shifted by 90°.

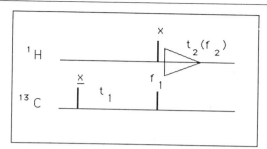

FIG. 2. Heteronuclear COSY sequence with proton detection. The simplest phase cycle is $f_1$: $+x, -x$; $f_2$: $+, -$.

*Basic Experiments.* In the early days of 2D NMR, all heteronuclear experiments were performed with detection of the $S$ spin. The simplest pulse sequence is given in Fig. 1. This yields a spectrum with antiphase multiplet structure in both frequency directions. No decoupling can be employed. The simplest heteronuclear correlated spectroscopy COSY pulse sequence with proton detection is outlined in Fig. 2. To our knowledge neither of these sequences has been applied to proteins. They seem to be of little interest as far as polarization transfer through one-bond coupling is concerned. However, the sequence of Fig. 2 or variations thereof may be attractive for heteronuclear long-range correlation where double-transfer techniques are not efficient and decoupling is not desirable.

*Decoupling.* Decoupling in either frequency dimension can simplify the appearance of the spectra and increase the sensitivity by collapsing the multiplet components. Decoupling requires that the multiplet components are in phase. This makes necessary insertion of dephasing and rephasing delays. Figure 3A shows the simplest heteronuclear COSY sequence with decoupling in both dimensions. The first 90°(H) pulse creates transverse proton magnetization which is in phase with respect to the heteronuclear coupling. Broad-band carbon decoupling[4] during $t_1$ prevents evolution under heteronuclear coupling and keeps the proton magnetization in phase. The $\tau$–180°(H, C)–$\tau$ period allows evolution under $J$ coupling but refocuses evolution under chemical shift. Thus the pair of 90° pulses can perform a coherence transfer. The following $\tau$–180°(H, C)–$\tau$ period achieves refocusing to in-phase magnetization so that decoupling can be applied during acquisition. The delays $\tau$ have to be tuned to $1/(4J)$. This means that these techniques can only be applied reasonably well for one-bond correlation or cases where all heteronuclear couplings have similar values. Alternatively, several subspectra have to be acquired which are tuned to different classes of coupling constants. Broad-band

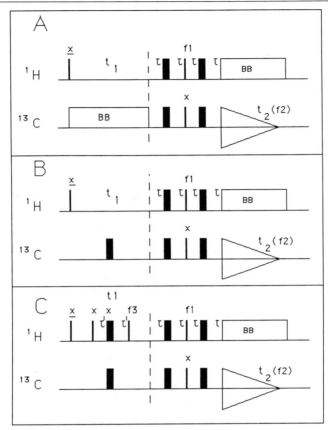

FIG. 3. Heteronuclear COSY sequences with broad-band decoupling. The thin and thick vertical bars symbolize 90° and 180° pulses, respectively. The delays $\tau$ are tuned to $1/(4J)$. The delays $\tau'$ are tuned to $1/(2J)$. Decoupling along $\omega_2$ is obtained by broad-band irradiation during $t_2$. Decoupling along $\omega_1$ can be achieved by (A) broad-band decoupling during $t_1$, or by (B) insertion of a 180° carbon pulse in the center of the $t_1$ period. Selective decoupling of one-bond couplings or long-range couplings can be achieved by insertion of a bilinear rotation decoupling (BIRD) sandwich with $f_3$ set to $+x$ or $-x$, respectively (C). The simplest phase cycle is $f_1$: $+x,-x$; $f_2$: $+,-$.

decoupling during $t_1$ by continuous wave (CW) irradiation may not be efficient enough to decouple a whole carbon or nitrogen spectrum and may also lead to sample heating. It is preferable to use decoupling by pulse trains, such as WALTZ-16.[17] Decoupling by CW irradiation or by continuous pulse trains can be used for weakly and strongly coupled spin

[17] A. J. Shaka, J. Keeler, and R. Freeman, *J. Magn. Reson.* **53**, 313 (1983).

systems equally well. Broad-band decoupling along $\omega_1$ of weakly coupled $I$-spin systems from $S$ spins can also be achieved by insertion of a $180°(S)$ pulse in the center of the $t_1$ period[4,18] (Fig. 3B). In spin systems where there is strong coupling between $I$ spins, a $180°(S)$ pulse may lead to coherence transfer between $I$ spins and produce undesired effects.[15] Selective decoupling of either one-bond or long-range coupling can be achieved by using a bilinear rotation decoupling (BIRD) pulse instead[19,20] (Fig. 3C). The delay $\tau'$ has to be set to $1/(2J)$ (one bond). The phase $f_4$ of the last $90°(H)$ pulse of the BIRD sandwich set to $+x$ or $-x$ results in decoupling of one-bond or long-range couplings, respectively. Broadband decoupling during detection ($t_2$) can be achieved by CW irradiation. Obviously the decoupling schemes of Fig. 3 can also be used in a reversed mode, i.e., for proton-detected experiments. In this case CW irradiation at the carbon or nitrogen spectrum during $t_2$ is not feasible if the whole spectrum should be covered. Decoupling during $t_2$ is usually achieved by rapid pulsing. The best known decoupling pulse sequences are among the WALTZ,[17] MLEV,[21] and GARP[22] families. At the moment, the GARP sequences provide the widest effective decoupling fields, but they are not yet available on all commercial spectrometers. WALTZ-16[17] decoupling is most commonly used.

*Double-INEPT Transfer.* The first proton-detected heteronuclear experiment on a protein was a double-transfer experiment, starting with proton polarization and ending with proton detection. It uses the INEPT transfer (insensitive nuclei enhancement by polarization transfer). The pulse sequence is shown in Fig. 4. The delays $\tau$ are tuned to $1/(4J)$. The $180°(H)$ pulse in the center of $t_1$ achieves decoupling along $\omega_1$, and broadband decoupling can also be used during $t_2$. The value of this sequence is comparable with the heteronuclear multiple-quantum COSY described below. The disadvantage compared to that is that two coherence transfer steps have to be employed which may reduce sensitivity, the advantage is that no homonuclear proton couplings are active during $t_1$ which might lead to mixed line shapes and partial cancellation of cross-peaks in heteronuclear multiple-quantum experiments described below. Thus this experiment may be useful for routine heteronuclear COSY spectra of proteins.

*Heteronuclear Multiple-Quantum COSY.* The heteronuclear multiple-quantum COSY is the most frequently used method for hetero-COSY of proteins in the moment. The first experiment of this type has been de-

[18] G. Bodenhausen and R. Freeman, *Chem. Phys. Lett.* **69**, 471 (1977).
[19] J. R. Garbow, D. P. Weitekamp, and A. Pines, *Chem. Phys. Lett.* **93**, 504 (1982).
[20] A. Bax, *J. Magn. Reson.* **53**, 517 (1983).
[21] M. H. Levitt, R. Freeman, and T. Frenkiel, *Adv. Magn. Reson.* **11**, 47 (1983).
[22] A. J. Shaka, P. B. Barker, and R. Freeman, *J. Magn. Reson.* **64**, 547 (1985).

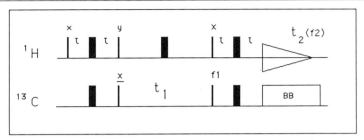

FIG. 4. Double INEPT transfer experiment. The simplest phase cycles $f_1$ and $f_2$ are the same as in Figs. 1–3.

scribed by Müller.[23] The pulse sequence of the experiment is shown in Fig. 5A. Prior to the first 90°(C) pulse we have heteronuclear antiphase magnetization $2I_xS_z$ which is converted to heteronuclear two-spin coherence $-2I_xS_y$ by the following 90°(C) pulse. The simultaneous 90°(H) pulse has the only effect to flip magnetization of protons not coupled to $^{13}$C back to the $z$ axis. The 180°(H) pulse in the center of $t_1$ refocuses the proton chemical shift evolution so that the heteronuclear two-quantum coherence evolves only under carbon chemical shift. The final 90°(C) pulse converts the heteronuclear two-quantum coherence back into proton antiphase magnetization $2I_xS_z$. This results in hetero-COSY spectra which are decoupled along $\omega_1$ but undecoupled and antiphase along $\omega_2$. Decoupling along $\omega_2$ can be achieved when the pulse sequence in Fig. 5B is used, which has an additional refocusing period prior to detection.[24,25] The heteronuclear multiple-quantum COSY techniques have the advantage over the double-INEPT experiment (Fig. 4) that no coherence transfer is employed, and they may thus have a higher sensitivity. They have the disadvantage, however, that $^1$H–$^1$H couplings are active during $t_1$. This leads to proton–proton multiplet splitting along $\omega_1$, and to mixed line shapes and partial canceling of cross-peaks, in particular for methylene groups. To overcome this problem, it has been suggested to apply a $z$ filter prior to acquisition.[12] This is a practical approach for isotope-enriched samples. When the $S$ spin is present only at natural abundance, however, the $z$ filter perturbs the elimination of the signals of protons which are not coupled to the NMR-active $S$ spin and results in a poor performance of the phase cycling schemes. Modifications of these experiments have been used where the 180°(H) pulse was omitted (Fig. 5C). This results in a heteronuclear double-quantum or zero-quantum evolution during $t_1$, ei-

[23] L. Müller, *J. Am. Chem. Soc.* **101**, 4481 (1979).
[24] M. R. Bendall, D. T. Pegg, and D. M. Doddrell, *J. Magn. Reson.* **52**, 81 (1983).
[25] A. Bax, R. H. Griffey, and B. L. Hawkins, *J. Magn. Reson.* **55**, 301 (1983).

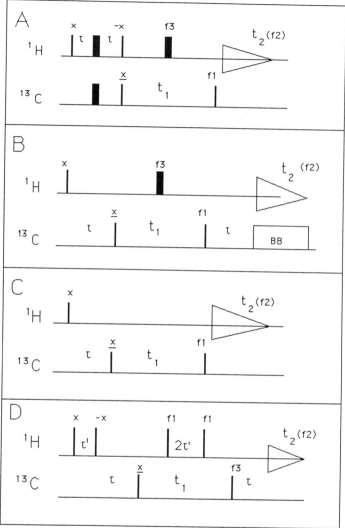

FIG. 5. Heteronuclear multiple-quantum correlation experiments. (A) Sequence of Mül-ler.[23] The delay $\tau$ is tuned to $1/(4J)$. Signals are detected in antiphase, and no decoupling is possible during $t_2$. Evolution of proton chemical shift is refocused by the 180° pulse in the center of $t_1$. (B) Sequence with refocusing of heteronuclear coupling.[24,25] The delay $\tau$ is set to $1/(2J)$. (C) Sequence without refocusing of heteronuclear coupling and proton chemical shift evolution. The simplest phase cycles are $f_1$: $x,-x$; $f_2$: $+,-,-,+$; $f_3$: $x, x, y, y, -x,$ $-x, -y, -y$. (D) Sequence with jump-return preparation and refocusing pulses for experiments in $H_2O$.[26]

ther of which can be selected by a proper phase cycling. Since the proton chemical shift evolution is not refocused, only absolute value Fourier transforms can be used in this case. Another modification of the sequence of Fig. 5A has been applied where the excitation pulse and the 180°(H) refocusing pulse are replaced by jump–return pulses[26] (Fig. 5D). The phase of the pulse pair in the center of the evolution period is cycled in the EXORCYCLE manner.[27] This is interesting when experiments have to be performed in $H_2O$, such as $^1H$–$^{15}N$ correlation, and the elimination of the solvent resonance is difficult.

   *Editing Techniques.* Editing techniques which discriminate between $CH$, $CH_2$, and $CH_3$ groups, such as DEPT[24] (distortionless enhancement by polarization transfer) and refocused INEPT, are common for 1D heteronuclear coherence transfer experiments of small organic molecules. Usually these experiments involve a coherence transfer from protons to carbons. Analogous techniques for editing proton-detected experiments have also been developed but little used. For proteins, however, neither of these methods has been employed so far. To cope with the low concentration of protein samples, editing in double-transfer experiments would be desirable as outlined in Fig. 6A and B. Figure 6A is a double-refocused INEPT experiment. The first $\tau$–180°(C, H)–$\tau$ period creates heteronuclear antiphase magnetization $2I_xS_z$. The delay $\tau$ has to be tuned to $1/(4J)_{HC}$. The following pair of 90° pulses performs a coherence transfer to produce $2I_zS_y$. The following $\tau$–180°(C, H)–$\tau$ period performs refocusing to in-phase carbon magnetization, $S_x$. The delay $\tau'$ may be set at $1/(8J)_{CH}$, $1/(4J)_{CH}$, or $3/(8J)_{CH}$ to achieve maximum, minimum, or zero intensity from different $CH_n$ groups. This refocusing is proportional to $\sin(\pi J2\tau')$ for CH groups, to $\sin(\pi J4\tau')$ for $CH_2$ groups, and to $\frac{3}{4}[\sin(\pi J2\tau') + \sin(\pi J6\tau')]$ for $CH_3$ groups. The $\tau''$–180°(C, H)–$\tau''$ period which follows the evolution performs defocusing to $2I_zS_y$ which is necessary to achieve a coherence transfer to the proton spins by the following pair of 90° pulses. This defocusing has the same characteristic dependence on $\tau''$ for CH, $CH_2$, and $CH_3$ groups as the refocusing during $\tau'$. The final $\tau$–180°(C, H)–$\tau$ period refocuses the antiphase proton magnetization $2I_xS_z$ to $I_y$, which is finally detected. The lengths of the delays $\tau'$ can be used for editing. If $\tau' = \tau'' = 1/(4J)_{CH}$, only CH signals will be observed. If $\tau' = \tau'' = 1/(8J)_{CH}$, CH, $CH_2$, and $CH_3$ signals will all be present and have the same sign. If $\tau' = 1/(8J)_{CH}$ and $\tau'' = 3/(8J)_{CH}$, CH, $CH_2$, and $CH_3$ signals will be present, but $CH_2$ signals will have the opposite sign. The editing with the double-DEPT sequence of Fig. 6B is analogous. It is achieved,

[26] V. Sklenář and A. Bax, *J. Magn. Reson.* **74,** 469 (1987).
[27] G. Bodenhausen, R. Freeman, and D. L. Turner, *J. Magn. Reson.* **27,** 511 (1977).

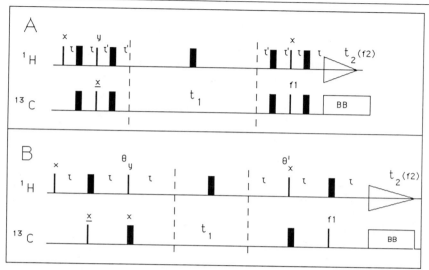

FIG. 6. (A) Double-refocused INEPT experiment. The delay $\tau$ is set to $1/(4J)$. The delays $\tau'$ and $\tau''$ may be tuned to select for CH, $CH_2$ or $CH_3$ groups, respectively (see text). (B) Double-DEPT experiment. The delay $\tau$ is set to $1/2J$. The flip angles $\theta$ and $\theta'$ can be tuned to select for CH, $CH_2$, or $CH_3$ groups, respectively (see text). The phase cycles $f_1$ and $f_2$ can be chosen as in Fig. 1.

however, by adjusting the width of the $\theta_1$ and $\theta_2$ pulses. $\theta$ pulses of 45, 90, and 135°, etc., have the same effect in the double-DEPT sequence as delays $\tau'$ and $\tau''$ of $1/(8J)_{CH}$, $1/(4J)_{CH}$, $3/(8J)_{CH}$, etc., in the INEPT sequence. To improve INEPT and DEPT experiments with a single transfer, various purge procedures have been used.[28] Purging of undesired signals in DEPT or INEPT experiments with a double transfer can also be achieved by $z$ filters.[29]

*Elimination of $^{12}CH$ Signals.* Heteronuclear experiments with proton detection require elimination of signals of protons which are not coupled to $^{13}C$, i.e., 99% of the signals. To rely only on phase cycling requires a very stable instrument, and significant differences of the performance can be observed between individual instruments, correlated with the laboratory environment and the design of the instrument. Depending on the performance of difference spectroscopy, it may be important to reduce the intensity of $^{12}CH$ signals by other means, in addition to phase cycling. This has been attempted either by selective alignment of $^{12}CH$ proton

[28] O. W. Sørensen and R. R. Ernst, *J. Magn. Reson.* **51**, 477 (1983).
[29] N. R. Nirmala and G. Wagner, *J. Am. Chem. Soc.* **110**, 7557 (1988).

signals along the $z$ axis,[23,30,31] or by selectively destroying $^{12}$CH signals by homospoil pulses while $^{13}$CH signals are saved as longitudinal $zz$ order[30,31] or by selective inversion of $^{12}$CH proton signals via a BIRD pulse sequence prior to the actual pulse sequence which is started when the relaxing $^{12}$CH proton signals are close to zero.[32] Methods have been proposed which eliminate $^{12}$CH signals by applying a spin lock at different instances of heteronuclear pulse sequences.[33]

## Heteronuclear RELAY Experiments

Heteronuclear relayed coherence transfer spectroscopy (RELAY) experiments transfer coherence between three nuclei of at least two different kinds, e.g., H and $X$, and cross-peaks connect nuclei which have no direct coupling but share a common coupling partner. The first heteronuclear RELAY experiments were described by Bolton and Bodenhausen.[34,35] These techniques can be denoted as H($\omega_1$)–H–$X$($\omega_2$) RELAY experiments. The first proton is frequency labeled during $t_1$, then coherence is transferred to a coupled proton, and afterward this is transferred to the $X$ nucleus, the frequency of which is recorded during detection. Different versions of such RELAY experiments are possible and have been tried, e.g., an H($\omega_1$)–$X$–H($\omega_2$) experiment[36,37] or an H–$X$($\omega_1$)–H–H($\omega_2$) experiment.[12,30,31,38] The latter type is of most use for protein NMR. Two versions are shown in Fig. 7. In the sequence of Fig. 7A[30,31] the $^{12}$CH proton signals are destroyed by a homospoil pulse (HS) while the $^{13}$CH signals are saved as longitudinal $zz$ order. The delays $\tau$ are tuned to $1/(4J)_{CH}$ while the delay $\tau'$ is tuned to achieve optimum proton–proton coherence transfer, i.e., approximately $1/(4J)_{HH}$. The final $90^\circ_y$(H) pulse achieves the H–H coherence transfer. The sequence uses a heteronuclear multiple-quantum evolution period as described in the section, Editing Techniques. A simpler version of this sequence is shown in Fig. 7B.[12] It does not attempt to eliminate $^{12}$CH proton signals. This sequence has successfully been applied to ($^{113}$Cd$_7$)metallothionein, where the cadmium was isotope enriched. The $90^\circ_y$(H) pulse can be replaced with a spin-lock

[30] D. Brühwiler and G. Wagner, J. Magn. Reson. **69**, 546 (1986).

[31] G. Wagner and D. Brühwiler, Biochemistry **25**, 5839 (1985).

[32] A. Bax and S. Subramanian, J. Magn. Reson. **67**, 565 (1986).

[33] G. Otting and K. Wüthrich, J. Magn. Reson. **76**, 569 (1988).

[34] P. H. Bolton and G. Bodenhausen, Chem. Phys. Lett. **89**, 139 (1982).

[35] P. H. Bolton, J. Magn. Reson. **48**, 336 (1982).

[36] D. Neuhaus, G. Wider, G. Wagner, and K. Wüthrich, J. Magn. Reson. **56**, 164 (1984).

[37] M. A. Delsuc, E. Guittet, N. Trotin, and J. Y. Lallemand, J. Magn. Reson. **56**, 163 (1984).

[38] P. H. Bolton, J. Magn. Reson. **62**, 143 (1985).

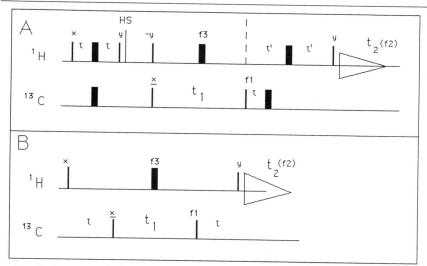

FIG. 7. Sequences for heteronuclear RELAY experiments with heteronuclear multiple-quantum evolution period. Sequence (A) is optimized for elimination of $^{12}$CH proton signals. $\tau$ is set to $1/(4J)_{CH}$, $\tau'$ is optimized for homonuclear coherence transfer. Sequence (B) is only useful when the isotope of the heteronucleus is enriched. $\tau$ is set to $1/(2J)_{CH}$. Phase cycles are the same as in Fig. 5.

sequence to achieve total correlation between the carbon and all protons which are coupled to the $^{13}$CH proton directly bound to the $X$ nucleus.[39] For correlations with heteronuclei of low natural abundance the latter sequence seems to be of lesser performance.

### Long-Range Correlation Experiments

To achieve long-range correlation between protons and other nuclei is of high interest since information about torsion angles could be achieved which is not accessible from homonuclear proton coupling constants. No successful experiment of this type has been reported for a protein so far. For sensitivity reasons only proton-detected experiments should be considered. The most promising experiment has been developed by Bax and Summers[40] (Fig. 8A). It is a heteronuclear multiple-quantum experiment similar to that of Fig. 5B but without refocusing period. The delay $\tau$ is tuned, however, to $1/(2J)_{\text{long range}}$. In addition, a 90°(C) pulse is inserted at

[39] L. Lerner and A. Bax, *J. Magn. Reson.* **69,** 375 (1986).
[40] A. Bax and M. F. Summers, *J. Am. Chem. Soc.* **108,** 2093 (1986).

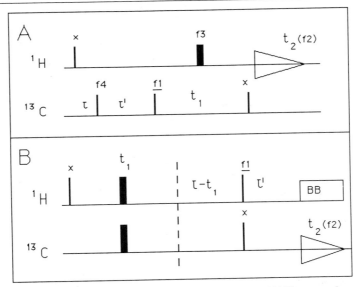

FIG. 8. Sequences for heteronuclear long-range correlation. (A) Heteronuclear multiple-quantum experiment with proton detection.[40] $\tau$ is set to $1/(4J)_{CH}$ for the one-bond coupling, $\tau'$ is optimized for heteronuclear long-range coupling. The phases $f_1$ to $f_3$ are the same as in Fig. 5. $f_4$ is phase alternated without changing the receiver phase. (B) COLOC experiment[41] which uses $S$-spin detection. $\tau$ and $\tau'$ are optimized for heteronuclear long-range coupling. $f_1$ and $f_2$ are phase cycled as in Fig. 1.

a time $1/(2J)_{\text{short range}}$ after the first proton pulse and is phase alternated while keeping the receiver phase constant. This eliminates cross-peaks of short-range correlations. The spectra have mixed line shape and only absolute value transforms should be made. Thus accurate measurements of the long-range coupling constants, the parameters of interest, are not possible. The best known heteronuclear 2D experiment for long-range correlation is correlation spectroscopy via long-range coupling (COLOC[41]) (Fig. 8B). It uses $X$ detection and is therefore not well suited for studies of proteins. However, proton-detected versions of this experiment may be developed. COLOC is a constant time experiment analogous to homonuclear experiments developed earlier.[42,43] The evolution time is fixed to a value around $1/(2J)_{\text{long range}}$. The delay $\tau_2$ could be omitted, and broad-band decoupling should not be applied in this case.

[41] H. Kessler, C. Griesinger, and K. Wagner, J. Am. Chem. Soc. 109, 6927 (1987).
[42] A. Bax and R. Freeman, J. Magn. Reson. 44, 542 (1983).
[43] M. Rance, G. Wagner, O. W. Sørensen, K. Wüthrich, and R. R. Ernst, J. Magn. Reson. 59, 250 (1984).

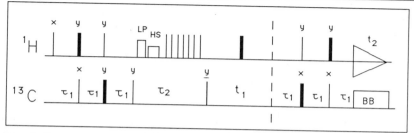

FIG. 9. Sequence for measurements of $^{13}$C relaxation times in 2D heteronuclear correlated experiments.[46] $\tau_1$ is tuned to $1/(2J)$, $\tau_2$ is the relaxation delay. LP (long pulse) and HS (homospoil pulse) have the purpose to eliminate residual transverse magnetization. The series of 90°(H) pulses during $\tau_2$ keep the proton magnetization saturated. The experiment is related to the 1D version of Skelnar et al.[45]

## Relaxation Time Measurements

One-dimensional pulse sequences for measurement of $^{13}$C relaxation times using proton detection have been reported recently.[44,45] They use either double-INEPT or double-DEPT transfer. To provide the necessary resolution for protein spectra, 2D versions of such experiments are mandatory. Figure 9 shows a pulse sequence for 2D relaxation time measurements which we are using for measurements of $^{13}$C relaxation rates in proteins.[46] With modern high-field NMR spectrometers such experiments can be performed with protein solutions with about a week of instrument time (24 hr/delay).

## Other Experiments

Among the first 2D NMR experiments applied to proteins was 2D $J$-resolved spectroscopy. Both homonuclear and heteronuclear experiments have been developed.[1,12] For proteins, $J$-resolved spectra have limited use since they lack resolution along the chemical shift axis, and only those resonances can be analyzed which are well resolved already in 1D spectra. Their main purpose, i.e., to measure coupling constants, can be achieved also in highly resolved COSY spectra. Another experiment of potential interest is a heteronuclear NOE experiment. It could provide information on mobility and on distances between hydrogens and other nuclei. Several techniques have been proposed in the literature.[47–49] For

[44] L. E. Kay, T. L. Jue, B. Bangerter, and P. C. Demou, *J. Magn. Reson.* **73**, 558 (1987).
[45] V. Sklenář, D. Torchia, and A. Bax, *J. Magn. Reson.* **73**, 375 (1987).
[46] N. R. Nirmala and G. Wagner, *J. Am. Chem. Soc.* **110**, 7557 (1988).
[47] K. E. Kövér and G. Batta, *Progr. NMR Spectrosc.* **19**, 223 (1987).
[48] N. Niccolai, C. Rossi, V. Brizzi, and W. A. Gibbons, *J. Am. Chem. Soc.* **106**, 5732 (1984).
[49] C. Yu and G. C. Levy, *J. Am. Chem. Soc.* **105**, 6994 (1983).

macromolecules, however, all heteronuclear NOE's are expected to be close to zero. It seems that only isotope enrichment will provide enough sensitivity to detect heteronuclear NOE's.

Applications for Studies of Protein Structure and Dynamics

The main impact of heteronuclear NMR experiments on studies of proteins seems to be in the analysis of dynamic processes. These can be studied by relaxation time measurements and by exchange spectroscopy. The only conformational parameters which might become accessible in the near future by heteronuclear experiments are heteronuclear coupling constants. Furthermore, heteronuclear NMR experiments may be important if they can be employed to resolve ambiguities in proton resonance assignments. To obtain such information requires assignments of the heteronuclei. Assignment of the spectra of carbon, nitrogen, or other nuclei bound to proteins most often relies on cross-correlation with already assigned proton spectra.

*Assignment of Carbon and Nitrogen Spectra*

A large fraction of the carbon resonances of basic pancreatic trypsin inhibitor (BPTI) were assigned using the heteronuclear COSY experiment of Fig. 5A and the hetero-RELAY experiment of Fig. 7A.[30,31] Figure 10 shows the region of the $C^\alpha H$ cross-peaks. The assignments relied on knowing the $\alpha$-proton resonance positions so that direct $C^\alpha H$ correlations could be established. If $\alpha$-proton resonances were degenerate, unique assignments could be made in most cases by identification of RELAY cross-peaks between $\alpha$-carbons and $\beta$-protons. A similarly complete set of assignments was obtained recently for the nitrogens of BPTI using the pulse sequence of Fig. 5C.[50] Obviously, assignments of quaternary carbons cannot be obtained by these methods. It may be possible that such assignments will be achieved via long-range couplings.

*Identification of Proton Spin Systems via Heteronuclear Experiments*

At present assignment of proton resonances is more important than assignment of carbon or nitrogen resonances since they can be used to measure distances via NOE experiments, to characterize torsion angles via coupling constants or to study dynamic effects. For analysis of NOESY spectra of proteins it is important that as many proton resonances as possible are assigned, otherwise NOE's to crowded spectral

[50] J. Glushka and D. Cowburn, *J. Am. Chem. Soc.* **109**, 7879 (1987).

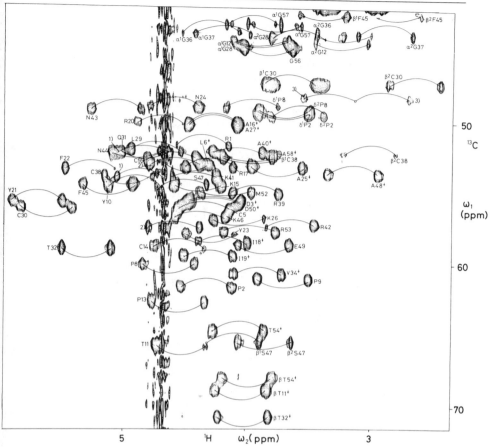

FIG. 10. Assignment of the $C^\alpha H$ groups in a heteronuclear correlated spectrum of basic pancreatic trypsin inhibitor (adapted from Ref. 31). The pulse sequence of Fig. 5A was used.

regions will sometimes remain ambiguous since one cannot be sure whether a certain cross-peak connects to a known resonance or to a signal which is not yet assigned. Heteronuclear experiments can serve to assign proton resonances which may have escaped detection by homonuclear experiments since they provide resolution in a third dimension. The usual way to assign proton spin systems[1] of amino acids is to search for NH-$\alpha$ or $\alpha$–$\beta$ cross-peaks in $^1$H COSY or double-quantum filtered COSY spectra (see this series, Vol. 177, [7] by Basus). If the active coupling is small compared to the line width these cross-peaks may have low intensity or even be absent. This situation has improved by introduction of TOCSY (homonuclear Hartmann–Hahn spectroscopy; HOHAHA)[51,52] spectra which do not suffer from cancellation of antiphase multiplet components (see [8] by Bax, this volume). However, often several residues may be degenerate with respect to the $\alpha$-protons and $\beta$-protons, and their identification by homonuclear experiments alone may be difficult. In the intensively studied BPTI, the spin systems of Arg-1 and Pro-13 could not be detected by homonuclear experiments alone.[53] In heteronuclear COSY experiments the one-bond coupling constants are always much larger than the line widths and can be refocused to in-phase mulitplet structure so that no cancellation occurs. A heteronuclear $^1H$–$^{13}C$ COSY yields, for example, one cross-peak for each residue except for glycines, which have two.[31] Furthermore, glycines $C^\alpha H$ cross-peaks can be identified by DEPT or INEPT editing (see above).

### Connectivities between Degenerate Protons

Homonuclear 2D proton experiments suffer sometimes from problems of degeneracy. COSY cross-peaks can never be established between degenerate resonances. This is a serious problem for identification of threonine spin systems where the $\alpha$- and $\beta$-protons are sometimes degenerate. A similar problem exists for aromatic side chains. If, for example, the $\delta$- and $\varepsilon$-protons of a tyrosine side chain are degenerate they will show neither a cross-peak nor a diagonal peak in a double-quantum filtered COSY. In a heteronuclear COSY, however, the $C^\delta H$ and $C^\varepsilon H$ cross-peaks are well resolved due to the largely different chemical shifts of the carbon resonances. Connectivities between degenerate $\delta$- and $\varepsilon$-protons, for example, can be established by heteronuclear RELAY experiments where carbon decoupling is omitted during detection. The pathway of magnetization transfer can be characterized as an $H^\delta$–$C^\delta(\omega_1)$–$H^\delta$–$H^\varepsilon(\omega_2)$ RELAY transfer. A cross-peak will appear between $C^\delta$ and $H^\varepsilon$. The position of this cross-peak would overlap with the direct $C^\delta H^\delta$ cross-peak if carbon decoupling would be applied. Without decoupling, however, this cross-peak falls in the empty space between the two multiplet components. This method has been demonstrated for connectivities between nearly degenerate $\alpha$- and $\beta$-protons of threonines in BPTI,[31] and the aromatic proton spectrum of this protein could be completely assigned only by using this technique.[54] The same principle could be used to establish NOE connectivities between degenerate protons which can never be mea-

[51] L. Braunschweiler and R. R. Ernst, *J. Magn. Reson.* **53**, 521 (1983).
[52] A. Bax and D. G. Davis, *J. Magn. Reson.* **65**, 355 (1985).
[53] G. Wagner and K. Wüthrich, *J. Mol. Biol.* **155**, 347 (1982).
[54] G. Wagner, D. Brühwiler, and K. Wüthrich, *J. Mol. Biol.* **196**, 227 (1987).

sured with homonuclear experiments. This may be particularly valuable for characterization of $\beta$-sheet and $\alpha$-helical secondary structures, which rely on observations of $d_{NN}$ and $d_{\alpha\alpha}$ connectivities, and NH and $\alpha$-proton resonances may in some cases be degenerate.

## Analysis of Exchange Phenomena

Intramolecular exchange processes between different conformational states seem to be common for proteins and may be important for certain aspects of protein function. NMR can be used to detect such slowly interconverting conformational substates and to characterize equilibrium constants and exchange rates. If such exchange processes are slow on the NMR time scale (exchange rates smaller than the chemical shift differences) they may cause the appearance of different sets of signals corresponding to the different conformational substates. If the substates have different energies the signals will have different intensities, and the intensity ratios will be temperature dependent in most cases. In favorable situations line shapes may change with temperature. The best known of such exchange phenomena are folding–unfolding processes, rotation of aromatic or other side chains, proline cis–trans isomerization, or protonation–deprotonation equilibria. The latter are mostly fast on the NMR time scale. Chemical exchange can be studied by homonuclear 2D exchange spectroscopy (NOESY)[55] (see Berkowitz and Balaban [16], this volume), homonuclear 2D $zz$ spectroscopy,[56] or 2D rotating frame NOE experiments (Camelspin, ROESY)[57,58] (see Brown and Farmer [11], this volume). Exchange cross-peaks in NOESY and ROESY experiments are often close to the diagonal, in homonuclear $zz$-spectra they are usually far off the diagonal but they may suffer from cancellation effects due to unresolved $J$-couplings. Equilibrium constants and exchange rates cannot be measured properly in NOESY or ROESY spectra since, for this purpose, diagonal peaks and cross-peaks would have to be integrated which are often overlapped by many other signals. Equilibrium constants and exchange rates could be measured in homonuclear $zz$ spectra if the coupling constants are the same in the different interconverting substates. In addition, it is necessary that a pair of coupled protons is undergoing exchange.

[55] J. Jeener, B. H. Meier, P. Bachmann, and R. R. Ernst, J. Chem. Phys. **71**, 4546 (1979).
[56] G. Wagner, G. Bodenhausen, N. Müller, M. Rance, O. W. Sørensen, R. R. Ernst, and K. Wüthrich, J. Am. Chem. Soc. **107**, 6440 (1985).
[57] A. A. Bothner-By, R. L. Stephens, J. Lee, C. D. Warren, and R. W. Jeanloz, J. Am. Chem. Soc. **106**, 811 (1984).
[58] A. Bax and D. G. Davis, J. Magn. Reson. **63**, 207 (1985).

Heteronuclear $zz$ experiments[59] provide additional unique information to characterize such intramolecular exchange processes. The role of the diagonal in the homonuclear exchange experiments is taken by the direct cross-peaks (as in homonuclear $zz$ spectroscopy). The intensity of the direct peaks and the cross-peaks is not biased by variation in the coupling constants, which are very similar for all CH groups, and do not depend on conformational parameters such as torsion angles. Thus, this technique can be used to measure rates and equilibrium constants for such exchange processes.

### Relaxation Time Measurements

$^{13}$C relaxation time measurements provide information about reorientational motions of CH bonds (overall rotation and internal reorientation) (see London [18], this volume). $^{13}$C relaxation in proteins seems to be dominated by dipole–dipole interaction, at least at the magnetic field strengths available today. The relaxation times depend on the inverse sixth power of the distance between the interacting nuclei and on the distribution of the frequencies of rotational motions which modulate the dipole–dipole interaction and induce nuclear transitions. For protonated carbons the contribution of the directly bound protons is dominant. Since the CH distance is fixed by the chemical bond, parameters about internal motions can be obtained by carbon relaxation times. The most straightforward analysis is possible for CH groups, while for CH$_3$ groups the interpretation is complicated by the methyl rotation. Relaxation time measurements on methine carbons in proteins with conventional 1D techniques are difficult to analyze because the resonances are heavily overlapped. Relaxation time measurements in heteronuclear C–H correlation experiments provide the necessary resolution,[31] and the high sensitivity of proton-detected heteronuclear experiments allows one to measure a series of 2D spectra, analogous to inversion recovery experiments. For a 20 m$M$ solution of a protein, such a series can be recorded in a week of instrument time. Ideally, the pulse sequence of Fig. 9 yields a series of spectra where the intensity of the cross-peaks recover from a value of $-1.0$ to $+0.25$ in units of the intensities in a normal proton-detected heteronuclear COSY experiment.[46] The motivation for such experiments is to identify a variation of relaxation times within the protein and to differentiate between regions of higher and lower mobility. Motions that are manifested in relaxation times are so fast that they can be simulated by molecular dynamics calculations. Thus relaxation time measurements provide one of the few experimental data sets to check and to optimize these computation procedures. Relaxation time measurements may also have implica-

[59] G. Montelione and G. Wagner, *J. Am. Chem. Soc.*, in press.

tions for interpretation of NOESY spectra. Ideally, they could lead to a characterization of the spectral density function of motions of C–H bond vectors in a protein. This would provide some help to interpret H–H NOE's in terms of distances. Obviously, the spectral density functions of motions of the H–H connecting vector should be known for a proper analysis of H–H NOE's in terms of distances. This cannot be obtained from carbon relaxation times. However, if there is a variation of carbon relaxation times found within a protein, one can use a local calibration of NOE's with known distances instead of a global calibration as it is done usually,[1] and distances may be characterized more accurately.

## Long-Range Couplings

Heteronuclear long-range couplings are a large, still unused resource of conformational parameters. Vicinal coupling constants $^3J(H^\beta, C')$, $^3J(H^\beta, N)$, and $^3J(H^\beta, H^\alpha)$ will provide the best means to characterize the torsion angle $\chi^1$ and to obtain stereospecific assignments of $\beta$-methylene protons. In a similar way, dihedral angles $\phi$ and $\psi$ may be characterized by vicinal homonuclear and heteronuclear coupling constants. Such measurements have been carried out with small peptides[41] where high concentrations can be obtained and line widths are smaller. For proteins such experiments are more difficult, but it can be foreseen that this resource will become available, in particular when isotope-enriched proteins are available.

## Conclusion

Heteronuclear experiments on proteins can provide useful complementary information to homonuclear proton NMR experiments. Such experiments have become practical since proton-detected heteronuclear experiments have become available. As this technology has been introduced only recently, the potential of heteronuclear techniques is certainly not exhausted yet, and a number of useful new heteronuclear experiments can be expected in the near future. At the moment the main attractions of heteronuclear experiments of proteins are to clarify and to complete proton spin system assignments, to study protein dynamics via exchange spectroscopy or relaxation time measurements, and to get additional conformation parameters by measuring heteronuclear long-range couplings.

## Acknowledgments

The preparation of this chapter was supported in part by a grant from the National Science Foundation (DMB-8616059).

## [6] Multiple-Quantum Nuclear Magnetic Resonance

*By* Mark Rance, Walter J. Chazin, Claudio Dalvit,
and Peter E. Wright

### Introduction

Multiple-quantum (MQ) spectroscopy offers significant advantages for analysis of the highly overlapped $^1$H NMR spectra of proteins. By combining multiple-quantum techniques with COSY and relayed coherence transfer experiments, it is possible to obtain essentially *complete* proton assignments for proteins of 100 or more amino acid residues. The availability of such extensive assignments makes it possible to utilize NMR at an unprecedented level of detail to determine protein conformation and dynamics in solution.

Two fundamental multiple-quantum experiments, multiple-quantum-filtered COSY[1,2] (MQF-COSY) and two-dimensional multiple-quantum (MQ) spectroscopy,[3] are of particular importance for making unambiguous $^1$H spin system assignments in proteins. To achieve this, acquisition of spectra in the phase-sensitive mode is virtually mandatory. With increasing reliance on multiple-relayed coherence transfer techniques to assign protein spectra, the multiple-quantum methods are becoming even more important, because it is not generally feasible to assign resonances unambiguously to specific side-chain protons, i.e., to discriminate between $C^\beta$, $C^\gamma$, and $C^\delta$ protons, solely on the basis of multiple-step relayed connectivities. This problem is most severe for the isotropic mixing (TOCSY or HOHAHA) experiment,[4,5] because there is a multiplicity of pathways for coherence transfer through an amino acid side chain and because the dynamics of the coherence transfer process are complex. It is therefore necessary to supplement these experiments with multiple-quantum experiments which have no ambiguities.

The most significant advantages of multiple-quantum experiments are summarized below.

1. MQ and MQF experiments can provide substantial simplification of both one- (1D) and two-dimensional (2D) spectra.

[1] U. Piantini, O. W. Sørensen, and R. R. Ernst, *J. Am. Chem. Soc.* **104**, 6800 (1982).
[2] A. J. Shaka and R. Freeman, *J. Magn. Reson.* **51**, 169 (1983).
[3] L. Braunschweiler, G. Bodenhausen, and R. R. Ernst, *Mol. Phys.* **48**, 535 (1983).
[4] L. Braunschweiler and R. R. Ernst, *J. Magn. Reson.* **53**, 521 (1983).
[5] A. Bax and D. G. Davis, *J. Magn. Reson.* **65**, 355 (1985).

2. MQ filters remove much of the dispersive contribution to the diagonal of COSY-type spectra and can greatly attenuate the amplitude of the diagonal peaks.

3. Multiplets in MQF-COSY spectra often have unique fine structure which provides increased resolving power in crowded regions plus information on the relative signs of the active couplings, thereby allowing distinction between geminal and vicinal couplings.

4. MQ spectra contain no diagonal (autocorrelation) peaks, thus allowing identification of nearly degenerate coupled spins.

5. Remote and combination peaks in MQ spectra provide a unique source of information for unambiguous spin system identification.

6. Most peaks in MQ spectra have a predominantly in-phase multiplet structure in the $\omega_1$ dimension, thus avoiding the loss of intensity due to self-cancellation of antiphase multiplets, which is characteristic of COSY spectra. This becomes increasingly important as the ratio of line width to active coupling increases.

7. The elimination and reduction of the diagonal peaks in MQ and MQF-COSY spectra, respectively, can significantly reduce the amount of $t_1$ noise in the 2D spectra.

Our aim in this chapter is to provide an introduction to multiple-quantum spectroscopy as it is applied to the study of proteins. There is insufficient space to cover all aspects of multiple-quantum spectroscopy; instead we restrict ourselves to a discussion of those subjects of greatest importance in applications to proteins. Practical aspects will be emphasized over theoretical concepts. The theory of multiple-quantum NMR spectroscopy has been dealt with in considerable detail elsewhere and the reader is referred to several excellent reviews.[3,6–9]

## Basic Theory

The general scheme for a multiple-quantum experiment involves four basic steps: (1) a preparation period, in which the desired order of coherence is generated; (2) an evolution period, during which the spin system evolves under the influence of the chemical shifts and scalar couplings; (3) a mixing period, during which there usually is some transfer of coherence; and (4) a detection period, where the NMR signal is recorded. For ease of discussion, we shall classify homonuclear multiple-quantum ex-

[6] G. Bodenhausen, *Prog. NMR Spectrosc.* **14,** 137 (1981).
[7] G. P. Drobny, *Annu. Rev. Phys. Chem.* **36,** 451 (1985).
[8] D. Weitekamp, *Adv. Magn. Reson.* **11,** 111 (1983).
[9] M. A. Thomas and A. Kumar, *J. Magn. Reson.* **54,** 319 (1983).

periments into three categories: (1) one-dimensional multiple-quantum filtration,[10,11] (2) two-dimensional correlated spectroscopy with multiple-quantum filtration,[1,2] and (3) two-dimensional multiple-quantum spectroscopy.[3,9]

Multiple-quantum filtration involves the creation of a desired order of multiple-quantum coherence (MQC) in a spin system, immediately followed by a transfer of this MQC to single-quantum coherences (SQC) for detection; all other coherence transfer pathways are eliminated, providing for a considerable simplification of the resulting 1D or 2D spectrum. In a 2D MQ experiment the MQC created is allowed to evolve for a time $t_1$ before transfer back to SQC; in this case a 2D Fourier transform results in a spectrum containing multiple-quantum frequencies in one dimension and single-quantum frequencies in the other.

There are several methods for the creation of multiple-quantum coherence.[12] The most generally applicable method is to apply a pair of nonselective 90° rf pulses separated in time by a suitable delay $\tau$. While selective excitation of specific MQC is difficult (but see Ref. 13), it is relatively easy to detect a particular order of MQC selectively using the different transformation properties of the various orders of MQC. MQC of order $p$ is $p$ times more sensitive to rf phase shifts than is SQC.[12] By a suitable combination of free-induction decays (FID's) collected with different phases of the multiple-quantum excitation pulses, it is possible to cancel out all of the signals originating from unwanted MQC and retain only the specific MQC of interest.

The degree of spectral simplification to be expected in multiple-quantum experiments (MQF or MQ) can be determined from the so-called coherence transfer selection rules identified by Braunschweiler et al.[3]: (1) multiple-quantum coherence involving a set of $q$ spins can only be transferred to or created from single-quantum transitions of a spin $A$ when the scalar couplings between $A$ and all $q$ spins (or $q - 1$ spins if $A$ belongs to the set of $q$ spins) are resolved; and (2) since couplings between magnetically equivalent nuclei are ineffective in the isotropic phase, MQC involving two or more equivalent nuclei cannot be transferred to or created from the single-quantum transitions of these equivalent nuclei. These rules, while not strictly valid in the presence of strong coupling or differential relaxation effects (see later), are in general extremely useful for interpretation of multiple-quantum spectra.

[10] A. Bax, R. Freeman, and S. P. Kempsell, *J. Am. Chem. Soc.* **102**, 4851 (1980).
[11] G. Bodenhausen and C. M. Dobson, *J. Magn. Reson.* **44**, 212 (1981).
[12] A. Wokaun and R. R. Ernst, *Chem. Phys. Lett.* **52**, 407 (1977).
[13] M. H. Levitt and R. R. Ernst, *J. Chem. Phys.* **83**, 3297 (1985); erratum: **84**, 4114 (1986).

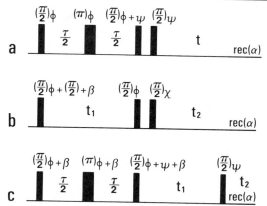

FIG. 1. Pulse sequences for 1D MQF (a), MQF-COSY (b), and MQ (c) experiments. The basic phase cycling for selective detection of $p$ quantum coherence consists of $2p$ steps, in which the phase $\phi = k\pi/p$, where $k = 0, 1, \ldots, 2p - 1$. The resulting free-induction decays are alternately added to and subtracted from computer memory. The phase of the receiver is designated $\alpha$. The phase $\psi$ is 0 for even-quantum and $\pi/2$ for odd-quantum excitation. In addition to the basic phase cycling, CYCLOPS[27] can be included by incrementing all pulse phases in steps of $\pi/2$ and appropriately routing the quadrature data in computer memory, giving an $8p$ step-phase cycle. In practice the $(\pi)_\phi$ refocusing pulse is often replaced by the composite pulse[28] $(\pi/2)_{\phi+(\pi/2)} - (\pi)_\phi - (\pi/2)_{\phi+(\pi/2)}$. Because the transfer of coherence between single and MQ transitions is particularly sensitive to rf pulse imperfections it may be desirable to employ more fully compensated pulse sequences.[29-31]

## Techniques of Multiple-Quantum Spectroscopy

### One-Dimensional Multiple-Quantum Filters

The basic pulse sequence for 1D multiple-quantum filtration[10] is shown in Fig. 1a. As discussed above, the first two 90° rf pulses create MQC; the 180° pulse is inserted in the middle of the preparation period in order to refocus the precession due to chemical shift and the inhomogeneity of the static magnetic field, allowing off-set independent excitation of the MQC. The third 90° pulse reconverts the MQC to SQC, which is subsequently detected. The phase $\phi$ of the multiple-quantum preparation pulses is cycled such that only the coherence transfer pathway[14] through the desired order of MQC is selected; for a $p$-quantum filter, the phase $\phi$ is cycled through the values $\phi = k\pi/p$, $k = 0, 1, 2, \ldots, 2p - 1$, and the resulting signals are alternately added and subtracted (receiver reference phase $\alpha = 0$). For $\psi = 0$ only even orders of MQC are created whereas for $\psi = \pi/2$

[14] G. Bodenhausen, H. Kogler, and R. R. Ernst, *J. Magn. Reson.* **58**, 370 (1984).

only odd orders of MQC are generated. A useful variation of this method is to store the signals recorded for each setting of the phases $\phi$ and $\psi$ separately instead of combining them during acquisition. Multiple-quantum-filtered spectra of different orders can then be generated from the same data set by taking different linear combinations of the separate FID's.[12,14] Symmetrical sequences[11,15,16] have been employed to obtain both 1D and 2D MQ spectra with in-phase multiplet structure in $\omega_2$ instead of the antiphase character which is inherent to using the sequences of Fig. 1. However, there will be a loss of intensity due to relaxation during the additional refocusing period, so the benefits of using symmetrical sequences need to be judged on an individual basis.

Spectral simplification via multiple-quantum filtration occurs because a $p$-quantum coherence can be generated only in a spin system containing $p$ or more spins. The appearance of a resonance in a $p$-quantum-filtered spectrum indicates that the spin in question has resolved couplings to at least $p - 1$ other spins. Larger values of $p$ offer the greatest simplification, but there is a significant loss in signal intensity for larger $p$ due to competing excitation of the more numerous lower order coherences.

### Two-Dimensional Multiple-Quantum-Filtered COSY

The pulse sequence for MQF-COSY[1,2] is shown in Fig. 1b. The first 90° pulse creates SQC, which is subsequently frequency labeled during the evolution period $t_1$, as in the conventional COSY[17,18] experiment. MQF-COSY differs from the conventional COSY in the coherence transfer pathway taken after the evolution period. In the COSY experiment, SQC is transferred from one spin to another; in MQF-COSY the coherence transfer is through a transient state of MQC. In the sequence of Fig. 1b, all orders of MQC are excited, and so the filter order is determined solely by the incrementation of the pulse phase $\phi$. The phase cycling of $\phi$ and $\alpha$ for a $p$-quantum filter is identical to that described above for the 1D MQF experiments. It is possible to generate 2D MQF-COSY spectra of all desired orders from the same data set.[19]

Quadrature detection in the $\omega_1$ dimension in a 2D MQF-COSY experiment is easily accomplished[20] via either the time-proportional phase incre-

[15] O. W. Sørensen, M. H. Levitt, and R. R. Ernst, *J. Magn. Reson.* **55**, 104 (1983).
[16] M. Rance, O. W. Sørensen, W. Leupin, H. Kogler, K. Wüthrich, and R. R. Ernst, *J. Magn. Reson.* **61**, 67 (1985).
[17] W. P. Aue, E. Bartholdi, and R. R. Ernst, *J. Chem. Phys.* **64**, 2229 (1976).
[18] A. Bax and R. Freeman, *J. Magn. Reson.* **44**, 542 (1981).
[19] R. Ramachandran, P. Darba, and L. R. Brown, *J. Magn. Reson.* **73**, 349 (1987).
[20] J. Keeler and D. Neuhaus, *J. Magn. Reson.* **63**, 454 (1985).

mentation (TPPI) technique[14,21-23] or the hypercomplex method.[24-26] In the TPPI method the phase $\beta$ in Fig. 1b is incremented in steps of $\pi/2$ for each incrementation of the variable delay $t_1$. The spectral width in the $\omega_1$ dimension is given by $1/(2\Delta t_1)$, where $\Delta t_1$ is the step size of the $t_1$ increment. The data are processed by first transforming the FID's $s(t_1, t_2)$ to give the complex result $S(t_1, \omega_2)$, including any phase correction which may be necessary in the $\omega_2$ dimension, and then performing a real Fourier transform with respect to $t_1$ of the absorption (real) part of $S(t_1, \omega_2)$ to give $S(\omega_1, \omega_2)$. By separately transforming the dispersion (imaginary) part of $S(t_1, \omega_2)$ it is possible to execute a posttransform phase correction of the 2D spectrum in either dimension. The hypercomplex method requires that for each value of $t_1$, two FID's are recorded with $\beta = 0$ and $\pi/2$; after the normal Fourier transformation of the data with respect to $t_2$, a complex data set is formed by setting the real part equal to the absorption component of the $S(t_1, \omega_2, \beta = 0)$ data and the imaginary part equal to the absorption component of the $S(t_1, \omega_2, \beta = \pi/2)$ data and then performing a complex Fourier transformation with respect to $t_1$. In this case the $\omega_1$ spectral width is given by $1/\Delta t_1$. As discussed by Keeler and Neuhaus,[20] the two methods for $\omega_1$ quadrature detection are formally equivalent; one practical difference, however, is that axial peaks (i.e., signals generated from longitudinal magnetization present during $t_1$) will appear along the bottom of a TPPI-generated spectrum but through the middle of a hyper-complex-generated spectrum.

The selection rules for observation of peaks in MQF-COSY spectra are as follow[1]: (1) the appearance of a diagonal peak in a $p$-quantum-filtered spectrum indicates that the spin in question has couplings to $p - 1$ other spins, and (2) the presence of a cross-peak in a $p$-quantum-filtered spectrum indicates a coupling between two spins which share a common set of at least $p$-2 additional coupling partners.

*Two-Dimensional Multiple-Quantum Spectroscopy*

The pulse sequence most often used for 2D MQ experiments is shown in Fig. 1c. The multiple-quantum preparation period is the same as in

[21] A. G. Redfield and S. D. Kunz, *J. Magn. Reson.* **19**, 250 (1975).

[22] G. Drobny, A. Pines, S. Sinton, D. P. Weitekamp, and D. Wemmer, *Symp. Faraday Soc.* **13**, 49 (1979).

[23] D. Marion and K. Wüthrich, *Biochem. Biophys. Res. Commun.* **113**, 967 (1983).

[24] L. Müller and R. R. Ernst, *Mol. Phys.* **38**, 963 (1979).

[25] D. J. States, R. A. Haberkorn, and D. J. Ruben, *J. Magn. Reson.* **48**, 286 (1982).

[26] M. Ohuchi, M. Hosono, K. Furihata, and H. Seto, *J. Magn. Reson.* **72**, 279 (1987).

[27] D. I. Hoult and R. E. Richards, *Proc. R. Soc. London, Ser. A* **344**, 311 (1975).

[28] M. H. Levitt and R. Freeman, *J. Magn. Reson.* **33**, 473 (1979).

the 1D multiple-quantum filter experiment. Quadrature detection in the $\omega_1$ dimension can be achieved by either the TPPI technique or the hypercomplex method, as for MQF-COSY; the required phase increment $\beta$ for each of the excitation pulses is $\pi/(2p)$ for a $p$-quantum spectrum,[14] since a $p$-quantum coherence is $p$ times more sensitive to rf phase shifts than is SQC. For MQ spectra of proteins dissolved in $^2H_2O$ we have often found it desirable to collect the spectra with no $\omega_1$ quadrature detection, since fewer experiments are then required for a given digital resolution in $\omega_1$. This produces spectra which are folded about $\omega_1 = 0$, but with a judicious choice of the transmitter frequency there is seldom a problem of undesirable overlaps or ambiguities as to the sign of the $\omega_1$ frequency.

The choice of the multiple-quantum excitation period $\tau$ is very important for optimal MQ excitation. For example, in an AX spin system the optimum value of $\tau$ for the generation of 2QC is $1/(2J_{AX})$ (in the weak coupling limit and ignoring relaxation). In an AMX spin system, there are six 2Q coherences and the efficiencies of their excitation have more complicated expressions. In general, the efficiency of excitation of each MQC in an arbitrary spin system will be some calculable function of $\tau$ and the relevant $J$ couplings, and will depend on the coupling topology of the spin system.[9]

The peaks which are present in a multiple-quantum spectrum can be classified into three general types: (1) direct peaks, which arise when MQC actively involving a given spin (i.e., that spin participates in the multiple-quantum transition) is transferred to SQC of that spin, (2) remote peaks, due to the transfer of MQC to SQC of a spin not actively involved in the multiple quantum transition, and (3) combination peaks, which arise from $q$-spin–$p$-quantum coherences.[3] The direct peaks are exactly analogous to the cross-peaks in a COSY or MQF-COSY spectrum. The remote peaks provide information similar to the relayed connectivities in a relayed coherence transfer spectrum,[32] but also give additional information concerning magnetically equivalent nuclei.

A schematic representation of a 2Q spectrum of an $AMX_2$ spin system is shown in Fig. 2.[33] The filled circles represent the direct connectivities if $J_{AX} = 0$. The open symbols represent peaks which could be observed if $J_{AX} \neq 0$. The direct peaks always occur in pairs, equidistant from the two-quantum pseudo-diagonal ($\omega_1 = 2\omega_2$). The remote peaks are indicated by

[29] M. H. Levitt, *Prog. NMR Spectrosc.* **18**, 61 (1986).
[30] S. Wimperis and G. Bodenhausen, *J. Magn. Reson.* **71**, 355 (1987).
[31] N. Müller, R. R. Ernst, and K. Wüthrich, *J. Am. Chem. Soc.* **108**, 6482 (1986).
[32] G. Eich, G. Bodenhausen, and R. R. Ernst, *J. Am. Chem. Soc.* **104**, 3731 (1982).
[33] O. W. Sørensen, Ph.D. Dissertation No. 7658, ETH Zürich (1984).

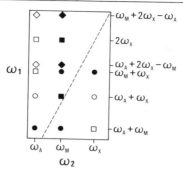

Fig. 2. Schematic diagram of a 2Q spectrum of an $AMX_2$ spin system, adapted from Ref. 33. Circles, squares, and diamonds indicate direct, remote, and combination peaks, respectively. The corresponding two-quantum frequencies are indicated at the side, and the dashed line is the two-quantum pseudo-diagonal ($\omega_1 = 2\omega_2$). Peaks indicated by open symbols will be observable only if $J_{AX} \neq 0$. An additional pair of combination peaks arising from the 4-spin–2Q-coherence $\omega_A + \omega_M + \omega_X - \omega_X$ are coincident with the direct peaks at $\omega_1 = \omega_A + \omega_M$, $\omega_2 = \omega_A$, $\omega_M$.

squares. A 2Q coherence between spins A and X can be created due to their mutual coupling to spin M, and this coherence can be transferred subsequently to SQC of spin M. Thus, a peak appears at $\omega_1 = \omega_A + \omega_X$, $\omega_2 = \omega_M$ and indicates that the fragments identified from the direct peaks, A–M and M–X, actually form an A–M–X system. A second remote peak arises from 2Q coherence between the magnetically equivalent X spins transferred to SQC of the M spin. This peak allows one to distinguish between A–M–X and A–M–$X_n$ ($n > 1$) spin systems. The diamonds in Fig. 2 are the combination peaks arising from 4-spin–2-quantum coherences. In each case, four spins are actively involved in the 2Q transition, with three spins flipping one way and the fourth flipping the opposite direction so that the net change in magnetic quantum number is two. For an $AMX_2$ spin system there are three unique 4-spin–2-quantum transitions. However, one ($\omega_1 = \omega_A + \omega_M + \omega_X - \omega_X$) is coincident with the AM direct peaks. The observation of combination peaks is useful for confirming assignments made from direct and remote connectivities.

Multiple-quantum spectroscopy does not always give rise to spectral simplification. A double-quantum spectrum is in some regards more complex than a COSY spectrum, due to the presence of remote and combination peaks. However, the absence of diagonal (autocorrelation) peaks represents a major advantage over COSY. In addition, the remote and combination peaks often appear in regions of a protein spectrum which are not particularly crowded and they are therefore invaluable for the unambiguous assignment of resonances.

Due to the smaller number of 3Q transitions compared to 2Q transitions for a given spin system, a 3Q spectrum is considerably simplified compared to a COSY or a 2Q spectrum. In our example of an $AMX_2$ spin system, only three peaks are observed for $J_{AX} = 0$, all at $\omega_2 = \omega_M$. One peak is located at $\omega_1 = \omega_A + \omega_M + \omega_X$, another at $\omega_1 = \omega_M + 2\omega_X$; both are direct peaks, since the M spin is actively involved in the 3Q coherences. A remote peak is located at $\omega_1 = \omega_A + 2\omega_X$. No peaks are expected at the $\omega_2$ frequencies of the A and X spins since they each have only one coupling partner. However, if $J_{AX} \neq 0$ then direct peaks may also be observed at $\omega_1 = \omega_A + \omega_M + \omega_X$, $\omega_2 = \omega_A$, $\omega_X$ and $\omega_1 = \omega_A + 2\omega_X$, $\omega_2 = \omega_A$ with a remote peak at $\omega_1 = \omega_M + 2\omega_X$, $\omega_2 = \omega_A$.

### Exceptions to Coherence Transfer Selection Rules

The coherence transfer selection rules,[1,3] which are normally used to determine the appearance of an MQF-COSY or MQ spectrum, are strictly valid only in the limit of weak coupling and in the absence of relaxation effects. Virtual coupling effects in strongly coupled spin systems will give rise to violations of these rules.[34] In addition, it has been shown that differential relaxation of degenerate transitions of magnetically equivalent methyl protons will allow MQC actively involving two or more of these spins to be created from and transferred to SQC of these spins.[34–38] This phenomenon leads to otherwise unexpected methyl resonances in pQF-COSY and pQ spectra ($p > 2$) and to unexpected combination lines in 2Q spectra (for example, the combination peaks in the upper left of Fig. 4). These relaxation effects are dependent on the motional correlation times and, while often seen in protein spectra, normally are not observable for small molecules.

### Flip-Angle Effects

Although an extensive discussion of flip-angle effects is beyond the scope of this chapter (see Ref. 39), some aspects should be discussed briefly. It is possible to determine the sign of a MQ frequency by employ-

[34] M. Rance and P. E. Wright, *Chem. Phys. Lett.* **124**, 572 (1986).
[35] M. Rance, C. Dalvit, and P. E. Wright, *Biochem. Biophys. Res. Commun.* **131**, 1094 (1985).
[36] M. Rance and P. E. Wright, *J. Magn. Reson.* **66**, 372 (1986).
[37] N. Müller, G. Bodenhausen, K. Wüthrich, and R. R. Ernst, *J. Magn. Reson.* **65**, 531 (1985).
[38] N. Müller, G. Bodenhausen, and R. R. Ernst, *J. Magn. Reson.* **75**, 297 (1987).
[39] N. Murali, Y. V. S. Ramakrishna, K. Chandrasekhar, M. A. Thomas, and A. Kumar, *Pramana* **23**, 547 (1984).

ing a non-$\pi/2$ flip angle for the last (detection) pulse in Fig. 1c[40]; another consequence of this method is that some multiplet components can have somewhat greater intensity compared to results obtained with a $\pi/2$ pulse. In addition, remote peaks in MQ spectra can be attenuated relative to direct peaks by using a non-$\pi/2$ pulse.[41] Thus, flip-angle effects appear to be useful for spectral simplification and may find other applications such as in automated spectral analysis using pattern recognition procedures.[42,43]

There are several drawbacks, however, to employing a non-$\pi/2$ detection pulse in MQ experiments with proteins, including (1) $\omega_1$ quadrature detection is not perfect since the "image" peaks are attenuated but not eliminated; (2) significant dispersive character is introduced to the line shapes, which degrades the achievable resolution; (3) the remote peaks are not completely suppressed, and thus may be confused with weak direct connectivities in absolute value spectra; and (4) the intensity of the direct peaks is reduced if a high degree of suppression of remote peaks is attempted. Finally we stress that in phase-sensitive MQ spectra of proteins, remote peaks are, rather than being confusing, an invaluable source of information for making unambiguous spin system assignments.

*Sensitivity*

The question of sensitivity in multiple-quantum spectroscopy is not straightforward because many factors, both theoretical and practical, must be considered. Some key factors are as follow: (1) the efficiency of excitation of the MQC of interest, (2) relaxation times of the MQC and SQC, (3) the extent of constructive and destructive interference among multiplet components, and (4) $t_1$ noise reduction.

There is insufficient space here for a detailed discussion of these points, but a few general comments are appropriate. Theoretically the 2QF-COSY experiment suffers a 2-fold loss of sensitivity compared to COSY. Despite this sensitivity loss, in practice the 2QF-COSY experiment usually results in a net improvement in the quality of spectra and can provide information that cannot be obtained in COSY. In theory the sensitivity of a 3QF-COSY spectrum for an AMX spin system is one-half that of 2QF-COSY. However, constructive and destructive interference effects among cross-peak multiplet components can lead, in favorable

[40] T. H. Mareci and R. Freeman, *J. Magn. Reson.* **48**, 158 (1982).
[41] T. H. Mareci and R. Freeman, *J. Magn. Reson.* **51**, 531 (1983).
[42] M. Novic, H. Oschkinat, P. Pfändler, and G. Bodenhausen, *J. Magn. Reson.* **73**, 493 (1987).
[43] J. Boyd and C. Redfield, *J. Magn. Reson.* **68**, 67 (1986).

cases, to peaks in a 3QF-COSY spectrum being more intense than the corresponding peaks in a 2QF-COSY spectrum.[31] Similar arguments hold for the corresponding MQ experiments.

Our practical experience has been that MQ spectroscopy often provides superior sensitivity compared to the corresponding MQF-COSY, and that as the molecular weight of the protein increases and lines become broader, the predominance of in-phase multiplet structure in the $\omega_1$ dimension can lead to the 2Q experiment having significantly better sensitivity than 2QF-COSY and even the COSY experiment.

### Applications

In the sections below we discuss various practical aspects of multiple-quantum techniques as applied to proteins.[36,44,45] As illustrations we show experimental data for two proteins, French bean plastocyanin (99 residues, $M_r$ ~10,500) and sperm whale myoglobin (153 residues, $M_r$ ~18,000). We stress that multiple-quantum techniques are applicable to proteins of molecular weight up to at least 18,000 and in fact become particularly advantageous as the molecular weight increases.[46]

### *Two- and Three-Quantum-Filtered COSY Experiments*

The 2QF-COSY experiment has become popular as a result of its two major advantages over COSY.[47] First, the diagonal peaks in 2QF-COSY spectra have antiphase multiplet structure instead of in phase, and thus experience self-cancellation to the same extent as do the cross-peaks. For larger proteins this results in a significant reduction in the size of the diagonal peaks, which often results in less $t_1$ noise. Second, the cross-peaks in 2QF-COSY spectra and the major contribution to the diagonal peaks have absorption line shapes in both dimensions. Elimination of the dispersive character of the diagonal peaks allows identification of cross-peaks lying immediately adjacent to the diagonal. These same advantages also pertain to higher order MQF-COSY spectra.

The 3QF-COSY experiment offers further advantages over 2QF-COSY, one of which is spectral simplification. The cross-peaks for spins which do not share at least one common coupling partner are eliminated from the 3QF-COSY spectrum, as are the diagonal peaks for spins which

[44] G. Wagner and E. R. P. Zuiderweg, *Biochem. Biophys. Res. Commun.* **113,** 854 (1983).
[45] J. Boyd, C. M. Dobson, and C. Redfield, *J. Magn. Reson.* **55,** 170 (1983).
[46] C. Dalvit and P. E. Wright, *J. Mol. Biol.* **194,** 313 (1987).
[47] M. Rance, O. W. Sørensen, G. Bodenhausen, G. Wagner, R. R. Ernst, and K. Wüthrich, *Biochem. Biophys. Res. Commun.* **117,** 479 (1983).

have less than two coupling partners. For example, the $C^\alpha H$–$C^\beta H$ cross-peak and the $C^\alpha H$ diagonal peak for valine residues are eliminated by three-quantum filtration and, in spectra recorded in $^2H_2O$, all peaks of glycine residues are eliminated (Fig. 3). In principal, one might expect all aromatic cross-peaks to be eliminated by three-quantum filtration of protein spectra if these spin systems are considered to be linear. In practice, relatively weak cross-peaks are often observed due to long-range coupling and strong coupling effects.

One of the most important applications of 3QF-COSY spectroscopy is for the analysis of connectivities between $C^\alpha H$ and $C^\beta H$ protons. In COSY or 2QF-COSY spectra it is common to observe only one of the $C^\alpha H$–$C^\beta H$ cross-peaks, due to the presence of one small and one large $^3J_{\alpha\beta}$ coupling constant in the $C^\alpha H$–$C^\beta H_2$ moiety. In the 3QF-COSY spectrum, however, both of the $C^\alpha H$–$C^\beta H$ cross-peaks are usually observed

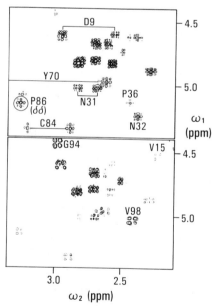

FIG. 3. Corresponding sections of 500-MHz phase-sensitive 2QF-COSY (lower panel) and 3QF-COSY (upper panel) spectra of French bean Cu(I) plastocyanin. The acquisition and data-processing parameters were identical, except that the 3QF-COSY spectrum was acquired with four times as many FID's per $t_1$ value. Spin systems whose cross-peaks are markedly stronger in the 3QF-COSY spectrum are labeled. Spin systems that do not pass through the three-quantum filter (glycine and those with linear coupling topology) are labeled in the 2QF-COSY spectrum. The signs of the multiplet components are inverted for the cross-peak arising from the geminal pair of $C^\delta$ protons of Pro-86 (circled).

because their intensity is less dependent on the magnitude of the small coupling constant.[31,36,43] In addition, the double antiphase multiplet structure gives superior resolving power for analysis of crowded regions of the spectrum. Finally, the relative signs of the multiplets provide information on the relative signs of the active couplings.[31] All of these points are illustrated in Fig. 3.

3QF-COSY (and also 3Q) spectra are also useful for the assignment of the $C^{\gamma}H_2$–$C^{\delta}H_3$ fragment of isoleucine spin systems. Although the multiplet structure of the methyl group resonance is quite distinctive, it is often difficult to precisely locate the $C^{\gamma}H_2$ resonances in COSY and 2QF-COSY spectra. Identification of these resonances is greatly simplified in 3QF-COSY (and 3Q) spectra because all other methyl resonances are greatly suppressed. (Note that weak peaks are usually observed for the other methyl groups due to differential relaxation effects.)

## Multiple-Quantum Spectroscopy

All MQ spectra are free of diagonal peaks (except in the presence of relaxation effects, as discussed above) and thus connectivities can be established between resonances which are nearly degenerate. This is particularly significant in larger proteins, where the many overlapping diagonal peaks often completely mask nearby cross-peaks. The strong diagonal peaks often cause significant $t_1$ noise in COSY spectra which may obscure important cross-peaks. In MQ spectra the $t_1$ noise is drastically reduced due to the absence of diagonal peaks. The limiting factor in the observation of peaks is often the level of thermal noise.

*Double-Quantum Spectroscopy.* Phase-sensitive 2Q spectroscopy offers several unique advantages for $^1$H NMR studies of proteins[44] and hence will be discussed separately. The direct connectivities are represented in a 2Q spectrum by pairs of peaks arranged symmetrically on opposite sides of the pseudo-diagonal ($\omega_1 = 2\omega_2$), while the remote connectivities give rise to isolated peaks. In principle, this difference should allow the direct and remote peaks to be distinguished. In practice, since protein spectra are highly complicated, it is frequently the case that only one of the pair of expected peaks characterizing direct connectivities is observed (Fig. 4), inviting possible erroneous assignments.[48] However, distinction between direct and remote peaks is always possible in phase-sensitive 2Q spectra (and MQ spectra in general) due to predictable variations in the phase and intensity of the peaks as a function of the multiple-quantum excitation period.[48] Other characteristics which help to distinguish the two type of connectivities are (1) when the direct peaks are

[48] C. Dalvit, M. Rance, and P. E. Wright, *J. Magn. Reson.* **69**, 356 (1986).

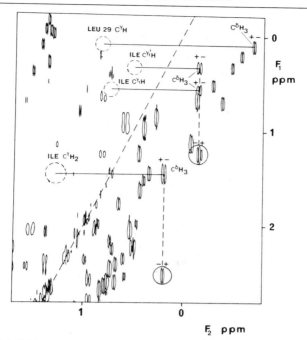

FIG. 4. High-field region of the phase-sensitive double-quantum spectrum of the CO complex of sperm whale myoglobin, with the MQ excitation period $\tau = 40$ msec. The sloping, dashed line represents the double-quantum pseudo-diagonal ($\omega_1 = 2\omega_2$). The direct connectivities are indicated by the horizontal lines for various leucine and isoleucine residues; in each case a peak is observed at the $\omega_2$ frequency of the methyl resonance but the other direct peak is missing (broken circles) due to its high multiplicity in the $\omega_2$ dimension and/or relaxation. The remote peaks at $\omega_2 = \omega_C \delta_{H3}$, $\omega_1 = \omega_C \gamma_H + \omega_C \gamma'_H$ for isoleucine are enclosed in solid circles. Both positive and negative contour levels are drawn. Differentiation of these remote peaks from the direct connectivities is possible due to their different phase characteristics and sign. The peaks on the pseudo-diagonal and the combination lines in the upper left region of the spectrum arise through violations of the coherence transfer selection rules. (From Dalvit *et al.*[48])

phased to be absorptive in $\omega_2$, the remote peaks have dispersive line shapes; and (2) the remote peaks have antiphase multiplet structure with respect to two couplings in the $\omega_2$ dimension, rather than to one coupling as is the case for the direct peaks. (In the $\omega_1$ dimension, many of the direct and remote peaks have predominantly in-phase absorptive line shapes, although there can be significant antiphase dispersive contributions due to couplings to passive spins.) The ease with which direct and remote peaks can be distinguished in phase-sensitive 2Q spectra is clear from Figs. 4 and 5. These figures also illustrate the importance of the remote peaks for

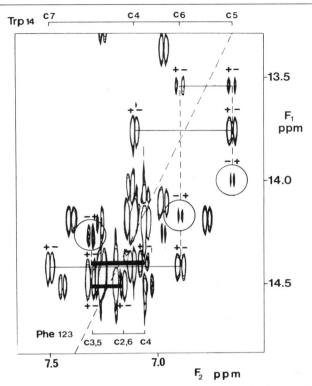

FIG. 5. A section of the aromatic region of the phase-sensitive double-quantum spectrum of the CO complex of myoglobin recorded with $\tau = 40$ msec. The sloping dashed line represents the double-quantum pseudo-diagonal ($\omega_1 = 2\omega_2$). The direct (solid lines) and remote connectivities (circled) for Trp-14 and for Phe-123 are indicated. For tryptophan the remote peaks at $\omega_2 = \omega_6$, $\omega_1 = \omega_7 + \omega_5$ and at $\omega_2 = \omega_5$, $\omega_1 = \omega_6 + \omega_4$ make assignment of the spin system straightforward. In the case of Phe-123 only one remote peak is shown at $\omega_2 = \omega_{3,5}$ and $\omega_1 = \omega_{2,6} + \omega_4$; the other, at $\omega_2 = \omega_4$ and $\omega_1 = 2\omega_{3,5}$ is very weak. (Adapted from Dalvit $et$ $al.$[48])

making unambiguous spin system assignments. Observation of remote peaks at $\omega_1 = \omega_\beta + \omega'_\beta$, $\omega_2 = \omega_\alpha$ in 2Q spectra are particularly useful for determining the complete set of $C^\alpha H$ and $C^\beta H$ connectivities in an unambiguous manner,[36] and for identification of the chemical shifts of tightly coupled and degenerate proton resonances that are often observed at the ends of long side chains (e.g., Ref. 49).

2Q spectroscopy can in principle provide the complete backbone fingerprint (NH–$C^\alpha H$ connectivities) of a protein in $^1H_2O$ solution.[50]

[49] W. J. Chazin, M. Rance, and P. E. Wright, *FEBS Lett.* **222**, 109 (1987).
[50] G. Otting and K. Wüthrich, *J. Magn. Reson.* **66**, 359 (1986).

Some of the NH–C$^\alpha$H cross-peaks are usually missing from COSY spectra recorded in $^1$H$_2$O due to saturation of C$^\alpha$H protons, located under the water peak, during solvent suppression. The observation of a complete fingerprint is possible in the 2Q spectrum since 2QC created from SQC of the NH is unaffected by saturation of the C$^\alpha$H. Thus a peak can be observed at $\omega_2 = \omega_{NH}$ and $\omega_1 = \omega_{NH} + \omega_{C^\alpha H}$ even if the C$^\alpha$H is completely saturated during the equilibration period. It is important to note that the water must not be continuously irradiated during the multiple-quantum excitation period, because this would "decouple" the underlying C$^\alpha$H resonances from their NH partners and no 2QC could be created.

2Q spectra recorded in $^1$H$_2$O are also of great value for unambiguous identification of glycine spin systems[44] since a remote peak is observed at $\omega_2 = \omega_{NH}$, $\omega_1 = \omega_\alpha + \omega_{\alpha'}$. These peaks are located in an otherwise unoccupied region of the spectrum, and thus are easily identified.

The 2Q experiment has been shown to be useful for establishing scalar connectivities by way of small long-range couplings.[51] By selecting an appropriate $\tau$ value, direct connectivities between the C$^\varepsilon$H and C$^\delta$H and between the C$^\delta$H and C$^\beta$H protons of histidine can be observed, even in rather large proteins (Fig. 6). The observation of C$^\delta$H–C$^\beta$H connectivities is important in the sequential assignment procedure since it allows unequivocal connection of the imidazole ring protons with the backbone protons of the same residue.

MQ spectroscopy, in particular 2Q spectroscopy, has a unique advantage in studies of spin systems undergoing certain types of exchange processes.[52] This arises from the fact that the frequency of most MQ transitions are insensitive to a mutual exchange of spins actively participating in the transitions and hence no exchange broadening occurs.

*Three-Quantum Spectroscopy.* One of the principal advantages of a 3Q spectrum is the high degree of simplification which is achieved in comparison to COSY, 2QF-COSY, 3QF-COSY, and 2Q spectra. This is illustrated in Fig. 7, where comparable regions of 2QF-COSY, 3QF-COSY, 2Q, and 3Q spectra of plastocyanin are shown. The C$^\alpha$H–C$^\beta$H$_2$ fragment of a serine residue is readily identified in the 3Q spectrum, which clearly demonstrates the advantage of spectral analysis based on the mutual correlation of three spins. While not identified in Fig. 7, the relative signs of the multiplet components make it easy to distinguish between geminal and vicinal couplings.[36]

The degree of spectral simplification which can be achieved in 3Q spectra is further illustrated in Fig. 8, which shows the aromatic region of plastocyanin. If one makes the approximation that all aromatic spin sys-

[51] C. Dalvit, P. E. Wright, and M. Rance, *J. Magn. Reson.* **71**, 539 (1987).
[52] M. Rance, *J. Am. Chem. Soc.* **110**, 1973 (1988).

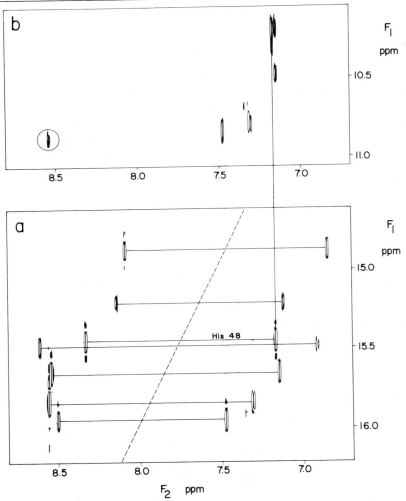

FIG. 6. Two regions of the phase-sensitive double-quantum spectrum of the CO complex of myoglobin obtained with $\tau = 100$ msec. The sloping dashed line in (a) represents the double-quantum pseudo-diagonal ($\omega_1 = 2\omega_2$); the peaks disposed symmetrically with respect to the diagonal and connected with solid lines represent direct connectivities between the $C^\varepsilon H$ and $C^\delta H$ resonances of the histidines. In (b) the peaks at the $\omega_2$ frequency of the $C^\delta H$ resonances originate from double-quantum coherence between the $C^\delta$ and $C^\beta$ protons. (From Dalvit *et al.*[51])

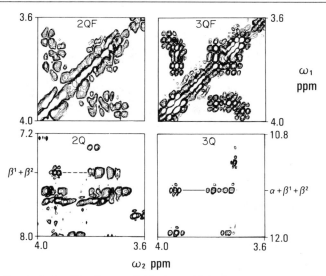

FIG. 7. Spectral simplification available in multiple-quantum spectra. Corresponding sections of 500-MHz phase-sensitive 2QF-COSY, 3QF-COSY, 2Q, and 3Q spectra of French bean Cu(I) plastocyanin showing COSY cross-peaks and 2Q and 3Q direct and remote peaks of Ser-23. The 2Q excitation period was 30 msec and the 3Q excitation period was 36 msec. Partial degeneracy of cross-peaks of several spin systems precludes an unambiguous analysis of the COSY spectra, whereas the two direct and one remote peak at $\omega_1 = \omega_\beta + \omega_{\beta'}$, $\omega_2 = \omega_\beta$, $\omega_{\beta'}$, $\omega_\alpha$ in the 2Q spectrum and the three direct peaks at $\omega_1 = \omega_\alpha + \omega_\beta + \omega_{\beta'}$, $\omega_2 = \omega_\alpha$, $\omega_\beta$, $\omega_{\beta'}$ in the 3Q spectrum are free from overlap and reveal the three resonance frequencies of the spin system.

tems are linear and ignores strong coupling effects, then only two 3Q peaks are expected for phenylalanine and tryptophan and none is expected for tyrosine or histidine. These assumptions are not strictly valid, but the resulting additional peaks are usually weak and are thus easily distinguished.

In practice it is often the case, particularly in 3Q spectra, that one of the direct peaks is not observed, perhaps due to the multiplicity in $\omega_2$ or to exchange effects which selectively broaden one resonance. In such cases no information is lost since the multiple-quantum frequencies can be employed to calculate the chemical shift of a spin actively involved in the MQC, even when no peak is observed at its $\omega_2$ frequency. For example if a $C^\alpha H$ resonance cannot be observed due to an intense noise band from the residual solvent signal but the 3Q peaks can be resolved at the $\omega_2$ frequencies of the two $C^\beta H$ resonances, then the $C^\alpha H$ chemical shift can be calculated.

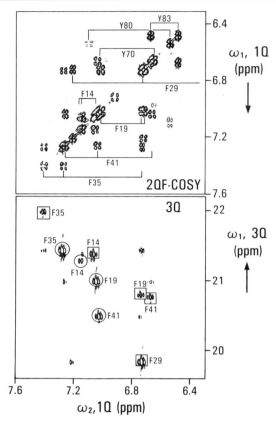

FIG. 8. The aromatic region of 500-MHz phase-sensitive 2QF-COSY and 3Q spectra acquired from a 10 m$M$ solution of French bean Cu(I) plastocyanin in $^2H_2O$. The excitation period in the 3Q spectrum was 36 msec. Circles and boxes in the lower panel identify 3Q direct peaks at $\omega_1 = \omega_\delta + \omega_\varepsilon + \omega_\zeta$, $\omega_2 = \omega_\varepsilon$ and $\omega_1 = 2\omega_\varepsilon + \omega_\zeta$, $\omega_2 = \omega_\zeta$, respectively. The aromatic region of the 3Q spectrum has been folded in the $\omega_1$ dimension, and therefore has an opposite sense relative to the 2QF-COSY spectrum as indicated by the arrows. Note that the tyrosine spin systems are effectively removed from the 3Q spectrum.

## Higher Order MQ and MQF-COSY Experiments

Even though the sensitivity of a 4QF-COSY experiment will be in theory approximately half that of a 3QF-COSY and 8-fold less than COSY, it is still feasible to perform such experiments on present high-field spectrometers in reasonable periods of time. Müller *et al.*[31] have described 4QF-COSY spectra of a protein, which are highly simplified since all two- and three-spin systems are eliminated. Four-quantum spectroscopy is also possible and in some cases may even be preferable to 4QF-COSY since many of the peaks in the 4QF-COSY spectrum are diagonal peaks which are usually uninformative, whereas all of the resonances in the 4Q spectrum provide information on multispin correlations.

Two-dimensional spectra involving MQC of order higher than four suffer from very low sensitivity due to extremely inefficient excitation of the MQC. However, it is feasible to obtain 1D multiple-quantum-filtered spectra with high-order filters. Five-quantum-filtered 1D spectra have been reported for plastocyanin (99 residues)[35] and we have successfully recorded 1D 5QF spectra for a protein of 146 residues (C. Dalvit and P. E. Wright, unpublished). Such spectra can be very useful for identifying specific spins, such as the $C^{\beta}H$ of threonines.

Although zero-quantum spectroscopy has a few unique features which distinguish it from higher order MQ experiments,[53,54] such as the insensitivity of 0QC to magnetic field inhomogeneity,[17] there are also some drawbacks which have limited its usefulness to date in studies of proteins.[55,56]

### Integrated Strategy for Spin System Identification

The multiple-quantum experiments described above, when used conjointly with multiple-step relay and TOCSY (HOHAHA) experiments, provide a powerful integrated approach to spin system assignment.[49,57,58] This integrated strategy utilizes relayed coherence transfer techniques to establish long-range connectivities from side-chain protons to (1) the backbone NH proton (or $C^{\alpha}H$ in the case of proline) or (2) the side-chain terminal protons, e.g., the $C^{\varepsilon}H_2$ protons of lysine. Since relayed experiments (including TOCSY) cannot reliably discriminate between $C^{\beta}H$, $C^{\beta'}H$, $C^{\gamma}H$, $C^{\gamma'}H$, etc., protons, proton-specific assignments are made by the systematic application of complementary MQF-COSY and MQ experiments. Multiple-quantum spectra also provide verification of connectivities identified via multistep coherence transfer experiments and are used to identify resonances for which relayed connectivities are either greatly attenuated or not observed at all due to unfavorable coupling topology and/or small coupling constants.

The connectivities for any given spin system are usually established independently in several different experiments. This redundancy is very

[53] G. Pouzard, S. Sukumar, and L. D. Hall, *J. Am. Chem. Soc.* **103,** 4209 (1981).
[54] L. D. Hall and T. J. Norwood, *J. Magn. Reson.* **69,** 391 (1986).
[55] L. Müller, *J. Magn. Reson.* **59,** 326 (1984).
[56] P. H. Bolton, *J. Magn. Reson.* **60,** 342 (1984).
[57] W. J. Chazin and P. E. Wright, *Biopolymers* **26,** 973 (1987).
[58] W. J. Chazin, M. Rance, and P. E. Wright, *J. Mol. Biol.* **202,** 603 (1988).

important for analysis of the extremely crowded spectra of proteins since it provides a series of cross-checks on the spin system assignments. Multiple-quantum techniques are invaluable, by themselves and in conjunction with complementary methods, for unambiguously assigning the complex $^1H$ NMR spectra of proteins and other biological macromolecules.

## Acknowledgments

We thank Prof. R. R. Ernst for providing a preprint of Ref. 38, Ms. L. Harvey for preparation of the manuscript, and the National Science Foundation (DMB-8517959 to M.R.) and the National Institutes of Health (NIH DK34909 and GM36643 to P.E.W.) for financial support.

## [7] Detection of Insensitive Nuclei

By Ad Bax, Steven W. Sparks, and Dennis A. Torchia

## Introduction

Because of their larger chemical shift dispersions, $^{13}C$ and $^{15}N$ NMR spectra are often better resolved than the corresponding $^1H$ spectra. It was this particular feature that stimulated numerous heteronuclear NMR studies of proteins in the 1970s. The goal of these studies was to use measurements of $^{13}C$ and $^{15}N$ chemical shifts and relaxation rates to obtain information about molecular structure and dynamics. Unfortunately, the low sensitivity of $^{13}C$ and $^{15}N$ NMR necessitated the use of large sample quantities, and assignment of the $^{13}C$ and $^{15}N$ spectra was difficult, relying heavily on off-resonance decoupling techniques (provided that the attached protons were resolved and assigned).

The assignment problem has been greatly simplified by two-dimensional (2D) heteronuclear correlation spectra, in which the two coordinates of each resonance are the chemical shifts of a proton and of its directly attached heteronucleus. If the $^1H$ spectrum has been assigned, the heteronuclear correlation spectrum can be used to assign the corresponding $^{13}C$ and $^{15}N$ spectra. Alternatively, characteristic shifts to certain $^{13}C$ and $^{15}N$ nuclei are useful for assigning the $^1H$ NMR spectrum. The early 2D heteronuclear correlation techniques[1-4] were only partially suc-

[1] A. A. Maudsley and R. R. Ernst, *Chem. Phys. Lett.* **50**, 368 (1977).

[2] A. A. Maudsley, L. Müller, and R. R. Ernst, *J. Magn. Reson.* **28**, 463 (1977).

[3] G. Bodenhausen and R. Freeman, *J. Magn. Reson.* **28**, 471 (1977).

[4] A. Bax and G. A. Morris, *J. Magn. Reson.* **42**, 501 (1981).

cessful for generating such correlation maps for proteins.[5-7] These methods relied on direct detection of the heteronucleus, the NMR amplitude of which was modulated by the frequency of its coupled proton(s). The major impediment of this approach was the low sensitivity of such methods, lower by at least a factor of three to five relative to a simple one-dimensional heteronuclear spectrum.

In more recently developed 2D correlation techniques the sensitive $^1$H signal is detected, modulated by the frequency of its heteronuclear coupling partner, $X$. For historical reasons, this approach is often referred to as reverse correlation. Relative to a regular one-dimensional $^1$H spectrum, the sensitivity of such a reverse-correlation map is decreased by the natural abundance of the heteronucleus (1.1% for $^{13}$C and 0.37% for $^{15}$N), but not by its magnetogyric ratio, $\gamma_X$. Maudsly et al.[2] pointed out the potential advantages of directly detecting the nucleus with the higher $\gamma$ more than a decade ago. Subsequent work by Müller and Ernst,[8] Müller,[9] Bodenhausen and Ruben,[10] Bendall et al.,[11] Redfield,[12] and Bax et al.[13] paved the way for applying such techniques in a practical fashion.

This chapter describes a number of 2D reverse-correlation experiments that yield high-sensitivity heteronuclear correlation spectra. The optimal pulse sequence depends on the particular application and on the spectrometer used. For example, for correlation through long-range coupling or through one-bond coupling, different pulse schemes have to be used. Certain pulse schemes provide better suppression of artifacts but yield poorer line shapes.

This chapter is not a comprehensive review of heteronuclear correlation spectroscopy but rather a brief guide of what one may expect from such methods and how to record optimal spectra.

## Sensitivity Gain from Reverse Correlation

There has been some confusion over what gain in sensitivity may be expected from the $^1$H-detected heteronuclear correlation techniques.

[5] T. M. Chan and J. L. Markley, J. Am. Chem. Soc. 104, 4010 (1982).
[6] J. L. Markley, W. M. Westler, T. M. Chan, C. L. Kojiro, and E. L. Ulrich, Fed. Proc., Fed. Am. Soc. Exp. Biol. 43, 2648 (1984).
[7] W. M. Westler, G. Ortiz-Polo, and J. L. Markley, J. Magn. Reson. 58, 354 (1984).
[8] L. Müller and R. R. Ernst, Mol. Phys. 38, 963 (1979).
[9] L. Müller, J. Am. Chem. Soc. 101, 4481 (1979).
[10] G. Bodenhausen and D. J. Ruben, Chem. Phys. Lett. 69, 185 (1980).
[11] M. R. Bendall, D. T. Pegg, and D. M. Dodrell, J. Magn. Reson. 52, 81 (1983).
[12] A. G. Redfield, Chem. Phys. Lett. 96, 537 (1983).
[13] A. Bax, R. H. Griffey, and B. L. Hawkins, J. Magn. Reson. 55, 301 (1983).

Numbers ranging from 10- to 1000-fold have been quoted in the literature. Here, the theoretical gain in sensitivity obtainable with the reverse-correlation method over the $X$-detected correlation technique will be briefly discussed.

The NMR sensitivity of a nucleus is proportional to $\gamma^{5/2}$.[14] Therefore, at a first glance, one might expect a gain in sensitivity of $(\gamma_H/\gamma_X)^{5/2}$ by directly detecting the proton instead of the $X$ nucleus. However, in the $X$-detected experiment the sensitivity is improved by a factor $\gamma_H/\gamma_X$ because of polarization transfer from $^1H$ to $X$. The expected gain in sensitivity is thus reduced to $(\gamma_H/\gamma_X)^{3/2}$. This number should be multiplied by the number of protons, $N$, directly attached to $X$; for the $X$-detected experiment, the amount of polarization transfer is nearly independent of $N$, whereas for reverse detection the detected signal is directly proportional to $N$. For $^{13}C$–H correlation of methyl groups, the gain in sensitivity therefore equals $3 \times (4)^{3/2} = 24$, for methylenes it is 16, and for methine sites it is 8. For $^{15}N$–H correlation of peptide amide resonances, the gain is about 30.

In the above discussion, the assumption has been made that $^1H$ and $X$ nucleus line widths are identical. For $X = {}^{13}C$, $^{15}N$, this assumption is generally incorrect for small molecules, where the $^1H$ resonance is often split by homonuclear $J$ couplings. For macromolecules, however, the line width often is dominated by the heteronuclear dipolar coupling and will be quite similar for protons and for the $X$ nuclei.

This entire sensitivity discussion has been restricted to the case where the quantity of sample is limited. If unlimited sample is available, the $X$-detection experiment can be performed in a large-diameter sample tube, improving its sensitivity significantly. Another practical consideration is that the so-called $t_1$ noise can be much worse in the reverse-correlation spectra. Finally, for successful reverse correlation it is essential to have some special hardware, including a reverse-detection probe.

### Pulse Schemes for Reverse Correlation

A large variety of different heteronuclear reverse-correlation schemes has been proposed in the literature. No attempt will be made to review all these methods here, but the advantages and limitations of a small selection of such schemes will be discussed. The key to successful reverse-correlation experiments on natural abundance samples is that they should contain relatively few $^1H$ pulses. This facilitates suppression of the much stronger resonances from protons not coupled to the heteronucleus. The best schemes rely on the principle of heteronuclear multiple-quantum

---

[14] D. I. Hoult and R. E. Richards, *J. Magn. Reson.* **24,** 71 (1976).

coherence. Five different pulse schemes for correlation through one-bond couplings are shown in Fig. 1.

*Zero- and Double-Quantum Correlation.* Scheme a in Fig. 1 is the simplest, and it was first developed for correlating the imino protons in tRNA with their attached $^{15}N$ nuclei.[15] Scheme a (Fig. 1) can be used with or without decoupling of the heteronucleus and it can easily be adapted to studies in $H_2O$ without presaturation by replacing the first $^1H$ pulse of Scheme a (Fig. 1) by one of the water-suppression schemes discussed by Hore.[16] This particular scheme, with the $^1H$ pulse replaced by a 2–1–4 Redfield pulse, was used by Glushka and Cowburn[17] for generating a high-quality $^1H$–$^{15}N$ shift correlation map of the amide resonances in basic pancreatic trypsin inhibitor (BPTI). Two disadvantages of Scheme a in Fig. 1 are that the acquired spectra cannot be phased to the absorption mode which necessitates the use of magnitude calculations in both dimensions of the 2D spectrum, decreasing resolution and sensitivity. Second, in Scheme a (Fig. 1) the detected $^1H$ signals are modulated by the zero- and double-quantum frequencies, corresponding to the sums and differences of the $^1H$ and $X$ nucleus offsets from their respective carriers. Hence, analysis is less convenient because the 2D spectrum is not a conventional correlation map with $^1H$ chemical shifts along one axis and the $X$ nucleus chemical shifts along the other axis. More serious is the fact that the $^1H$ signals are modulated by both zero- and double-quantum frequencies, i.e., the sensitivity of the experiment is decreased by $\sqrt{2}$ relative to the case where the $^1H$ signals are modulated by only a single frequency, the $X$ nucleus chemical shift. Therefore, resolution and sensitivity of Scheme a (Fig. 1) are far from optimal, but its strong points are (1) the easy $H_2O$ suppression and (2) a minimum amount of $t_1$ noise.

*Constant-Time Heteronuclear Correlation.* An interesting and underused variation of Scheme a (Fig. 1) has been developed by Müller *et al.*,[18] and is sketched in Scheme b (Fig. 1). Scheme b (Fig. 1) also employs a single $^1H$ excitation pulse, but it uses a constant duration of the evolution period through which an $X$ nucleus 180° pulse is shifted in a stepwise fashion. Scheme b (Fig. 1) has several advantages over Scheme a (Fig. 1): Water suppression is even easier with Scheme b (Fig. 1) since the $^1H$ signal sampling is further removed from this pulse. Therefore, a relatively high receiver gain setting can be used, even for concentrated samples. In practice, Scheme b (Fig. 1) also gives the best suppression of signals not

[15] R. H. Griffey, C. D. Poulter, A. Bax, B. L. Hawkins, Z. Yamaizumi, and S. Nishimura, *Proc. Natl. Acad. Sci. U.S.A.* **80,** 5895 (1983).
[16] P. J. Hore, this volume [3].
[17] J. Glushka and D. Cowburn, *J. Am. Chem. Soc.* **109,** 7879 (1987).
[18] L. Müller, R. A. Schiksnis, and S. J. Opella, *J. Magn. Reson.* **66,** 379 (1986).

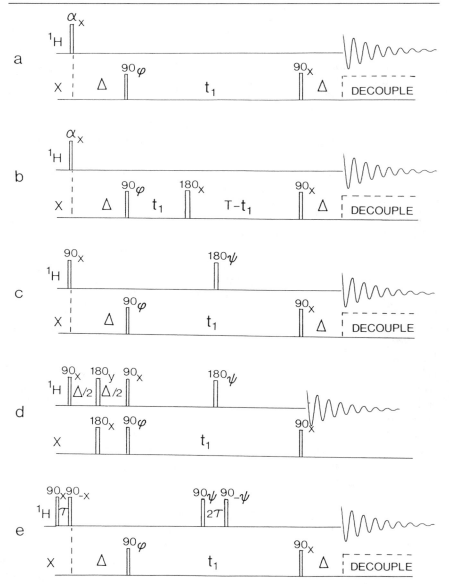

FIG. 1. Pulse schemes for heteronuclear correlation of protons with their directly attached $X$ nuclei. All schemes employ $^1H$ detection. (a) Zero- and double-quantum correlation (absolute value). (b) Constant-time chemical shift correlation (absolute value). (c) HMQC chemical shift correlation (phase sensitive). (d) Flip-back chemical shift correlation (phase sensitive). (e) HMQC correlation in $H_2O$ without presaturation (phase sensitive). For Schemes (a) and (b), the following phase cycling can be used (with TPPI incrementation of

coupled to the $X$ nucleus. For a given time, $T$, between the 90° $X$ pulses, the acquisition time in the $t_1$ dimension runs from $-T$ to $+T$, providing high resolution in the $F_1$ dimension. The acquired $^1H$ signals are modulated by the $X$ nucleus chemical shift and not by the multiple-quantum frequencies (even while the "magnetization" exists as zero- and double-quantum coherence during the time $T$). Hence, the $^1H$ signals are modulated by only a single frequency, increasing sensitivity relative to Scheme a (Fig. 1). Finally, the spectrum conveniently displays $X$ nucleus chemical shift frequencies along the $F_1$ axis. For best results, spectra should be acquired in the hypercomplex (or TPPI) fashion and the magnitude spectrum should be calculated in both dimensions. For digital filtering, a pseudo-echo filter is recommended in the $t_1$ dimension, with the signal at the edges of the $t_1$ time domain (at $t_1 = +T$) attenuated by about a factor of three relative to the center. In $f_2$, conservative filtering using, for example, a cosine bell, cosine-squared bell, or 60°-shifted sine (squared) bell is recommended. In all respects, this sequence is superior to Scheme a of Fig. 1 and it is surprising that this method has not found more application.

*Heteronuclear Multiple-Quantum Correlation (HMQC).* Scheme c (Fig. 1) is a simple modification of Scheme a (Fig. 1), where the zero- and double-quantum frequencies now are interchanged by the 180° $^1H$ pulse, applied at the center of the evolution period. This has the net effect of eliminating the $^1H$ chemical shift component from the multiple-quantum frequency. The final spectrum therefore has the appearance of a regular heteronuclear chemical shift correlation map, with $^1H$ chemical shifts along one axis and the $X$ nucleus shift along the other axis. Neglecting the effect of small homonuclear couplings, this method permits the recording of pure absorption spectra, offering the highest possible resolution. Scheme c (Fig. 1) is often referred to as the HMQC (heteronuclear multiple-quantum correlation) or the "forbidden echo"[19] technique. A modified version[20] has been widely applied to the study of small molecules, and several applications of the HMQC scheme to natural abundance protein

[19] R. H. Griffey, A. G. Redfield, R. E. Loomis, and F. W. Dahlquist, *Biochemistry* **24**, 817 (1985).

[20] A. Bax and S. Subramanian, *J. Magn. Reson.* **67**, 565 (1986).

---

$\phi$): $\phi = x, -x$; Acq $= x, -x$. For Schemes (c), (d), and (e): $\phi = x, -x$; $\psi = x, x, y, y, -x, -x,$ $-y, -y$; Acq $= x, -x, -x, x, x, -x, -x, x$. To avoid poor cancellation of non-$X$-coupled protons caused by imperfect steady state $z$ magnetization (too short a delay time between scans), the phase of the final 90° $X$ pulse and the receiver phase should be inverted after the eight-step cycle.[38]

studies have appeared.[21,22] Problems with the HMQC scheme are (1) suppression of signals from protons not coupled to $X$ becomes more difficult by the addition of the 180° pulse, and (2) that recording the spectrum in $H_2O$ solution is also more difficult [see discussion of Scheme e (Fig. 1) below]. The suppression of signals from protons not coupled to $X$ improves with the square root of the number of scans and also is easier for relatively broad resonances. For proteins, the number of scans per $t_1$ value needed for sufficient signal to noise is usually quite large, on our Nicolet and Bruker spectrometers, so that suppression of signals not coupled to $X$ does not present any practical problems in the study of macromolecules, provided that proper precautions are taken (e.g., no sample spinning).

As an example, Scheme c (Fig. 1) is applied to a sample of BPTI in $D_2O$, $p^2H$ 6.6, 7 m$M$, 70 m$M$ NaCl, 35°. Figure 2 shows the $C_\alpha$ region of the HMQC spectrum, recorded at 600 MHz. The total measuring time was 11 hr. This spectrum was recorded with the same level for the $^{13}C$ pulses and the $^{13}C$ decoupling (3.3-kHz rf field), using WALTZ16 decoupling modulation.[23] Generating this 3.3-kHz rf field required about 3-W rf power, sufficiently high that sample heating and associated lock signal deterioration became significant. Therefore, the data acquisition time $(t_2)$ (and the decoupling duration) was limited to 80 msec. WALTZ modulation with a 3.3-kHz rf field provides sufficiently good decoupling over an 8-kHz band width. A better choice for broad-band $^{13}C$ decoupling is to use the GARP modulation scheme,[24] covering nearly twice this band width with the same amount of rf power. However, to effectively excite this wider band width with the 90° pulses of the HMQC scheme, power switching between the $^{13}C$ pulses and the $^{13}C$ decoupling becomes essential, a feature not available on our spectrometer when the spectrum of Fig. 2 was recorded.

Most experimental work to date consistently avoids heteronuclear decoupling during $^1H$ data acquisition. However, it should be noted that heteronuclear decoupling doubles the signal-to-noise ratio (provided appropriate rf filtering is applied) and reduces signal overlap. Line shapes are otherwise unaffected.

*Flip-Back Heteronuclear Correlation.* If for instrumental limitations broad-band $X$ nucleus decoupling is impossible, the original Scheme d (Fig. 1) proposed by Müller[9] may be preferable. In Scheme d (Fig. 1), the double $^1H/X$ 180° pulse eliminates the effects of $^1H$ offset during the first

[21] V. Sklenář and A. Bax, *J. Magn. Reson.* **71**, 379 (1987).
[22] A. Bax and L. Lerner, *Science* **232**, 960 (1986).
[23] A. J. Shaka, J. Keeler, T. Frenkiel, and R. Freeman, *J. Magn. Reson.* **52**, 335 (1983).
[24] A. J. Shaka, P. B. Barker, and R. Freeman, *J. Magn. Reson.* **64**, 547 (1985).

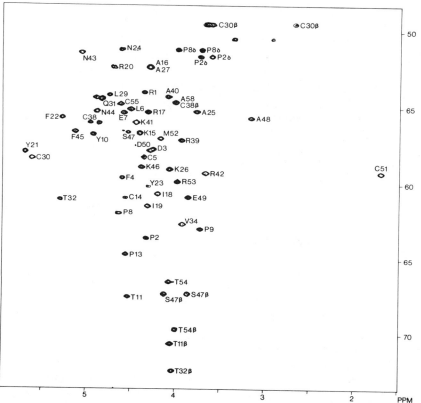

FIG. 2. $C_\alpha$ region of the $^1H$–$^{13}C$ shift correlation spectrum recorded at 600-MHz $^1H$ frequency, for a sample of 20 mg natural abundance BPTI in 0.5 ml $D_2O$. The measuring time was 11 hr. Assignments are taken from Wagner and Bruewihler.[25] Acquisition times in the $t_1$ and $t_2$ dimensions were 50 and 70 msec, respectively. Sine bell digital filtering (45° shifted) and zero filling were used in both dimensions. TPPI type phase cycling was used.

delay, $\Delta$, but leaves the heteronuclear coupling intact. The second $^1H$ 90° pulse flips the magnetization from protons not coupled to $X$ to the $-z$ axis; magnetization from protons coupled to $X$ is in antiphase along the $\pm x$ axis at this point in time and is converted into heteronuclear zero- and double-quantum coherence. The 180° $^1H$ pulse, applied at the midpoint of the evolution period, serves the same function as in Scheme c (Fig. 1), but has the additional effect of turning magnetization from protons not coupled to $X$ back to the $+z$ axis. Spin diffusion then causes the rapid recovery of the longitudinal magnetization of the $X$-coupled spins. Two potential advantages of Scheme d (Fig. 1) are (1) that a somewhat faster repetition rate can be used and (2) that (at least in principle) only signal from $X$-coupled protons reaches the receiver. This latter property decreases dynamic range problems and therefore can increase sensitivity of the experiment. In our experience, the flip back does *not* improve the cancellation of signals from protons not coupled to $X$. High-quality spectra using the flip-back scheme have been reported by Wagner and Bruewihler[25] (natural abundance $^{13}$C of BPTI) and by Stockman et al.[26] ($^{15}$N-labeled flavodoxin).

Scheme d (Fig. 1) is less useful if $X$ nucleus decoupling during data acquisition is required. $^1$H chemical shifts are in phase at the end of the evolution period, whereas the decoupling should be started a time $1/(2J_{XH})$ later. In principle another set of double 180° pulses could be inserted between the final 90° $X$ pulse and the start of data acquisition. In practice, if one wants to record decoupled spectra, Scheme c (Fig. 1) is probably preferable.

The heteronuclear coupled spectra, obtained with Scheme d (Fig. 1), are lower in signal-to-noise ratio by a factor of two relative to the decoupled spectra of Scheme c (Fig. 1) and show twice the number of resonances, i.e., an increased chance of overlap. In practice, the increased complexity of the spectrum may also have advantages. Several authors[17,25] state that the characteristic antiphase doublet pattern facilitates recognition of such heteronuclear correlations and may make it easier to distinguish them from artifacts and $t_1$ noise.

Very recently, Otting and Wuethrich[27] have proposed interesting modifications to Scheme d (Fig. 1) that would considerably alleviate the problem of suppressing signals from protons not coupled to $X$, with or without $X$ nucleus decoupling. We have not yet had the opportunity to test the performance of these sequences, but results presented by these authors suggest that these schemes function quite well.

*The 1–1 Echo Scheme.* Correlating the backbone amide protons with their attached $^{15}$N nuclei can be particularly valuable. It greatly improves the resolution in this region of the spectrum and the $^{15}$N chemical shifts may contain structural information. Moreover, for cloned proteins, it is relatively easy to selectively $^{15}$N-label certain types of amino acids, facilitating the assignment process and permitting the recording of edited NOESY and HOHAHA spectra.[19,28–31] As a first step, after $^{15}$N incorpora-

[25] G. Wagner and D. Bruewihler, *Biochemistry* **25**, 5839 (1986).

[26] B. J. Stockman, W. M. Westler, E. S. Mooberry, and J. L. Markley, *Biochemistry* **27**, 136 (1988).

[27] G. Otting and K. Wuethrich, *J. Magn. Reson.* **76**, 569 (1988).

[28] A. Bax and M. Weiss, *J. Magn. Reson.* **71**, 571 (1987).

[29] R. H. Griffey and A. G. Redfield, *Q. Rev. Biophys.* **19**, 51 (1987).

tion in a protein, it is useful to record an $^1$H–$^{15}$N correlation map. This usually gives much clearer results than a simple spin-echo difference spectrum[32] where partial overlap and low-level $^{15}$N-labeling via transamination might be difficult to spot. To ensure that no amide resonances are lost due to presaturation, it is desirable to record these types of spectra with a sequence that avoids excitation of the water resonance, making presaturation unnecessary. A simple scheme for doing this is to replace the 90 and 180° $^1$H pulses in Scheme c (Fig. 1) by jump-and-return 1–1[16] pulses (Scheme e, Fig. 1).[19] However, it should be noted that the 1–1 pulse at the center of the evolution period, which serves as a refocusing pulse, only works well over a relatively narrow frequency band. To obtain pure phase correlation spectra, phase cycling of this refocusing pulse unit therefore is essential.[33]

As an example, Fig. 3 shows the $^{15}$N–$^1$H correlation spectrum obtained for the protein staphylococcal nuclease, complexed with pdTp and $Ca^{2+}$. Leucine, and to a lesser degree serine, were $^{15}$N labeled. Spectra were recorded at 600 MHz, 35°, pH 7.4, 1.5 m$M$, 100 m$M$ NaCl. Because of the $^{15}$N labeling, high-sensitivity spectra can be obtained in a very short period of time. The spectrum of Fig. 3 was recorded in about 45 min, a minimum time dictated by the required phase cycling and the number of $t_1$ increments needed. The sample was also labeled with $^{13}$C in the carbonyl position of lysine residues, giving rise to partially resolved doublet structures for Leu-7, Leu-25, Leu-137, and Ser-128, each of which are preceded by a lysine residue. At contour levels lower than shown, a large number of additional correlations become visible which probably correspond to glycine residues that carry an $^{15}$N label derived from serine.

### Heteronuclear Relay Experiments

One-bond heteronuclear correlations of isotopically labeled proteins are very sensitive experiments, comparable to the one-dimensional $^1$H spectrum. Therefore, it is relatively straightforward to extend this type of experiment by combining it with NOESY, COSY, or HOHAHA.[25,27–31] Of the several dozen different pulse schemes available for these purposes, we discuss a single example: a combination of HOHAHA and heteronu-

[30] M. Rance, P. E. Wright, B. A. Messerle, and L. D. Field, *J. Am. Chem. Soc.* **109**, 1591 (1987).
[31] H. Senn, G. Otting, and K. Wuethrich, *J. Am. Chem. Soc.* **109**, 1090 (1987); S. W. Fesik, R. T. Gampe, and T. W. Rockway, *J. Magn. Reson.* **74**, 366 (1987).
[32] R. Freeman, T. H. Mareci, and G. A. Morris, *J. Magn. Reson.* **42**, 341 (1981).
[33] V. Sklenář and A. Bax, *J. Magn. Reson.* **74**, 469 (1987).

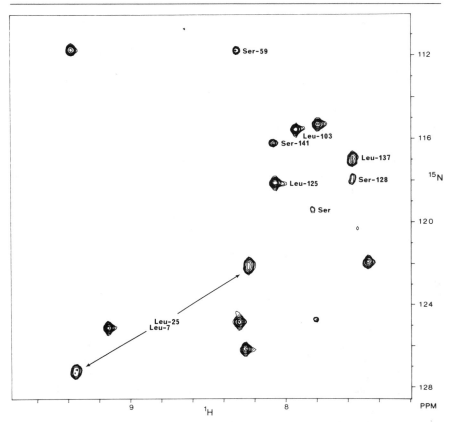

FIG. 3. $^1H-^{15}N$ shift correlation spectrum of labeled staphylococcal nuclease, recorded at 600 MHz. The spectrum was recorded with Scheme e of Fig. 1, using eight scans per $t_1$ value, preceded by two dummy scans. A 200 × 1024 data matrix was recorded, corresponding to acquisition times of 40 and 50 msec in the $t_1$ and $t_2$ dimension, respectively. The total measuring time was 45 min, but it should be noted that nearly one-half of this time was overhead for dummy scans and for writing the data to disk.

clear correlation. One of the possible pulse schemes for this purpose is sketched in Fig. 4. This sequence combines in a straightforward manner the pulse scheme (Scheme c in Fig. 1) with HOHAHA type mixing.[34,35] At the beginning of the MLEV17 mixing, the $^1H$ magnetization is aligned along the $y$ axis and modulated in amplitude by the $X$ spin chemical shift. The MLEV17 mixing scheme then redistributes the magnetization of the

[34] L. Braunschweiler and R. R. Ernst, *J. Magn. Reson.* **53**, 521 (1983).
[35] A. Bax and D. G. Davis, *J. Magn. Reson.* **65**, 355 (1985).

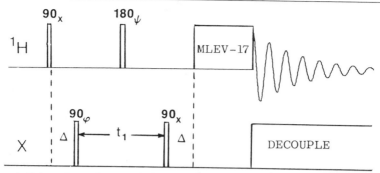

FIG. 4. Pulse scheme for correlating entire ¹H spin systems with a labeled X nucleus. Either the WALTZ or the MLEV scheme can be used for the homonuclear mixing (for details, see Bax [8], this volume). Phase cycling as in Fig. 1.

X-coupled proton over all its coupling partners. So, in the final 2D spectrum, all coupled protons will be modulated by the frequency of the X nucleus.

As an example, Fig. 5 shows the $^{13}C-^1H-^1H$ relay spectrum of $[^{13}C_\beta]$ Ala-labeled staphylococcal nuclease, recorded at 500 MHz, 42°, 1.5 mM in a 5-mm sample tube. The level of $^{13}C$ labeling was about 20%. The duration of the MLEV17 mixing period was set to 30 msec and no trim pulses[35] were used. At the beginning of the MLEV17 mixing, only magnetization from the methyl protons is modulated by the chemical shift of the $^{13}C$. The MLEV17 transfers magnetization from the methyl protons to the $C_\alpha$ proton, resulting in correlations between $C_\alpha H$ and $^{13}C_\beta$. This spectrum provides significant simplification compared to a simple direct 2D HOHAHA spectrum (compared with Fig. 5, Bax [8], this volume). Future developments are expected where such experiments are performed in a three-dimensional fashion,[36] permitting the use of much more extensive labeling without introducing spectral overlap.

## Correlation via Small Couplings

There is, of course, no fundamental difference between correlation via direct or via long-range couplings. However, there are large practical differences in how to optimize the experiment and what pulse sequence to choose. Problems in correlating chemical shifts of protons and heteronuclei via long-range couplings are that (1) the size of the couplings show large variations, (2) the heteronuclear couplings are typically of the same

[36] H. Oschkinat, C. Griesinger, P. J. Kraulis, O. W. Sorensen, R. R. Ernst, A. M. Gronenborn, and G. M. Clore, *Nature (London)* **332,** 374 (1988).

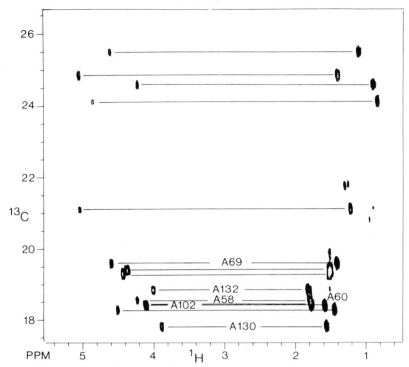

FIG. 5. Example of an HOHAHA $^{13}$C relay spectrum recorded with the scheme of Fig. 4, using an MLEV-17 mixing scheme with a total duration of 30 msec. The sample of staphylococcal nuclease was about 20% [$^{13}C_\beta$]Ala. For every methyl group, the corresponding C$^\alpha$H is clearly observed. The spectrum was recorded at 500 MHz, 42°, using a total measuring time of 12 hr. Resonances not connected by horizontal bars originate from natural abundance signals.

order of magnitude as the homonuclear $^1$H–$^1$H couplings, and (3) in macromolecules the heteronuclear long-range couplings are often smaller than the natural line widths of the proton resonances. As a consequence, in macromolecules the sensitivity of long-range correlation experiments is reduced dramatically relative to the correlation through one-bond couplings, described above. For the study of proteins at low concentrations, isotopic labeling is therefore always essential.

Our discussion is limited to one sequence that yields long-range correlation spectra, although it should be realized that this particular sequence is not necessarily the best for all applications. Its pulse scheme is

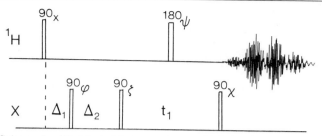

FIG. 6. Scheme for $^1$H-detected heteronuclear multiple-bond correlation (HMBC) via multiple-quantum coherence. The delay, $\Delta_1$, and the first $90_\phi$ X pulse serve to suppress direct X–H correlations (J filter[37]). For correlation to X nuclei that do not have directly attached protons, $\Delta_1$ should be set to zero and the $90_\phi$ pulse may be omitted. The phase cycling (with TPPI incrementation of $\zeta$) is $\psi = x, x, y, y, -x, -x, -y, -y; \zeta = x, -x; \chi = x, x, x, x, x, x, x, x, -x, -x, -x, -x, -x, -x, -x, -x$; Acq = $x, -x, -x, x, x, -x, -x, x, -x, x, x, -x, -x, x, x, -x$. The phase cycling of $\chi$ serves to eliminate effects of changes in the steady-state $z$ magnetization,[38] introduced by the phase cycling of $\psi$. After 16 steps, the phase $\phi$ may be inverted (J filter) without changing the receiver phase.

sketched in Fig. 6,[37,38] and inspection shows that this scheme is nothing but a slightly modified version of Scheme c (Fig. 1). This sequence, first applied to long-range $^1$H–$^{13}$C correlation in coenzyme $B_{12}$,[39] is known under the name HMBC, for heteronuclear multiple-bond correlation. The first (optional) $90°$ X pulse serves as a 1D J filter,[37] to eliminate one-bond correlations from the 2D spectrum. The second $90°$ pulse, applied after another delay $\Delta_2$, creates the multiple-bond multiple-quantum coherence. The $180°$ pulse removes the $^1$H chemical shift contribution and the final $90°$ X pulse converts the multiple-quantum coherence back into antiphase $^1$H magnetization. No X decoupling is applied during data acquisition.

Because of homonuclear $^1$H–$^1$H couplings it is not possible to obtain absorption mode spectra in the $F_2$ dimension of the 2D spectrum. However, in the $F_1$ dimension the data are simply modulated in amplitude by the X chemical shift and an absorption mode representation in this dimension can be obtained.[40] Because the data are in antiphase at the beginning of the detection period, $t_2$, the time domain signal in the $t_2$ dimension starts at zero; it is a sine function that is rapidly damped by the short $T_2$ of the protons. For sensitivity purposes it is best to apply a matched filter to

[37] H. Kogler, O. W. Sorensen, G. Bodenhausen, and R. R. Ernst, *J. Magn. Reson.* **55**, 157 (1983).
[38] J. Cavanagh and J. Keeler, *J. Magn. Reson.* **77**, 612 (1988).
[39] A. Bax and M. F. Summers, *J. Am. Chem. Soc.* **108**, 2093 (1986).
[40] A. Bax and D. Marion, *J. Magn. Reson.* **78**, 186 (1988).

these data, i.e., in the $t_2$ dimension a nonshifted sine bell is a suitable function if the data acquisition time in the $t_2$ dimension is set to about $2$–$3 \times T_2$.

As an example, the scheme of Fig. 6 is applied to a sample of staphylococcal nuclease complexed with pdTp and calcium. Fourteen milligrams of the complex (18 kDa) was dissolved in 0.5 ml $D_2O$ and spectra were recorded at 600 MHz, 35°, $p^2H$ 7.4. Figure 7 shows the long-range $^1H$–$^{13}C$ correlation spectrum for a sample where threonine residues are labeled

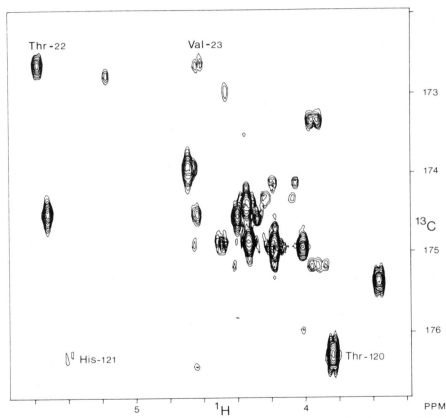

FIG. 7. Six hundred megahertz $^1H$–$^{13}C$ correlation of staphylococcal nuclease, recorded with the scheme of Fig. 6. The protein was labeled in the carbonyl position of the 11 Thr residues. Experimental details: Acquisition times in the $t_1$ and $t_2$ dimensions, 40 and 102 msec; sine bell filter in $t_2$, 60° shifted sine bell in $t_1$, $\Delta_1 = 0$, $\Delta_2 = 33$ msec; total measuring time 14 hr; absorption mode in $F_1$, absolute value in $F_2$. Because more than 11 $^{13}C$ chemical shifts are observed in this spectrum, it is suspected that some of the low-intensity correlations originate from natural abundance or low-level $^{13}C$-labeled amino acids other than Thr.

with $^{13}$C in the carbonyl position. A large number of cross-peaks can be seen, corresponding to the 11 different threonine residues present in the protein. The intensities of the correlations vary dramatically, the difference between the highest and lowest contour level in Fig. 7 is a factor of 48. These intensity differences reflect the different sizes of the long-range couplings and the differences in line width of the $C_\alpha H$ resonances. In practice, only two- and three-bond couplings can be sufficiently large to yield observable correlations. The size of $^3J_{CH}$ depends strongly on conformation and up to four correlations (three for threonine) in principle can be observed for a single carbonyl resonance (two $C_\alpha$ protons and up to two $C_\beta$ protons for most amino acids). However, in practice many of these possible correlations have too low an intensity to be observable in proteins of the size of staphylococcal nuclease. This is unfortunate because otherwise this type of correlation would be extremely valuable for obtaining sequential assignments. Two examples of such sequential assignment are labeled in Fig. 7; both the $C_\alpha$ protons of Thr-22 and Val-23 show a correlation to the carbonyl of Thr-22 and similarly, His-121 $C_\alpha H$ and Thr-120 $C_\alpha H$ show connectivity to the carbonyl of Thr-120. Of course, it would be of major interest to also correlate the NH resonances with the labeled carbonyls. However, for the staphylococcal nuclease complex, the $T_2$ values of the NH resonances were too short (11–13 msec) to permit this type of correlation to be observed.

The example shown here is only one of the many different applications of the $^1$H-detected methodology. Other very interesting applications of long-range heteronuclear correlation in proteins concern the detection of metal nuclei (Cd,[41,42] Hg,[43] Pt) and phosphorus.[44] In cases where the $T_1$ of the heteronucleus is shorter than the $T_1$ of the protons but its $T_2$ is not shorter than the $T_2$ of the protons, a different sort of approach, not based on multiple-quantum coherence may be favorable.[45]

Discussion

The heteronuclear two-dimensional experiments discussed in this chapter show particular promise for alleviating assignment problems in proteins. Although the one-bond correlation techniques, at least in princi-

[41] M. H. Frey, G. Wagner, M. Vasak, O. W. Sorensen, D. Neuhaus, E. Woergoetter, J. H. Kaegi, R. R. Ernst, and K. Wuethrich, *J. Am. Chem. Soc.* **107**, 6847 (1985).
[42] J. D. Otvos, H. R. Engeseth, and S. Wehrli, *J. Magn. Reson.* **61**, 579 (1985); D. Live, I. M. Armitage, D. C. Dalgarno, and D. Cowburn, *J. Am. Chem. Soc.* **107**, 1775 (1985).
[43] D. A. Vidusek, M. F. Roberts, and G. Bodenhausen, *J. Am. Chem. Soc.* **104**, 5452 (1982).
[44] D. H. Live and D. E. Edmonson, *J. Am. Chem. Soc.* **110**, 4468 (1988).
[45] V. Sklenář, H. Miyashiro, G. Zon, H. T. Miles, and A. Bax, *FEBS Lett.* **208**, 94 (1986).

ple, could be applied to natural abundance samples of small proteins, the most promising area of application is for $^{13}$C- and $^{15}$N-labeled proteins that are too large for straightforward analysis. These techniques then can be combined with NOESY and HOHAHA methods to obtain spectra of reduced complexity. Alternatively, three-dimensional NMR techniques that are based on combining HOHAHA or NOESY with heteronuclear correlation are expected to become of major practical relevance for NMR analysis of proteins. Although the use of one-bond heteronuclear correlations is largely limited to solving assignment problems, the multiple-bond correlations also carry structural information since the intensities of the correlations reflect the size of the heteronuclear $J$ couplings.

For the methods discussed in this chapter it is essential to have access to a so-called "inverse probehead," with the $^1$H observe coil close to the sample (for the highest possible sensitivity) and the decoupler coil on the outside. Despite the fact that the regular so-called "dual probe" may function quite well for regular proton observation, its sensitivity for the inverse correlation experiments is dramatically lower. In our experience the inverse probe shows the same sensitivity (within 10%) and line shape as the regular $^1$H-dedicated probehead and, as a result, in our laboratory we typically leave the inverse probehead in the magnet for months at a time, saving instrument time and reducing the possibility of damage. All experiments are relatively "risk free," provided that the system is protected from an overdose of $X$ nucleus decoupling power.

Currently, almost all spectrometers (even new ones) are designed to directly detect heteronuclei, and at best, inverse detection options have been added as an afterthought. We expect this situation will change during the next few years, and application of the heteronuclear correlation techniques may then become as straightforward as the present recording of COSY and NOESY spectra.

---

[46] L. Lerner and A. Bax, *J. Magn. Reson.* **69**, 375 (1986).

## [8] Homonuclear Hartmann–Hahn Experiments

*By* AD BAX

### Introduction

The principles and applications of homonuclear correlation experiments have been discussed by Markley.[1] In this chapter one of these techniques, known as HOHAHA or TOCSY, will be discussed in more detail, with particular emphasis on the experimental details required for recording optimal spectra on biological samples. Total correlation spectroscopy (TOCSY) was first proposed by Braunschweiler and Ernst.[2] This experiment has particular potential for protein NMR studies since it permits correlation of all protons within a scalar coupling network. Hence, a complete subspectrum can be obtained for every amino acid, making resonance assignments considerably easier. However, the TOCSY experiment never gained much popularity, mainly because of the limited band width that could be covered with the mixing schemes used in the original pulse sequence. We accidentally rediscovered this type of magnetization transfer when analyzing artifacts[3] occurring in rotating frame NOE (ROESY[4,5]) spectra. It was shown that these artifacts are caused by a homonuclear Hartmann–Hahn (HOHAHA) effect that could be described quantitatively by mathematical expressions derived for the heteronuclear case by Müller and Ernst[6] and Chingas *et al.*[7] Magnetization transfer via the Hartmann–Hahn effect occurs only when the difference in the absolute magnitudes of the effective rf fields experienced by two coupled spins is smaller than the scalar interaction between the two spins. The effective rf field is the vector sum of the resonance offset, $\delta$, and the applied rf field. Its magnitude is therefore strongly dependent on the resonance offset.

By considering the spurious magnetization transfer in the ROESY

[1] J. L. Markley, this volume [2].
[2] L. Braunschweiler and R. R. Ernst, *J. Magn. Reson.* **53,** 521 (1983).
[3] A. Bax and D. G. Davis, *J. Magn. Reson.* **63,** 207 (1985).
[4] A. A. Bothner-By, R. L. Stephens, J. T. Lee, C. D. Warren, and R. W. Jeanloz, *J. Am. Chem. Soc.* **106,** 811 (1984).
[5] L. R. Brown and B. T. Farmer II, this volume [11].
[6] L. Müller and R. R. Ernst, *Mol. Phys.* **38,** 963 (1979).
[7] G. C. Chingas, A. N. Garroway, R. D. Bertrand, and W. B. Moniz, *J. Chem. Phys.* **74,** 127 (1981).

experiment as a case of homonuclear Hartmann–Hahn cross-polarization, it was relatively straightforward to develop methods that would overcome the effects of rf offset. A series of different mixing schemes is now available and their advantages and disadvantages will be discussed. As will be shown, the HOHAHA experiments are very efficient at establishing direct (COSY[8]) and indirect (RELAY[9]) types of spin–spin connectivities. Of particular importance for the study of proteins is the fact that the efficiency of HOHAHA magnetization transfer is less sensitive to short transverse relaxation times than the alternative COSY and RELAY type experiments. As will be demonstrated here, methods for recording the HOHAHA spectra in $H_2O$ solution are especially useful for connecting the side-chain protons with the usually better resolved amide resonances. At presence, most commercial spectrometers require some special hardware for producing the rf fields that have to be generated during the mixing period. Therefore, a brief discussion of the hardware requirements of the HOHAHA experiment is also included.

### Principles of Homonuclear Hartmann–Hahn Spectroscopy

A detailed theoretical analysis of Hartmann–Hahn cross-polarization is beyond the scope of this chapter. For background information on heteronuclear Hartmann–Hahn cross-polarization in liquids the reader is referred to the papers by Müller and Ernst[6] and Chingas et al.[7] For analogous descriptions of the homonuclear case, the reader could consult a variety of papers.[2,3,10–14] Here, the principles will be briefly reiterated to provide some insight into selecting the optimal parameters for a particular experiment.

As pointed out by Braunschweiler and Ernst, the magnetization of a spin, $I$, under isotropic mixing conditions is periodically converted into magnetization of a second spin, $S$, provided that $I$ and $S$ are scalar cou-

---

[8] K. Nagayama, A. Kumar, K. Wuethrich, and R. R. Ernst, *J. Magn. Reson.* **40,** 321 (1980); A. Bax and R. Freeman, *ibid.* **44,** 542 (1981).

[9] G. Eich, G. Bodenhausen, and R. R. Ernst, *J. Am. Chem. Soc.* **104,** 3732 (1982).

[10] A. Bax and D. G. Davis, *in* "Advanced Magnetic Resonance Techniques in Systems of High Molecular Complexity" (N. Niccolai and G. Valensin, eds.), pp. 21–48. Birkhaeuser, Basel, 1986.

[11] J. S. Waugh, *J. Magn. Reson.* **68,** 189 (1986).

[12] R. Bazzo and J. Boyd, *J. Magn. Reson.* **75,** 452 (1987).

[13] R. R. Ernst, G. Bodenhausen, and A. Wokaun, "Principles of Nuclear Magnetic Resonance in One and Two Dimensions," p. 444. Oxford Univ. Press (Clarendon), London and New York, 1987.

[14] A. Bax, *J. Magn. Reson.* **77,** 134 (1988).

pled. Mathematically, this is described by the expression

$$I_x \xrightarrow{\ J\tau IS\ } I_x[1 + \cos(2\pi J\tau)]/2 + S_x[1 - \cos(2\pi J\tau)]/2$$
$$+(I_y S_z - I_z S_y) \sin(2\pi J\tau) \quad (1)$$

where $\tau$ is the duration of the mixing period. $I_x$ and $S_x$ denote the $x$ components of transverse $I$ and $S$ spin magnetization, and similarly the indices $y$ and $z$ refer to the $y$ and $z$ components of magnetization in the regular rotating frame. This expression shows how the $x$ component of $I$ spin magnetization is converted periodically (with period $1/J$) into the $x$ component of $S$ spin magnetization. Similarly, during isotropic mixing, $I_y$ is periodically interconverted with $S_y$ and likewise $I_z$ with $S_z$. If the magnetization transfer takes place during the mixing period of a two-dimensional (2D) experiment (Fig. 1), this generates off-diagonal cross-peaks in the 2D spectrum. Equation (1) shows that for two coupled spins all $I$ spin magnetization can be transferred to $S$ (in the absence of relaxation) if the duration of the mixing period, $\tau$, is set to $1/(2J)$. The diagonal peaks in this case would disappear from the 2D spectrum and *all* intensity would be present in the cross-peaks. For systems consisting of more than two spins, an analytical expression analogous to Eq. (1) for the transfer of magnetization becomes much more complicated,[15,16] and complete transfer of magnetization from the diagonal to the cross-peaks is generally not possible. Moreover, in protein studies short relaxation times prohibit the use of long mixing periods needed for transferring the maximal amount of magnetization to the cross peaks.

Equation (1) indicates that in-phase $I$ spin magnetization ($I_x$) can be transferred to in-phase $S$ spin magnetization. This is very similar to, for example, the NOESY experiment in that one can record the spectrum in the absorption mode, with all cross-peaks having the same sign as diagonal peaks. However, more careful inspection of Eq. (1) shows that $I_x$ magnetization can also be transferred into $I_y S_z$ and $I_z S_y$ terms. These latter terms denote antiphase magnetization components; for example, $I_z S_y$ refers to $S$ spin transverse magnetization along the $\pm y$ axis, i.e., the magnetization vector that corresponds to the $S$ spin doublet component with spin $I$ in the $\alpha$ spin state points along the $y$ axis and the second doublet component points along the $-y$ axis. Because the line widths in protein spectra often are large relative to the $J$ coupling, the overlapping antiphase multiplet components largely cancel one another and the spectrum still appears to be absorptive. Glycine residues can be an exception to this rule because their doublet components often are partially resolved

[15] N. Chandrakumar, *J. Magn. Reson.* **71**, 322 (1987).
[16] N. Chandrakumar and S. Subramanian, *J. Magn. Reson.* **62**, 346 (1985).

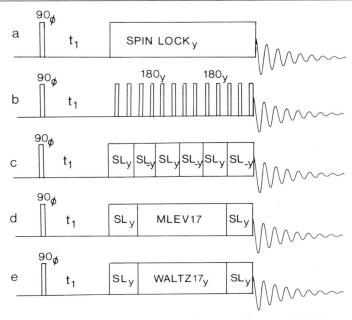

FIG. 1. Pulse schemes for 2D homonuclear Hartmann–Hahn and TOCSY experiments. (a) Using a spin-lock field of constant phase, requiring excessively strong rf fields for cross-polarization over a wide band width, but possibly useful for restricting magnetization transfer to spins that are close in chemical shift. (b) The original TOCSY experiment,[2] using a series of closely spaced 180° pulses to accomplish magnetization transfer. The shortest possible 180° pulse width (or composite 180° pulses[13]) should be used and pulse spacing should be less than 1/(2SW), where SW is the spectral width in hertz. (c) Magnetization transfer using a phase-alternated spin-lock scheme. (d) Mixing using an MLEV17 cycle, preceded and followed by trim pulses (typically 1–2 msec each). The MLEV17$_y$ mixing consists of an integral number of repetitions of the following sequence: $ABBA$ $BBAA$ $BAAB$ $AABB$ $60_y$, where $A = 90_x 180_y 90_x$ and $B = 90_{-x} 180_{-y} 90_{-x}$. (e) Mixing using the WALTZ17$_y$ cycle, consisting of an integral number of repetitions of $ABBA - \alpha_y$, where $A = 270_{-x} 360_x 180_{-x} 270_x 90_{-x} 180_x 360_{-x} 180_x 270_{-x}$ and $B = 270_x 360_{-x} 180_x 270_{-x} 90_x 180_{-x} 360_x 180_{-x} 270_x$; the flip angle $\alpha$ is adjusted between 0 and 90°. The phase cycling used is $\phi = x, -x$ and Acq $= x, -x$ for TPPI type experiments and $\phi = x, y, -x, -y$ and Acq $= x, x, -x, -x$ for hypercomplex data acquisition. In addition, CYCLOPS may be added by incrementing all phases ($\phi$, trim pulses, spin lock or MLEV or WALTZ and Acq) by 90° after completion of the short phase cycle.

(large coupling with no passive spins if the NH proton has been exchanged for $^2$H) and the phase distortion of the cross-peak between the geminal protons then can be observed.

Equation (1) is only valid if all chemical shift and rf terms of the Hamiltonian have identical values for each of the two spins. For example, if the $J$-coupled spins $I$ and $S$ have identical chemical shifts, periodic oscillation of magnetization between spins $I$ and $S$ occurs without any rf field being present (infinitely strong coupling case). In general, spins $I$ and $S$ have different chemical shifts and the purpose of the rf irradiation is to eliminate the difference in the chemical shifts between the two protons. This then would create a situation where the spins would become infinitely strongly coupled, resulting in periodic transfer of magnetization between the two coupled spins. One simple way to accomplish this is the application of a strong rf field (Fig. 1a). We will consider the effective rf fields experienced by the two coupled spins. Each effective rf field corresponds to the vector sum of the applied rf field (along the $y$ axis) and the resonance offset vector (along the $z$ axis) for that particular spin (Fig. 2). The magnitude of the effective rf field, $\nu_{\text{eff}}$, for a particular spin with resonance offset $\delta$ from the carrier is then given by

$$\nu_{\text{eff}} = (\delta^2 + \nu^2)^{1/2} \tag{2}$$

where $\nu$ is the nominal rf field strength (in frequency units). For cases where the rf field strength is much larger than the offset ($\nu \gg \delta$) the difference, $\Delta$, in effective field strengths for two spins $I$ and $S$ is

$$\Delta = \nu_{\text{effI}} - \nu_{\text{effS}} \approx (\delta_I^2 - \delta_S^2)/2\nu \tag{3}$$

A condition for Hartmann–Hahn transfer is that $\Delta \ll J$. As can be seen from Eq. (3), one way to obtain this is to use a very strong rf field. Alternatively, the carrier can be put exactly halfway between the two

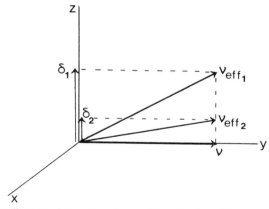

FIG. 2. Vector diagram of the orientations of the effective rf fields in the rotating frame, experienced by two spins with offsets $\delta_1$ and $\delta_2$, when an rf field of strength $\nu$ is applied along the $y$ axis.

resonances, such that $\delta_I = -\delta_S$. A second point worth noting is that the effective $J$ coupling during spin-lock conditions is reduced because the two effective fields point in different directions. This slightly slows down the rate of magnetization transfer between coupled spins.[11,12,14] In practice, the method of Fig. 1a is not suitable for wide-band cross-polarization: an rf field of several megahertz would be needed to cover a 10-ppm band width at 500-MHz $^1$H frequency! In addition, transverse NOE would also give rise to intense cross-peaks which might cancel the Hartmann–Hahn transfers which are of opposite sign. The same is true for the mixing scheme used in the original TOCSY experiment (Fig. 1b). Although the power required for wide-band homonuclear cross-polarization is significantly reduced with this method relative to continuous irradiation, transverse NOE's are also present and the band width that can be covered effectively is still quite narrow.

### Mixing Based on Composite Pulse Decoupling Schemes

Effective methods for wide-band cross-polarization in proteins must satisfy three criteria: (1) the effects of transverse NOE's must be minimized; (2) minimal rf power must be used during the cross-polarization to avoid sample heating; and (3) they must be easy to use in a routine fashion. Below, two sequences that satisfy these criteria will be discussed briefly.

Arguments presented by Waugh on the theory of spin decoupling[17] suggested to us that sequences effective at broad-band heteronuclear decoupling might be useful for obtaining wide-band homonuclear cross-polarization at low rf power. The first improvement we made on the scheme of Fig. 1a was to modulate the phase of the spin-lock field (Fig. 1c), the analog of square-wave heteronuclear decoupling.[18] Much more effective decoupling schemes have been developed by Freeman, Levitt, and co-workers.[19,20] In a first attempt, we tried to use their MLEV16 scheme[21] for wide-band homonuclear cross-polarization. The MLEV16 composite pulse sequence consists of a large number of 90 and 180° pulses that are phase shifted relative to one another, as described in the caption of Fig. 1. The first problem encountered was that this scheme preserves both the $x$ and the $y$ component of transverse magnetization present at the end of the evolution period. However, the relaxation rates of these two

[17] J. S. Waugh, *J. Magn. Reson.* **50**, 30 (1982).
[18] J. B. Grutzner and A. E. Santini, *J. Magn. Reson.* **19**, 178 (1975).
[19] T. A. Frenkiel, M. H. Levitt, and R. Freeman, *Adv. Magn. Reson.* **11**, 47 (1983).
[20] A. J. Shaka and J. Keeler, *Prog. NMR Spectrosc.* **19**, 47 (1987).
[21] M. H. Levitt, R. Freeman, and T. A. Frenkiel, *J. Magn. Reson.* **47**, 328 (1982).

components differ from one another during the MLEV16 irradiation (*vide infra*), resulting in quadrature artifacts and phase distortions. As discussed later, trim pulses before and after the mixing period can be used to eliminate either the $x$ or the $y$ component. The MLEV16 scheme is supposed to accomplish isotropic mixing of the protons over a wide band width, which means that the magnetization vector of any isolated spin in the rotating frame ends up in its starting orientation after a single MLEV16 pulse cycle has been completed. As calculated by Waugh,[11] even in theory the MLEV 16 sequence is not quite perfect at doing this. In practice, small errors in phase shifting and pulse width aggravate this imperfection. So it can happen that $y$ magnetization present at the end of the evolution period finishes along the $z$ axis at the end of the mixing period (after many MLEV16 cycles), and remains undetected during the $t_2$ period of the 2D experiment. This then results in phase and amplitude distortions in the 2D spectrum. Our solution to this problem was to add a seventeenth pulse ($60_y^\circ$) which prevents the $y$ spin magnetization from rotating away from the $y$ axis during subsequent MLEV16 cycles. Note that the addition of the seventeenth pulse severely reduces the band width over which this MLEV17 sequence[22] provides efficient homonuclear cross-polarization (or heteronuclear decoupling). However, its beneficial effects are to ensure that pure-phase spectra free of anomalous amplitude distortions can be recorded. It also can be shown that the addition of this seventeenth pulse makes the mixing nonisotropic, which eliminates magnetization (sensitivity) loss that otherwise would be caused by the generation of multiple-quantum coherence.[11,23]

As mentioned earlier, in principle both the $x$ and the $y$ components of transverse magnetization are preserved during the mixing period. The undesirable $x$ component can be eliminated by the application of so-called trim pulses before and after the mixing period. Because of the severe inhomogeneity of the rf field, any magnetization not parallel to the effective field vector (the $y$ axis in Fig. 1d) is rapidly defocused. The effect of any nondefocused $x$ magnetization that can give rise to a weak anti-diagonal in the 2D spectrum can be eliminated by coadding the results of two 2D experiments recorded under identical conditions except that the second experiment uses trim pulses that are 90° longer than the first experiment.[23]

Use of the MLEV17 sequence for protein structure determination has been pioneered by Clore, Gronenborn, and co-workers,[24] and it is now

[22] A. Bax and D. G. Davis, *J. Magn. Reson.* **65,** 355 (1985).
[23] A. Bax, *Isr. J. Chem.* **28,** in press (1989).
[24] G. M. Clore, S. R. Martin, and A. M. Gronenborn, *J. Mol. Biol.* **191,** 553 (1986).

used routinely in a number of laboratories. The MLEV17 sequence is quite forgiving; the sequence is not particularly sensitive to the calibration of hardware rf phase shifts or balancing of the amplitudes of the phase-shifted pulses. Exact calibration of the pulse widths also is not very critical. For effective cross-polarization over a 10-ppm band width at 500 MHz (assuming the worst case of an NH proton removed by 5 ppm from the carrier and $C_\alpha H$ on resonance, with a $J$ coupling of 4 Hz), an rf field strength of about 10 kHz is needed. Typically this requires about 3- to 5-W rf power. Even if this power is applied for relatively short mixing period durations of 30–60 msec, this can result in significant sample heating effects. The inhomogeneous sample heating may cause deterioration of the lock signal and an increase in $t_1$ noise.

Mixing based on the WALTZ16[25] decoupling cycle can be slightly more efficient than the MLEV17 scheme, providing a reduction in rf power by up to about 40%. However, our computer simulations and experimental results suggest that this type of mixing is more critical with respect to exact balancing of the amplitudes of the two phase-shifted rf transmit channels. On some of the newer types of NMR instruments (which often use digital rf phase shifting) this presents no particular problem and the WALTZ-based mixing scheme (Fig. 1e) is then preferable over the MLEV17 scheme. For similar reasons as for the MLEV17 scheme, a seventeenth pulse may be added at the end of each WALTZ16 cycle. In practice, on our Bruker AM600 spectrometer, a length of 30–60° for this seventeenth pulse is sufficient. To determine what duration is needed for the seventeenth pulse, the following simple experiment should be conducted. Use a compound with well-resolved resonances spread over at least 8 ppm. Record the spectrum that corresponds to the first increment of the 2D experiment. First use a short mixing time (one WALTZ17 cycle) followed by a 3-msec trim pulse and determine what phase parameters are needed for phasing the 1D spectrum to the absorption mode. Next, using a 75-msec mixing period, remove the final trim pulse, and reduce the width of the seventeenth pulse down from 180° until phase distortions become apparent in the spectrum (when phased with the parameters for the short mixing period with trim pulse). The seventeenth pulse should be set to the minimal width for which the spectrum still has an absorption mode ($\pm 15°$) appearance. All spectra shown in this chapter have been recorded with the scheme of Fig. 1e, using a 60° flip angle for the seventeenth pulse.

### Reducing Relaxation during Mixing

In the scheme of Fig. 1a, the spin-locked magnetization decays with a time constant $T_{1\rho}$, which for proteins is identical to $T_2$. In the scheme of

Fig. 1d, the $y$ magnetization present at the start of the MLEV17 sequence goes through a trajectory where it spends one-half its time along the $z$ axis (relaxing with $T_1$) and one-half its time in the transverse plane where it relaxes with $T_2$. Because for macromolecules, $^1H$ $T_1$ values typically are much larger than $T_2$ values, the magnetization decay is reduced by almost a factor of two relative to Schemes a–c in Fig. 1.[22]

Similarly, during the WALTZ mixing scheme of Fig. 1e, magnetization perpendicular to the axis along which the WALTZ pulses are applied (the $x$ axis in Fig. 1e) relaxes by almost a factor of two more slowly relative to magnetization parallel to that axis. Note that the advantage in relaxation only applies to in-phase spin-locked magnetization; the anti-phase $I_yS_z$ and $I_zS_y$ terms, essential in the HOHAHA process, are not reduced in relaxation rate.

### Interference from NOE Effects

As mentioned earlier, transverse NOE effects can interfere with the Hartmann–Hahn effects. Since, under spin-locked conditions, the NOE effect is always positive, NOE cross-peaks are of opposite sign to the diagonal. During the MLEV and WALTZ sequences, however, the "spin-locked" magnetization spends half its time along the $z$ axis, where the NOE is negative for macromolecules. Although the spin-locked NOE effect is larger than the regular NOE effect, the fact that they are of opposite sign strongly reduces the net NOE effect. This minimizes reduction of HOHAHA cross-peak intensities in cases where both HOHAHA and NOE effects are present. Note that for the WALTZ sequence the NOE effects would persist if the phase of the WALTZ pulses were changed by 90° with respect to the trim pulses.

### Applications of HOHAHA to Peptides and Proteins

The power of the HOHAHA experiment to solve spectral assignment problems (cf. Basus [7], Vol. 177, this series) is probably most clearly demonstrated for peptides and proteins by correlating amide protons with $C_\alpha$ and side-chain protons. As an example, Fig. 3 shows the complete 2D correlation spectrum for the antimicrobial peptide, magainin 2 (23 amino acids).[26] The spectrum was recorded in a mixture of trifluoroethanol (TFE) and $H_2O$ (25/75, v/v) using Scheme e of Fig. 1, with presaturation of the $H_2O$ resonance prior to the evolution period. Because of the acidic sample conditions used (pH 4 prior to addition of TFE) the amide proton exchange is sufficiently slow that presaturation does not obliterate any of

[25] A. J. Shaka, J. Keeler, T. A. Frenkiel, and R. Freeman, *J. Magn. Reson.* **52**, 335 (1983).

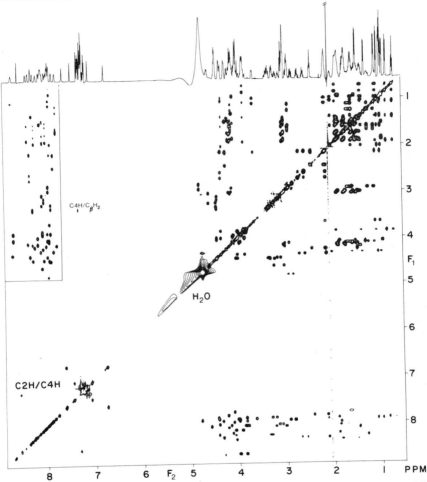

FIG. 3. Two-dimensional HOHAHA spectrum of the peptide magainin 2, 15 mg in 0.375 ml $H_2O$/0.125 ml trifluoroethanol, 27°. The spectrum was recorded at 600 MHz, using a WALTZ17 mixing sequence with $\alpha = 60°$. The 90° pulse width was 35 $\mu$sec, duration of the WALTZ irradiation was 75 msec, and the trim pulse duration was 2 msec. Presaturation of the $H_2O$ resonance was used during the recycling delay. Data were acquired in the TPPI mode and the size of the acquired data matrix was 700 × 2048, corresponding to acquisition times of 42 and 120 msec in the $t_1$ and $t_2$ dimension, respectively. Eight scans, preceded by two dummy scans, were recorded per $t_1$ value, and the total measuring time was 3.5 hr. Zero filling was used to yield a digital resolution of 4 Hz in both dimensions.

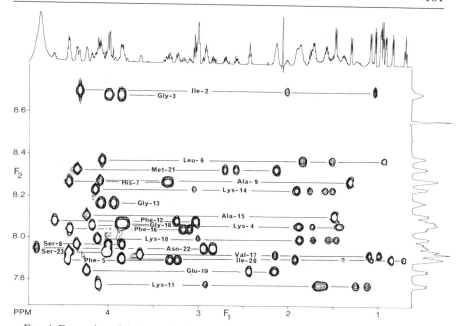

Fig. 4. Expansion of the boxed region of Fig. 3, showing the amide to side-chain connectivities. With the exceptions of Gly-1 and Ile-2, for each amide proton nearly all side-chain resonances connected directly or indirectly via scalar couplings to the $C_\alpha$–H resonance are observed, despite relatively small NH–$C_\alpha$H $J$ couplings ($\alpha$ helix).

the exchangeable peptide protons. Under these conditions the peptide adopts an $\alpha$-helical conformation,[27] and consequently the NH–$C_\alpha$H $J$ couplings are relatively small (4–5 Hz). Nevertheless, for the mixing time used in this experiment (79 msec) a large proportion of the amide magnetization has been transferred to the aliphatic protons. An expansion of the boxed region of Fig. 3 is shown in Fig. 4, showing the complete coupled spin systems of the side chains. For example, all four lysine side chains can be clearly identified, correlating the NH proton directly with the terminal $C_\varepsilon H_2$ protons. Similarly, the complete spin system of Ile-20 can be recognized immediately. Analogous connectivities for Ile-2 are considerably weaker because of the broadening of the NH resonance. No connectivities between the aromatic ring protons of the phenylalanine residues and the aliphatic protons are seen, although small couplings between $C_\delta$ and $C_\beta$ protons have been reported for such residues. In contrast, the isolated C2H and C4H protons of the histidine residue show intense cor-

[26] M. Zasloff, *Proc. Natl. Acad. Sci. U.S.A.* **84**, 5449 (1987).

relations to each other and C4H shows a correlation to the $C_\beta$ protons. In our experience, this latter connectivity is usually not observable in molecules significantly larger than the magainin peptide. Note also that the cross-peak for transfer from $C_\beta H_2$ ($F_1$) to C4H ($F_2$) is much stronger than the cross-peak corresponding to transfer in the opposite direction. This is caused by too short a delay time between scans: by waiting insufficiently long the slowly relaxing C4H protons have less magnetization available for transfer to the $C_\beta H_2$ protons than vice versa. This type of asymmetry artifact caused by too-short delay times can be seen in many types of 2D experiments.

Some distortion of the 2D spectrum of Fig. 3 in the region of the residual water is seen. This residual water always tends to be a much larger problem in HOHAHA experiments in comparison with, for example, NOESY or COSY methods. The reason lies in the relatively long trim pulses used. Water magnetization that is relatively far removed from the receiver coil and is present in a very poor homogeneous region of the magnetic field is excited by these long pulses, giving a significant response in the 2D spectrum. As discussed in the next section, an alternative method that does not use $^1H$ presaturation will automatically avoid excitation of this spurious $H_2O$ signal.

### Tracing Connectivities along the Aliphatic Side Chains

For proteins larger than 60 or 70 amino acids, the $T_2$ ($T_{1_\rho}$) values of the amide protons often become extremely short (<20 msec) and the long mixing periods needed for connecting the amide resonances with all side-chain resonances necessarily lead to very low signal to noise in the final spectrum. However, connectivities to the $C_\alpha$ and $C_\beta$ protons often are visible. In this case it can be very useful to record a second HOHAHA spectrum in $D_2O$ solution. Typically, magnetization propagates very rapidly along the aliphatic side chains, and a mixing time of 35–50 msec is often sufficient to observe all possible connectivities. As a result, the spectrum becomes very crowded; and to distinguish direct from indirect connectivities, it may be necessary to record HOHAHA spectra for at least two mixing times or to compare the spectrum with an absorption-mode COSY spectrum.

The $T_2$ values of the aliphatic resonances often are substantially longer than those of the amide protons and excellent spectra can be obtained for quite large proteins. Figure 5 shows the aromatic and aliphatic regions of staphylococcal nuclease, complexed with pdTp and calcium (18 kDa).

[27] D. Marion, M. Zasloff, and A. Bax, *FEBS. Lett.* **227**, 21 (1988).

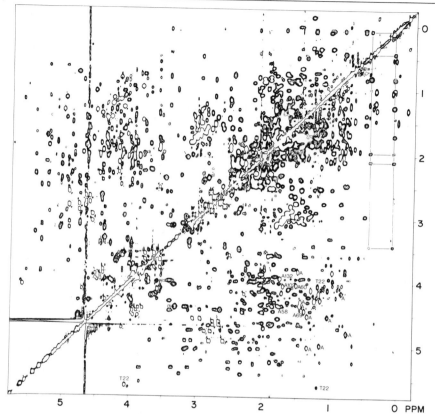

FIG. 5. Two-dimensional HOHAHA spectrum of the aliphatic region of a staphylococcal nuclease/pdTp/Ca$^{2+}$ complex (18 kDa), 1.5 m$M$ in D$_2$O, 100 m$M$ NaCl, pD 7.4. The spectrum was recorded with a mixing time of 35 msec and the measuring time was 6 hr. Duration of the trim pulses was 1.5 msec and the 90° pulse width was 28 $\mu$sec. Labeled cross-peaks follow the assignments of Torchia et al.[28]

Some of the assignments, based on isotopic labeling studies,[28] are indicated in the figure. The phase distortion observed for the intense resonance of the residual HDO resonance is largely due to the finite duration of the trim pulses. As mentioned earlier, coadding data recorded with two different durations of the trim pulses can remove this type of artifact.

The relative intensities of resonances from nonmobile methylene resonances can become very low if the strongest resolution enhancement is

[28] D. A. Torchia, S. W. Sparks, and A. Bax, Biochemistry 27, 5135 (1988).

used. For this reason it may be useful to process the data twice, using different line-broadening functions. The same, of course, is also true for processing the data from NOESY and COSY spectra.

## Recording HOHAHA Spectra without Presaturation

As mentioned above, presaturation of $H_2O$ when recording HOHAHA spectra can lead to serious problems caused by water from outside the receiver coil. Moreover, presaturation obliterates a band of $C_\alpha$ protons resonating close to the $H_2O$ resonance. In addition, when working at or near physiological pH, the amide exchange rates for many of the residues are such that $H_2O$ presaturation also saturates the amide resonances, which in turn may partially saturate other protons that have a strong NOE interaction with these amide protons. For these reasons, it may be better to record the HOHAHA spectrum without presaturation, by using one of the nonexcitation water suppression schemes. One such scheme[29] has been demonstrated to function well for the HOHAHA method. On our Bruker AM600 spectrometer we prefer to use a slight variation of this method which has a more favorable excitation profile. Its pulse sequence is sketched in Fig. 6. Essentially, this is the same sequence as the one of Fig. 1e, with the following minor modifications: After the mixing sequence, the spin-locked magnetization is stored along the $z$ axis by means of the $90_\psi$ flip-back pulse. Subsequently, the $z$ magnetization is "read" by a conventional 1–1 "jump-and-return"[30] sequence. One has to prevent the water resonance from getting inverted, which would lead to severe radiation damping problems. Therefore, the first excitation pulse is cycled only in a two-step (not the regular four-step) fashion. Below, the procedure typically followed during the setup of this experiment is described briefly. First, the 90° pulse is determined, simply by taking one-half the pulse width of a 180° pulse or one-fourth the width of a 360° pulse. Next, the scheme of Fig. 6 is executed with the pulse widths of the final 1–1 sequence set to zero. The width of the flip-back pulse is then finely adjusted to minimize the amount of $H_2O$ signal. The 1–1 sequence is then reinstated and the receiver gain is adjusted to avoid overload.

Typically at 500 or 600 MHz, a 90° pulse width of 30 or 25 $\mu$sec will be optimal for coverage of a 10-ppm spectral width. More power will result in too much sample heating, causing lock problems and $t_1$ noise. Also, to minimize the effect of unavoidable rf heating, the sample should be

[29] A. Bax, V. Sklenář, A. M. Gronenborn, and G. M. Clore, *J. Am. Chem. Soc.* **109**, 6511 (1987).
[30] P. Plateau and M. Guéron, *J. Am. Chem. Soc.* **104**, 7310 (1982).

FIG. 6. Pulse scheme of the HOHAHA experiment for recording spectra in $H_2O$ solution without the use of presaturation. This scheme has additional pulses at the end of the mixing period (compare to Fig. 1e). The $90_\psi$ pulse flips the spin-locked magnetization back to the $z$ axis. An additional saturation pulse (not shown) may be applied at the end of data acquisition.[29] Phase cycling used on our Bruker AM-600 spectrometer is as follows: $\phi = -45°, 135°$; $\psi = -x, x; \xi = x, x, y, y, -x, -x, -y, -y$; Acq $= x, x, y, y, -x, -x, -y, -y$. On successive $t_1$ increments the phase $\phi$ is incremented (from odd to even numbered spectra) or decremented by 90° (from even to odd). This procedure prevents the water resonance from getting inverted and still permits the suppression of axial peaks and the recording of 2D quadrature spectra. Similar procedures can be used for recording the data in the hypercomplex format.[29]

brought to a steady state by starting the 2D experiment for about 5 min, immediately followed by a restart of the real experiment.

The procedure described above is demonstrated for a sample of 4.5 m$M$ hen egg white lysozyme (14 kDa), pH 4.2, 36°, 80 m$M$ NaCl. The fingerprint region of the HOHAHA spectrum, obtained with a 36-msec mixing time is shown in Fig. 7. Complete assignments for this protein, based on very careful double-quantum-filtered COSY and RELAY experiments, were given recently by Redfield and Dobson.[31] Comparison of their double-quantum-filtered COSY spectrum with the spectrum of Fig. 7 indicates that the resolutions in the fingerprint regions of the two spectra are quite similar, despite the 20% higher field strength used for the HOHAHA experiment. However, the lowest contour level in the HOHAHA spectrum is far above the thermal noise level, so much stronger resolution enhancement and higher digital resolution probably could be used to further enhance spectral resolution. Baseline correction of the final 2D spectrum was not used but would enhance the appearance of the spectrum significantly, particularly in the lower right-hand corner of the spectrum closest to the $H_2O$ resonance.

Because of the relatively short mixing period used (36 msec), the amount of relay observed in the spectrum of Fig. 7 is relatively limited. However, NH cross-peaks to both the $C_\alpha H$ and the $C_\beta H$ resonances are present for most residues. For Val-2, the entire spin system is clearly visible. The amount of relay is significantly larger than observed for other Val residues, perhaps because this amino acid at the N-terminus may have increased mobility of the side chain. $NH–C_\beta H_3$ connectivities for

[31] C. Redfield and C. M. Dobson, *Biochemistry* **27**, 122 (1988).

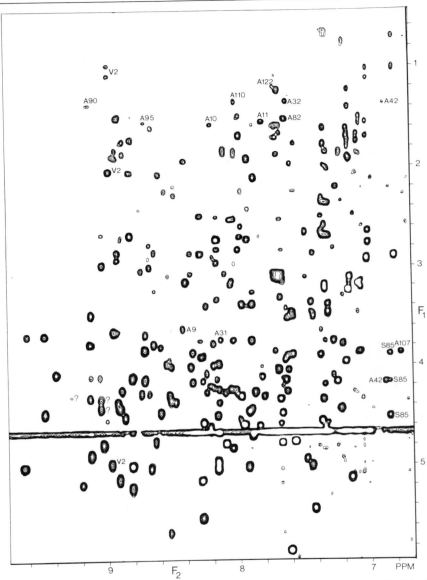

FIG. 7. Two-dimensional HOHAHA spectrum of hen egg white lysozyme, 4.5 mM, pH 4.2, 75 mM NaCl, 35°. The spectrum was recorded with the scheme of Fig. 6, using a mixing time of 35 msec, 24-$\mu$sec 90° pulse width, 800 × 2048 data matrix, acquisition times of 40 and 102 msec in $t_1$ and $t_2$ dimension, respectively. Sixteen scans were recorded per $t_1$ value and the total measuring time was 7 hr. Assignments are taken from Redfield and Dobson.[31] No baseline correction procedure of the frequency domain spectra was used and the lowest

the alanine residues are labeled in Fig. 1. At the contour level shown, these relay connectivities are absent for Ala-9, Ala-31, and Ala-107. At lower contour levels these connectivities are also observed, but it is not clear why the relay intensities would be weaker for these residues, considering that the coupling between $C_\alpha$ and $C_\beta$ protons is always 7 Hz. Three unidentified cross-peaks, not present in the COSY spectrum of Ref. 30 (near 9.1/4.4 ppm), are marked "?."

## Discussion

Appropriate hardware is required for the recording of high-quality HOHAHA spectra. First, an rf amplifier at $^1$H frequencies is needed that can produce the power needed for a 25- to 35-$\mu$sec 90° pulse (typically 2–5 W) and that can produce this amount of rf power without any droop for at least 50 msec. Attenuating the high "observe pulse power" is usually not an adequate solution and may result in burned attenuators or in a collapse of the power supply of the amplifier. A linear or near-linear amplifier as is often used for $^1$H decoupling is better suited for the HOHAHA irradiation. To keep the complexity of the 2D experiment to a minimum, we use the same power for pulses, trim pulses, and mixing sequences. A slight loss in sensitivity and a small linear phase error in the $F_1$ dimension resulting from rf offset effects at these medium levels of rf power cause no major problems in practice. If more effective mixing sequences are developed in the future, it may become necessary to switch between high-power pulses and low-power mixing for their application, as has been proposed for the ROESY experiment.[32] At present, there appears no need for such power switching.

If one wants to use the $^1$H decoupler amplifier for generating rf pulses, the decoupler rf must be phase coherent with the receiver. Although in principle this latter requirement can be avoided by using an interesting $z$-filtered version of the experiment,[33] in practice this approach causes extra $t_1$ noise and with it we have been unable to generate spectra of a quality similar to that obtained with the sequence described in this chapter.

[32] C. Griesinger and R. R. Ernst, *J. Magn. Reson.* **75**, 261 (1987).
[33] M. Rance, *J. Magn. Reson.* **74**, 557 (1987).

contour level is far above the thermal noise, but close to the baseline distortions caused by incomplete water suppression and by frequency-dependent phase correction used in the $F_1$ dimension. Three of the 12 NH–$C_\beta H_3$ relay connectivities for Ala residues (A31, A9, A107) are not observed at this contour level, despite intense NH–$C_\alpha$H cross-peaks. Resonances marked "?" were not observed in the COSY spectrum.

Of the various mixing schemes discussed here, the MLEV17 and WALTZ17 schemes appear to be most suitable for protein studies. All of the earlier schemes (Fig. 1a–c) give strong additional NOE cross-peaks which can be very confusing and which also can decrease the sensitivity of the experiment significantly in cases where NOE and HOHAHA effects compete. Comparison of the MLEV17 and WALTZ17 schemes reveals a few differences. From a practical point of view, the MLEV17 mixing scheme appears to be less sensitive to imperfections in the phase-shifting hardware. However, the band width that can be covered with MLEV17 for a given amount of rf power is about 25% smaller. WALTZ17 mixing has the intrinsic disadvantage that the effective size of the coupling during mixing (and thus the transfer rate) is reduced more than it is for the MLEV sequence. For offsets used in practice (smaller than $\nu_{rf}/2$), this effect is small. The maximal band width covered by the WALTZ17 sequence is about $\pm 0.4\nu_{rf}$ and for MLEV17, about $\pm 0.33\nu_{rf}$. For most diamagnetic proteins this allows effective cross-polarization over the entire spectral range with only a few watts of rf power. Overall, if spectrometer hardware is sufficiently good, the WALTZ sequence appears preferable over the MLEV mixing scheme.

Two-dimensional HOHAHA experiments are particularly useful for delineating individual spin systems in proteins. The method provides high-sensitivity, high-resolution spectra that contain a wealth of relayed connectivity information. There are two reasons why HOHAHA spectroscopy may be more sensitive than conventional COSY-type experiments. First, because of the in-phase nature of the transfer of magnetization, cancellation of absorptive components within a cross-multiplet does not occur. Second, no magnetization is lost to the generation of spurious multiple-quantum coherence, a major source of sensitivity loss in conventional RELAY experiments. Resolution in the fingerprint region of double-quantum-filtered COSY and HOHAHA spectra is quite similar; although in our experience, the sensitivity of the HOHAHA spectra is typically much higher. For studying spectral regions close to the diagonal, the HOHAHA spectrum usually offers the best appearance because of the relatively low intensity of the diagonal.

### Acknowledgment

I thank Laura Lerner for many useful comments during the preparation of this manuscript and Rolf Tschudin for technical assistance in modifying spectrometer hardware. The nuclease sample was kindly provided by Dennis Torchia and Steven Sparks, and the magainin-2 sample by Michael Zasloff. I also wish to acknowledge stimulating discussions with Marius Clore and Angela Gronenborn.

# [9] Two-Dimensional Nuclear Overhauser Effect: Complete Relaxation Matrix Analysis

*By* BRANDAN A. BORGIAS and THOMAS L. JAMES

## Introduction

The use of two-dimensional nuclear Overhauser effect (2D NOE) spectroscopy in the structural analysis of macromolecules has become widespread in recent years.[1] The use of distances obtained from 2D NOE spectra as constraints in either molecular dynamics[2] or distance geometry calculations[3] has resulted in protein structures in solution that are compatible with X-ray crystal structures. Obtaining accurate distances from 2D NOE intensities is complicated, however, by the nonlinear relationship between the time development of cross-peak intensity and the distance separating the two protons correlated by the cross-peak. Approaches to obtaining the best possible distances from the 2D NOE experiment are the focus of this chapter.

Several reports have appeared describing protocols for obtaining distances quantifiably from intensities.[4-7] The simplest method, called ISPA (isolated spin-pair approximation), assumes that the time course of the intensities can be linearized under suitable conditions (i.e., short mixing times and short correlation times).[4] Another method, which we will call the DIRECT method, is feasible when enough of the intensities (both diagonal and cross-peak) can be assigned and accurately measured so that the complete relaxation network is well represented by the intensities.[5,6] The intensities are then back transformed directly into distances by solving for the set of relaxation equations which are consistent with the intensities. Finally, there are iterative methods, which refine a structure to optimize the match between calculated and experimental intensities.[7,8] One such method is being developed in our laboratory and is called COMATOSE (complete matrix analysis torsion optimized structure).[7] This method is an extension of CORMA (complete relaxation matrix anal-

---

[1] K. Wüthrich, this series, Vol. 177 [6].
[2] R. Scheek, R. Kaptein, and W. F. van Gunsteren, this series, Vol. 177 [10].
[3] I. D. Kuntz, J. F. Thompson, and C. M. Oshiro, this series, Vol. 177 [9].
[4] A. M. Gronenborn and G. M. Clore, *Prog. Nucl. Magn. Reson. Spectrosc.* **17**, (1985).
[5] E. T. Olejniczak, R. T. Gampe, Jr., and S. W. Fesik, *J. Magn. Reson.* **67**, 28 (1986).
[6] P. A. Mirau, *28th Exp. NMR Conf. 1987* Abstr. No. WK12.
[7] B. A. Borgias and T. L. James, *J. Magn. Reson.* **79**, 493 (1988).
[8] J.-F. Lefèvre, A. N. Lane, and O. Jardetzky, *Biochemistry* **26**, 5076 (1987).

METHODS IN ENZYMOLOGY, VOL. 176

ysis) for calculating 2D NOE intensities from an assumed molecular structure.[9,10]

As will be elaborated below, the ISPA method incurs serious errors if quantitative distances are required, although for many applications it may be suitable. The major source of error in estimating distances from 2D NOE intensities arises from the occurrence of spin diffusion, which becomes important in systems with long correlation times (i.e., $\omega\tau_c \geq 1$). Under the influence of spin diffusion, a cross-peak between distant protons gains intensity which (1) depends primarily on the network of protons linking the correlated protons, and (2) bears little relation to the distance between them. The ISPA method is incapable of accounting for spin diffusion. The DIRECT method works well in principle, and incorporates the effects of spin diffusion, but it is frequently the case that many cross-peaks and diagonal peaks are poorly resolved. Hence, the intensity matrix used in analysis will necessarily incorporate many errors, and the resulting distance matrix will consequently also be in error. COMATOSE can work under the experimental conditions most commonly obtained, namely incomplete resolution of all peaks and random errors in intensity measurement. However, it consumes considerably more computer time since it is an iterative process.

### The Structure/Intensity Relationship in 2D NOE Spectroscopy

Relaxation processes leading to the 2D NOE spectrum include both simple internuclear (e.g., $^1H-^1H$) relaxation *and* effects due to interacting pairs, triplets, etc., of nuclear spins, i.e., cross-correlation effects.[11,12] We will limit our discussion to only the former since it has been shown that cross-correlation effects contribute negligibly to the 2D NOE intensities generated during mixing times that yield the most usable data.[12] The relaxation network for the macromolecule is described by a set of equations which are an extension of Solomon's equations describing the dipolar relaxation process in a two-spin system.[13] The time course of the magnetization during the mixing period of the 2D NOE experiment is described by the system of equations[14]:

[9] J. W. Keepers and T. L. James, *J. Magn. Reson.* **57**, 404 (1984).
[10] B. A. Borgias and T. L. James, "Complete Relaxation Matrix Analysis (CORMA)." University of California, San Francisco, 1987.
[11] L. Werbelow and D. M. Grant, *Adv. Magn. Reson.* **9**, 189 (1978).
[12] T. E. Bull, *J. Magn. Reson.* **72**, 397 (1987).
[13] I. Solomon, *Phys. Rev.* **99**, 559 (1955).
[14] S. Macura and R. R. Ernst, *Mol. Phys.* **41**, 95 (1980).

$$\partial M/\partial t = -RM \tag{1}$$

where $\mathbf{M}$ is the magnetization vector describing the deviation from thermal equilibrium ($\mathbf{M} = \mathbf{M}_z - \mathbf{M}_0$), and $\mathbf{R}$ is the matrix describing the complete dipole–dipole relaxation network. The diagonal elements of the rate matrix are simply the longitudinal relaxation rates ($\rho_i$), while the off-diagonal elements are the cross-relaxation rates ($\sigma_{ij}$).

$$R_{ii} \equiv \rho_i = 2(n_i - 1)(W_1^{ii} + W_2^{ii}) + \sum_{j \neq i} n_j(W_0^{ij} + 2W_1^{ij} + W_2^{ij}) + R_{1i} \tag{2a}$$

$$R_{ij} \equiv \sigma_{ij} = n_i(W_2^{ij} - W_0^{ij}) \tag{2b}$$

Here $n_i$ is the number of equivalent spins in a group, such as a methyl rotor, and the zero-, single-, and double-transition probabilities $W_n^{ij}$ are given (for isotropic random reorientation of the molecule) by

$$W_0^{ij} = q\tau_c/r_{ij}^6 \tag{3a}$$

$$W_1^{ij} = 1.5(q\tau_c/r_{ij}^6)\{1/[1 + (\omega\tau_c)^2]\} \tag{3b}$$

$$W_2^{ij} = 6(q\tau_c/r_{ij}^6)\{1/[1 + 4(\omega\tau_c)^2]\} \tag{3c}$$

where $q = 0.1\gamma^4\hbar^2$. The term $R_{1i}$ represents external sources of relaxation such as paramagnetic impurities or spin labels, and will be ignored in the following discussion. Inspection of Eqs. (2a,b) and (3a–c) reveals the $1/r^6$ distance dependence of the relaxation rates. Equation (1) has the familiar solution

$$\mathbf{M}(\tau_m) = \mathbf{a}(\tau_m)\mathbf{M}(0) = e^{-\mathbf{R}\tau_m}\mathbf{M}(0) \tag{4}$$

where $\mathbf{a}$ is the matrix of so-called mixing coefficients which are proportional to the measured 2D NOE intensities. Note that the exponential of Eq. (4) *cannot* be calculated directly by performing a term-by-term exponentiation of the rate matrix [i.e., $a_{ij}(\tau_m) \neq e^{-R_{ij}\tau_m}$]. A simple approach to calculating the intensities or distances is to recast the exponential into a series expansion:

$$\mathbf{a}(\tau_m) = e^{-\mathbf{R}\tau_m} \approx \mathbf{1} - \mathbf{R}\tau_m + \tfrac{1}{2}\mathbf{R}^2\tau_m^2 - \cdots + [(-1)^n/n!]\mathbf{R}^n\tau_m^n + \cdots \tag{5}$$

Truncation of the series expansion in Eq. (5) after the second term results in a linear relationship between the measured intensities and the relaxation rate constants. This yields an easy approach to calculating distances provided the intensities are obtained with a sufficiently short mixing time.

## Applying ISPA to Obtain Distances

The truncated series yields a linear relationship between the mixing coefficients at a specified mixing time and the relaxation rates, and consequently a quick way of estimating distances. This method has proved to be useful in generating protein structures that are consistent with X-ray crystal structures with RMS (root mean square) deviations in backbone distances of less than 2 Å. Typically, an internuclear distance $r_{ij}$ is estimated from a cross-peak intensity $a_{ij}$ by making reference to a fixed distance $r_{ref}$ in the molecule and its corresponding intensity $a_{ref}$. Distances are calculated according to Eq. (6):

$$r_{ij} = r_{ref}(a_{ref}/a_{ij})^{1/6} \qquad (6)$$

It is assumed that the correlation time for the distance to be measured is the same as the correlation time for the reference distance. The effectiveness of this method in obtaining approximate distances and, ultimately, protein structures is a consequence of its restricted use to long-range interactions between protons on residues separated by several intervening aminoacyl residues. This provides important secondary and tertiary structure information without specifying details of local structure. Such details are typically resolved by minimizing the energy of the structure which has been loosely defined by the 2D NOE constraints. The utility of using ISPA distances for analysis of nucleic acid structure is less certain, however. Typically, only cross-peaks between adjacent nucleotides are observed, so errors in distance estimates can be significant in the final analysis.

If approximate ISPA distances are desired, it is still important to restrict the analysis to data collected in the time domain where the approximation is valid. The question of how short a mixing time is needed for the truncation of Eq. (5) to be reasonably accurate has not been generally answered, although mixing times typically used are frequently in the 50- to 100-msec range. Is this short enough? That question can be answered by running through the series expansion term by term for a representative case.

Intensities were calculated according to Eq. (5) for protons in B-DNA assuming a correlation time of 4 nsec (at 500 MHz, $\omega\tau_c = 12.6$), and mixing times of 10, 50, 100, 150, and 200 msec. The results are given in Table I and Fig. 1. At 10 msec, intensities were in error by only 5–15%. However, intensities for distances between 1.8 and 2.3 Å were found to be overestimated by 20–90% at 50 msec. The intensities for these cross-peaks are expected to be dominated by direct relaxation. Intensities for longer distances, where spin diffusion is expected to make a significant contribution, were found to be underestimated by 50–70%. Moreover, simply increasing the number of terms in the expansion proved to be a poor route to obtain intensities. Figure 1 shows the fluctuations typical of a range of distances at a 200-msec mixing time. We found that the series

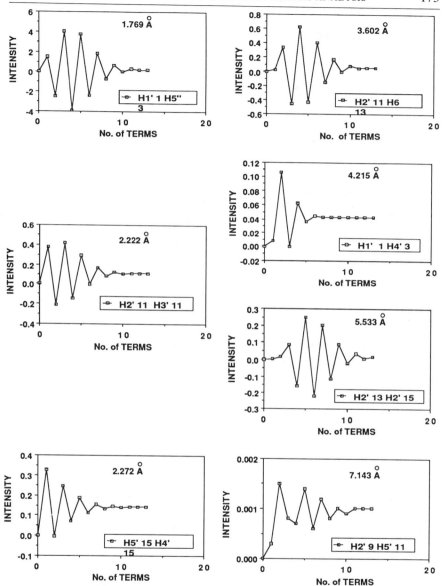

FIG. 1. Plots showing the convergence of intensities calculated using a term-by-term series summation of Eq. (5). Calculations assume a mixing time of 200 msec, a correlation time of 4 nsec, and an operating frequency of 500 MHz ($\omega\tau_c = 12.6$). Note that the intensity scales are not the same. The distances and specific proton–proton pair for the intensity are indicated.

TABLE I
CONVERGENCE OF SERIES EXPANSION FOR 2D NOE
INTENSITY CALCULATION

| $\tau_m{}^b$ | $N^c$ | Error[a] in single-term approximation | |
| --- | --- | --- | --- |
| | | $r_{ij} = 1.77$ to $2.27$ Å | $r_{ij} = 3.60$ to $7.15$ Å |
| 10 | 3 | +5 to 15% | −0 to −30% |
| 50 | 4 | +30 to 90% | −45 to −70% |
| 100 | 7 | +60 to 205% | −50 to −80% |
| 150 | 10 | +90 to 410% | −60 to −90% |
| 200 | 13 | +130 to 570% | −60 to −90% |

[a] Error is defined as $(I_1 - I_N)/I_N$, where $I_1$ is the single-term intensity, and $I_N$ is the intensity calculated after $N$ terms. Reported in these columns are the ranges of errors encountered for several distances as indicated. Intensities for short distances ($\leq 3.0$ Å) are typically overestimated by the single-term approximation. For longer distances the intensities are typically underestimated. The range of errors quoted here actually corresponds to errors for specific distances of 1.77, 2.22, 2.27, 3.60, 4.22, 5.53, and 7.15 Å.

[b] Mixing time for calculation in milliseconds. Correlation time is assumed to be 4 nsec, and the operating frequency is 500 MHz ($\omega\tau_c = 12.6$).

[c] Number of terms in series expansion required to achieve less than a 5% deviation in intensities averaged over all calculated intensities.

converges slowly, requiring three terms at 10 msec and as many as 13 terms to converge for a 200-msec experiment with a correlation time of 4 nsec (see Table I). The trade-off between performing matrix diagonalization (see below) and series summation occurs at about three terms. Hence only for very short mixing times would a series expansion ever be useful.

## Calculating Intensities by CORMA

A more expeditious approach to calculating intensities is to take advantage of linear algebra and the simplifications which arise from working with the characteristic eigenvalues and eigenvectors of a matrix. The rate matrix $\mathbf{R}$ can be represented by a product of matrices: $\mathbf{R} = \chi\lambda\chi^T$ where $\chi$ is the unitary matrix of orthonormal eigenvectors ($\chi^{-1} = \chi^T$), and $\lambda$ is the diagonal matrix of eigenvalues. The utility of making this transformation is that, since $\lambda$ is diagonal, the series expansion for its exponential (and consequently that of the mixing coefficient matrix) collapses.

$$\mathbf{a} = 1 - \chi\lambda\chi^T\tau_m + \tfrac{1}{2}\chi\lambda\chi^T\chi\lambda\chi^T\tau_m^2 - \cdots \quad (7a)$$
$$\mathbf{a} = \chi e^{-\lambda\tau_m}\chi^T \quad (7b)$$

This calculation allows one to readily calculate all the cross-peak intensities for a proposed structural model. Then comparison between calculated and measured intensities allows a determination as to the validity of the model structure. We have developed a program for performing this calculation, named CORMA, which is described in the Appendix at the end of this volume and available upon request from the authors.[9] Figure 2

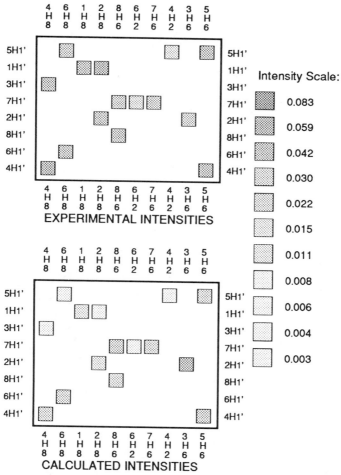

FIG. 2. Schematic intensities for the aromatic-H1′ region in [d(GGTATACC)]₂ as determined by CORMA. The atoms were ordered according to chemical shift. Experimental intensities were obtained at 500 MHz with a mixing time of 250 msec. Calculated intensities assume a correlation time of 4 nsec and a B–D–B conformation. Only these peaks which were unambiguously assigned and measured with high confidence are shown in these plots.

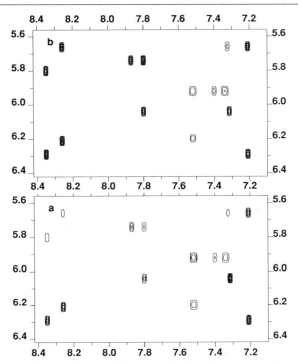

FIG. 3. Contour plots of the intensities shown in Fig. 2 as generated by software developed by H. Widmer. CORMA calculated intensities are shown in (a), experimental intensities are shown in (b). Line shapes for both plots are calculated assuming a natural line width of 5 Hz, and acquisition times of 0.15 sec in both $t_1$ and $t_2$, a 45° phase-shifted sine-squared apodization was applied in both domains. Coupling constants were assigned to model the splitting in the experimentally obtained spectrum (not shown). Peak assignments are as shown in Fig. 2 and correspond (in ppm) to: A4–H8 8.35, A6–H8 8.26, G1–H8 7.87, G2–H8 7.80, C8–H6 7.52, A6–H2 7.40, C7–H6 7.34, A4–H2 7.33, T3–H6 7.26, T5–H6 7.21, A4–H1' 6.29, A6–H1' 6.21, C8–H1' 6.20, G2–H1' 6.04, C7–H1' 5.92, T3–H1' 5.80, G1–H1' 5.74, T5–H1' 5.66. Displaying the intensities in this manner results in greater apparent contrast than the simple gray-scale representation of Fig. 2. However, the validity of this display is subject to the selection of appropriate coupling constants and the assumption that all peaks have the same natural line width.

is a schematic representation of intensities (generated by the program CORMA) for the aromatic-H1' region of [d(GGTATACC)]$_2$ in a B–D–B conformation in comparison with experimentally obtained intensities. An alternative representation, using contours and true chemical shift axes as generated by the program LINSHA,[15] is shown in Fig. 3.

[15] H. Widmer and K. Wüthrich, *J. Magn. Reson.* **70,** 270 (1986).

An alternative to the matrix diagonalization and intensity calculation according to Eq. (7a,b) is to perform a numerical integration of Eq. (1).[8] We do not have any experience with this approach, nor an estimate of the time required to perform the calculation.

## DIRECT Calculation of Distances

The central intensity/distance relationship described above can be used in the reverse direction to obtain distance estimates directly from complete, accurate intensity sets without having to propose an initial model. Rearrangement of Eq. (4) gives

$$\{-\ln[\mathbf{a}(\tau_m)/\mathbf{a}(0)]\}/\tau_m = \mathbf{R} \tag{8}$$

This approach has been described by others,[5,6] and we have made our own evaluation of it.[7] Under ideal circumstances, it is without doubt the best way to arrive at accurate distance estimates. However, the typical 2D NOE experiment does not yield all the information necessary for the DIRECT calculation to work well. Usually the diagonal peaks are not resolved, so estimates of these intensities can be in serious error. Further, several cross-peaks will also be unresolved or obscured by proximity to the diagonal peaks. As a consequence, the intensity matrix will incorporate many errors so that the resulting relaxation matrix will not represent the spatial arrangement of protons leading to the intensities.

## Iterative Structure Refinement

The exact calculation of intensities[7,8] have been incorporated according to Eq. (1) into iterative structure refinement programs. We have chosen to minimize the function:

$$E = \sum_i (I_i^{obs} - kI_i^{calc})^2 \tag{9}$$

where $k$ is a constant scaling factor applied to all the calculated intensities to normalize the sums of the calculated and observed intensities. Alternatively, one could use a weighted function:[8]

$$E^2 = \sum_i (I_i^{obs} - I_i^{calc})^2/\sigma_i^2 \tag{10}$$

where $\sigma_i^2$ is the standard deviation of the intensity estimated from the signal-to-noise ratio of the measured intensity $I_i^{obs}$. The use of the weighting function is prompted by the fact that smaller cross-peaks can easily have uncertainties which approach the magnitude of the intensity. However, this has the effect of focusing the refinement on only the shortest distances.

The next issue is the selection of structural parameters for optimization. Due to the limited number of observable cross-peaks ($\sim$5–10 per proton), it is important to reduce the degrees of freedom of the structure. In our COMATOSE program,[7] we have focused on the torsion angles and pseudorotation phases and amplitudes which effect changes in proton–proton distances. The pseudorotation phase angle P and amplitude $\theta_{max}$ are defined using the expressions developed by Altona and Sundaralingam.[16] The torsion angles ($\theta_i$) in the sugar ring are defined by

$$\theta_j = \theta_{max} \cos[P + (j - 2)4\pi/5] \qquad j = 0, \cdots, 4 \qquad (11a)$$

and

$$\tan P = (\theta_4 + \theta_1 - \theta_3 - \theta_0)/\{2\theta_2[\sin(\pi/5) + \sin(2\pi/5)]\} \qquad (11b)$$

gives the relationship between the phase and the individual torsion angles. Then, to reduce the number of coordinate changes required for changes in chain torsion angles, we treat each nucleotide (or aminoacyl residue) as a freely floating entity described by three Cartesian coordinates and three orientation angles. This avoids the calculations necessary to propagate torsion angle changes along the full length of the chain. Thus we arrive at 10 parameters per nucleotide (6–10 per aminoacyl residue). Alternatively,[8] the relative orientation of the nucleotides with respect to each other could be defined using the helical parameters: helical twist $\theta_t$, base pair roll $\theta_R$, propeller twist $\theta_p$, base-pair tilt t, helix displacement D, and local pitch h. Figure 4 shows a plot of COMATOSE-refined proton–proton distances against the corresponding ideal distances in the case of a single strand of DNA, d(GGTATACC).

### Matrix Analysis of Relaxation for Discerning Geometry of an Aqueous Structure—MARDIGRAS

A potentially quite useful extension of the DIRECT calculation of distances is iterative relaxation matrix analysis (IRMA) proposed by Kaptein.[17] We have developed a variant of IRMA termed MARDIGRAS. The general scheme is shown in Fig. 5. The critical feature of MARDIGRAS (or IRMA) is the generation of the augmented intensity matrix which contains all the scaled experimental intensities and the intensities calculated from a suitable model structure (obtained from model building, distance geometry, etc.). By solving the augmented intensity matrix for distances, one generates a distance set which will be in reasonable agreement with the model, but which is also partially restrained by experi-

[16] C. Altona and M. Sundaralingam, *J. Am. Chem. Soc.* **94**, 8205 (1972).
[17] R. Boelens, T. M. G. Konig, and R. Kaptein, *J. Mol. Struct.* **173**, 299 (1988).

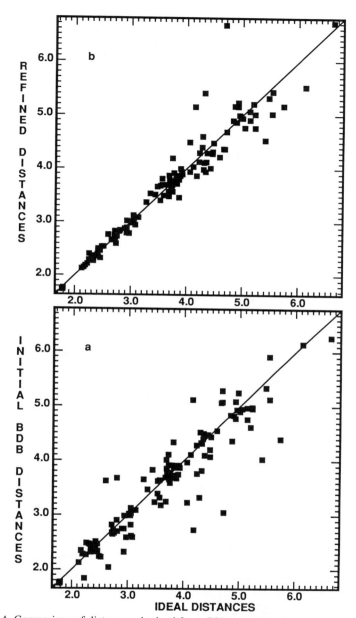

FIG. 4. Comparison of distances obtained from COMATOSE refinement with ideal distances. Intensities were calculated by CORMA for d(GGTATACC) in energy-minimized BDB-DNA conformation assuming an isotropic correlation time of 4 nsec and a mixing time of 250 msec. The starting conformation was an idealized BDB-DNA conformation prior to energy minimization. This conformation has the GG and CC segments in standard Arnott-B conformation and the TATA segment in a wrinkled-D conformation. (a) Prior to COMATOSE refinement. (b) After COMATOSE refinement.

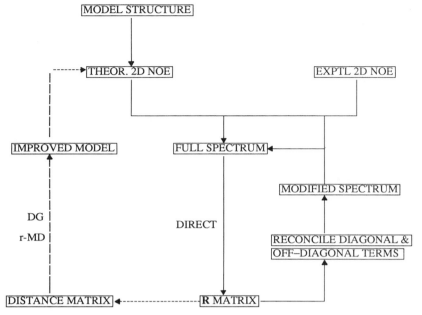

Fig. 5. Schematic diagram of matrix analysis of relaxation for discerning geometry of an aqueous structure (MARDIGRAS). A model structure is used to generate a theoretical 2D NOE spectrum (using CORMA). Wherever possible, experimental intensities are substituted into the theoretical spectrum to yield a full spectrum suitable for direct solution of the rate and distance matrices. This process incorporates all the effects of network relaxation and spin diffusion and will be relatively accurate, depending on how close the model is to the true structure. An augmented matrix or modified spectrum utilizes known distance constraints and required internal consistency between diagonal and off-diagonal elements. To improve the structure, distance geometry (DG) or restrained molecular dynamics (r-MD) can be used to generate an improved model which incorporates the distance restraints generated by the previous pass of the intensity/distance calculation.

mental intensities. By iterating through the cycle of structure generation via DG or MD on one branch and CORMA-type calculations on the other, one should eventually reach a self-consistent structure which incorporates all structural information inherent in the 2D NOE intensities. The variation introduced in MARDIGRAS, and shown on the lower right in Fig. 5, utilizes requirements for internal consistency in the relaxation matrix itself rather than iterating through the computer time-consuming restrained MD or DG procedures after a single pass through the relaxation matrix. With MARDIGRAS, constrained distances are used to give relaxation matrix intensities as a first approximation, and diagonal and off-diagonal elements are required to be consistent. This yields a modified spectrum which is then recombined with the experimental 2D NOE data to yield a new augmented matrix.

We have made some simple tests of the accuracy of the distances derived from the augmented intensity matrix for use in MARDIGRAS. The target structure was based on X-ray coordinates of bovine pancreatic trypsin inhibitor (BPTI) to which we appropriately added protons. CORMA, with a realistic random noise level of ±0.003, was used to generate the "experimental" spectrum. For one initial model we generated a randomized structure in which the proton coordinates had a root-mean-square shift of 3 Å relative to the ideal structure. The short (≤5 Å) distances generated from the augmented intensity matrix were within a few percentages of their ideal values. Longer distances tended to reflect the roundoff error of the weakest intensities, or the random distribution of the model. Note that the starting structure in this test tended to have severe distance errors for intraresidue distances, but on the whole was approximately correct in terms of the long-range structure. To test the sensitivity of the algorithm to starting structure, we generated an extended chain structure based on the BPTI primary sequence. As seen in Fig. 6, initial distances varied considerably from ideal X-ray distances. This initial structure has the short-range intraresidue distances approximately correct, while the long-range distances are in gross error. But with successive cycles through MARDIGRAS, distances improved dramatically. The resulting distances could then be used as either distance geometry or molecular dynamics input.

## Conformational Multiplicity and Motion

All the preceding has tacitly assumed a rigid structure undergoing isotropic tumbling. In fact, however, the molecule may undergo anisotropic motion, and have internal librational motion which may allow the molecule to sample a range of related conformations. This is akin to the problem of thermal motion in crystallography, with the exception that in NMR there is a strongly skewed relationship between the intensities and the distances: the intensities are approximately proportional to the inverse-sixth power of the distances. Thus, an intensity representing the average over all available conformations will underrepresent long distances. So it is possible that distances obtained from analysis of 2D NOE intensities will be inconsistent with any physically realistic structure. It is possible to incorporate anisotropic motion into the calculation of spectral densities, but this can be done only at the expense of an increase in the number of parameters undergoing the fitting procedure. Behling et al.[17] have approached this problem in a stepwise manner in which distance

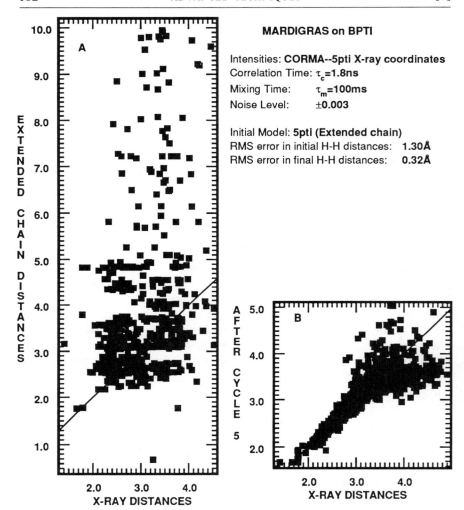

FIG. 6. Comparison of initial interproton distances from extended chain BPTI sequence with those of the "true" X-ray structure, and comparison of the interproton distances following MARDIGRAS refinement with those of the true X-ray structure. In (A) only those distances in the extended chain model less than 10 Å are shown. The actual range of distances extends to 182 Å. In (B) all calculated distances corresponding to input intensities are shown.

constraints derived according to the ISPA approximation were adjusted periodically during energy minimization to account for the presumed anisotropic motion of the proton–proton vector.[18]

## Conclusions

With experimentally obtainable 2D NOE intensities it is possible to arrive at target distances suitable as constraints in other structure generation programs, such as distance geometry, molecular dynamics, and molecular mechanics. Depending on the application, distances obtained with very little computational effort (i.e., using the ISPA approximation) may be suitable for defining important long-range distances which substantially define the tertiary structure of the macromolecule under scrutiny. For more accurate determination of distances, necessary for elucidation of detailed local structure, e.g., in DNA helices, it becomes necessary to incorporate spin diffusion and relaxation contributions from the complete network of all the interacting nuclei. In some cases (i.e., very well-resolved spectra) it may be possible to obtain distances directly from the intensities according to Eq. (8). However, it will typically be the case with larger macromolecular systems, such as small proteins and DNA fragments, that insufficient information will be available from the 2D NOE spectrum to successfully perform the transformation. In these cases, an iterative structure refinement algorithm which optimizes the fit between calculated and observed intensities may be called for. Moreover, it is important that the calculated intensities take into account the effects of network relaxation and spin diffusion.

In any study, it is important to bear in mind that the measured NOE's are dependent not only on the static distances defining the network of interacting protons, but also on the motional characteristics of the molecule. In general, any intensity measured in the 2D NOE experiment will be an ensemble average over all the conformations available to the molecule. Consequently, the distances so obtained will be weighted toward the short end of the range because of the $1/r^6$ dependence of the relaxation rate on distance.

[18] R. W. Behling, S. N. Rao, P. K. Kollman, and D. R. Kearns, *Biochemistry* **26**, 4674 (1987).

# [10] Selective Relaxation Techniques

By Neri Niccolai and Claudio Rossi

The pairwise dipole–dipole interactions which occur among $I = \frac{1}{2}$ nuclei may dominate nuclear relaxation, provided that suitably short internuclear distances between the magnetic moments in the molecular framework and the motion of the internuclear vectors are typical of biomolecules in solution.[1-3] In complex spin systems, experimental confirmation of the relevance of dipolar relaxation was obtained mainly from measurements of homo- and heteronuclear Overhauser effects and spin–lattice relaxation rate determinations.[4] The observation of dipolar connectivities can provide powerful information on molecular structure and dynamics in solution since the dipolar interaction between two nuclei, $I$ and $S$, yields a direct relaxation contribution, $R_{IS}$, and a cross-relaxation contribution, $\sigma_{IS}$, if both $I$ and $S$ nuclei are simultaneously excited.

$$R_{IS} = \frac{1}{10} \frac{\hbar^2 \gamma_I^2 \gamma_S^2}{r_{IS}^6} \left[ \frac{3\tau_c}{1 + \omega_I^2 \tau_c^2} + \frac{6\tau_c}{1 + (\omega_I + \omega_S)^2 \tau_c^2} + \frac{\tau_c}{1 + (\omega_I - \omega_S)^2 \tau_c^2} \right] \quad (1)$$

$$\sigma_{IS} = \frac{1}{10} \frac{\hbar^2 \gamma_I^2 \gamma_S^2}{r_{IS}^6} \left[ \frac{6\tau_c}{1 + (\omega_I + \omega_S)^2 \tau_c^2} - \frac{\tau_c}{1 + (\omega_I - \omega_S)^2 \tau_c^2} \right] \quad (2)$$

where $\hbar$ is the reduced Plank's constant, $\gamma_I$, $\gamma_S$, $\omega_I$, and $\omega_S$ are the magnetogyric and Larmor frequencies of nuclei $I$ and $S$, $r_{IS}$ is the internuclear distance, and $\tau_c$ is the effective correlation time which modulates the $I$–$S$ internuclear vector. From Eq. (1) and (2), it is apparent that $R_{IS}$ and $\sigma_{IS}$ can be directly interpreted in terms of molecular conformations (via $r_{IS}$) and motions (via $\tau_c$).[5]

In multispin systems, within the approximation of independent pairwise interactions, the spin–lattice relaxation rate of an excited nucleus $i$ interacting with neighboring $j$ nuclei, at their thermal equilibrium, can be described by Eq. (3):

$$R_i = \sum_{i \neq j} R_{ij} + R^* \quad (3)$$

[1] A. Abragam, "The Principles of Nuclear Magnetism." Oxford Univ. Press (Clarendon), London and New York, 1961.
[2] F. Bloch, *Phys. Rev.* **105**, 1206 (1957).
[3] A. G. Redfield, *IBM J. Res. Dev.* **1**, 19 (1957).
[4] J. H. Noggle and R. E. Schirmer, "The Nuclear Overhauser Effect." Academic Press, New York, 1971.
[5] I. Solomon, *Phys. Rev.* **99**, 559 (1955).

where each $R_{ij}$ is the contribution described by Eq. (1), and $R^*$ takes into account nondipolar relaxation processes. If the nuclear spin of $j$ is $M_j \neq \frac{1}{2}$, the right-handside of Eq. (1) must be divided by $m_j(m_j + 1)$. The modified equation has been used for the analysis of relaxation data of nitrogen-bound protons, where the $^{14}N-^1H$ interaction may yield a significant contribution to the observed relaxation of backbone amide protons of peptides and proteins.[6]

If the spin $i$ is excited together with one or more of the $j$ dipolar-coupled nuclei, then cross-relaxation occurs and the following relation holds,

$$R_i = \sum_{i \neq j} R_{ij} + R^* + \sum_{i \neq j} \sigma_{ij} \qquad (4)$$

where the $\sigma$ term is described by Eq. (2).

In nonselective spin–lattice relaxation experiments on proton nuclei, therefore, the measured relaxation rate is the sum of many terms related to all the direct and $i–j$ cross-relaxation contributions. Thus, only limited information on molecular conformation and dynamics can be obtained from nonselective proton relaxation data since several different internuclear distances and correlation times determine the experiment rate. No structural information can be obtained from $^{13}C$ nonselective relaxation of protonated carbons since the one-bond C–H interaction dominates the relaxation process, while a conformational analysis of the relaxation of quaternary carbons is complicated by the multiplicity of long-range C–H interactions.[7,8]

$^1H$ and $^{13}C$ nuclei can be considered to be limiting examples of complexity for dipolar relaxation analysis and, in both cases, the possibility of using spin–lattice relaxation rates for conformational investigations is based on the ability to dissect the complex relaxation process of a multispin system. In principle, internuclear distances can be measured for biomolecules in solution provided that (1) they have a single predominant structure or a limited number of conformers and (2) single $R_{ij}$ or $\sigma_{ij}$ can be experimentally measured. This can be achieved by selective spin–lattice relaxation techniques and by combining selective relaxation rates and nuclear Overhauser effects (NOE) as described below. It should be noted here that, in order to make structural use of relaxation spectroscopy in dealing with polypeptides, backbone nuclei such $^1H_\alpha$, $N^1H$, and $^{13}C{=}O$ of the peptide group, and side-chain nuclei of aromatic rings are most suit-

[6] H. E. Bleich, K. R. K. Easwaran, and J. A. Glasel, *J. Magn. Reson.* **31**, 517 (1978).
[7] A. Allerhand, D. Doddrell, and R. Komorski, *J. Chem. Phys.* **55**, 189 (1971).
[8] L. G. Werbelow and D. M. Grant, *J. Chem. Phys.* **63**, 544 (1975).

able ones since their spin–lattice relaxation is very sensitive to a magnetic environment which is dependent on conformational features.

The selective excitation of single protons simplifies the nuclear relaxation pathway and can be performed by a variety of techniques.[9] If the experimental conditions are such that the selective relaxation process of a nucleus involved in dipole–dipole interactions is described by the above reported equations, simple and reliable methods of calculation of relaxation parameters from selective relaxation measurements can be used, and structural and/or dynamic information can be obtained.

### Frequency-Selective Pulsed Excitation

As summarized in Table I,[10–12] a suitable excitation envelope can be adjusted (1) by adjusting the pulse amplitude and width, (2) by using long pulses, and (3) by using tailored "soft" pulses. Good selectivity of the excitation pulse is a preliminary requirement for the further analysis of relaxation data. The most suitable pulse technique will depend on the specific characteristic of the spectrometer and the system to be investigated. Because weaker pulses with longer duration provide greater selectivity, two main problems arise in principle if spin–lattice and spin–spin relaxation processes during the excitation are no longer negligible: (1) complete inversion by an initial 180° selective pulse may not be possible and (2) the recovery curves obtained using the nonselective 90° monitoring pulse may be distorted. The effects of incomplete inversion can be accounted for by normalizing the recovery curves to include an initial pulse of flip angle less than 180°. Those due to relaxation during the pulse can be safely ignored for relaxation times 20–50 times longer than the excitation pulse. In molecular systems in which closely spaced resonances and short relaxation times are present, the DANTE excitation technique or the one-dimensional version of the NOESY experiment[13] should be used.

[9] C. J. Turner, *Prog. NMR Spectrosc.* **16**, 311 (1984).
[10] R. Freeman, H. D. W. Hill, B. L. Tomlinson, and L. D. Hall, *J. Chem. Phys.* **61**, 4466 (1974).
[11] T. R. Brown and S. Ogawa, *Proc. Natl. Acad. Sci. U.S.A.* **74**, 3627 (1977).
[12] G. A. Morris and R. Freeman, *J. Magn. Reson.* **29**, 433 (1978).
[13] G. Robinson, P. W. Kuchel, B. E. Chapman, D. M. Doddrell, and M. G. Irving, *J. Magn. Reson.* **63**, 314 (1985).

TABLE I

PULSE SEQUENCES USED FOR SELECTIVE PROTON INVERSION

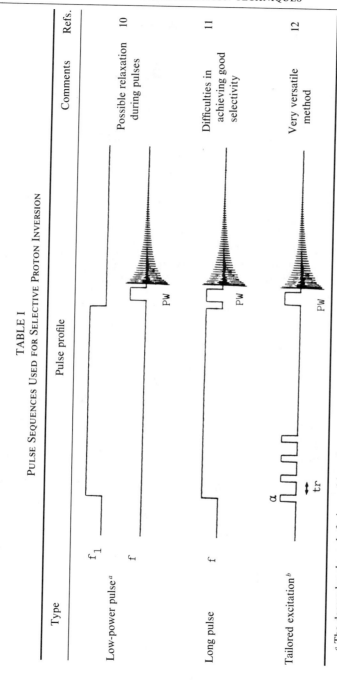

| Type | Pulse profile | Comments | Refs. |
|---|---|---|---|
| Low-power pulse[a] | $f_1$ / $f$ PW | Possible relaxation during pulses | 10 |
| Long pulse | $f$ PW | Difficulties in achieving good selectivity | 11 |
| Tailored excitation[b] | $\alpha$ / $t_r$ PW | Very versatile method | 12 |

[a] The decoupler channel, $f_1$, is used in order to generate the selective pulse.
[b] The selective pulse consists of a train of short pulses with a duration $\alpha$, separated by short intervals $t_r$. The interval $t_r$ is calculated from the distance of the carrier frequency $\nu_0$ from the signal to be selectively excited. $t_r = 1/\Delta\nu$.

Spin–Lattice Relaxation of Selectively Excited Nuclei

## $^1H$ Relaxation

In proton spin systems, the nonselective inversion recovery, $180°$–$\tau$–$90°$ pulse sequence gives partially relaxed spectra where it is possible to follow the evolution of peak intensities determined by the spin–lattice relaxation of the individual $^1H$ nuclei of the investigated molecule. In this case, all the $i$–$j$ dipole–dipole interactions contribute to the overall relaxation process with direct and cross-relaxation terms, and the spin–lattice relaxation is described by Eq. (4). It has been shown by Freeman and co-workers[10] that the cross-relaxation can be negligible provided that the $180°$ pulse selectively excites single-proton resonances. In particular, if the initial relaxation is monitored after the selective population inversion of proton $i$, i.e., short delays are used between the selective and the nonselective observing $90°$ pulse, the initial recovery rate of the magnetization is described by Eq. (3). At longer delay times, cross-relaxation becomes significant, as can be observed from the intensity changes of the resonances of protons dipolarly coupled to the excited proton. It follows that the importance of selective excitation techniques in nuclear relaxation spectroscopy is 2-fold: simplification of the proton relaxation process may be achieved and selective dipole–dipole interactions can be observed.[10,14,15]

The initial selective relaxation rate $R_i(SE)$ is a fundamental parameter which can be used for obtaining information on (1) relaxation mechanisms, (2) molecular dynamics, and (3) molecular conformations.

1. In molecular systems which satisfy the extreme narrowing condition $(\omega_0\tau_c)^2 \ll 1$, for proton $i$, the ratio between the nonselective relaxation, $R_i(NS)$, and $R_i(SE)$ is diagnostic of the efficiency of the dipole–dipole relaxation mechanism since this ratio, $F_i$, is equal to 1.5 for a relaxation pathway completely dominated by the $i$–$j$ proton dipolar interactions.

2. As shown in Fig. 1, $F_i$ also depends on $\omega_0\tau_c$ and, hence, for rigid isotropically reorienting molecules, motions can be quantified from the latter parameter.[16]

3. The combined use of $R_i(SE)$ and homonuclear Overhauser effects,

[14] L. D. Hall and H. D. W. Hill, *J. Am. Chem. Soc.* **98**, 1269 (1976).
[15] N. Niccolai, M. P. Miles, S. P. Hehir, and W. A. Gibbons, *J. Am. Chem. Soc.* **100**, 6528 (1978).
[16] N. Niccolai, M. P. Miles, and W. A. Gibbons, *Biochem. Biophys. Res. Commun.* **91**, 157 (1979).

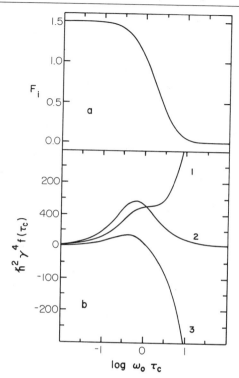

FIG. 1. (a) Graph of $F = R_i(NS)/R_i(i)$ versus $\log \omega_0\tau_c$. (b) Graph of $R_i(1)$, $R_i(ij)$ (2), and $\sigma_{ij}$ (3) for a two-spin system; (1), (2), and (3) are readily generalized for proton $i$, involved in multiple interactions with several protons, $i \neq j$.

$\mathrm{NOE}_i(j)$, yields an absolute evaluation of $\sigma_{ij}$ rates since, neglecting cross-polarization effects, the following relation holds[17]

$$\mathrm{NOE}_i(j) = \sigma_{ij}/R_i(\mathrm{SE}) \tag{5}$$

where $\mathrm{NOE}_i(j)$ is the Overhauser effect observed on proton $i$ upon selective excitation of proton $j$.

The calculation of the cross-relaxation rate relative to a single $i$–$j$ dipolar interaction is also possible by the simultaneous double excitation of protons $i$ and $j$. If selectivity in the double excitation can be technically achieved (see Fig. 2), the relaxation process is driven by the biselective

[17] C. R. Jones, C. T. Sikakana, S. Hehir, M. C. Kuo, and W. A. Gibbons, *Biophys. J.* **24**, 815 (1978).

relaxation rate, $R_{i,j}(BS)$, where

$$R_{i,j}(BS) = R_{i,j}(SE) + \sigma_{ij} \qquad (6)$$

It should be noted that the initial conditions for the evaluation of $R_{i,j}(BS)$'s are critically important because the peak intensities observed in the partially relaxed spectra are determined by the relaxation process and the time-dependent $i-j$ NOE.

On the basis of Eq. (2), the $i-j$ internuclear distance can be calculated from the $\sigma_{ij}$ parameter if the correlation time which modulates the $i-j$ interaction is known (see Fig. 3). It is evident that the choice of a suitable effective $\tau_c$ is of primary importance in the structural analysis of cross-relaxation rates, and different methods are available, as discussed in the section Choice of Effective Correlation Time. In the case that internal motions are present within the molecular backbone or the overall reorien-

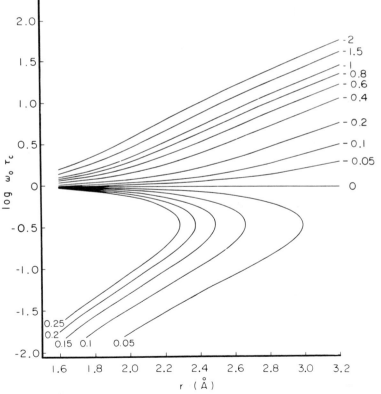

FIG. 2. Isosigma contours for $\log \omega_0 \tau_c$ vs interproton distances calculated from Eq. (2); the selected $\sigma$ values are expressed in $sec^{-1}$.

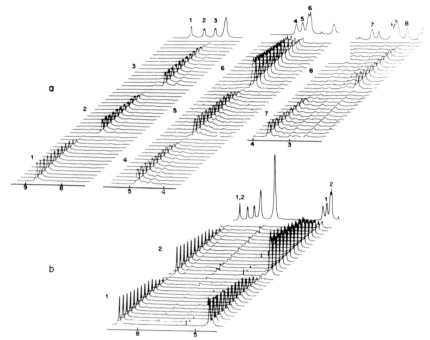

FIG. 3. (a) Monoselective and (b) biselective excitation partially relaxed spectra recorded in the ($M_0$–$M_z$) mode of gramicidin S (10 m$M$) in DMSO-$d_6$ at 26°. In (a), selective inversion recovery experiments are shown for Phe NH (1), Orn NH (2), Leu NH (3), Orn H (4), Leu H (5), Phe H (6), Pro $H_1$ (7), and Pro $H_2$ (8); in (b), biselective experiments are shown for Phe NH–Leu H (1) and Phe NH–Phe H (2).

tation is not fully isotropic, the accurate evaluation of the correlation time which is effective on a particular $i$–$j$ vector is not straightforward.[18] However, the sixth power dependence of the experimental relaxation parameters on $r_{ij}$ makes the accuracy of the $\tau_c$ to be used in the calculation of internuclear distances less critical.

## $^{13}C$ Relaxation

Spin–lattice relaxation rates of $^{13}C$ nuclei have been extensively used for obtaining information on the structure and dynamics of biopolymers in solution.[19] These nuclei have a low natural abundance and hence their

[18] D. E. Woessner, *J. Chem. Phys.* **42**, 1855 (1965).
[19] G. C. Levy, ed., "Topics in Carbon-13 NMR Spectroscopy," Vol. 2. Wiley, New York, 1976.

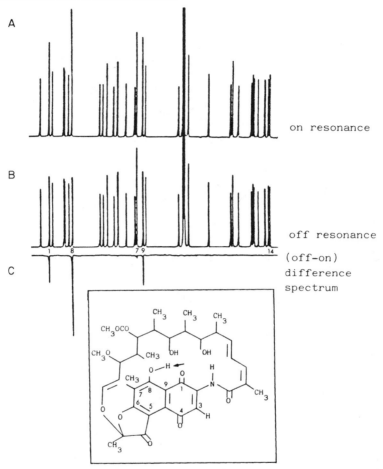

A

on resonance

B

off resonance

(off-on)
C                                        difference
spectrum

FIG. 4. $^{13}C$ spectra of $0.5$ mol·$dm^{-3}$ rifamycin in $CDCl_3$ recorded on an XL-200 Varian spectrometer at 23°: (A) $^{13}C$ spectrum after irradiation of the phenolic hydroxyl proton; (B) as (A) but with the decoupler set "off-resonance." (C) Difference spectrum (B − A). The presaturation selective pulse on the phenolic hydroxyl proton had a 10-sec duration and 0.5-W power. During the acquisition of the FID's the decoupler was on with a power of 4 W in the broad-band mode. The structure of rifamycin S is shown.

relaxation measurements are often limited by the solute concentration. In spite of this sensitivity problem, natural abundance $^{13}C$ relaxation studies present many advantages over proton studies. The large dispersion of carbon chemical shifts makes it more practical to determine the relaxation of individual nuclei, and the removal of the multiplet structure due to continuous proton broad-band decoupling yields relaxation recoveries which approach single exponential behavior. Cross-relaxation among the carbon nuclei is negligible, due to their natural abundance, and Eq. (3), previously used for analyzing the $^1$H selective relaxation, also describes the nonselective $^{13}$C relaxation rates. It follows that the latter relaxation parameter may give useful information on relaxation mechanisms, molecular dynamics, and structures from a combined analysis with other $^{13}$C relaxation data such as selective $\{^1H\}^{13}$C-NOE's (see Fig. 4),[20] biselective,[21] $R_C(H_i)$, and polyselective relaxation rates, $R_C(H_{BB})$.[22]

The experimental procedures for measuring these three parameters are shown in Table II. Thus, proton–carbon cross-relaxation rates, $\sigma_{CH}$, can be obtained from Eq. (7) and experimental parameters:

$$\sigma_{CH} = \{^1H\}^{13}C\text{-NOE}_{CH} \times R_c \tag{7}$$

Then the proton–carbon internuclear distance $r_{CH}$ can be calculated provided that $\tau_c$ is independently known and Eq. (2) is used (see Fig. 5). The same heteronuclear cross-relaxation rate can be calculated from the difference $[R_C(H_i) - R_C]$, while the effectiveness of the intramolecular $^1$H–$^{13}$C dipolar interactions and the molecular motion can be probed from the $F_C$ parameter, defined as

$$F_C = R_C(H_{BB})/R_C \tag{8}$$

and $F_C$ is equal to 2.99 if carbon nuclei have a relaxation pathway dominated by the interactions with neighboring protons and if the molecular reorientations are fast in the NMR time scale.

## Choice of Effective Correlation Time

Assumptions such that the geminal H–H interaction in CH$_2$ and CH$_3$ groups is predominant for the nonselective relaxation of methyl and methene protons allow the calculation of $\tau_c$ for these interactions if the geminal distance is known and independent of conformation. Calibration interproton distances are available in backbone and side-chain moieties of a polypeptide, and internuclear vectors such as H$_\alpha$–H$_\alpha$ of glycine, H$_\varepsilon$–H$_\delta$ of tyrosine, and H$_\delta$–H$_\delta$ of proline, and calculated cross-relaxation rates, have been used for the evaluation of $\tau_c$.

The correlation times of C–H vectors can be calculated from nonselective relaxation measurements of $^{13}$C-protonated nuclei. From the latter

[20] N. Niccolai, C. Rossi, V. Brizzi, and W. A. Gibbons, *J. Am. Chem. Soc.* **106,** 5732 (1984).
[21] C. Rossi, *J. Chem. Phys.* **84,** 6581 (1986).
[22] C. Rossi, N. Marchettini, L. Pogliani, F. Laschi, and N. Niccolai, *Chem. Phys. Lett.* **136,** 506 (1987).

TABLE II

PULSE SEQUENCES USED FOR SELECTIVE PROTON–CARBON EXPERIMENTS

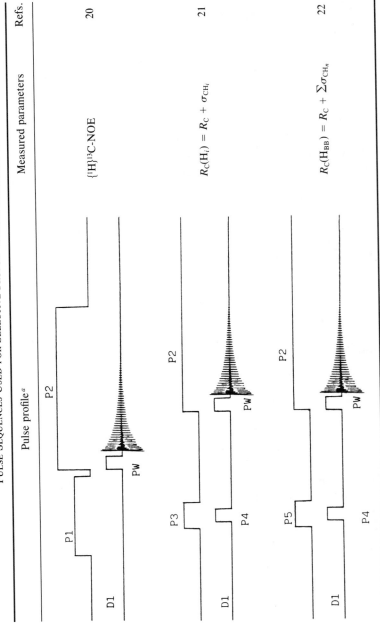

| Pulse profile[a] | Measured parameters | Refs. |
|---|---|---|
| | $\{^1\mathrm{H}\}^{13}\mathrm{C}\text{-NOE}$ | 20 |
| | $R_\mathrm{C}(\mathrm{H}_i) = R_\mathrm{C} + \sigma_{\mathrm{CH}_i}$ | 21 |
| | $R_\mathrm{C}(\mathrm{H_{BB}}) = R_\mathrm{C} + \sum \sigma_{\mathrm{CH}_n}$ | 22 |

[a] P1, Selective proton pulse; P2, broad-band proton decoupling; P3, selective 180° proton pulse; P4, nonselective 180° carbon pulse; P5, nonselective 180° proton pulse; PW, 90° carbon pulse; D1, relaxation delay.

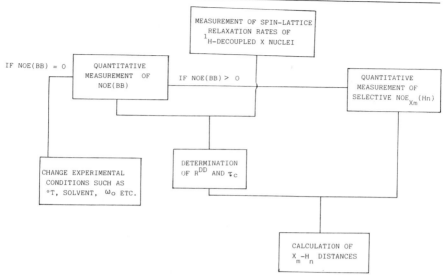

FIG. 5. The strategy for obtaining information on molecular structure and dynamics from combined analysis of selective NOE's and spin–lattice relaxation rates.

relaxation rates, backbone and side-chain dynamics can be monitored, since the one-bond C–H dipolar interaction dominates the relaxation pathway of carbon nuclei.[6,7] $F_i$ and $F_C$ can be used as probes of molecular motion, if the intramolecular dipolar relaxation mechanism is assumed to be dominant. Geminal proton–carbon dipolar connectivities can also be interpreted in terms of the dynamics of individual molecular sites. All these independent 1D relaxation techniques for delineating Brownian motion and internal flexibilities can offer a reliable basis for the structural analysis of dipolar interactions.

## Analysis of Selective Relaxation Data

Following a single or multiple selective excitation, peak intensities from partially relaxed spectra can be measured; initial rates can be obtained provided that the experiment is correctly carried out and theoretical problems do not quench the dipolar relaxation strategy. Two important aspects still have to be considered: the role of the scalar coupling between spin $i$ and spin $j$, and spin-diffusional mechanisms of nuclear magnetization transfer within the biopolymer.

The scalar coupling effects on nuclear relaxation have been thoroughly investigated.[23] For each particular scalar coupled system, equa-

[23] R. L. Vold and R. R. Vold, *Prog. NMR Spectrosc.* **19**, 79 (1978).

tions have been derived for the relaxation process. It can be concluded that strongly coupled spin systems are not suitable for selective relaxation approaches because of technical and theoretical complexities. The equations reported in the previous sections hold if the chemical shift difference between nuclei $i$ and $j$, $\Delta$, is much larger than their scalar coupling constant $J$; $\Delta/j > 10$ is a reasonable safety condition[23] generally met for the backbone nuclei with which we are concerned.

The size, shape, and aggregational state in solution determine the dynamics of the molecule under selective relaxation analysis. If molecular motions are such that the $\omega_0\tau_c \gg 1$ limit holds, then spin diffusion might be effective and this aspect should be taken into account in the quantification of molecular structures and motions.[24]

The evolution of $M_{zi}$, the magnetization of the $i$ nucleus dipolar coupled to a group of other $j$ spins, follows Eq. (9).

$$d\langle M_{zi}\rangle/dt = -R_i(\langle M_{zi}\rangle - M_{0i}) - \sum_{i\neq j} \sigma_{ij}(\langle M_{zi}\rangle - M_{zj}) \tag{9}$$

where $R_i$ values are the direct relaxation rates of Eq. (3). The quantities of Eq. (9) have simple experimental interpretations since $\langle M_{zi}\rangle$ is proportional to the total integrated intensity of the NMR signal of spin $i$ and $M_{0i}$ is the value of $\langle M_{zi}\rangle$ at thermal equilibrium.

Calculations of the relaxation rates are then performed by measuring the peak intensity of the excited nucleus $i$, $I_{zi}$, in the selective partially relaxed spectra obtained at different delay times between the selective 180° pulse and the nonselective 90° pulse.

An excellent review is available for the analysis of selective relaxation data[25] and only general remarks will be made here:

1. The recovery curve of a spin $i$ may follow single exponential behavior as in the case of $^{13}C$ relaxation measured under conditions of continuous broad-band proton decoupling. The same behavior is expected for single selective $^1H$ relaxation, provided that the influence of cross-relaxation during the experiment is negligible or considered by measuring initial rates in the semilog plot of the reduced intensities $[(I_{0i} - I_{zi})/2I_{0i}]$ vs delay times. In both cases, the experimentally derived $R_i$ values can be used with good confidence for the study of molecular conformation and dynamics.

2. After a multiple selective 180° pulse or when the cross-relaxation is as fast as the direct relaxation (spin-diffusion limit) the analysis of relaxation data is more complex and multiexponential fitting to the experimental

24 A. Kalk and J. J. C. Berendsen, *J. Magn. Reson.* **24**, 343 (1976).
25 I. D. Campbell, C. M. Dobson, R. G. Ratcliffe, and R. J. P. Williams, *J. Magn. Reson.* **29**, 397 (1978).

relaxation curve must be performed. This complex procedure can yield misleading results if the curve fitting is complicated by the scattering of experimental data.

It should also be noted that whenever the initial rate estimate becomes critically important, as in the latter cases, the effects of imperfect 180° pulses are more significant because of residual transverse magnetization, if peak intensities are sampled at very short delay times between the selective and nonselective pulses.

### Applications of Selective Relaxation Techniques

#### Homonuclear Intramolecular Dipolar Connectivities

Many papers have been published in which selective excitation techniques have been used in the evaluation of cross-relaxation rates for obtaining internuclear distances and correlation times in amino acids,[15] peptides,[16] natural products,[26] and carbohydrates.[11] Calibration distances have been widely used for calculating molecular motion in systems with restricted internal flexibility or the latter have been considered on the basis of rotamer population analysis.[27]

For the above reasons, time-dependent NOE's and the combined use of proton–proton Overhauser enhancements and single selective relaxation rates seems to yield the most reproducible and accurate distance measurements. In the era of two-dimensional NMR, these spin–lattice relaxation methods have diminished in importance but, within the limits of experimental conditions, are still unmatchable for quantitative structural determinations.

#### Heteronuclear Intramolecular Dipolar Connectivities

Many studies on proton–carbon dipolar couplings have been performed with 1D and 2D relaxation techniques, as recently reviewed.[28] The relaxation behavior of quaternary carbons in natural products,[29] peptides,[30] and other organic molecules has been shown to be a powerful tool for defining proton microenvironments. The combined use of heteronu-

[26] N. Niccolai, A. K. Schnoes, and W. A. Gibbons, *J. Am. Chem. Soc.* **102**, 1513 (1980).
[27] C. Rossi, L. Pogliani, F. Laschi, and N. Niccolai, *J. Chem. Soc., Faraday Trans. 1* **79**, 2955 (1983).
[28] K. E. Kövér and G. Batta, *Prog. NMR Spectrosc.* **19**, 223 (1987).
[29] N. Niccolai, C. Rossi, P. Mascagni, W. A. Gibbons, and V. Brizzi, *J. Chem. Soc., Perkin Trans. 2* p. 239 (1985).
[30] N. Niccolai, C. Rossi, P. Mascagni, P. Neri, and W. A. Gibbons, *Biochem. Biophys. Res. Commun.* **124**, 739 (1984).

clear NOE's and $^{13}C$ spin–lattice relaxation has been proposed for the simultaneous determination of acceptor–donor groups of hydrogen bonds.[20] It has also been observed that the proton–carbon dipolar interactions which depend on molecular conformation are generally very weak and not detectable via conventional heteronuclear NOESY spectra.[31] An alternative approach for measuring proton–carbon dipolar interaction is offered by the combined use of selective and nonselective $^{13}C$ spin–lattice relaxation rates.[22]

### Intermolecular Solute–Solvent Dipolar Interactions

Selective excitation of solvent molecules in the presence of proteins and cellular systems has also been used to investigate the magnetic interactions of water protons with those present on the surface of macrosystems. From a comparison of the selective and nonselective spin–lattice relaxation rates of water protons in solutions of bovine serum albumin, cross-relaxation has been found between solvent and protein nuclei.[32] This experiment confirmed that the generally observed water proton relaxation rate enhancement due to the presence of macromolecules in solution is derived not only from dynamic changes in solvent molecules but also from dipolar intermolecular connectivities. Similarly, the selective relaxation behavior of water protons in the presence of red blood cells was studied at different temperatures and isotopic H/D dilutions.[33] A strong dipolar coupling between solvent and surface exchangeable protons was observed. It was suggested that solvent selective excitation techniques are more suitable for detecting proton relaxation enhancements in the presence of macrosystems than nonselective spin–lattice relaxation measurements.

### Intermolecular Solute–Solute Dipolar Interactions

The formation of intermolecular adducts can be followed by selective relaxation measurements. It has been shown that the ligand–receptor interaction induces large changes in the direct relaxation contributions of the proton signals of the ligand. The reduced mobility of the bound ligand is reflected in the experimental relaxation parameter $R(SE)$, provided that fast exchange conditions hold, so that

$$R_i(SE)_{exp} = p_f R_i(SE)_f + p_b R(SE)_b \qquad (10)$$

[31] N. Niccolai, A. Prugnola, P. Mascagni, C. Rossi, L. Pogliani, and W. A. Gibbons, *Spectrosc. Lett.* **20,** 307 (1987).
[32] G. Valensin and N. Niccolai, *Chem. Phys. Lett.* **79,** 47 (1981).
[33] N. Niccolai, L. Pogliani, and C. Rossi, *Chem. Phys. Lett.* **110,** 294 (1984).

where $f$ and $b$ subscripts refer to the free and bound environments and $p$'s are the fractions of ligand molecules in the two states.[34] Selective relaxation rates may therefore be used to calculate association constants for this kind of interaction.[35]

Intermolecular proton–carbon dipolar couplings were also detected by using selective heteronuclear techniques in model systems.[36]

[34] G. Valensin, T. Kushnir, and G. Navon, *J. Magn. Reson.* **46**, 23 (1982).
[35] I. Barni-Comparini, E. Gaggelli, N. Marchettini, and G. Valensin, *Biophys. J.* **48**, 247 (1985).
[36] C. Rossi, N. Niccolai, and F. Laschi, *J. Phys. Chem.* **91**, 3903 (1987).

# [11] Rotating-Frame Nuclear Overhauser Effect

*By* L. R. BROWN and B. T. FARMER II

## Introduction

Cross-relaxation in the laboratory frame has become a very popular method for investigating interatomic distances in macromolecules and, hence, for obtaining structural and conformational information.[1,2] One reason for the popularity of this method is the availability of the two-dimensional (2D) NMR experiment, NOESY, which allows all spatially proximal pairs of hydrogen atoms to be detected simultaneously. The pulse sequence for this experiment (Fig. 1A) creates spin magnetizations which, at the outset of the mixing time $\tau_m$, are aligned parallel with the external, static magnetic field (Fig. 1A). Because the spins have different oscillation frequencies during the evolution period $t_1$, the spin magnetizations are not in thermal equilibrium at the beginning of $\tau_m$ and net magnetization transfer by cross-relaxation can take place between pairs of spatially proximal spins. The consequence is that a single laboratory-frame 2D cross-relaxation (NOESY) spectrum contains a detailed description of all the short distances in a molecular conformation.

It is also possible to measure cross-relaxation rates in the rotating frame. This experiment was originally proposed by Bothner-By *et al.*[3] and

[1] J. H. Noggle and R. E. Schirmer, "The Nuclear Overhauser Effect." Academic Press, New York, 1971.
[2] K. Wüthrich, "NMR of Proteins and Nucleic Acids." Wiley, New York, 1986.
[3] A. A. Bothner-By, R. L. Stephens, J. Lee, C. D. Warren, and R. W. Jeanloz, *J. Am. Chem. Soc.* **106**, 811 (1984).

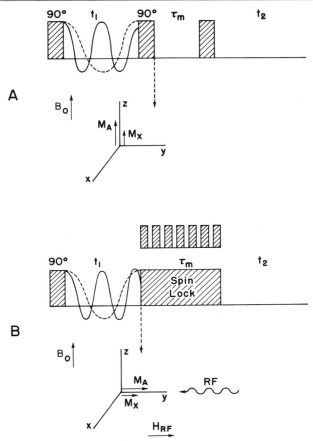

FIG. 1. Experimental schemes for measurement of two-dimensional cross-relaxation NMR spectra. (A) Laboratory-frame cross-relaxation. The second 90° pulse creates longitudinal magnetization of unequal magnitude for spins $A$ and $X$, which cross-relax during the fixed mixing time $\tau_m$ and are detected during $t_2$ following the third 90° pulse. (B) Rotating-frame cross-relaxation. At the end of $t_1$ the spin-lock rf field "locks" the $y$ component of the transverse magnetization along the $y$ axis of the rotating frame. Spins $A$ and $X$ cross-relax during the mixing period $\tau_m$ for which the spin-lock rf field is applied and are then detected during $t_2$. The spin-locking rf field may consist of either continuous wave irradiation or pulsed irradiation.

termed CAMELSPIN, but was subsequently named ROESY,[4] i.e., rotating-frame Overhauser effect spectroscopy. In the two-dimensional version of ROESY (Fig. 1B), at the end of the evolution period $t_1$ a strong, on-resonance radio frequency (rf) field is applied to the spins in a direction

[4] A. Bax and D. G. Davis, *J. Magn. Reson.* **63,** 207 (1985).

$90°$ to the external magnetic field $B_0$. This causes the spin magnetization to become "spin locked" parallel to the rf magnetic field (Fig. 1B). Because of their different oscillation frequencies during $t_1$, different spins have different amounts of magnetization aligned with the rf field. Net transfer of magnetization can then take place between spatially proximal spins, but with the spin magnetization perpendicular to the external, static magnetic field. As with NOESY, the two-dimensional ROESY experiment allows all pairs of spatially proximal spins to be detected simultaneously.

In ROESY experiments, cross-relaxation takes place perpendicular to the external, static magnetic field and is therefore dependent on spin–spin relaxation processes.[5,6] In contrast, in NOESY experiments cross-relaxation takes place parallel to the external, static magnetic field and is dependent on spin–lattice relaxation processes. Consequently, cross-relaxation in the rotating frame has a different dependence on molecular motion than cross-relaxation in the laboratory frame.[7] This results in three characteristics which make cross-relaxation in the rotating frame an attractive complement to, or even a replacement for, measurements of cross-relaxation in the laboratory frame.

1. The cross-relaxation rate in the rotating frame, $\sigma^r$, is always positive. In contrast, the laboratory-frame cross-relaxation rate, $\sigma^n$, becomes zero for rigid body isotropic motion such that $\omega_0 \tau_c = \sqrt{5}/2$, where $\omega_0$ is the Larmor frequency and $\tau_c$ is the correlation time for isotropic molecular tumbling. This means that measurement of cross-relaxation rates in the rotating frame is particularly attractive for moderately sized molecules where $\sigma^n \approx 0$ and the NOESY experiment therefore shows low sensitivity.

2. For isotropic, rigid body motion in the slow motional regime, i.e., $\omega_0 \tau_c \gg 1$, cross-relaxation in the laboratory frame is an energy-conserving process[8] and this leads to substantial spin diffusion. In contrast, for slow molecular motion cross-relaxation in the rotating frame is tempered by dipolar relaxation,[7] thereby attenuating spin diffusion. This means that more accurate measurements of cross-relaxation rates, and hence better descriptions of molecular structure, may be obtainable from measurements in the rotating frame.

3. The different dependence of $\sigma^n$ and $\sigma^r$ on molecular motion makes the measurement of both cross-relaxation rates an attractive method to study molecular motion. In most previous studies of molecular structure

[5] I. Solomon, *Phys. Rev.* **99**, 559 (1955).
[6] G. P. Jones, *Phys. Rev.* **148**, 332 (1966).
[7] B. T. Farmer II, S. Macura, and L. R. Brown, *J. Magn. Reson.* **80**, 1 (1988).
[8] S. Macura and R. R. Ernst, *Mol. Phys.* **41**, 95 (1980).

using $\sigma^n$, it has implicitly been assumed that rigid body, isotropic motion occurs. If both $\sigma^n$ and $\sigma^r$ are measured, it is possible to detect when this assumption is not justified and, therefore, when extreme care is warranted for interpretation of cross-relaxation rates in terms of interatomic distances. In favorable cases, determination of both $\sigma^n$ and $\sigma^r$ may allow the nature of the molecular motion to be determined.

To take advantage of the above characteristics of rotating-frame cross-relaxation, it is necessary to obtain quantitative measurements and interpretations of rotating-frame cross-relaxation rates. There are several aspects to this. (1) Transfer of magnetization in the rotating frame by mechanisms other than cross-relaxation must be accounted for or suppressed. (2) To obtain quantitative measurements of $\sigma^r$, one must have the necessary instrumental capabilities and appropriate procedures for extraction of reliable cross-relaxation rates from NMR spectra. (3) The resulting cross-relaxation rates must be quantitatively interpreted in terms of molecular parameters, including both interatomic distances and molecular motion.

In the following sections the three situations, in which measurements of $\sigma^r$ offer special advantages, are first discussed and then the experimental requirements are considered.

### Small and Moderately Sized Molecules

For rigid body, isotropic motion, Fig. 2 shows the relative sensitivity, $\xi$, of NOESY compared to ROESY. For small molecules with $\tau_c \gtrsim 5 \times 10^{-10}$ the two experiments show equal sensitivity and the same cross-relaxation rate ($\sigma^n = \sigma^r$). It is well known that in the fast motion regimen NOESY is not an effective experiment and the same will be true of ROESY. For rapid molecular tumbling, a steady-state NOE measurement will generally be the most effective experiment. For medium-sized molecules such that $\omega_0\tau_c \approx \sqrt{5}/2$, i.e., $\sigma^n \approx 0$, ROESY is much more sensitive than NOESY and measurements of $\sigma^r$ are the method of choice.

An experimental example of this is shown for polymyxin B (Fig. 3) in Fig. 4. In both ROESY (Fig. 4A) and NOESY (Fig. 4B), cross-peaks are observed between the $\delta$ hydrogens of the aromatic ring of Phe-6 and the $\alpha$, $\beta$, and $\beta'$ hydrogens of the same residue. Note that in the ROESY spectrum the cross-peaks have the opposite sign of the diagonal peak. This will always be true for ROESY cross-relaxation cross-peaks since $\sigma^r$ is always positive. In contrast, the NOESY cross-peaks show the same sign as the diagonal peak. This indicates that $\sigma^n$ is negative, i.e., $\omega_0\tau_c > \sqrt{5}/2$. As expected for a molecule of the size of polymyxin B, ROESY shows about a 3-fold better signal-to-noise ratio than NOESY.

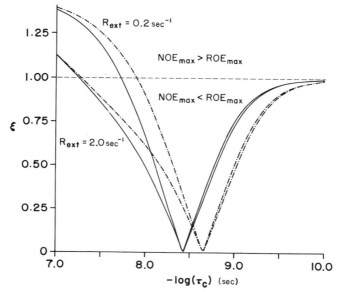

FIG. 2. The relative sensitivity, $\xi$, of NOESY compared to ROESY for an $AX$ spin system undergoing rigid body, isotropic motion with the correlation time, $\tau_c$, for molecular rotation. $\xi$ is defined as $|a_c^n(\tau_{opt}^n)/a_c^r(\tau_{opt}^r)|$ where $a_c^{n,\,r}(\tau_{opt}^{n,\,r})$ are the maximum cross-peak intensities in NOESY and ROESY. Curves are shown for the external relaxation rates $R_{ext} = 0.2$ and 2.0 sec$^{-1}$ at Larmor frequencies of 300 MHz (—) and 500 MHz (-·-). The dashed line at $\xi = 1$ corresponds to equal sensitivity of NOESY and ROESY. (From Farmer *et al.*[7])

The cross-relaxation rates in the laboratory frame and the rotating frame are both simply related to the interatomic distance, $r_{ij}$, between two hydrogen atoms $i$ and $j$:

$$\sigma_{ij}^{n,\,r} = q_{ij}f^{n,\,r}(\omega_0\tau_c)r_{ij}^{-6} \tag{1}$$

where $q_{ij} = 0.1(h\gamma_i\gamma_j/2\pi)^2(\mu_0/4\pi)^2 = 5.688 \times 10^{10}$ sec$^{-2}$ Å$^6$ is a constant and both

$$f^n(\omega_0\tau_c) = \{[6\tau_c/(1 + 4\omega_0^2\tau_c^2)] - \tau_c\} \tag{2}$$

and

$$f^r(\omega_0\tau_c) = \{[3\tau_c/(1 + \omega_0^2\tau_c^2)] + 2\tau_c\} \tag{3}$$

are simple functions of the Larmor frequency, $\omega_0$, and the correlation time for isotropic molecular tumbling, $\tau_c$. If $\tau_c$ is known, Eq. (1) can be

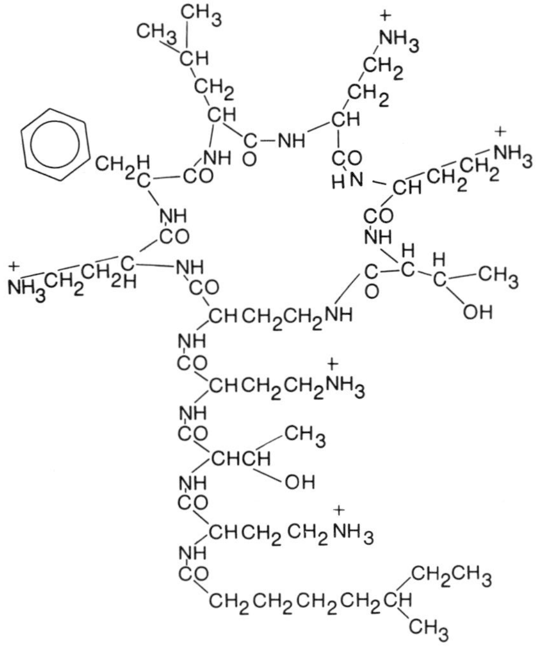

FIG. 3. The structure of polymyxin B.

used to obtain absolute interatomic distances from either $\sigma^n$ or $\sigma^r$, subject to the assumptions discussed by Borgias and James ([9], this volume). If $\tau_c$ is not known, relative interatomic distances can be obtained from

$$\sigma_{ij}^{n,\,r}/\sigma_{kl}^{n,\,r} = (r_{kl}/r_{ij})^6 \tag{4}$$

Equation (4) can in turn be used to obtain unknown distances if $\sigma^n$ or $\sigma^r$ can be measured for one interatomic distance which is known, e.g., the distance between two hydrogens of a methylene group.

Equation (1) has been widely used to obtain interatomic distances from measurements of $\sigma^n$. However, it must be remembered that this equation is strictly valid only for rigid body, isotropic motion and that if such motion does not prevail, fallacious interatomic distances may be obtained. In the case of polymyxin B, it is quite clear that the assumption of rigid body, isotropic motion is not justifiable. As shown in Fig. 4C, in the NOESY experiment the cross-peak between the $\alpha$ and $\beta$ hydrogens of the fatty acid tail shows the opposite sign to the diagonal peak, implying that $\omega_0 \tau_c < \sqrt{5}/2$ and $\sigma^n > 0$. This is inconsistent with the cross-peaks for

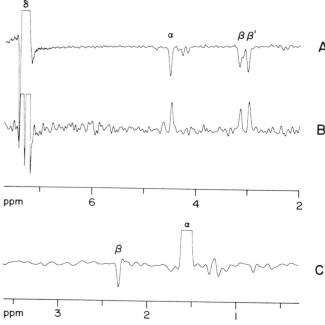

FIG. 4. Cross-sections through two-dimensional cross-relaxation NMR spectra for polymyxin B in $H_2O$. (A) ROESY. Cross-peaks are shown between the $\delta$ hydrogens of the aromatic ring of phenylalanine-6 and the $\alpha$, $\beta$, and $\beta'$ hydrogens of the same residue. (B) NOESY. The same cross-peaks as in (A) are shown. (C) NOESY. Cross-peaks are shown between the $\alpha$ and $\beta$ hydrogens of the 6-methyloctanoic acid.

Phe-6 and is clear evidence that rigid body, isotropic motion does not pertain to this molecule. Because $\sigma^r$ is always positive and all cross-peaks are consequently negative, this clear evidence for differential mobility could not be observed in the ROESY spectrum.

Both $\sigma^n$ and $\sigma^r$ are highly sensitive to internal and/or anisotropic motion for overall tumbling times which are near the Larmor frequency, i.e., for medium-sized molecules. This is illustrated in Fig. 5 for overall tumbling times such that $\omega_0\tau_c = 0.5$ and $2.0$ and internal rotation characterized by free diffusion with a diffusion constant, $D_{internal}$, such that $D_{internal}/D_{global} = 3$ or $9$. Depending on the angle $\phi$ between the axis for internal rotation and the interatomic vector between the two spins, both $\sigma^n$ and $\sigma^r$ can be changed manyfold. This means that completely fallacious estimates of interatomic distances can be obtained despite the $r_{ij}^{-6}$ dependence of $\sigma^n$ and $\sigma^r$. Similar complications can be caused by anisotropic motion of a rigid body. The effect of internal and/or anisotropic

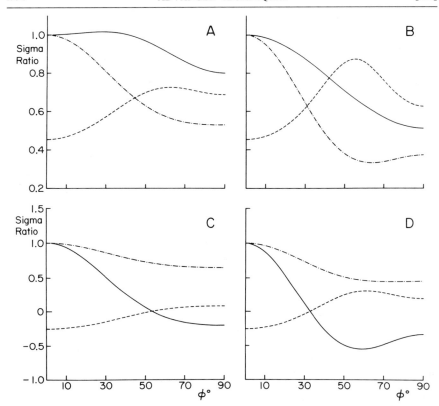

FIG. 5. The ratios $\sigma^n/\sigma^n_{iso}$ (—), $\sigma^r/\sigma^r_{iso}$ (-·-), and $\sigma^n/\sigma^r$ (---) for an $AX$ spin system undergoing isotropic global motion with diffusion constant $D_{global}$ and free internal rotation with diffusion constant $D_{internal}$ as a function of the angle $\phi$ between the axis of internal rotation and the $AX$ interatomic vector. $\sigma^{n,r}_{iso}$ denotes the cross-relaxation rate for isotropic global motion with diffusion constant $D_{global}$ and no internal motion. (A) $D_{global}/D_{internal} = 1/3$. (B) $D_{global}/D_{internal} = 1/9$. (C) $D_{global}/D_{internal} = 1/3$. (D) $D_{global}/D_{internal} = 1/9$. Plots (A) and (B) correspond to $\omega_0\tau_c = 0.5$ and plots (C) and (D) to $\omega_0\tau_c = 2.0$.

motion is greater than for macromolecules, which means that special care is needed when using rotating-frame cross-relaxation rates to study structures of medium-sized molecules. Because NOESY is an insensitive experiment in this motional regime, it appears that the most productive approach will be the measurement of $\sigma^r$ at different field strengths. Indeed, the influence of internal and/or anisotropic motion on cross-relaxation rates for medium-sized molecules is so strong that structural interpretations should only be regarded as quantitatively reliable if measurements have been made at different magnetic field strengths and, if necessary, internal and/or anisotropic motion has been explicitly included in the interpretation of the cross-relaxation rates.

## Macromolecules

For isotropic, rigid body motion which is slow ($\omega_0 \tau_c \gg 1$), in the laboratory frame the leakage relaxation for two spins with a dipole–dipole interaction depends only on external relaxation.[8] Because the exchange of magnetization between two spins is therefore energy conserving and highly efficient, spin diffusion is a serious problem in NOESY experiments on macromolecules. In the rotating frame and with slow molecular motion, the dipole–dipole interaction between two spins does contribute to the leakage relaxation.[7] This has two consequences. First, leakage relaxation among magnetically equivalent spins will attenuate cross-relaxation to a further spin. Second, spin diffusion is strongly attenuated for rotating-frame cross-relaxation.

Figure 6 shows the relative sensitivity for NOESY compared to ROESY for cross-relaxation within an $AX_3$ spin system as a function of the overall rotational rate ($\tau_c$) and the diffusion constant ($D_{int}$) for internal

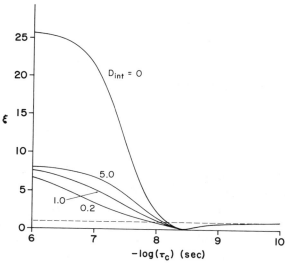

FIG. 6. The relative sensitivity, $\xi$, for NOESY compared to ROESY for an $AX_3$ spin system undergoing isotropic global motion with correlation time $\tau_c$ and with the indicated free diffusion constant, $D_{int}$, for rotation of the equivalent $X_3$ spins. $D_{int}$ is given in gigahertz. The $X_3$ spins have the geometry of the hydrogens of a methyl group and the $A$ spin is located on the axis of rotation of the methyl group. (From Farmer *et al.*[7])

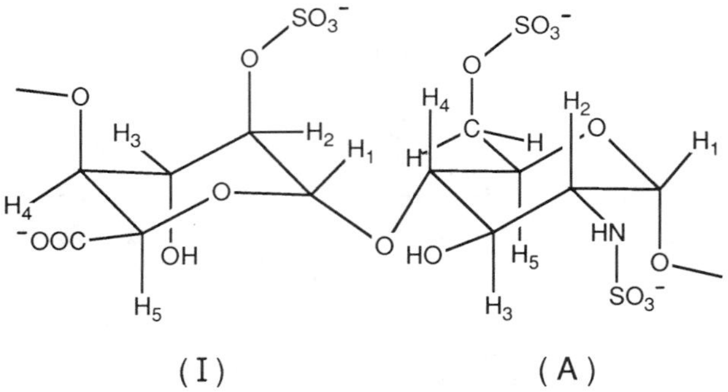

FIG. 7. The structure of the major disaccharide component of heparin. (From Farmer et al.[10])

rotation of the $X_3$ group. For overall rotation such that $\omega_0 \tau_c \gtrsim 5$ and no internal motion, NOESY is much more sensitive than ROESY. This is because the leakage relaxation among the $X_3$ spins in the ROESY experiment strongly attenuates cross-relaxation to the $A$ spin. If the $X_3$ spins rotate internally, the contribution of leakage relaxation among the $X_3$ spins is reduced, but NOESY remains the more sensitive experiment (Fig. 6). For large macromolecules, detection of cross-relaxation to a group of magnetically equivalent spins will in general be more efficient with NOESY than with ROESY. The above characteristics also mean that the presence of magnetically equivalent spins and the nature of their motion may have to be taken explicitly into account when using rotating-frame cross-relaxation rates to estimate interatomic distances in macromolecules.

Because $\sigma^r$ is always positive, cross-relaxation in the rotating frame will give cross-peaks with sign $(-1)^m$,[9,10] where $m$ is the number of transfers in the cross-relaxation pathway (the diagonal peaks are assumed to be positive). For example, for a cross-relaxation pathway $A \rightarrow B \rightarrow C$, the $AB$ cross-peak will be negative and the $AC$ cross-peak positive. Because of the strong attenuation of spin diffusion in ROESY experiments on macromolecules, it is unlikely that more than one relay of magnetization will be observed. Qualitatively this means that any positive cross-relaxation cross-peaks in ROESY must arise from spin diffusion and that all negative cross-peaks are likely to be due to direct, one-step cross-relaxation.

It has been suggested that the sign alternation characteristic of ROESY experiments offers a major advantage over NOESY experi-

[9] A. Bax, V. Sklenář, and M. F. Summers, J. Magn. Reson. **70**, 327 (1986).
[10] B. T. Farmer II, S. Macura, and L. R. Brown, J. Magn. Reson. **72**, 347 (1987).

ments,[9] where the effects of spin diffusion can generally only be detected by recording several experiments with different mixing times, including short mixing times where the signal-to-noise ratio is poor. While it is true that positive cross-relaxation cross-peaks in ROESY can only arise from spin diffusion, the presence of sign alternation necessitates care in the quantitative interpretation of ROESY spectra. For heparin (Fig. 7), Fig. 8 shows the intensity of the A1 to I5 ROESY cross-peak as a function of the mixing time. Initially the cross-peak shows negative intensity due to direct cross-relaxation between A1 and I5. At longer mixing times, a relay pathway, most likely A1 → I4 → I5 or A1 → I3 → I5, begins to dominate and the cross-peak passes through zero to become positive. If a single mixing time had been used, the A1 to I5 cross-relaxation might be interpreted in any of three ways: (1) direct cross-relaxation ($\tau_m \approx 60$ msec; note that a poor estimate of $\sigma^r$ would be obtained), (2) no cross-relaxation ($\tau_m \approx 80$ msec), or (3) relayed cross-relaxation ($\tau_m \approx 100$ msec). A ROESY spectrum with a short mixing time should be recorded even when only a qualitative interpretation is needed. This type of three-spin effect is well known from laboratory-frame cross-relaxation measurements on small molecules,[1] where $\sigma^n$ is also positive, and it is likely to be commonplace in ROESY experiments. This means that for ROESY, as for NOESY, quantitative measurement of cross-relaxation rates will require

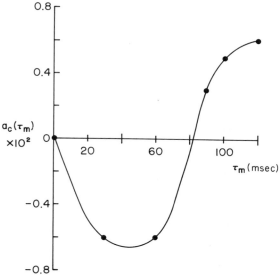

FIG. 8. The intensity of the A1–I5 cross-peak intensity versus the mixing time $\tau_m$ for ROESY spectra of heparin in $^2H_2O$. (From Farmer et al.[10])

recording a series of spectra at different mixing times, including very short mixing times.

## Molecular Motion

For rigid body, isotropic molecular motion, Fig. 9 shows that the ratio $\sigma_{ij}^n/\sigma_{ij}^r$ is independent of the interatomic distance $r_{ij}$ and depends only on the Larmor frequency $\omega_0$ and the correlation time $\tau_c$ [see Eqs. (1) to (3)]. This provides a necessary, but not sufficient test (see below) for such motion, i.e., for rigid body, isotropic motion the ratio $\sigma^n/\sigma^r$ must be the same for all pairs of hydrogen atoms in the molecule under study. If this ratio is not constant for all pairs, then the assumption of rigid body, isotropic motion is unwarranted and considerable care should be taken when interpreting cross-relaxation rates in terms of interatomic distances.

Quantitative expressions have recently been derived for $\sigma^n$ and $\sigma^r$ for axially symmetric molecules, e.g., DNA, and for globular molecules with one degree of internal rotational freedom.[7] The results obtained suggest that by measuring both $\sigma^n$ and $\sigma^r$, angular orientations of interatomic vectors relative to the diffusional symmetry axis can be obtained for axially symmetric molecules. This would be very complementary infor-

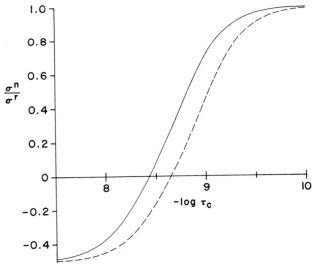

FIG. 9. The ratio $\sigma^n/\sigma^r$ for rigid body isotropic motion of an $AX$ spin system as a function of the correlation time $\tau_c$ for Larmor frequencies of 300 MHz (—) and 500 MHz (- - -). (From Farmer et al.[7])

mation to the estimates of interatomic distances which are usually obtained from cross-relaxation rates. For molecules which have isotropic overall motion and internal rotation characterized by free diffusion, these equations were used to show the cross-relaxation rates observed for medium-sized molecules can be very sensitive to internal motion (Fig. 5). For macromolecules with rotational diffusion constants such that $D_{global} \gg \omega_0 \gg D_{internal}$, the observed cross-relaxation rates will be given by[7]

$$\sigma^{n, r} = \tfrac{1}{4}(3 \cos^2 \phi - 1)^2 \sigma_{iso}^{n, r} = S^2(\phi)\sigma_{iso}^{n, r} \tag{5}$$

where $\sigma_{iso}^{n, r}$ is the cross-relaxation rate in the absence of internal motion and $\phi$ is the angle between the internal rotation axis and the interatomic vector. In this case both $\sigma^n$ and $\sigma^r$ are reduced according to the order parameter $S^2(\phi)$, but the ratio $\sigma^n/\sigma^r$ is independent of internal motion. This is analogous to the model-free interpretation of $^1H-^{13}C$ dipolar relaxation of Lipari and Szabo[11] and suggests that for fast internal motion with more than one degree of freedom, $\sigma^n$ and $\sigma^r$ can be interpreted in terms of an order parameter. Since $\sigma^n/\sigma^r$ is independent of the internal motion, this order parameter could be determined by measurement of both $\sigma^n$ and $\sigma^r$ provided the interatomic distance is known.

## Artifactual Contributions to ROESY Spectra

In laboratory-frame NMR experiments it is well known that magnetization transfer can occur by both coherent and incoherent processes. Examples of incoherent processes, i.e., processes in which phase coherence is not retained, include cross-relaxation, which depends on transfer of magnetization via relaxation processes, and chemical exchange, which depends on mass transfer, e.g., a chemical equilibrium, to cause magnetization transfer. In isotropic liquids, the most important coherent transfer mechanism is the scalar ($J$) coupling. Since the same chemical and physical processes will exist, it is no surprise that analogous transfer processes exist for rotating-frame NMR experiments.

Table I shows the characteristics of cross- and diagonal peaks for magnetization transfers in the rotating frame. In principle, cross-relaxation cross-peaks can be distinguished by the fact that these are the only cross-peaks which show negative absorption. In practice the simple classification of peak types based on the characteristics in Table I generally breaks down for two reasons. First, the same cross-peak may have contributions from more than one transfer mechanism, e.g., within scalar coupled spin networks it is quite usual to have a cross-peak which shows

[11] G. Lipari and A. Szabo, *J. Am. Chem. Soc.* **104**, 4546 (1982).

TABLE I

INTENSITY TYPES IN ROTATING-FRAME MAGNETIZATION
TRANSFER EXPERIMENTS

| Peak type | Signal intensity |
|-----------|------------------|
| No transfer | |
|   Diagonal | Positive absorption |
| Single transfer | |
|   Cross-relaxation | Negative absorption |
|   Homonuclear Hartmann–Hahn | Positive absorption |
|   COSY | Antiphase dispersion |
|   Chemical exchange | Positive absorption |

contributions from cross-relaxation, homonuclear Hartmann–Hahn transfer, and COSY-type transfer. Second, because substantial mixing times (ca. 10–200 msec) are used in ROESY experiments, it is not unusual to have multistep magnetization transfers. In this case the nature of the cross-peak becomes a product of the cross-peak types for the individual steps. For example, transfer by cross-relaxation followed by homonuclear Hartmann–Hahn transfer, which is also known as isotropic mixing, gives a cross-peak of the type (−absorption)(+absorption) = (−absorption) which is indistinguishable from a single-step cross-relaxation transfer. It has already been shown that this type of multistep transfer can lead to incorrect structural interpretations if not recognized.[10,12] An example of multistep cross-relaxation transfer has already been given above (Fig. 8).

In the context of measurement of cross-relaxation in the rotating frame, the other transfer mechanisms represent artifacts which it would be desirable to suppress. At present there are no known rotating-frame cross-relaxation experiments for suppression of chemical exchange cross-peaks. However, chemical exchange is not likely to be encountered very often. COSY-type cross-peaks can in principle be suppressed by means of differential resonance offset for two spins, rf inhomogeneity, and the exact choice of both the mixing time and the spin-lock field strength.[4] It may be difficult, however, to satisfy these conditions for all the resonances in a spectrum. Fortunately, COSY-type cross-peaks are usually easily recognizable by their antiphase character and are not a serious problem in quantitative analysis of phase-sensitive ROESY spectra. The most troublesome artifactual transfer mechanism is homonuclear Hartmann–Hahn transfer.

[12] D. Neuhaus and J. Keeler, *J. Magn. Reson.* **68,** 568 (1986).

Suppression of Homonuclear Hartmann–Hahn Transfer

Cross-relaxation in the rotating frame shows only weak dependence on resonance offset effects, whereas homonuclear Hartmann–Hahn transfer is strongly dependent on resonance offset (see also Bax [8], this volume).[4,12-14] This means that by judicious choice of the carrier frequency and field strength of the rf spin-lock irradiation, it is possible to measure cross-relaxation transfer while suppressing homonuclear Hartmann–Hahn transfer. To suppress homonuclear Hartmann–Hahn transfer, it is suggested that the transmitter carrier frequency $\omega_c$ be placed in a region of the spectrum so that the following two criteria are met: (1) $\min[(\{\omega\} - \omega_c)/\gamma B_1] \geq 0.1$, where $B_1$ is the spin-locking field strength in gauss and $\{\omega\}$ is the set of precessional frequencies (in rad/sec) defining all observable resonances in the spectrum; and (2) $\min[(\omega_i - \omega_j)/\gamma B_1] \geq 0.2$ for all directly scalar coupled spins $i$ and $j$. It is of course necessary to retain spin locking to achieve transfer by cross relaxation. For spin $i$ *on resonance,* this requires that the tilt angle, $\theta_j$, of spin $j$ relative to the external magnetic field, which is defined by $\tan \theta_j = \gamma B_{sl}/(\omega_j - \omega_c)$, be such that $\theta_j > 60°$.[14] Provided that $\theta_i - \theta_j$ is not too large, one can achieve spin locking for values of $\theta$ much less that 60°. Ernst and co-workers have recently exploited values of $\theta < 90°$ to investigate dipolar cross-correlation in the rotating frame.[14a] This method of suppressing homonuclear Hartmann–Hahn transfer has limitations because it is not always possible to satisfy all of the above conditions, especially for scalar coupled spins which have very similar chemical shifts, e.g., the $\beta$ and $\beta'$ hydrogens of amino acid residues. In addition, this method places high demands on the phase and amplitude stability of the continuous wave rf spin-locking field.

An alternate method for suppression of homonuclear Hartmann–Hahn transfer has recently been suggested by Kessler et al.[15] This involves pulsed rf irradiation of the form

$$[\beta - \tau]_n = \text{pulsed rf irradiation} \qquad (6)$$

where $\beta$ represents a pulse of tip angle $\beta$, $\tau$ a delay, and $n$ represents the number of times this basic sequence is repeated during the mixing time. There are essentially two potential benefits to using pulsed rf irradiation during the mixing time. First, the suppression of homonuclear Hartmann–Hahn transfer has been claimed,[15] although this has been recently dis-

---

[13] D. G. Davis, *J. Am. Chem. Soc.* **109**, 3471 (1987).
[14] B. T. Farmer II and L. R. Brown, *J. Magn. Reson.* **72**, 197 (1987).
[14a] C. Griesinger and R. R. Ernst, *Chem. Phys. Lett.* **152**, 239 (1988).
[15] H. Kessler, C. Griesinger, R. Kerssebaum, K. Wagner, and R. R. Ernst, *J. Am. Chem. Soc.* **109**, 607 (1987).

puted by Bax.[16] The resonance offset dependence of homonuclear Hartmann–Hahn transfer in the pulsed spin-lock experiment is essentially equivalent to that for the average spin-lock field over the period $(\beta - \tau)$,[16] provided that $\beta \ll 180°$. Second, the demand placed on the amplifier power supply for the spin-lock rf field can be much less, thereby allowing the transmitter to serve as the rf source for both the initial $90°_{-x}$ pulse (Fig. 1) and the spin lock, which is an advantage on many commercial spectrometers.

To adjust the effective field strength in the pulsed method, either the rf amplifier output or the $\tau/t_\beta$ ratio in Eq. (6) can be increased or decreased, where $t_\beta$ is the time required to achieve the $\beta$ pulse. The effective spin-lock field strength is given approximately by $B_\beta[t_\beta/(t_\beta + \tau)]$ where $B_\beta$ is the field strength of the $\beta$ pulse. As the spectral width increases, due either to the characteristics of the sample or to the use of a higher static magnetic field strength, it will be necessary to decrease $\tau/t_\beta$ in order to achieve spin locking of resonances in the spectral wings. Since $\tau/t_\beta = 8$–10 and $\beta = 30°$ seems to be sufficient for most nonparamagnetic biological macromolecules, many transmitter amplifier power supplies that would not stand up to the duty cycle imposed by the continuous wave (CW) spin lock will be sufficient for the pulsed spin lock. The pulsed spin lock has the potential disadvantages that sample heating will be greater and that the spectrometer must be capable of generating the fast pulses for the spin lock.

Recently, it has been suggested that the frequency of the spin lock be swept during the spin-lock period.[17] Because ROESY is considerably less sensitive to resonance offset effects than is homonuclear Hartmann–Hahn transfer, this method preferentially emphasizes transfer by rotating frame cross-relaxation. Instrumentally, the method requires the ability to shift the carrier frequency of the spin lock relatively rapidly, reproducibly, and without discontinuities of phase.

### Resonance-Offset Effects

Resonance offset affects the sensitivity and quantitation of ROESY spectra by modifying the cross-relaxation rate and by decreasing the amount of spin-locked magnetization that is observable in $t_2$. In the presence of resonance offset, the effective cross-relaxation rate, $\sigma^e_{ij}$, is[13,14,18]

[16] A. Bax, *J. Magn. Reson.* **77**, 134 (1988).
[17] J. Cavanagh and J. Keeler, *J. Magn. Reson.* **80**, 186 (1988).
[18] C. Griesinger and R. R. Ernst, *J. Magn. Reson.* **75**, 261 (1987).

TABLE II
FACTORS FOR SCALING OF OBSERVED ROESY
PEAK INTENSITY

| Peak type | Scaling factor[a] |
|---|---|
| CW spin lock | |
| $M(i, j)$ (cross peak) | $(\sin^2 \theta_i \sin^2 \theta_j)^{-1}$ |
| $M(i, i)$ (diagonal peak) | $(\sin^2 \theta_i)^{-1}$ |
| CW spin lock bracketed by 90° pulses[b] | |
| $M(i, j)$ (cross peak) | $(\sin \theta_i \sin \theta_j)^{-1}$ |
| $M(i, i)$ (diagonal peak) | 1 |

[a] The tilt angles $\theta_i$ and $\theta_j$ are defined by $\tan \theta_{i,j} = \gamma \beta_1 / (\omega_{i,j} - \omega_c)$ where $\omega_c$ is the rf carrier frequency for the spin lock.
[b] Reference 18.

$$\sigma_{ij}^e(\theta_i, \theta_j) = n_i q_{ij}(\sigma_{ij}^r \sin \theta_i \sin \theta_j + \sigma_{ij}^n \cos \theta_i \cos \theta_j) \tag{7}$$

where $\theta_i$ and $\theta_j$ are the tilt angles for spins $i$ and $j$ and $\sigma_{ij}^r$ is the rotating-frame cross-relaxation rate in the absence of resonance offset. From Eq. (7) it is clear that an accurate value for $\sigma_{ij}^r$ can only be obtained from $\sigma_{ij}^e$ if either spin $i$ or spin $j$ is on resonance, i.e., $\cos \theta_i$ or $\cos \theta_j = 0$. However, for small resonance offsets such that $\sin \theta_i \sin \theta_j \gg \cos \theta_i \cos \theta_j$, it is possible to make an approximate correction for the effect of resonance offset on the observed cross-relaxation rate $\sigma_{ij}^e$. It is also necessary to correct for the proportion of transverse magnetization which is parallel to the spin-lock field and for the proportion of the spin-locked magnetization which will be in the transverse plane, and hence, observable during $t_2$. To extract a value for $\sigma^r$ from a ROESY cross-peak build-up curve, it is necessary to correct both cross- and diagonal-peak intensities with the factors shown in Table II. The factor $\sin^2 \theta_i \sin^2 \theta_j$ can lead to substantial attenuation of the cross-peak intensities. Some improvement can be obtained by applying hard 90° pulses immediately before and after the CW spin lock (Table II).[18]

## Analysis of Build-Up Curves

Because cross-relaxation in the rotating frame is analogous to cross-relaxation in the laboratory frame, cross-relaxation rates from ROESY data can in principle be extracted by the same methods used for NOESY data, namely, (1) initial slopes, (2) the full spin-pair approximation, or (3) a full matrix calculation (Borgias and James [9], this volume).[19–21] There are as yet few quantitative results in the literature. However, theoretical considerations (L. R. Brown and B. T. Farmer II, unpublished results) suggest that spin diffusion and cross-relaxation among equivalent spins will compromise methods (1) and (2) above to a greater extent in ROESY than for NOESY. A full matrix calculation may be more appropriate for determination of cross-relaxation rates for ROESY data. In NOESY, this method has had limited success due both to the extensive propagation of magnetization by spin diffusion and to the indistinguishability of direct and indirect transfers based on the sign of the transferred magnetization. Both of these factors are less of a problem for ROESY experiments.

[19] J. W. Keepers and T. L. James, *J. Magn. Reson.* **57**, 404 (1984).
[20] S. Macura, B. T. Farmer II, and L. R. Brown, *J. Magn. Reson.* **70**, 493 (1986).
[21] R. R. Ernst, G. Bodenhausen, and A. Wokaun, "Principles of Nuclear Magnetic Resonance in One and Two Dimensions." Oxford Univ. Press (Clarendon), London and New York, 1987.

## [12] Modern Spectrum Analysis in Nuclear Magnetic Resonance: Alternatives to the Fourier Transform

### By JEFFREY C. HOCH

### Introduction

The discrete Fourier transform (DFT) is virtually synonymous with spectrum analysis in modern NMR spectroscopy, with good reason. The Wiener–Khinchine theorem provides a theoretical link between the autocorrelation function of a time series (the free induction decay) and its power spectrum. Fast Fourier transform algorithms allow rapid and robust computation of the DFT. The DFT is only an approximation to the continuous Fourier transform, however, and techniques for improving the spectral estimate obtained using the DFT, such as the use of window functions and zero filling, have been employed since the earliest days of pulsed NMR.

Spectral estimates which avoid some of the shortcomings of the DFT, in particular the difficulty of obtaining high-resolution spectral estimates from short data records, have been used in fields other than NMR spectroscopy for more than two decades. Two recent developments are responsible for the migration of these techniques from fields such as geophysics, radio astronomy, and speech recognition to NMR spectroscopy. One is the decreasing reliance by NMR spectroscopists on instrument manufacturers for data-processing software. The other is that the cost of computing has declined so dramatically over the past decade that it is no longer impractical to consider spectral estimates which take as much as two orders of magnitude more time to compute than the DFT.

The purpose of this chapter is to provide an introduction to some new non-DFT spectral estimates. We begin with a review of some properties of the DFT, with an emphasis on shortcomings which motivate the search for new spectral estimates. The next section provides a description of the theoretical basis of some of the new methods of spectral analysis. The discussion is not rigorous, but seeks to place the methods on a plausible (if not firm) footing. Finally, the methods described are applied to NMR data. Application to a common set of NMR data permits an overview of the relative strengths of the various methods from the standpoint of the sensitivity, resolution, dynamic range, resistance to data contamination, and the computation cost of the spectral estimates. One important caveat applies: since the spectral estimates are nonlinear, the resulting spectra can depend dramatically on the data being analyzed and the values of adjustable parameters used by the methods. The comparisons made here must therefore be viewed as illustrative. It is hoped, however, that they may serve as a practical guide to the issues involved in selecting a method for the spectral analysis of real experimental data. It must also be emphasized that the author did not invent any of the methods described here. Used prudently, these new methods promise to push back the bounds on sensitivity and resolution placed on NMR by traditional methods of spectral analysis based on the DFT, and in the processes enhance the power of NMR as a technique for studying biologically important molecules.

## Motivation: DFT as a Spectral Estimate

The DFT is an approximation to the continuous Fourier transform, and the validity of the approximation depends on the data being analyzed. Differences between the two are a result of *sampling* (discretization) and *truncation* (finite data length).

Part of the difficulty lies in the fact that the DFT implicitly treats the finite time series as periodic, with period $NT$, where $N$ is the number of samples and $T$ is the time interval between samples. If we consider the expansion of a sampled time series $X$ in terms of a finite number of Fourier components (the inverse DFT),

$$x(kT) = (1/N) \sum_{n=0}^{N-1} X[n/(NT)]e^{i2\pi nk/N} \tag{1}$$

where $X[n/(NT)]$ are the Fourier components, it is clear that $x[(k + N)T]$ is equal to $x(kT)$, since $e^{i2\pi n} = 1$ when $n$ is an integer. The discontinuities thus implicit in the DFT analysis of finite time series which are not periodic with period $NT$ lead to additional frequency components in the frequency domain, a phenomenon termed *leakage*. These frequency components are due to the convolution of $\sin f/f$, the Fourier transform of a square wave representing the finite sampling interval, with the spectrum of the time series.

Another difficulty associated with the DFT stems from the relationship between the number of data samples and the digital frequency resolution,[1] $\Delta f = 1/NT$, where $\Delta f$ is the spacing between points in the frequency domain. To achieve high digital frequency resolution, it is desirable to collect as long a data record as possible. In pulsed NMR experiments, however, the signal of interest invariably diminishes with time. Thus collecting long data records to achieve high resolution rapidly leads to a case of diminishing returns by degrading the signal-to-noise ratio ($S/N$) of the resulting spectrum. One technique for overcoming this degradation is to collect data for a short period of time and then append zeros to the data. This process also introduces discontinuities into the time series, resulting in leakage. The leakage can be severe if the signal of interest has not decayed sufficiently before the end of the sampled interval. Weighting functions (alternately referred to as apodization, window, or taper functions) can be applied to the time data to minimize the extent of leakage by reducing the intensity of the signal near the end of the sampled interval, although generally with an accompanying loss of resolution due to broadening of the spectral components. A variety of weighting functions exists, each function fulfilling a different goal (leakage reduction, resolution enhancement, sensitivity enhancement) or striking a compromise between various goals.

### Justification: Other Estimates

#### *Burg Maximum Entropy Method*

An intuitive notion about how a spectral estimate might be improved is the following: if the digital resolution of a DFT spectral estimate can be

---

[1] The term digital frequency resolution is used to refer to the frequency difference between adjacent points in a discrete spectral estimate, and should be distinguished from the resolution of the spectral estimate, which is a measure of the ability of a spectral estimate to distinguish between closely spaced frequency components. The digital resolution places a lower bound on the resolution of a spectral estimate.

improved by appending zeros to the end of the record, then the leakage caused by the discontinuity could be reduced by extrapolating the data record in a more reasonable fashion. Even the eye can provide a more reasonable extrapolation of a decaying sinusoid than appending zeroes. It is partly from the consideration of just these sorts of notions about spectral analysis that led John Burg to propose, in 1967, a formal solution to the problem of extrapolating a time series beyond its measured values.[2] Burg observed that the requirement that the $N + 1$ by $N + 1$ autocorrelation matrix

$$
\begin{bmatrix}
\phi_0 & \phi_1 & \cdots & \phi_N \\
\phi_1 & \phi_0 & & \\
\vdots & & \ddots & \\
\phi_N & & & \phi_0
\end{bmatrix}
\tag{2}
$$

must be positive semidefinite, and consequently must have a nonnegative determinant, places constraints on the values which can be taken on by $\phi_N$ when the first $N$ autocorrelation values ($\phi_0, \phi_1, \cdots, \phi_{N-1}$) are known. Since the determinant depends quadratically on $\phi_N$, there are two values for which the determinant is zero. The allowed values for $\phi_N$ are these two values and all values in between. Burg went on to compare this prescription for extending the autocorrelation function with the results obtained by asking what power spectrum has the maximum *entropy* consistent with the known values of the autocorrelation function. Shannon[3] first introduced the concept of entropy as a measure of information, or more precisely, a lack of information. Burg maximized the entropy measure

$$
H = \int \log S(f) df
\tag{3}
$$

where $S(f)$ is the spectral density at frequency $f$, subject to the constraint

$$
\int S(f) \cos(2\pi f \tau \Delta t) df = \phi(\tau)
\tag{4}
$$

for $\tau = 0$ to $N$, the $N + 1$ known values of the autocorrelation function. The solution to the resulting variational problem is

$$
S(f) = (P_{K+1}/W)/\left| 1 + \sum_{n=1}^{N} a_{n+1} e^{-i2\pi f n \Delta t} \right|^2
\tag{5}
$$

[2] J. P. Burg, *Proc. 37th Meet. Soc. Explor. Geophys., 1967;* reprinted *in* "Modern Spectrum Analysis" (D. G. Childers, ed.), p. 34. IEEE Press, New York, 1978.
[3] C. Shannon, *Bell Syst. Tech. J.* **27**, 379 (1948).

where $P_{K+1}$ and $a_{n+1}$ are determined from

$$
\begin{bmatrix}
\phi_0 & \phi_1 & \cdots & \phi_N \\
\phi_1 & \phi_0 & & \vdots \\
\vdots & & \phi_0 & \phi_1 \\
\phi_N & \cdots & \phi_1 & \phi_0
\end{bmatrix}
\begin{bmatrix}
1 \\
a_2 \\
a_3 \\
\vdots \\
a_{K+1}
\end{bmatrix}
=
\begin{bmatrix}
P_{K+1} \\
0 \\
0 \\
\vdots \\
0
\end{bmatrix}
\tag{6}
$$

The $a$ values correspond to the coefficients of a $K + 1$ term prediction error filter, or upon change of sign, a linear prediction (LP) filter:

$$
\phi_n = \sum_{j=1}^{N} \phi_{n-j}(-a_j)
\tag{7}
$$

A number of efficient algorithms for determining the prediction filter coefficients were known previously, so the estimation of the maximum entropy spectrum is straightforward. In addition to iterative algorithms due to Andersen[4] and Levinson,[5] the LP coefficients may be determined by least-squares methods.

It should be noted that the form for the entropy used by Burg differs from the Shannon information entropy

$$
H = \int S(f)\log S(f)df
\tag{8}
$$

The first formal statement of the maximum entropy principle in data analysis is usually attributed to Jaynes,[6] who used Eq. (8) for the entropy. Skilling and Gull,[7] have discussed the appropriateness of the two expressions for the entropy.

The Burg maximum entropy spectrum was derived for real data. The extension to complex data (consisting of real, imaginary pairs) is straightforward. Since the method deals with the autocorrelation function, however, the method is limited to obtaining the power spectrum, and is not capable of yielding phase-sensitive spectral estimates. Although derived for *stationary* time series, there appears to be no barrier to its application to decaying sinusoids. Some authors have applied an exponentially increasing weighting function to the data prior to computation of the LP coefficients in order to satisfy the stationary criterion, however.

[4] N. Andersen, *Geophysics* **9**, 69 (1970).
[5] N. Levinson, *J. Math. Phys.* **25**, 261 (1947).
[6] E. Jaynes, *IEEE Trans. Syst., Sci., Cybernet.* **SSC-4**, 227 (1968).
[7] J. Skilling and S. Gull, *SIAM-AMS Proc.* **14**, 167 (1984).

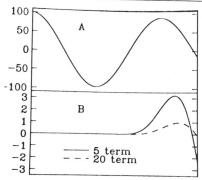

FIG. 1. Extrapolation error using linear prediction: (A) 256-point decaying sinusoid. (B) extrapolation error resulting from computing a linear prediction filter using the first 128 points of (A) and using the linear prediction filter to extrapolate from 128 to 256 points.

A practical problem with the method is that of the choice of the order $K$ of the linear prediction filter. The proper choice of $K$ depends on the number of frequency components, the $S/N$ ratio of the data, and the dynamic range. Although a number of criteria have been suggested for determining $K$,[8] a number slightly larger than the *a priori* estimate of the number of significant frequency components is frequently used. When the range in intensity between the weakest and strongest components is large, a value of $K$ much larger than the number of components present may be required. Linear prediction using a finite number of coefficients is only approximately valid for decaying sinusoids. Figure 1 illustrates the error incurred by extrapolation of a 256-point decaying sinusoid to 512 points using 5-term and 20-term linear prediction filters. Linear prediction seeks to minimize the mean square prediction error. When the dynamic range is large, the error is dominated by the largest components. Until the linear prediction filter order is large enough to make the prediction error associated with the strong component comparable to the intensity of the weak component, the filter will not accurately reflect the presence of the weak components, if at all. This can manifest itself as missing components, and as phase distortions in the LPZ spectral estimate (discussed in the following section).

The choice of $K$ can profoundly affect the appearance of the spectrum. Too small a value leads to a spectrum which is overly smooth. Too large a value yields spurious peaks. A related problem is that of "spontaneous

---

[8] H. Akaike, *IEEE Trans. Autom. Control* **AC-19,** 716 (1974).

splitting,"[9] in which a spectral component splits into two when $K$ is increased slightly.

## LPZ

A spectral estimate closely related to the Burg maximum entropy spectrum is LPZ, so named because it uses linear prediction and the $Z$ transform.[10-12] Derived by Tang and Norris, LPZ uses complex linear prediction to extrapolate a finite time series, and provides a closed form for the infinite sum corresponding to the DFT (or $Z$ transform) of the extrapolated time series. A similar spectral estimate has been derived by Ni and Scheraga.[13]

The linear prediction-extrapolated FID, $\hat{y}$, is given by

$$\hat{y}_i = \sum_{j=1}^{K} a_j \hat{y}_{i-j} \tag{9}$$

for $i > m$ (where $m$ is the number of data samples), $y$ is the sampled FID, and $K$ is the length of the linear prediction filter used to model the data. The spectrum of $\hat{y}$ is given by the one-sided $Z$ transform

$$S(z) = \sum_{n=0}^{\infty} \hat{y}_n z^{-n} \tag{10}$$

where $z$ is $e^{-i2\pi\omega t}$. Using linear prediction this becomes

$$S(z) = \sum_{k=0}^{\infty} \left( \sum_{j=1}^{K} a_j \hat{y}_{kj} \right) z^{-k}$$

$$= \sum_{j=1}^{K} \sum_{k=0}^{\infty} a_j \hat{y}_{k-j} z^k$$

$$= \sum_{j=1}^{K} \left( a_j \sum_{k=j}^{\infty} \hat{y}_i z^{k+j} \right) \tag{11}$$

$$= \sum_{j=1}^{K} \left( a_j z^j \sum_{k=j}^{\infty} \hat{y}_k z^k \right)$$

[9] P. Fougere, in "Maximum Entropy and Bayesian Methods in Inverse Problems" (C. Smith and W. Grandy, eds.), p. 303. Reidel, Dordrecht, Netherlands, 1985.

[10] J. Tang and J. Norris, J. Chem. Phys. **84**, 5210 (1986).

[11] J. Tang and J. Norris, J. Chem. Phys. Lett. **131**, 252 (1986).

[12] J. Tang and J. Norris, J. Magn. Reson. **69**, 180 (1986).

[13] F. Ni and H. Scheraga, J. Magn. Reson. **70**, 506 (1986).

$$= \sum_{j=1}^{K} \left[ a_j z^j \left( S(z) = \sum_{k=0}^{j-1} \hat{y}_k z^k \right) \right]$$

Define $G(k) = -\sum_{j=1}^{k} a_j z^j$. Then

$$S(z) = -G(K)S(z) - \sum_{j=1}^{K} \left( a_j z^j \sum_{k=0}^{j-1} \hat{y}_k z^k \right)$$

$$= -G(K)S(z) - \sum_{k=0}^{K-1} \sum_{j=i+1}^{K} a_j \hat{y}_k z^{k+j} \tag{12}$$

$$= -G(K)S(z) + \sum_{k=0}^{K-1} \hat{y}_k z^k [G(K) - G(k)]$$

Rearranging yields

$$S(z) = \{1/[1 + G(K)]\} \sum_{k=0}^{K-1} \hat{y}_k z^k [G(K) - G(k)] = H(z)/[1 + G(z)] \tag{13}$$

where $H(z) = -\sum_{m=1}^{K} b_m z^m$ and $b_m = \sum_{n=m}^{K} a_n \hat{y}_{n-m}$.

The sums required in the evaluation of $H(z)$ and $G(z)$ can be computed using the FFT. As does the Burg MEM spectral estimate, the LPZ spectrum depends markedly on the choice of $K$.

## Parametric Spectrum Analysis Using Linear Prediction: LPSVD, HSVD

What constitutes a spectrum? We customarily think of a plot of spectrum intensity as a function of frequency. Ultimately, however, we are interested in the parameters characterizing the significant components of the spectrum: frequency, amplitude, phase, and width. These parameters take on physical meaning when they are interpreted in terms of a model, which in high-resolution NMR spectroscopy typically consists of decaying sinusoids plus noise. Unfortunately, even if the number of decaying sinusoids present is known *a priori,* the problem of simultaneously determining the parameters characterizing the sinusoids by fitting the model function to the time-domain data is a notoriously ill posed problem. The nonlinear dependence of the model on the frequencies and decay rates is the principal source of difficulties.

A procedure recently applied to the analysis of noisy speech signals[14] has led to the development of several related robust procedures for deter-

[14] R. Kumaresan and D. Tufts, *IEEE Trans. Acoust., Speech, Signal Process.* **ASSP-30,** 833 (1982).

mining the model parameters for NMR data.[15,16] The principal feature of these methods is the determination of the frequencies and decay times from the zeroes of a complex polynomial formed from linear prediction coefficients fit to the FID. When the FID consists of $M$ exponentially decaying sinusoids in noise

$$y_n = \sum_{k=1}^{M} c_k e^{s_k n} + \varepsilon_n \tag{14}$$

where $c_k$ values are the (complex) amplitudes, $s_k = -\alpha_k + i2\pi f_k$, with decay constants $\alpha_k$ and frequency $f_k$, and $\varepsilon_n$ the noise, it has been shown that the complex polynomial

$$1 + a_1 z^{-1} + a_2 z^{-2} + \cdots + a_K z^{-K} \tag{15}$$

where the $a_i$ values are linear prediction coefficients fit to the FID, has zeroes at $z = e^{s_k}$ if $K$ is chosen to have a value between $M$ and $N - M$, where $N$ is the number of data samples.

Although the method does not directly provide an estimate of $M$, it provides a method for distinguishing some zeroes of Eq. (15) which are due to noise from those due to signal if $K$ is chosen larger than $M$. When the linear prediction coefficients are computed for the reversed time series, the $M$ signal zeroes fall outside the unit circle, corresponding to exponentially growing components. Zeroes falling inside the circle correspond to exponentially decaying components (for the time-reversed data), which can be attributed to noise.

Once the frequencies and decay constants of the significant components have been determined, the amplitudes and phases (or equivalently the complex amplitudes) can be determined by least-squares fit to the FID. The final result of the procedure is a table of frequencies, amplitudes, decay times, and phases of the significant spectral components. A traditional "spectrum" can be constructed by Fourier transforming a synthetic FID based on these parameters. The principal steps of the procedure are (1) least-squares determination of the complex linear prediction coefficients, (2) determining the complex roots of Eq. (15), (3) "filtering" out noise roots based on their location relative to the unit circle, (4) computation of the frequencies and decay times from the significant roots of Eq. (15), and (5) least-squares determination of the amplitudes and phases.

[15] H. Barkuijsen, R. de Beer, W. Bovée, and D. Van Ormondt, *J. Magn. Reson.* **61**, 465 (1985).

[16] J. Tang, C. Lin, M. Bowman, and J. Norris, *J. Magn. Reson.* **62**, 167 (1985).

The LPSVD method[15] derives its name from the method used to solve the linear least-squares problem (singular value decomposition). A related method (LPQRD)[16] uses Householder QR decomposition.[17] Both methods require the determination of the zeroes of a complex polynomial with an order comparable to the number of lines in the spectrum. A related method which obviates the need for polynomial root finding has been developed recently by Barkhuijsen et al.[18,19] Called HSVD because it uses singular value decomposition of a Hankel matrix formed from the data, diagonalization of a matrix formed from the eigenvectors of the signal singular values is used to determine the frequencies and decay times. The amplitudes and phases are determined by a subsequent least-squares fit, as in LPSVD.

The number of linear prediction coefficients determined using LPSVD or HSVD does not significantly affect the resulting spectrum, provided that the number exceeds the number of significant spectral components. The spectrum depends more critically on the "reduced" order, the number of singular values deemed significant which are included in the final least-squares step to determine the amplitudes and phases. Underestimation of the reduced order can lead to biased parameter estimates (in addition to missing peaks), while overestimation can introduce spurious peaks. The Burg MEM, LPZ, and LPSVD implementations used in this work require the experimenter to provide the linear prediction order, and for LPSVD the reduced order. The HSVD implementation used requires the linear prediction order and a starting value for the reduced order, which should be smaller than the number of peaks expected. The reduced order is increased until the root-mean-square deviation between the data and the HSVD fit is smaller than the (estimated) noise.

*Maximum Entropy Reconstruction*

In the Burg maximum entropy procedure the inverse Fourier transform of the spectrum estimate is taken to be exactly equal to the measured autocorrelation function over the sampled interval. A more general approach is to admit the possibility of errors in the measured data

$$d_i = \hat{d}_i + \varepsilon_i \qquad (16)$$

where $d_i$ values are the measured data, $\hat{d}_i$ the error-free signal, and $\varepsilon_i$ the errors. The spectrum we seek is that of $\hat{d}$, rather than $d$. Computing this

[17] C. L. Lawson and R. J. Hanson, "Solving Least Squares Problems." Englewood Cliffs, New Jersey, Prentice-Hall, 1974.
[18] H. Barkuijsen, R. de Beer, and D. Van Ormondt, *J. Magn. Reson.* **73**, 553 (1987).
[19] H. Yan and J. C. Gore, *J. Magn. Reson.* **80**, 324 (1988).

spectrum is equivalent to reconstructing $\hat{d}$ from the noisy measurements $d$. One possible reconstruction is that spectrum having maximum entropy which is consistent with the measured data. More concretely, we seek $\hat{f}$, and equivalently mock data $\hat{d}_k = (1/N) \sum_{n=0}^{N-1} \hat{f}_n e^{i2\pi nk/N}$, where $N$ is the number of frequencies at which $\hat{f}$ is to be determined, such that some measure $H$ of the entropy of $\hat{f}$ is a maximum and a constraint statistic $C$ which measures the discrepancy between $d$ and $\hat{d}$ is less than or equal to some value $C_0$. This constrained optimization problem can be converted to an unconstrained optimization through the introduction of a Lagrange multiplier. The objective function

$$O = H - \lambda C \tag{17}$$

is maximized, with $\lambda$ chosen so that the constraint is satisfied. This general strategy was applied by Wernecke and D'Addario[20] to the problem of Fourier synthesis in radio astronomy.

The appropriate form of the entropy measure has been the subject of considerable debate. The debate has focused primarily on the two forms

$$H_1 = \sum_{i=1}^{N} \log \hat{f}_i$$

and

$$H_2 = \sum_{i=1}^{N} \hat{f}_i \log \hat{f}_i \tag{18}$$

Both can be viewed as smoothness indicators. Skilling and Gull[7] have discussed the relative merits of the two forms. The form of the constraint statistic $C$ depends on assumptions about the errors $\varepsilon_i$. If the errors are assumed to be independent, then the constraint

$$C = \sum_{i=1}^{M} (1/\sigma_i^2)|d_i - \hat{d}_i|^2 = M \tag{19}$$

is appropriate, where $\sigma_i^2$ is the variance of the $i$th measurement and $M$ is the total number of measurements. In practice the variances for each of the measurements are generally assumed to be equal.

There is no known formal solution to this optimization problem, for either entropy measure. The problem must therefore be solved numerically. The size of typical NMR problems dictates the use of a first order method, such as steepest descent or conjugate gradiate, which require $O(N)$ storage locations, as opposed to second order methods which uti-

[20] S. J. Wernecke and L. R. D'Addario, *IEEE Trans. Comput.* **C-26**, 351 (1977).

lize the Hessian matrix of second derivatives and require $O(N^2)$ storage locations.[21]

The implementation of maximum entropy reconstruction used in the examples (the "Cambridge" algorithm; see Appendix at end of this volume) employs a modified descent (or ascent) strategy, using a family of search directions, rather than one.[22,23] Careful control of the step size along these search directions plays a crucial role in obtaining a stable algorithm. In addition to an estimate of the noise ($\sigma_i$), the method requires a default value which serves as a starting point for the iterative search and represents the value toward which an element of the spectrum $f_i$ tends when it is not constrained by the data ($\partial f_i/\partial C = 0$). The default value may be viewed as the average noise level; smaller values yield greater noise reduction in the reconstructed spectrum.

A problem with the use of entropy in the objective function is that it is defined only for positive values of $f_i$. In order to use maximum entropy to reconstruct spectra containing negative components, Laue et al.[24] have proposed reconstructing the positive and negative components separately,

$$\hat{f} = \hat{f}^+ - \hat{f}^-$$
(20)

where $\hat{f}^+$ and $\hat{f}^-$ are constrained to positive values. Alternatively, one could maximize the entropy of the power spectrum,[25]

$$H_3 = \sum_{i=1}^{N} |\hat{f}_i|^2 \log|\hat{f}_i|^2$$
(21)

While this entropy measure is defined for negative values of $\hat{f}_i$, the first derivative is singular at zero.

In addition to being noise cognizant, maximum entropy reconstruction is highly flexible since the measured data $d_i$ need not be collected at uniform intervals. This ability to deal with "missing" data, simply by excluding it from the computation of the constraint statistic, allows the method to be used when the data are not collected at uniform intervals. Barna et al.[26] have shown that different sampling schemes can yield

[21] W. Press, B. Flannery, S. Teukolsky, and W. Vetterling, "Numerical Recipes." Cambridge Univ. Press, London and New York, 1986.
[22] J. Skilling and R. Bryan, Mon. Not. R. Astron. Soc. 211, 111 (1984).
[23] S. Sibisi, J. Skilling, R. Brereton, E. Laue, and J. Staunton, Nature (London) 311, 446 (1984).
[24] E. Laue, J. Skilling, and J. Staunton, J. Magn. Reson. 63, 418 (1985).
[25] K. M. Wright and P. S. Belton, Mol. Phys. 58, 485 (1986).
[26] J. Barna, E. Laue, M. Mayger, J. Skilling, and W. Worrall, J. Magn. Reson. 73, 769 (1987).

higher sensitivity than uniform sampling for the same number of measurements. A further advantage of maximum entropy reconstruction is that it makes no assumptions about line shapes, and consequently can be applied without modification to Gaussian lines, powder spectra, or imaging data. Prior information, for example, in the form of expected line shapes and decay rates, can substantially improve maximum entropy reconstructions, however.

### Application: Estimates Compared

In this section the results of the spectral estimates described in the previous section are applied to NMR data. Examples are chosen to illustrate a number of characteristics of the spectral estimates, including sensitivity, resolution, and resistance to data contamination. Experimental data were recorded at 400 MHz on a JEOL GX-400 spectrometer.

The implementations of the Burg MEM and LPZ methods used are part of the Rowland NMR Toolkit (see Appendix at end of this volume). For both of these methods there are two parameters to be provided: the number of input points used to compute the complex linear prediction coefficients, and the order of the linear prediction filter. The LPSVD and HSVD implementations used were written by H. Barkhuijsen and R. de Beer of Delft University of Technology. For both of these methods there are three parameters: the number of data points used to compute the LP coefficients, the number of LP coefficients, and the "reduced" order, which corresponds to the number of significant roots to Eq. (15) expected. In the examples which follow the number of LP coefficients computed is always taken to be $0.75*N$, where $N$ is the number of input points used to compute the LP coefficients. Maximum entropy reconstructions were performed using an implementation of the "Cambridge algorithm" purchased from Maximum Entropy Data Consultants. The parameters used by this method are the number of data points to use in computing the constraint statistic, the estimated error in the measurements, a "default" value, toward which points in the spectrum tend if the derivative with respect to the constraint statistic vanishes, and an optional estimate of the decay times of the spectral components. Information on the availability of the computer programs is given in the Appendix at the end of this volume.

### Sensitivity

Figure 2 illustrates spectra obtained for ethylbenzene. The spectrum in Fig. 2A was computed using the DFT from data acquired using a 90° flip angle pulse. The remaining spectra shown in Fig. 2 were computed from

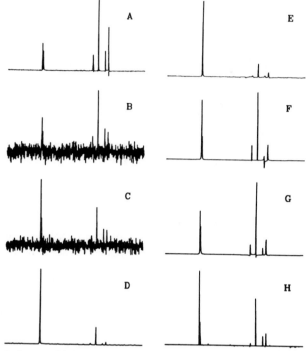

FIG. 2. Relative sensitivity of some spectral estimates: (A) 1024 (complex)-point spectrum obtained for a "standard" sample of 0.1% ethylbenzene in deuterochloroform, obtained from Wilmad, using a 90° pulse. The data for this and subsequent figures were obtained at 400 MHz on a JEOL GX-400 spectrometer. The outermost of the four lines at high field are the triplet and quartet of the ethyl group; the splitting is not observed at this digital resolution. The inner lines are due to impurities; these lines were present in several "standard" samples obtained from Wilmad. (B) Spectrum obtained for the same sample as (A), but using a 7° observe pulse. (C) FT spectrum using the same data as (B), but with exponential weighting corresponding to a 10-Hz Lorenzian line broadening applied prior to Fourier transformation. (D) Burg maximum entropy; 512 input points, linear prediction order 50. (E) LPZ: 512 input points, linear prediction order 50. (F) LPSVD: 512 input points, linear prediction order 384, reduced order 6. (G) HSVD: 512 input points, linear prediction order 384, reduced order 6. (H) Maximum entropy reconstruction: 512 input points, 40 iterations.

data acquired using a 7° pulse. The spectra in Figs. 2B and C were computed using the DFT: Figure 2B using 1024 complex data points, Fig. 2C using 512 complex points zero extended to 1024, and a 10-Hz Lorentzian line broadening applied.

The remaining spectra in Fig. 2 were evaluated at 1024 frequency values using the spectral estimates described in the section Maximum

Entropy Reconstruction with the following parameters: Fig. 2D, Burg MEM, 512 input points, 20-term LP filter; Fig. 2E, LPZ, 512 input points, 20-term LP filter; Fig. 2F, LPSVD, 512 input points, reduced order 6; Fig. 2G, HSVD 512 input points, reduced order 6; Fig. 2H, maximum entropy reconstruction, 1024 points used to compute $C$, $\sigma$ 0.01, default 0.001, 12 iterations.

Note the phase distortion of one of the peaks obtained using LPSVD. With LPSVD, using 256 inputs points rather than 512 points to compute the LP coefficients results in a spectrum missing a component. Using a reduced order of greater than six results in spurious peaks comparable in magnitude to the real peaks. HSVD performs somewhat better, but also exhibits spurious peaks when a reduced order greater than six is used. In contrast, maximum entropy reconstruction is "noise cognizant," admitting noise in the least structured manner possible.

The conventional means of quantifying sensitivity, the signal-to-noise ratio, is not useful for comparing the sensitivity of the spectral estimates, since (with the exception of maximum entropy reconstruction) noise in the data is not manifested as noise in the spectrum, but rather as spurious peaks or distortions of peaks. All the non-DFT spectral estimates yield spectra which appear to attain better noise suppression than a windowed DFT, without the concomitant line broadening. Appearances can be deceiving, however. For the Burg MEM and LPZ methods, using increasingly higher order linear prediction filters results in spurious peaks which are usually significantly smaller in magnitude than the real peaks. Using increasingly larger reduced order with the LPSVD and HSVD methods can lead to spurious peaks with magnitudes comparable to the real peaks, although it is sometimes possible to distinguish these spurious peaks on the basis of phase and line width. These properties of non-DFT spectral estimates heighten the importance of the distinction between sensitivity and signal-to-noise ratio.

## Resolution

Resolution refers to the ability to distinguish closely spaced components in a spectrum. For a discrete spectral estimate, digital resolution is the spacing between samples in the frequency domain. For parametric spectral estimates such as LPSVD, the digital resolution used to display the spectrum graphically is a matter of choice, and is independent of the ability of the method to distinguish closely spaced components. For the other methods, however, the digital resolution places a lower limit on the resolving power of the method. For experiments involving very wide spectral dispersion or short data records, this limitation can be severe.

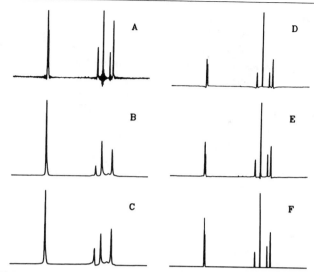

FIG. 3. High digital resolution spectral estimates from short data records. (A) FFT of the first 256 points of the data from Fig. 2A zero filled to 1024 points. (B) Burg MEM: 256 input points, linear prediction order 30. (C) LPZ: 256 input points, linear prediction order 30. (D) LPSVD: 256 input points, linear prediction order 192, reduced order 8. (E) HSVD: 256 point input points, linear prediction order 192, reduced order 8. (F) Maximum entropy reconstruction: 256 input points, 12 iterations.

The digital resolution of a DFT spectral estimate can be increased by appending zeroes to the data, with the drawback that the resulting spectrum is the convolution of the Fourier transform of the square data "window" ($\sin f/f$) with the desired spectrum. Presented in Fig. 3 are examples of the spectra produced by the methods described in the section Maximum Entropy Reconstruction using short data records obtained for ethylbenzene. In each case the spectrum was evaluated at 1024 frequency values using 256 (complex) data points. All of the non-DFT methods are free of "leakage." The resolution obtained using the simple linear prediction methods (Burg MEM and LPZ) is lower than that obtained with the hybrid parametric methods (LPSVD and HSVD) or maximum entropy reconstruction, which can be seen by comparing the low-field (aromatic) multiplet. Window functions can be used with non-DFT spectral estimates to improve resolution, as shown in the next section.

*Nonlinearity*

Nonlinearity of the non-DFT spectral estimates manifests itself in different ways and to different extents. The success of difference spectroscopy depends critically on the linearity of the spectral estimate. Figure 4 compares spectra obtained for the sum of two synthetic FID's with the sum of the spectra obtained for the individual FID's. The DFT is a linear spectral estimate, so the two means of obtaining the spectrum are equivalent. The non-DFT spectral estimates all exhibit intensity differences between the two means of computing the spectra. The Burg and LPZ spectra (LPZ not shown) both exhibit additional splitting in the spectrum obtained for the sum of the FID's. The intensity difference exhibited by maximum entropy reconstruction can be minimized by constraining the mock data obtained from the reconstructed spectrum more closely to the input data, at the expense of noise suppression, however.

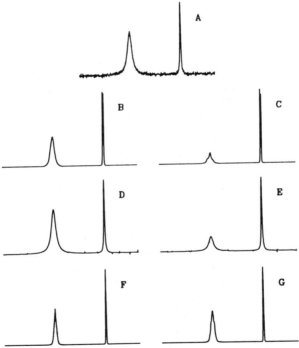

Fig. 4. Nonlinearity of the spectral estimates, illustrated by comparing the spectrum computed for the sum of two synthetic FID's with the sum of the spectra computed for the individual FID's. (A) FFT of the sum of the two FID's. (B) Sum of the Burg MEM spectra of the FID's. (C) Burg MEM spectrum of the sum of the FID's, using 512 input points and a linear prediction order of 20. (D) Sum of the HSVD spectra. (E) HSVD spectrum of the sum of the FID's, using 512 input points, linear prediction order 384, and reduced order 10. (F) Sum of the maximum entropy reconstructions. (G) Maximum entropy reconstructions of the sum of the FID's, 10 iterations.

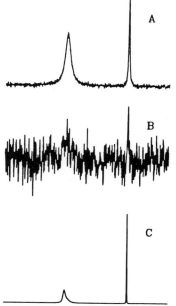

FIG. 5. Nonlinearity illustrated through deconvolution. (A) FFT of a synthetic FID. (B) FFT of the same data as in (A) but with a $-10$-Hz Lorentzian line broadening applied. (C) Burg MEM spectrum of the same data, using 512 input points and a linear prediction order of 4.

Nonlinearity is not necessarily bad. In fact, it is the nonlinearity of the non-DFT spectral estimates which affords them the ability to simultaneously enhance resolution and sensitivity compared to a DFT spectral estimate. Figure 5 illustrates a simple example of the use of an increasing exponential window function to achieve resolution enhancement. For linear methods, this results in amplification of the noise. A nonlinear spectral estimate, such as Burg MEM, discriminates between random noise and coherent signal.

Related to the nonlinearity of the spectral estimates is the question of intensity bias. Although this is an aspect that needs further investigation, the following general statements can be made. The Burg maximum entropy method yields biased intensities, but the integrated areas of spectral components tend to be accurate. Intensities using hybrid linear prediction methods, such as LPSVD, tend to be accurate. Neither simple nor hybrid linear prediction methods are terribly robust, however, and these statements characterize their behavior when they are not producing spurious components. Maximum entropy reconstruction usually yields biased intensities. The bias can be "tuned" by adjusting the relative weight applied to the constraint and entropy parts of the objective function. By tightly constraining the reconstructed spectrum to fit the data the bias can be minimized, although at the expense of noise suppression.

### Robustness

A spectral estimate is robust if it resists data contamination. Possible sources in NMR include transmitter breakthrough, receiver overload, and probe ring down. As an example, the first four points of a synthetic FID were manually "contaminated." The original FID contained components at $-500$ and 1000 Hz. Figure 6A and B illustrate the DFT spectra obtained for the original and contaminated data, respectively. The contaminated spectrum has a rolling baseline and a broad component at zero frequency, but the frequencies and relative intensities of the two original components are unaffected. The spectra computed for the contaminated data using the non-DFT spectral estimates are illustrated in Fig. 6C–H. The Burg maximum entropy (Fig. 6C) and LPZ (Fig. 6D) spectra exhibit some minor spurious components, but more importantly the frequencies of the major components are significantly biased. The principal components appear at $-526.6$ and 995.6 Hz in the Burg MEM spectrum and $-532.5$ and 989.7 Hz in the LPZ spectrum. In the HSVD (Fig. 6E) and LPSVD (Fig. 6F) spectra the frequencies and relative intensities of the major spectral components are accurate. The HSVD spectral estimate converged with a reduced order of eight, which accounts for the spurious structure. A reduced order of three was used in computing the LPSVD spectrum; using a reduced order of eight results in numerous sharp spurious components (not shown). In the maximum entropy reconstruction (Fig. 6G) the frequencies are accurate, but the relative intensities are biased. Figure 6G also illustrates the nonlinear scaling which occurs in maximum entropy reconstruction: noise near the top of the broad component is not suppressed as well as noise elsewhere in the spectrum because it is considerably higher than the "default" value. By constraining the maximum entropy reconstruction more tightly to the data the intensity bias can be diminished, although at the expense of sensitivity enhancement.

The examples given in Fig. 6 illustrate the kinds of artifacts that can be encountered. Prudence dictates that non-DFT spectral estimates should be compared with a DFT spectrum for the same data. Judicious choice of parameters for the non-DFT spectral estimates can dramatically improve the quality of the spectra (Fig. 6C–H). When data contamination is a potential problem, eliminating the first few data points may be essential.

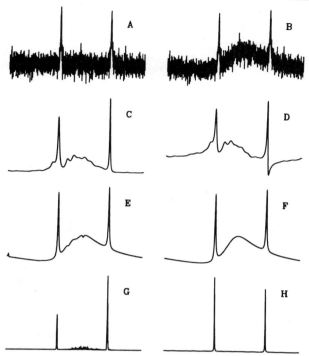

FIG. 6. Resistance of the spectral estimates to data contamination. (A) FFT of a synthetic FID. (B) FFT of the same data as in (A) but with the first four points "contaminated." The remaining spectral estimates are computed using this data. (C) Burg MEM: 512 input points, linear prediction order 20. (D) LPZ: 512 input points, linear prediction order 20. (E) HSVD: 256 input points, linear prediction order 192, reduced order 8. (F) LPSVD: 256 input points, linear prediction order 192, reduced order 3. (G) Maximum entropy reconstruction: 512 input points, 20 iterations. (H) Burg MEM, skipping the first four data points: 508 input points, linear prediction order 20.

The problems which have been described when linear prediction methods are applied to noisy data are probably due to nonideality of the noise realization, since the methods successfully suppress synthetic Gaussian noise added to a data set.

## Computing Time

The time required to compute the non-DFT spectra estimates depends on both the number of data points and the number of frequencies at which the spectrum is to be evaluated. Table I lists approximate central processor unit (CPU) times on a VAX 11/780 with a floating point accelerator for various data and spectrum sizes.

TABLE I
VAX 11/780 COMPUTING TIME (SECONDS)

| Method | Input points | Order[a] | Spectrum size | |
|---|---|---|---|---|
| | | | 1024 | 2048 |
| FFT | | | 0.3 | 0.7 |
| Burg MEM | 512 | 20 | 2.1 | 3.4 |
| | 512 | 40 | 4.1 | 5.7 |
| | 256 | 20 | 1.6 | 2.5 |
| LPZ | 512 | 20 | 2.8 | 4.3 |
| | 512 | 40 | 5.3 | 7.9 |
| | 256 | 20 | 2.2 | 3.6 |
| LPSVD | 512 | 20 | 424 | 427 |
| | 512 | 40 | 431 | 450 |
| | 256 | 20 | 90 | 83 |
| HSVD | 512 | 20 | 565 | 576 |
| | 512 | 40 | 665 | 665 |
| | 256 | 20 | 87 | 87 |
| ME reconstruction, | 512 | | 61 | 122 |
| 20 iterations | 1024 | | 62 | 124 |

[a] For Burg MEM and LPZ, order lists the linear prediction filter order. For LPSVD and HSVD order is the reduced order; the linear prediction filter order for both these methods is fixed at 0.75 times the number of input points.

## Two-Dimensional Data

Application of non-DFT methods to two-dimensional data is, in principle, straightforward. Since the computational cost of the Burg MEM, LPZ, LPSVD, and HSVD methods increases with spectral complexity, in practical applications they may be limited to use in the $t_1$ (column) dimension following Fourier transformation in the $t_2$ (row) dimension. In addition, long data records can be acquired in the $t_2$ dimension at little or no expense in overall acquisition time, since relaxation times govern the rate at which pulse sequences can be repeated. In contrast, overall acquisition time increases linearly with the length of the data record in the $t_1$ dimension. Consequently the benefits of the ability of the non-DFT spectral estimates to achieve high-resolution spectra from short data records without the truncation artifacts typifying zero-extended data accrue most significantly in the $t_1$ dimension.

The computational cost of maximum entropy reconstruction is independent of the complexity of the spectra, so in practice it could be applied to either dimension, or both. Maximum entropy reconstruction can be applied to both dimensions in two ways. It can be applied sequentially by first reconstructing the spectra of rows in the data matrix followed by columns, or it can be applied to both dimensions simultaneously (or all dimensions of higher dimensional spectra!). The constrained optimization problem is independent of the dimensionality of the data. It merely seeks that spectrum, having an arbitrary number of dimensions, which has the maximum entropy among those that are consistent with the available data. The objective function, consisting of the entropy and a constraint, is determined by summing over both (or all) dimensions. Although the feasibility of this approach has been demonstrated,[27] as a practical matter the computational resources required border on the prohibitive. Descent methods such as conjugate gradients require on the order of a few times $N$ storage locations, where $N$ is the number of points comprising the spectrum; the Cambridge algorithm requires at least $4N$ storage locations (reconstruction of spectra containing negative components adds another factor of two). In addition, since the method requires conversion from frequency (spectrum) to time (data) space, all four quadrants of hypercomplex spectra must be stored (eliminating a factor of two space savings which can be achieved by eliminating two of the four quadrants when column spectra are computed subsequent to row spectra). A moderate resolution phase-sensitive COSY spectrum consisting of 2048 (complex) points in both dimensions requires 64 megabytes of storage for all 4 quadrants of the hypercomplex spectrum (with a 32-bit word length) or 512 megabytes in order to carry out maximum entropy reconstruction in both dimensions simultaneously. Although disk storage sufficient to accommodate these requirements is readily available, the alternate (and repeated) row and column access of the spectrum matrix required by the conversion between spectrum and data space would result in prohibitively long computational times unless carried out on a computer with on the order of 512 megabytes of semiconductor memory. This remains the realm of supercomputers.

For the immediate future, non-DFT spectral estimates can be applied to the row or column dimensions of two-dimensional data independently. Although the computational requirements for applying these methods to the $t_1$ (column) dimension, for example, of two-dimensional data fall within the range of computers available to many NMR spectroscopists, there are other practical difficulties. For the Burg MEM and LPZ methods, the appropriate order of the linear prediction filter (and the reduced order for LPSVD and HSVD) depends on the number and relative intensity of spectral components, which varies from column to column. For the

[27] E. Laue, M. Mayger, J. Skilling, and J. Staunton, *J. Magn. Reson.* **68**, 14 (1986).

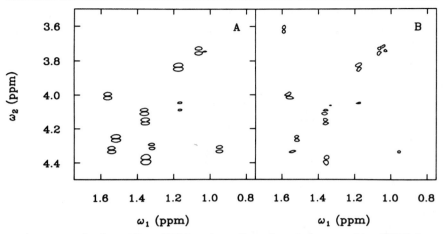

FIG. 7. Application of Burg MEM to the $t_1$ dimension of phase-sensitive COSY data, illustrated using the $\alpha$-methyl region of data obtained for hen egg white lysozyme. The data consisted of 1024 (complex) points in $t_2$ and 512 $t_1$ values, zero filled to 2048 in both dimensions. (A) Power spectrum computed following complex FFT applied to $t_1$. (B) Power spectrum obtained applying Burg MEM to $t_1$, 256 input points, linear prediction order 15.

Burg MEM and LPZ methods this can result in small changes in peak widths and intensities from column to column, producing oddly shaped peaks in two dimensions when the same order linear prediction filter is computed for each column. Figure 7 illustrates results using Burg MEM; LPZ results in similarly odd shapes.

Used judiciously, hybrid linear prediction methods, such as LPSVD, have been applied successfully to two-dimensional data. Schussheim and Cowburn[28] applied "image processing" to the parametric results to enhance the resulting spectra. Johnson et al.[29] improved the parameter estimates obtained using LPSVD by simultaneously optimizing all four parameters for each spectral component. When applying LPSVD to $t_2$, they achieved significant time savings by computing the frequencies and decay rates only once, computing only the phases and amplitudes for subsequent rows. Gesmar and Led[30] have applied linear prediction to both dimensions of simulated data.

These reports notwithstanding, the application of linear prediction methods to two-dimensional data presents a number of difficulties. The methods are not extremely robust, and they require an *a priori* estimate of

[28] A. Schussheim and D. Cowburn, *J. Magn. Reson.* **71,** 371 (1987).
[29] B. Johnson, J. Malikayil, and I. Armitage, *J. Magn. Reson.* **76,** 352 (1988).
[30] H. Gesmar and J. Led, *J. Magn. Reson.* **76,** 575 (1988).

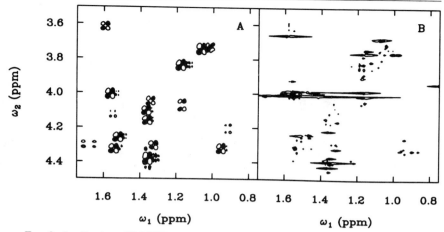

FIG. 8. Application of LPSVD to the $t_1$ dimension of phase-sensitive COSY data, using the same data as Fig. 7. Negative contours are filled. (A) Windowed FFT applied in both dimensions. (B) LPSVD applied to $t_1$, using 512 input points: linear prediction order 384, reduced order 15.

the number of significant spectral components. The presence of nonideal noise or an incorrect estimate of the number of frequency components can result in biased results or spurious peaks. Those methods which rely on rooting a complex polynomial are limited to relatively simple systems. They would not be appropriate, for example, applied to the $t_2$ dimension of an $^1$H COSY data of a macromolecule. Figure 8 illustrates the results that can be obtained applying LPSVD to the $t_1$ dimension of phase-sensitive COSY data for lysozyme from hen egg white. Figure 8A illustrates the $\alpha$-methyl cross-peak region obtained using windowed DFT's in both dimensions. Figure 8B illustrates the same region obtained substituting LPSVD for the Fourier transformation in the $t_1$ dimension.

Maximum entropy reconstruction is more robust than the linear prediction methods, and does not require an *a priori* estimate of the number of significant spectral components. Barna and Laue[31] have successfully applied maximum entropy reconstruction to the $t_1$ dimension of two-dimensional data using the same estimate of the noise level for each column. In the absence of severe $t_1$ noise, the noise level for each column should be approximately the same. Figure 9B illustrates results obtained applying maximum entropy reconstruction to the linearly sampled $t_1$ domain of phase-sensitive COSY data for lysozyme. Barna and Laue reported results for both linear and exponential sampling in $t_1$. Comparison

---

[31] J. Barna and E. Laue, *J. Magn. Reson.* **75**, 384 (1987).

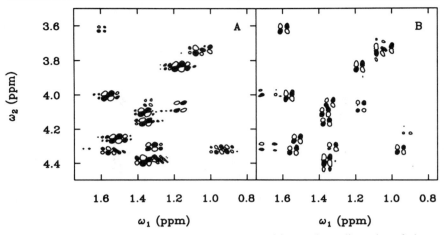

FIG. 9. Application of maximum entropy reconstruction to the $t_1$ dimension of phase-sensitive COSY data, using the same data as Fig. 7. (A) Windowed FFT applied in both dimensions (as in Fig. 8A), but using only 256 $t_1$ values, zero filled to 2048. (B) Maximum entropy reconstruction applied to $t_1$, using 256 $t_1$ values.

of Figs. 9A and B demonstrates the dramatic improvement over the DFT that can be obtained when maximum entropy reconstruction is applied to short data sets. Note particularly the elimination of truncation artifacts in the $\omega_1$ dimension. Comparison of Figs. 9B and 8A demonstrates that results obtained applying maximum entropy reconstructions to 256 $t_1$ values are comparable or better than results obtained applying DFT to 512 $t_1$ values. The spectrum illustrated in Fig. 9B required 4.5 hr of CPU time on a VAX 11/780; the spectrum in Fig. 8B required 26 hr.

### Overview

The simple linear prediction methods, Burg MEM and LPZ, are simple to implement and easy to use. Neither method is terribly robust. Burg MEM, as a power spectrum estimate, is limited to spectra containing positive peaks. LPZ, while it yields a complex spectrum estimate, can suffer from phase distortions when the spectrum is complex or the dynamic range high. The form of these spectral estimates makes them ill suited to arbitrary nonsymmetric line shapes, so they are unlikely to benefit the study of solids or magnetic resonance imaging.

LPSVD and HSVD, though more difficult to implement, are more robust. Since HSVD is not constrained by the difficulties of finding the zeroes of a complex polynomial, it effectively supercedes LPSVD. In principle these methods can be applied to arbitrary line shapes by adapting the model appropriately. An attractive feature of these methods is that they yield numerical estimates of the parameters characterizing the spectral components.

Maximum entropy reconstruction, while based on a simple idea, requires a complex optimizer for practical application. The optimizer used in the Cambridge algorithm consists of many thousands of lines of code. Despite this complexity, maximum entropy reconstruction affords a generality and flexibility which make it particularly attractive. An implementation suitable for reconstructing phase-sensitive spectra containing both positive and negative peaks is sufficiently general that it can be applied without modification to arbitrary line shapes and imaging data. This generality extends to data in one, two, or higher dimensions. The method can be "tuned" between the extremes of faithful intensity reproduction with low noise rejection and high noise rejection at the expense of intensity distortion. The very nature of maximum entropy reconstruction indicates that the method is highly resistant to spurious peaks: the reconstructed spectrum contains the minimum amount of structure necessary to satisfy the data constraints.

The application of these spectral estimates to NMR data remains an active area of research, particularly applied to two-dimensional data. Each of the methods described here warrants a place in the repetoire of methods available to practicing NMR spectroscopists. Increasing demands on sensitivity and resolution placed by spectroscopists on their data, and the increasing power and decreasing cost of computational resources, are likely to propel these methods into routine use along side traditional DFT methods. Finally, the nonlinear nature of these methods endows them with power which is equally susceptible to use and abuse.

## Acknowledgments

I thank R. de Beer of the Technical University of Delft for graciously providing code implementing LPSVD and HSVD. I thank J. Skilling, E. Laue, and J. Barna of the Cambridge Maximum Entropy Group for numerous discussions concerning maximum entropy reconstructions. Thanks are also due to members of the Rowland Institute for Science: to M. Burns, C. Shaefer, and A. Stern for numerous helpful discussions and for help writing software, to P. Connolly for collecting the lysozyme COSY data, and to M. Nilsson, J. Scarpetti, and L. Pew for assistance preparing the manuscript.

## [13] Solid-State Nuclear Magnetic Resonance Structural Studies of Proteins

*By* S. J. Opella and P. L. Stewart

### Introduction

This volume contains several chapters describing applications of solid-state NMR spectroscopy to the study of proteins. The results presented in these chapters are a direct consequence of developments in the instrumentation, methods, and theories used in NMR studies of solids. These developments began with the first demonstrations of high-resolution solid-state NMR spectroscopy.[1,2] The resolution and sensitivity of the experiments are now sufficient to obtain spectroscopic data from individual residues of proteins.

### Solid-State NMR Spectroscopy

Solid-state NMR spectroscopy brings several unique capabilities to the study of proteins. Most importantly, solid-state NMR spectroscopy enables the study of proteins that are immobile on NMR time scales ($10^3$–$10^6$ Hz) because they are in the crystalline or amorphous solid state or because they reorient slowly in solution as a result of their being part of a molecular aggregate. This means that a wide variety of biological systems may be characterized at atomic resolution, including large ordered systems, such as nucleoprotein or membrane–protein complexes. These systems are difficult or impossible to study by X-ray diffraction, because of its requirement for single-crystal samples, or by solution NMR spectroscopy, because of its requirements for low molecular weight and rapidly reorienting proteins. While both X-ray diffraction and solid-state NMR spectroscopy are well suited for studying single crystals of proteins, solid-state NMR spectroscopy has the advantage of being able to examine uniaxially oriented samples with equal facility as single crystals.[3] This relaxes the requirements on sample preparation considerably, since many biological systems that cannot be crystallized can be oriented by mechanical or magnetic methods. In addition, some solid-state NMR experiments can utilize unoriented solutions or polycrystalline powders.

[1] J. S. Waugh, L. M. Huber, and U. Haeberlen, *Phys. Rev. Lett.* **20**, 180 (1968).
[2] A. Pines, M. G. Gibby, and J. S. Waugh, *J. Chem. Phys.* **59**, 569 (1973).
[3] S. J. Opella and J. S. Waugh, *J. Chem. Phys.* **66**, 4919 (1977).

Solid-state NMR spectroscopy is highly effective at describing the structure[4] and dynamics[5] of proteins. A thorough solid-state NMR study of a protein enables the time-average relative positions and the intramolecular motions of individual sites to be described, providing extraordinary opportunities for understanding in detail the structures, dynamics, and mechanism of action of enzymes. In most studies of protein dynamics, unoriented samples are utilized and the line shapes of motionally averaged powder patterns from individual backbone and side-chain sites are analyzed.[5] This approach to studying molecular dynamics is particularly valuable because local motions can be readily characterized over a wide range of time scales in the absence of overall molecular reorientation. In addition, this approach has the advantage of relying on separate spectroscopic measurements to describe the types of motions and the frequencies of motions, since the line shapes are used to characterize the directions and amplitudes of the motions and relaxation parameters are used to characterize the frequencies of the motions.

This chapter describes the progress that has been made in developing an approach to determining structures of proteins[4,6–11] and peptides[12–16] in oriented samples with solid-state NMR spectroscopy. This approach relies on the spectral simplifications that result from sample orientation, with spin interactions at individual sites generally yielding single lines characterized by their resonance frequency or doublets characterized by the magnitude of the splitting. The observed values of the frequencies and splittings depend on the orientations of the principal axes of the spin-interaction tensors present at each site relative to the direction of the applied magnetic field. Therefore, the magnetic field of the NMR spectrometer defines a frame of reference. Molecular structural information is available because the orientations of many spin-interaction tensors in the molecular frame of reference have been established. Since molecular

[4] S. J. Opella, P. L. Stewart, and K. G. Valentine, *Q. Rev. Biophys.* **19**, 7 (1987).
[5] S. J. Opella, this series, Vol. 17, p. 327.
[6] E. Oldfield and T. M. Rothgeb, *J. Am. Chem. Soc.* **102**, 3635 (1980).
[7] T. M. Rothgeb and E. Oldfield, *J. Biol. Chem.* **256**, 1432 (1981).
[8] R. W. K. Lee and E. Oldfield, *J. Biol. Chem.* **257**, 5023 (1982).
[9] T. A. Cross and S. J. Opella, *J. Am. Chem. Soc.* **105**, 306 (1983).
[10] T. A. Cross and S. J. Opella, *J. Mol. Biol.* **182**, 367 (1985).
[11] P. L. Stewart, K. G. Valentine, and S. J. Opella, *J. Magn. Reson.* **71**, 45 (1987).
[12] R. Tycko, P. L. Stewart, and S. J. Opella, *J. Am. Chem. Soc.* **108**, 5419 (1986).
[13] P. L. Stewart, R. Tycko, and S. J. Opella, *Faraday Trans. 1* **24**, 3803 (1988).
[14] B. A. Cosnell, F. Separovic, A. J. Baldassi, and R. Smith, *Biophys. J.* **53**, 67 (1988).
[15] L. Nicholson, F. Moll, T. Mixon, P. LoGrasso, J. Lay, and T. A. Cross, *Biochemistry* **26**, 6621 (1988).
[16] J. H. Davis, *Biochemistry* **27**, 428 (1988).

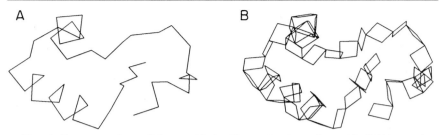

FIG. 1. Representations of the peptide backbone structure of crambin.[17] (A) α-carbon atoms connected by vectors. (B) Peptide planes outlined.

structures can be described on the basis of angles, given standard bond lengths, and geometries, it is possible to determine the structure of a protein with a sufficient number of spectroscopic measurements.[4]

*Protein Structure*

The pertinent features of protein backbone structure are illustrated with the small 46-residue protein crambin. A vector drawing of the positions of the α-carbon atoms of crambin determined by X-ray diffraction[17] is presented in Fig. 1A. This representation of the protein structure clearly shows the compact folding of the α-helical and β-sheet secondary structure elements. Proteins are polymers of amino acids joined with peptide bonds and all aspects of protein structure are strongly influenced by the chemical and geometric properties of the peptide linkage, especially since each peptide bond is part of a rigid planar grouping of atoms. The backbone of a protein consists of linked planes with quite regular geometry. The structure of crambin shown in Fig. 1B has each peptide plane drawn as a rectangle and is equivalent to the representation of the structure shown in Fig. 1A.

The relative orientations of the peptide planes in a protein are conveniently described with the torsion angles $\Phi$ and $\Psi$, which are defined in Fig. 2. The backbone torsion angles for crambin are plotted on a Ramachandran conformational energy diagram[18] in Fig. 3. The low-energy regions of the plot are designated for alanyl-like residues and correspond to the α-helical, β-sheet, and left-handed turn regions. All of the $\Phi$, $\Psi$ angle pairs of crambin are in low-energy regions, although three linkages involving glycine residues are involved in tight turns and have conformations which would be unfavorable for any other residue type. The com-

[17] W. Hendrickson and M. Teeter, *Nature (London)* **290**, 107 (1981).
[18] G. N. Ramachandran and V. Sasisekharan, *Adv. Protein Chem.* **23**, 283 (1968).

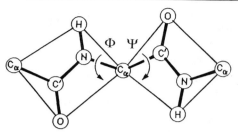

FIG. 2. Two peptide planes sharing a common $\alpha$-carbon atom. The rotation axis of the torsion angles $\Phi$ and $\Psi$ are defined with the arrows.

plete collection of $\Phi$, $\Psi$ pairs plotted in Fig. 3 fully characterizes the backbone structure of crambin.

## Methods

### Samples

There are two general requirements for the samples in structural studies. First, the sites of interest must be immobile on the time scales defined by the relevant nuclear spin interactions, ranging from about $10^3$ Hz for chemical shift interactions to $10^6$ Hz for nuclear quadrupole interactions. However, it is possible in favorable cases to deal with sites undergoing

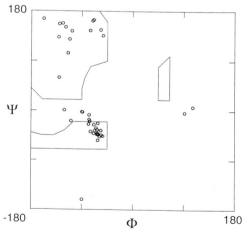

FIG. 3. Ramachandran conformational plots of $\Phi$ versus $\Psi$ angles. Each circle represents one $\Phi$, $\Psi$ pair of crambin.[17]

well-defined limited motional averaging, such as the 3-fold reorientation of methyl groups. Suitable immobile sites are present in protein crystals and in supramolecular structures in solution, such as nucleoprotein or membrane–protein complexes.

The second requirement is that the samples have at least one direction of orientation. In single crystals any arbitrary orientation of the sample can be studied. In uniaxially oriented samples, the direction of orientation must be parallel to that of the applied magnetic field.

*Instrumentation*

The experiments require a spectrometer optimized for solid-state NMR spectroscopy. This means that several channels of radio frequency (rf) irradiation must be delivered to a probe capable of dealing with high power without breakdown. All of the experiments benefit from using high magnetic fields. Spectrometers with adequate pulse programming capability, phase switching and stability, and rf power for solid-state NMR spectroscopy are generally home built. In many of these spectrometers, the high rf performance has been obtained at some cost in sensitivity and magnetic field homogeneity and stability. These trade-offs cannot be made in spectrometers used for solid-state NMR experiments on proteins, because it is always difficult to obtain, even in isotopically labeled samples, sufficient sensitivity to observe individual sites and it is necessary to have high intrinsic resolution for discrimination between similar sites in a protein. Many experiments require long-term signal averaging and, therefore, both the rf and magnetic field of the spectrometer must be highly stable.

## Determination of Peptide Backbone Structure

*Nuclear Spin Interactions*

The dipole–dipole, chemical shift anisotropy, and quadrupole interactions in polypeptide backbone sites are all highly anisotropic. The spin-interaction tensors have been determined in model peptides, yielding the orientations and magnitudes of the principal axes. These interactions can be used as sources of orientational information about peptide planes.[4] The dipole–dipole interaction between two spins depends on the distance between the two spins as well as the orientation of the internuclear vector with respect to the direction of the applied magnetic field; therefore this spin interaction provides spatial as well as angular information. In contrast, the chemical shift and quadrupole interactions, which reflect local electronic properties, provide only orientational information.

The dipole–dipole coupling between two spins is the simplest interaction to describe. The coupling of an amide nitrogen or an $\alpha$ carbon to a directly bonded proton, nitrogen, or carbon in a peptide linkage behaves very much like the interaction of an isolated spin pair. This coupling gives a spectrum that is a doublet when the nonobserved coupled spin is $S = \frac{1}{2}$. In an oriented sample, the frequency splitting of the doublet ($\Delta\nu_D$) has the angular dependence described by Eq. (1).

$$\Delta\nu_D = \nu_{\parallel}(3 \cos^2 \theta - 1) \tag{1}$$

$\theta$ is the angle between the direction of the applied magnetic field and the bond vector connecting the two nuclei and $2\nu_{\parallel}$ is the maximal frequency splitting for the dipolar coupling which is observed when the internuclear vector is parallel to the field direction. $\nu_{\parallel}$ can be readily calculated from the relevant gyromagnetic ratios, $\gamma_i$, and the distance between the two nuclei, $r$, using Eq. (2). There is excellent agreement between calculated and experimental values of $\nu_{\parallel}$ for $^{13}C$–$^{15}N$ and $^{13}C$–$^{13}C$ interactions; however, there are significant discrepancies between bond lengths determined from dipolar couplings and bond lengths determined by X-ray or neutron diffraction when one of the bonded atoms is hydrogen.[19] Therefore, it is necessary to use C–H or N–H bond lengths based on NMR measurements, while C–N and C–C bond lengths determined by diffraction methods can be used.

$$\nu_{\parallel} = h\gamma_A\gamma_B(1/r^3) \tag{2}$$

$^{13}C$ and $^{15}N$ are spin $S = \frac{1}{2}$ nuclei and the chemical shift anisotropy has an important influence on the spectra from these sites. Both chemical shift and quadrupole interactions are present at $^2H$ and $^{14}N$ sites which are spin $S = 1$ nuclei. However, because of the relative magnitudes of the energies of the two interactions, the spectral features are generally dominated by the effects of the quadrupole interaction. For $^{14}N$ nuclei in peptide bonds the magnitude of the quadrupole interaction is so large that even the second order quadrupole shifts are greater than the chemical shifts.

In $^{13}C$ and $^{15}N$ NMR experiments proton decoupling is essential to obtain high-resolution spectra which display only the effects of the chemical shift interaction. The chemical shift tensors for the $^{13}C$ sites[20–22] and

[19] J. E. Roberts, G. S. Harbison, M. G. Manowitz, J. Herzfeld, and R. G. Griffin, *J. Am. Chem. Soc.* **109**, 4163 (1987).

[20] R. E. Stark, L. W. Jelinski, D. J. Ruben, D. A. Torchia, and R. G. Griffin, *J. Magn. Reson.* **55**, 266 (1983).

[21] T. G. Oas, C. J. Hartzell, T. J. McMahon, G. P. Drobny, and F. W. Dahlquist, *J. Am. Chem. Soc.* **109**, 5956 (1987).

[22] C. J. Hartzell, M. Whitfield, T. G. Oas, and G. P. Drobny, *J. Am. Chem. Soc.* **109**, 5967 (1987).

[15]N sites[22-24] in peptide bonds have been determined for several model peptides. In oriented samples, the chemical shift interaction gives a single resonance line for each site with a characteristic frequency for each orientation. In a polycrystalline sample, where all orientations are present, the spectra take the form of powder patterns[25] which have a line shape and breadth characterized by the frequencies of the discontinuities, $\sigma_{ii}$. The observed resonance frequency for an oriented sample, $\nu_{cs}$, is given by Eq. (3) where $\theta$ and $\phi$ are the angles relating the principal axes of the chemical shift tensor to the magnetic field direction.

$$\nu_{cs} = \sigma_{11} \cos^2 \phi \sin^2 \theta + \sigma_{22} \sin^2 \phi \sin^2 \theta + \sigma_{33} \cos^2 \theta \qquad (3)$$

The quadrupole interaction of the spin $S = 1$, [14]N amide site is useful for determining angles in solid-state NMR studies and has several important advantages over the other observable interactions.[12-14] [14]N is over 99% naturally abundant, thus no isotopic labeling is required and sensitivity is high; the large quadrupole interaction provides high spectral resolution, with roughly a ratio of $10^3$ for total spectral width to line width; and second order effects are observable and yield additional angular information. The [14]N quadrupole tensor has been well characterized for model peptides.[26-28] Like the chemical shift tensor, it is nonaxially symmetric and thus there is a dependence on two angles, $\alpha$ and $\beta$, which describe the orientation of the principal axes of the tensor with respect to the magnetic field vector. Because [14]N is an $S = 1$ nucleus there are two fundamental $\Delta m = 1$ transitions and the spectra from oriented samples are doublets centered at the [14]N Larmor frequency. The observed splitting between the two frequencies gives angular information described by Eq. (4):

$$\Delta\nu_Q = (3/4)(e^2qQ/h)[3 \cos^2 \beta - 1 + \eta \sin^2 \beta \cos 2\alpha] \qquad (4)$$

where $e^2qQ/h$ is the quadrupole coupling constant and $\eta$ is the asymmetry parameter.

Additional orientationally dependent measurements can be made in the [14]N overtone NMR spectrum[11-14] where the weakly allowed overtone, or $\Delta m = 2$, transitions are directly excited and detected near twice the

[23] G. S. Harbison, L. W. Jelinski, R. E. Stark, D. A. Torchia, J. Herzfeld, and R. G. Griffin, *J. Magn. Reson.* **60**, 79 (1984).

[24] T. G. Oas, C. J. Hartzell, F. W. Dahlquist, and G. P. Drobny, *J. Am. Chem. Soc.* **109**, 5962 (1987).

[25] U. Haeberlen, "High Resolution NMR in Solids: Selective Averaging." Academic Press, New York, 1976.

[26] R. E. Stark, R. A. Haberkorn, and R. G. Griffin, *J. Chem. Phys.* **68**, 1996 (1978).

[27] G. F. Sadiq, S. G. Greenbaum, and P. J. Brey, *Org. Magn. Reson.* **17**, 191 (1981).

[28] R. Tycko and S. J. Opella, *J. Chem. Phys.* **86**, 1761 (1987).

Larmor frequency.[28,29] The observed resonance shift is twice that observed as a small perturbation in the fundamental spectrum and is a combination of second order shift and, to a lesser extent, chemical shift. The orientational dependence of twice the second order shift, $2\nu_Q(2)$, is given by Eq. (5):

$$
\begin{aligned}
2\nu_Q(2) = (1/\nu_0)(e^2qQ/h)^2\{&[9 \sin^2 \beta - (27/4)\sin^4 \beta] \\
&+ \eta \cos(2\alpha)[-3 \sin^2 \beta + (9/2)\sin^4 \beta] \\
&+ \eta^2[1 - (3/4)\cos^2(2\alpha)\sin^4 \beta]\}
\end{aligned} \tag{5}
$$

where $\nu_0$ is the $^{14}$N Larmor frequency and the angles $\alpha$ and $\beta$ are the same as for the fundamental measurement. The observed overtone shift also has a contribution of twice the chemical shift. The amide nitrogen chemical shift interaction may be approximated as axially symmetric and $\nu_{cs}$ may be rewritten in terms of $\alpha$ and $\beta$ as shown in Eq. (6).

$$
\nu_{cs} = \sigma_\| \sin^2 \beta \cos^2 \alpha + \sigma_\perp(1 - \sin^2 \beta \cos^2 \alpha) \tag{6}
$$

$\sigma_\|$ and $\sigma_\perp$ are the principal elements of an axially symmetric interaction.

Another angularly dependent measurement possible in $^{14}$N overtone spectroscopy is the relative nutation frequency. This is a measure of how allowed the transition is and it has its own characteristic angular dependence on the angles $\alpha$ and $\beta$. The nutation frequency is determined relative to that for a nearby fully allowed fundamental transition (deuterium is convenient) with the same rf power. The orientational dependence of the relative overtone nutation frequency, $\nu_{nut}^{(rel)}$, is

$$
\begin{aligned}
n_{nut}^{(rel)} = (\gamma_N/2\gamma_H\nu_0)(e^2qQ/h)\{&9 \sin^2 \beta \cos^2 \beta - 6\eta \cos(2\alpha)\sin^2 \beta \cos^2 \beta \\
&+ \eta^2[\sin^2 \beta - \cos^2(2\alpha)\sin^4 \beta]\}^{1/2}
\end{aligned} \tag{7}
$$

The combination of fundamental and overtone $^{14}$N NMR spectroscopy provides flexibility in both the experimental and interpretive aspects. In particular, $^{14}$N overtone NMR spectroscopy has the advantage over fundamental NMR spectroscopy of having a total spectral width of approximately 120 kHz (for amide nitrogens in a 5.9-T field) rather than the 4.8-MHz spectral width of fundamental $^{14}$N NMR. This greatly enhances the practicality of the spectroscopy. On the other hand, fundamental $^{14}$N spectra have the advantages of higher sensitivity and resolution. The combination of multiple independent structural measurements from overtone and fundamental $^{14}$N NMR spectra improves the accuracy and precision of the results.

$^2$H quadrupole interactions can also be useful for determining peptide backbone structure. For a single deuteron bonded to a carbon, the quad-

[29] R. Tycko and S. J. Opella, *J. Am. Chem. Soc.* **108**, 3531 (1986).

rupole tensor is axially symmetric with the largest principal component along the C–$^2$H bond.[30] Thus, the observed quadrupole splitting depends on a single angle. By specifically replacing $C_\alpha$ protons with deuterons, it is possible to measure the $^2$H quadrupole splitting. The angular dependence of the splitting is described by Eq. (8).

$$\Delta\nu_Q^{C-D} = (3/4)(e^2qQ/h)(3\cos^2\beta - 1) \tag{8}$$

$e^2qQ/h$ is the quadrupole coupling constant, which is approximately 180 kHz, and $\beta$ is the angle between the $C_\alpha$–$^2$H bond and the magnetic field. This angle is the same as would be found from the $^1$H–$^{13}C_\alpha$ dipole–dipole splitting and both measurements have the same angular dependence. Measurements of $^2$H quadrupole splittings for deuterons bonded to the $\beta$ and other side-chain carbons may also be useful, as may the $^1$H–$^{13}C_\beta$ dipole–dipole splittings. If there is motion in the side chain, the interaction will be averaged and the motional averaging must be taken into account in the analysis.

There is an important distinction between the angular dependence of the dipole–dipole coupling and that of the asymmetric chemical shift and quadrupole interactions. The dipolar interaction depends on the single angle between the $\nu_\parallel$ axis of the interaction and the applied magnetic field, because this interaction is axially symmetric. By contrast, the chemical shift and quadrupole interactions are generally nonaxially symmetric and their spectral frequencies depend on two angles. It is equally straightforward to calculate the spectral frequencies for an arbitrary orientation for both axially symmetric and nonaxially symmetric interactions. However, there is a large difference in the interpretation of the spectral frequencies in terms of orientation. Obviously the measurement of a spectrum from an axially symmetric interaction, such as a dipolar splitting, gives two or at most four values for a single angle. A nonaxially symmetric interaction has many possible combinations of two angles that are consistent with any given measurement. This distinction is important and will be demonstrated graphically in the next section.

For all of the spin interactions, the spectrum corresponding to an arbitrary orientation of a peptide plane, and hence any orientation of the spin-interaction tensor, can be readily calculated. The appropriate angles need to be determined from geometric considerations and then used along with the established magnitudes and orientations of the principal elements of the spin interactions of interest. In some cases it is possible to verify structural features by directly comparing experimental and calculated spectra. However, the actual determination of a structure requires the

[30] C. P. Slichter, "Principles of Magnetic Resonance." Springer-Verlag, New York, 1978.

extraction of angular information from the experimental measurements alone.

### Angular Restrictions from Experimental Measurements

The experimentally measured resonance frequencies and splittings are used to establish restrictions on the orientations of the spin-interaction tensors, and hence the orientations of the peptide planes, that are consistent with the chemical structure of the peptide plane.[4,11,12] In some cases, especially for those spin interactions with axial symmetry, such as dipole–dipole couplings, it is possible to directly determine the angles of bonds relative to the direction of the applied magnetic field. However, a general method that is useful for all spin interactions relies on representing the experimentally derived restrictions on orientation as areas in plots of two angles. This enables errors in the experimental measurements, as well as uncertainties in the magnitudes and orientations of the principal elements of the second rank tensors describing the spin interactions, to be directly evaluated.

There are a number of heteronuclear dipole–dipole couplings that can be measured in peptide groups.[4,10] These include interactions between the hydrogen and directly bonded nitrogen and between the nitrogen and $\alpha$ carbon or carbonyl carbon. These interactions are axially symmetric, readily calculated, and there is little uncertainty over the orientation and magnitude of the interaction, therefore they are reliable sources of angular information and can be used to illustrate the method. The angles between the N–H and N–C$'$ bonds and the direction of the applied magnetic field are shown for a peptide plane in Fig. 4. Also defined are the polar angles $\alpha$ and $\beta$, which describe the orientation of the molecular axis system with respect to the magnetic field vector.

Figure 5A presents a calculated spectrum that is a doublet with a splitting, $\Delta\nu_D$, corresponding to a heteronuclear $^{15}$N–$^{1}$H dipole–dipole coupling of 17.8 kHz. The angular dependence of this interaction is described by Eq. (1) where $\nu_\parallel$ is 9.93 kHz from an N–H bond length of 1.07 Å. The magnitude of the splitting as a function of $\theta^{\text{H–N}}$, the angle between the N–H bond and the magnetic field, is plotted in Fig. 5B. Using this plot, the experimental measurements indicate that the N–H bond is tilted by 15 or 165° from a direction parallel to the magnetic field. The possibility of the angle being 15 or 165° demonstrates the presence of ambiguities in the angles derived from NMR measurements.

The use of heteronuclear dipole–dipole couplings to determine the orientations of peptide planes can be demonstrated with plots of $\theta^{\text{H–N}}$ vs $\theta^{\text{C'–N}}$, the angle between the C–N bond and the magnetic field, where the

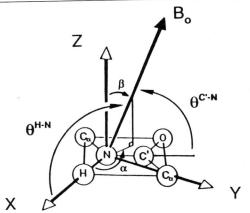

FIG. 4. A peptide plane with the molecular axis system, $X$, $Y$, $Z$, defined. The angles $\alpha$, $\beta$, $\theta^{H-N}$, and $\theta^{C'-N}$ are shown for an arbitrary direction of the magnetic field vector, $B_0$.

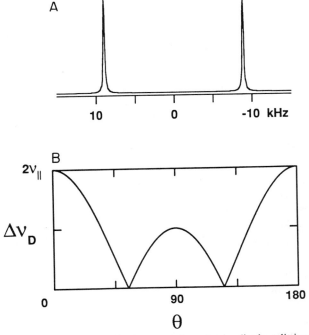

FIG. 5. (A) Calculated spectrum for heteronuclear dipole–dipole splitting of 17.8 kHz. (B) Plot of the dipole–dipole splitting, $\Delta\nu_D$, versus the angle $\theta$.

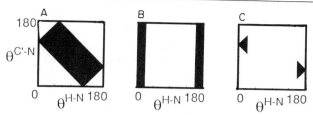

FIG. 6. Allowed combinations of angles between the C'–N bond and $B_0$ and between the N–H bond and $B_0$. (A) Combinations consistent with the stereochemistry of a peptide plane. (B) Combinations consistent with an N–H dipolar splitting of 17.8 kHz without regard for chemical or stereochemical constraints. (C) Combinations consistent with the conditions in the (A) and (B) restriction plots.

dark areas represent those combinations of $\theta^{N-H}$ and $\theta^{C'-N}$ that are consistent with the experimental data and the chemical properties of the peptide group. The precision of the determination of the angles is affected by experimental errors and uncertainties in the bond length which affects the magnitude of the maximal dipolar splitting represented by $\nu_\parallel$. In favorable cases $\nu_\parallel$ can be measured independently for an individual site, removing this constant as a source of uncertainty. Considering the practical problems of dealing with all of the sites in proteins, it is generally necessary to use an average value for the bond lengths and maximal dipolar splittings.

Figures 6 and 7 present plots of $\theta^{H-N}$ vs $\theta^{C'-N}$ to illustrate the roles of errors, uncertainties, and the chemical properties of the peptide group in translating experimental measurements into angular restrictions. The plot in Fig. 6B shows the restrictions placed on the orientation of an arbitrary N–H bond by the finding of $\Delta\nu_D = 17.8$ kHz, including the effects of experimental error and uncertainty in the bond length. If these latter

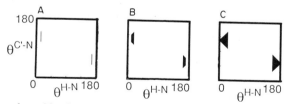

FIG. 7. Allowed combinations of angles between the C'–N bond and $B_0$ and between the N–H bond and $B_0$. (A) Combinations consistent with an N–H dipolar splitting of 17.8 kHz for an amide nitrogen in a peptide plane with an N–H bond length of 1.07 Å. (B) Same as (A) except allowing for experimental error, $\Delta\nu_D = 17.8 \pm 0.9$ kHz. (C) Same as (B) except allowing for variation in the bond length, $r = 1.07 \pm 0.04$ Å, as well as experimental error. Same as the plot in Fig. 6C.

considerations were not included, then there would be two infinitely thin lines at 15 and 165° instead of the broad bands in the plot.

The plot in Fig. 6A shows the restrictions on possible combinations of $\theta^{C'-N}$ and $\theta^{H-N}$ for a peptide plane. This representation clearly shows that the stereochemistry and rigidity of the peptide group by themselves provide a substantial reduction in the possible combinations of angles. A major strength of this approach to determining orientations is that angular restrictions from multiple sources can be combined. This is shown in Fig. 6C, where the restrictions from the single dipole–dipole splitting, including experimental error and uncertainty of the bond length, are combined with those of the stereochemistry of the peptide plane. The net result is that a very limited combination of angles are consistent with even this one NMR measurement.

The influences of errors and uncertainties on the combinations of angles consistent with the single measurement are illustrated in the plots in Fig. 7. The plot in Fig. 7C is the same as that in Fig. 6C, where the breadth and shape of the dark areas are determined by the errors and uncertainties for this particular measurement. The number of dark areas is determined by the symmetry properties of the angular dependence of the nuclear spin interaction. The plot in Fig. 7A shows the small areas, actually infinitely thin lines, that would be consistent with the single measured dipolar splitting if there was no error in the measurement and if the bond length was known precisely. The plot in Fig. 7B shows the areas that would be allowed taking into account experimental error, but having an established bond length.

The plots in Figs. 6 and 7 illustrate the nonlinear effects of errors and uncertainties on the combinations of angles that are consistent with a single experimental measurement. They clearly demonstrate the dramatic restrictions on possible orientations of a peptide plane that result from a single measurement by taking into account the chemical properties of the peptide plane. These plots also show the existence of ambiguities that arise in going from experimental data to angular restrictions. These ambiguities are simple for dipole–dipole couplings, but can become quite complex for other spin interactions.

Additional experimental measurements serve to reduce the allowed combinations of angles even further. This is demonstrated with the plots in Fig. 8. The plot in Fig. 8A is the same as in Fig. 7C for the restrictions from the single N–H dipolar coupling. The plot in Fig. 8B shows the angular restrictions from a C'–N dipolar coupling of 250 Hz, including the effects of experimental errors and bond length uncertainties. The plot in Fig. 8C then shows the results of combining the restrictions resulting from

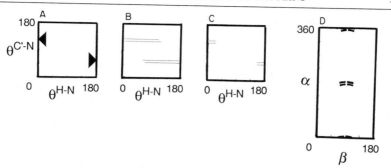

FIG. 8. Angular restrictions represented in $\theta^{H-N}$, $\theta^{C'-N}$, and $\alpha$, $\beta$ plots. (A) The restriction on $\theta^{H-N}$ and $\theta^{C'-N}$ from a single N–H dipole coupling. (B) The restriction on $\theta^{H-N}$ and $\theta^{C'-N}$ from a single C'–N dipole coupling. (C) The restriction on $\theta^{H-N}$ and $\theta^{C'-N}$ from both measurements. (D) The restriction on $\alpha$ and $\beta$ from both measurements.

the two independent experimental measurements on a single peptide plane.

Figure 8D converts the combined angular restrictions as displayed in the $\theta^{H-N}$ vs $\theta^{C'-N}$ plot in Fig. 8C to a more general $\alpha$ vs $\beta$ plot. No additional information is needed in going between these two representations, just the definition of the molecular axis systems shown in Fig. 4. It is realistic to determine the orientations of peptide planes to within a few degrees with experimental NMR data. The complexity in solid-state NMR protein structure determination arises from the ambiguities. The $\alpha$, $\beta$ plot in Fig. 8D shows that there are eight peptide plane orientations consistent with the two measured NMR parameters. The number of possible orientations can often be reduced with additional experimental measurements. However, even given many measurements for a single plane there would still be several possible orientations. Figure 9 shows the eight orientations of the peptide plane that are represented in the $\alpha$, $\beta$ plot in Fig. 8D.

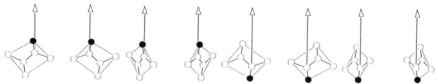

FIG. 9. The eight peptide plane orientations consistent with the allowed regions on the $\alpha$, $\beta$ plot in Fig. 8D. The dark circle represents the hydrogen attached to the nitrogen. The arrow represents the magnetic field vector.

The same type of analysis can be performed on the peptide planes in a protein. This yields the possible orientations of the peptide planes with respect to the direction of the magnetic field vector. In order to determine the structure of the protein, the orientations of the individual planes relative to a common vector must be converted into relative orientations of adjacent peptide planes. This requires a computer program to perform the necessary coordinate transformations and to select among the symmetry-related possibilities. The coordinate sets of the possible orientations for each peptide plane form the input to a computer program called TOTLINK,[4,11,31] which evaluates all possible combinations of plane orientations determined by solid-state NMR measurements and selects for the most chemically reasonable peptide backbone structures. The selection criteria are in the following order: (1) secondary structural elements including $\alpha$ helix, 3/10 helix, $\beta$ sheet, and type I, II, and III turns, (2) coarse Ramachandran $\Phi$, $\Psi$ energy values for Ala-like residues, glycyl residues, *trans*-prolyl residues, and non-glycyl residues before a proline or dihedral angle deviation from standard $\Phi$, $\Psi$ values, (3) agreement with disulfide bond information if given, and (4) agreement with $H-C_\alpha-B_0$ angles if given. A second program, EVAL, may be used to further select structures based on their overall size and shape. TOTLINK can vary all of the backbone bond angles and the torsion angle $\omega$ within narrow but adjustable limits. This is to ensure that possible structures are not missed because standard peptide planes were assumed and also to coarsely minimize the energy at every linkage.

At every planar linkage TOTLINK performs a transformation on the coordinates of the second plane in order to form a correct chemical linkage at the $C_\alpha$ atom, and finds the torsion angles, $\Phi$ and $\Psi$, which describe the relative orientation of the two planes. The transformation involves translating the coordinates of the second plane so that the coordinates for the $C_\alpha$ atom common to both planes coincide, and rotating the second plane around the magnetic field vector until a correct $N-C_\alpha-C'$ bond angle is formed. There are four possible outcomes: zero, one, two, or an infinite number of chemically correct linkages may be found. The most common outcome is two possible linkages. The second most common outcome is that no chemically correct linkages are possible, with the other two outcomes occurring very rarely. Figure 10 depicts the most common situation, where there are two chemically correct ways to link together two adjacent plane orientations. Notice that the two possible linked structures have different $\Phi$, $\Psi$ values. The reason there are usually two ways to link together two peptide planes is that there is cylindrical symmetry in

31 P. L. Stewart and S. J. Opella, unpublished results.

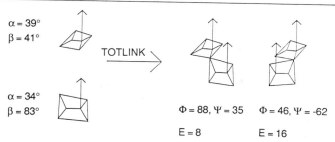

α = 39°
β = 41°
                    TOTLINK

α = 34°
β = 83°

Φ = 88, Ψ = 35     Φ = 46, Ψ = -62

E = 8              E = 16

FIG. 10. Linkage of two peptide planes by TOTLINK. There are two possible chemically correct structures. Their Φ, Ψ pairs and coarse Ramachandran energy values are given.

the orientational information. Solid-state NMR measurements yield only two polar angles, which describe the orientation of a plane with respect to one vector, the magnetic field vector. In order for the plane orientation to be completely specified with respect to a three-dimensional laboratory axis system, a third polar angle would be required.

The correct way to link together two adjacent planes can usually be distinguished if the angle between the $C_\alpha$–H or the $C_\alpha$–$C_\beta$ bond and the magnetic field vector, $B_0$, is known. This does not work for glycine residues since they have achiral $C_\alpha$ carbons and occasionally it does not work for nonglycine residues if the H–$C_\alpha$–$B_0$ and $C_\beta$–$C_\alpha$–$B_0$ angles happen to be similar. TOTLINK can evaluate agreement with H–$C_\alpha$–$B_0$ angles if they are given; however, if experimentally determined angles are used there will be ambiguities and error ranges to consider and this information may not be all that useful for selecting between possible linkages.

If H–$C_\beta$–$B_0$ angles are not given, then for a long peptide backbone there are a large number of possible structures which can be generated even if the orientation of every plane is known uniquely with respect to the magnetic field vector, i.e., each plane has a unique $\alpha, \beta$ pair. The maximum possible number of structures in this case is $2^{(n-1)}$ where $n$ is the number of peptide planes. (In the rare event that there are an infinite number of ways to link two planes together, TOTLINK selects for one possibility based on lowest Ramachandran energy.) The number of possible backbone structures is also greatly increased as the number of possible orientations for each plane increases. Fortunately, a large number of these possible structures are not chemically or biochemically reasonable. TOTLINK deals with the exponentially growing number of possible structures by evaluating sections of four peptide planes at a time.

When there are more than two peptide planes, TOTLINK must trace a "tree" structure of possible plane orientations. A small sample tree for three peptide planes, each with two possible orientations, is diagrammed

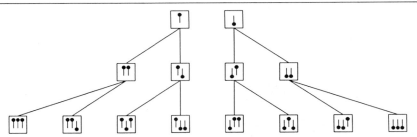

FIG. 11. A tree diagram showing the paths which must be tested for three peptide planes, each with two different orientations.[4] The two different vector orientations represent the two different plane orientations.

in Fig. 11. The first "branch" in the tree consists of starting with the first possible orientation for the first peptide plane and testing it with the first possible orientation of the second plane. Then the first possible orientation of plane two is tested with the first possible orientation of plane three. If, while tracing a branch, TOTLINK comes upon two plane orientations which cannot be linked together in a chemically correct manner this is considered a dead end and the current branch is stopped and a new branch initiated. A branch may also be stopped if the energy total for the lowest energy structure in the branch exceeds a specified limit.

After tracing all possible combinations of plane orientations in each section, the possible structures for that section are evaluated on the basis of maximum secondary structure and then lowest Ramachandran energy. The structures are then ranked and only the lowest energy ones are saved. The more possible plane orientations there are to consider, the higher the risk that the correct structure will not be saved at this point. After the second section has been evaluated, the saved structures for sections one and two are joined if they have the same orientation for their common plane and then the complete structures are evaluated and ranked on the basis of secondary structure and Ramachandran energy. Additional sections are evaluated and added on in the same manner.

*Backbone Structure of a Peptide in a Single Crystal*

The structure of the peptide *N*-acetyl-L-valyl-L-leucine (NAVL) has been determined with solid-state NMR experiments of a single-crystal sample.[13] Since $^{14}$N NMR spectroscopy was used, no isotopic labeling was necessary. Figure 12 shows the $^{14}$N NMR spectra at 8.4 T of a single crystal of NAVL at an arbitrary orientation with respect to $B_0$. The overtone lines are near twice the Larmor frequency, 51.703 MHz, and the fundamental lines are centered around the Larmor frequency of

A

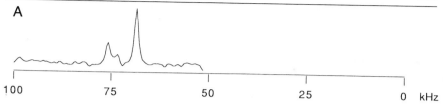

| | | | | |
|---|---|---|---|---|
| 100 | 75 | 50 | 25 | 0   kHz |

B

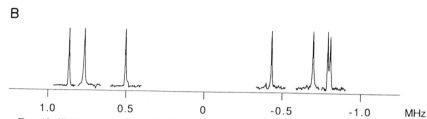

| | | | | |
|---|---|---|---|---|
| 1.0 | 0.5 | 0 | -0.5 | -1.0   MHz |

FIG. 12. $^{14}$N NMR spectra at 8.4 T of a single crystal of NAVL at an arbitrary orientation with respect to $B_0$. (A) The overtone spectrum shifted from twice the Larmor frequency, 51.703 MHz. (B) The fundamental spectrum centered at the Larmor frequency, 26.852 MHz.

26.852 MHz. Since there are two magnetically inequivalent molecules in the unit cell, four overtone lines are expected; however, one of the lines is not observed because of its near-zero transition moment. There should also be eight fundamental transitions, or four quadrupole splittings. The two highest frequency lines in the fundamental spectrum are very close in frequency and are not resolved, although they are resolved in a two-dimensional separated local field spectra. The fundamental spectrum is not symmetric about the Larmor frequency because of second order quadrupole effects.

Two-dimensional separated local field experiments for each of the fundamental $^{14}$N resonances were performed to determine the N–H dipolar couplings.[32] An overtone nutation frequency measurement[29] was made on the two stronger overtone lines. Large errors were placed on the nutation frequency measurements to account for a slightly different crystal orientation in this experiment. The relative nutation frequencies were calculated from the measured overtone nutation frequencies and the measured $^2$H nutation frequency with the same rf power. The relative nutation frequencies of the two weak overtone lines had to be estimated on the basis of their intensity and were given large error ranges. Table I lists all of the experimental $^{14}$N NMR measurements for the NAVL single crystal.

[32] R. K. Hester, J. C. Ackerman, B. L. Neff, and J. S. Waugh, *Phys. Rev. Lett.* **36**, 1081 (1976).

TABLE I

FUNDAMENTAL AND OVERTONE $^{14}$N NMR MEASUREMENTS FOR SINGLE CRYSTAL OF
$N$-ACETYL-L-VALYL-L-LEUCINE

| Parameter | Molecule 1 | | Molecule 2 | |
|---|---|---|---|---|
| | Val plane | Leu plane | Val plane | Leu plane |
| $^{14}$N quadrupole splitting | 1.676 MHz | 1.657 MHz | 0.923 MHz | 1.455 MHz |
| $^{14}$N overtone shift | 69 kHz | 74 kHz | 76 kHz | 82 kHz[a] |
| $^{1}$H–$^{14}$N dipole splitting (fundamental)[b] | 0.0 kHz | 8.0 kHz | 10.8 kHz | 15.9 kHz |
| Overtone relative nutation frequency[c] | 0.057 | 0.010[d] | 0.022 | 0.005[d] |

[a] Twice the shift observed in fundamental spectrum.
[b] Average measurement.
[c] From a second crystal in a similar orientation.
[d] Estimated from overtone resonance.

The measurements are analyzed in terms of restrictions on the angles $\alpha$ and $\beta$. Each measurement is first interpreted separately as an $\alpha$, $\beta$ plot showing all of the orientations which are allowed, then as an $\alpha$, $\beta$ plot with the combined restrictions from all of the measurements is made. In this way, it is possible to observe the mathematically complex regions where overlaps in the restrictions occur. In the analysis for one peptide plane of NAVL no overlap in the combined restriction plot was found until the orientations of the $^{14}$N quadrupole and N–H dipole tensors were rotated by 5° around their assumed positions. Experimental error and uncertainty in the tensor values are also included in the calculations resulting in angular restriction plots.

The angular restriction plots for the individual measurements from one of the four peptide planes in the unit cell of the NAVL crystal are shown in Fig. 13. First, the restriction from the largest quadrupole splitting, 1.676 MHz, is analyzed using the quadrupole constants determined for the valine amide chemical site and the $\alpha$, $\beta$ plot is shown in Fig. 13A. The second measurement analyzed in Fig. 13B is the shift observed in the overtone spectrum. Since the chemical shift effect is much smaller than the second order quadrupole effects, only an estimated axially symmetric amide chemical shift tensor is used without varying its parameters or orientation. The $^{14}$N–$^{1}$H dipole–dipole splitting is the third measurement analyzed using the $\alpha$, $\beta$ plot shown in Fig. 13C. The overtone relative nutation frequency has quite a different dependence on $\alpha$ and $\beta$ than the

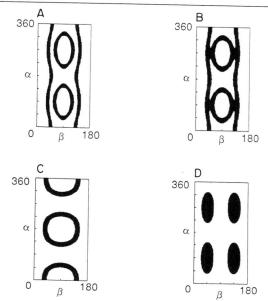

FIG. 13. Angular restrictions shown as $\alpha$, $\beta$ plots from each of the $^{14}$N NMR measurements for the Val plane of molecule 1 of NAVL. (A) Quadrupole splitting, (B) overtone shift, (C) N–H dipole splitting, and (D) overtone relative nutation frequency. Experimental error, variation in tensor values, and variation in tensor orientation are included.

other measurements, as shown in Fig. 13D, and thus is very useful, even if it can only be roughly estimated from the intensity of the overtone line.

The combined angular restrictions from these four measurements is shown in Fig. 14A. The same analysis was done for the other peptide planes. The combined restriction plot for the second plane of molecule 1

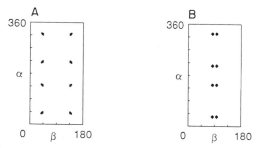

FIG. 14. Combined angular restrictions shown as $\alpha$, $\beta$ plots for the two peptide planes in molecule 1 of NAVL: (A) valine plane; (B) leucine plane.

is shown in Fig. 14B. Both plots show eight possible symmetry-related plane orientations. Eight is a common number of allowed possibilities since there are eight symmetry-related octants on the complete $\alpha$, $\beta$ plot for each of the $^{14}N$ NMR measurements used. Coordinate sets are generated for each possible plane orientation. Figure 15 shows all of the plane orientations used for molecule 1 in the structural analysis. Only four of the eight possible orientations for the Val plane are used, since the second four are related to the first four by changing only the direction of the magnetic field vector. Since this is also the case for the eight orientations of the Leu plane, including all eight for both planes would yield duplicate final structures.

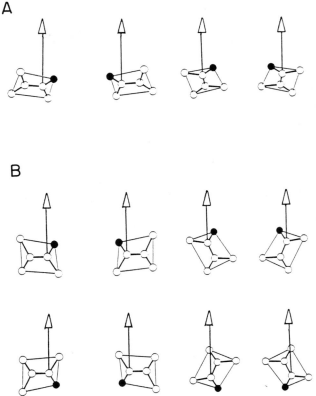

FIG. 15. The possible orientations used in the structural analysis for the valine (A) and leucine (B) peptide planes in molecule 1 of NAVL. The dark circle represents the hydrogen attached to the nitrogen. The arrow represents the magnetic field vector.

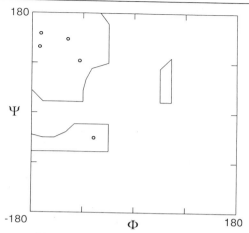

FIG. 16. The five possible average low-energy NMR backbone structures for NAVL shown on a Ramachandran conformational plot.

Each possible orientation for the Val plane shown in Fig. 15A must be tested with each possible orientation for the Leu plane shown in Fig. 15B. This gives 32 possible combinations of peptide plane orientations to evaluate. Each of these combinations may yield up to two chemically correct structures. However, most of the possible structures have $\Phi$, $\Psi$ torsion angles which fall on high-energy regions of a Ramachandran plot. Seven possible structures on the lowest energy regions of a Ramachandran plot were generated by the computer program TOTLINK. The same analysis was done for molecule 2 and seven different possible low-energy structures were generated.

It is known from the structure determined by X-ray diffraction[33] that the two molecules in the unit cell have the same molecular structure. Thus the correct structure should appear in the TOTLINK output for both molecules. By comparing the seven selected structures for both molecules, allowing for some experimental deviation between the $\Phi$, $\Psi$ values, and averaging the close $\Phi$, $\Psi$ pairs, five average NMR structures were obtained. Figure 16 is a Ramachandran plot showing the $\Phi$, $\Psi$ pairs for the five average NMR structures and coarse outlines of the three low-energy regions on an alanyl-like Ramachandran plot. One of the five structures ($\Phi = -117$, $\Psi = 133$) was obtained by averaging three $\Phi$, $\Psi$ pairs [($-106$, 127), ($-135$, 122), and ($-112$, 151)]. The variation in these three $\Phi$, $\Psi$ pairs can easily be explained by the error in $\alpha$ and $\beta$ and by not knowing

[33] P. J. Carroll, P. L. Stewart, and S. J. Opella, *Acta Crystallogr.*, in press.

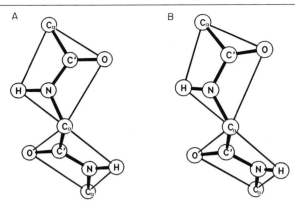

FIG. 17. Comparison of the best NMR structure (A) and the X-ray crystal structure (B) of NAVL. Only the two peptide planes are shown.

the exact peptide plane geometries. The torsion angles found for the same crystal form of NAVL by x-ray diffraction[33] are $\Phi = -118$ and $\Psi = 132$. The peptide planes in the best NMR structure and the x-ray structure are compared in Fig. 17. The dihedral angle deviation (DHAD) for $\Phi$ and $\Psi$ between the two structures is 1°.

## Structure of Two Peptide Planes in a Uniaxially Oriented Protein

The orientations of adjacent peptide planes in a protein may be determined in the same manner as demonstrated for a model peptide.[4] An analysis of two peptide planes in the coat protein of the filamentous bacteriophage fd is described here as an example. fd virus is composed of 2700 copies of the coat protein subunit arranged cylindrically around a single-stranded circle of DNA. Under appropriate conditions, a concentrated solution of the virus will orient spontaneously in a strong magnetic field. The long filamentous bacteriophage particles orient with the filament axis along the magnetic field direction. All of the protein subunits have the same orientation with respect to the magnetic field vector, $B_0$, because they are arranged symmetrically about the filament axis.

Biosynthetic labeling of individual amino acid types in the 50-residue coat protein enable solid-state NMR measurements such as $^{15}N-{}^1H$ and $^{13}C'-{}^{15}N$ dipole–dipole splittings to be made.[10] Assignments to specific backbone sites are made either by doubly labeling two adjacent residues with $^{13}C$ and $^{15}N$, by observing cross-peaks between nearby residues in spin-exchange experiments,[34] or changing residue types through site-spe-

[34] T. A. Cross, M. H. Frey, and S. J. Opella, *J. Am. Chem. Soc.* **105,** 7471 (1983).

TABLE II
$^{15}$N NMR MEASUREMENTS FOR TWO PEPTIDE
PLANES IN fd COAT PROTEIN

| Dipole splitting | Leu-41 (kHz) | Phe-42 (kHz) |
|---|---|---|
| $^1$H–$^{15}$N | 15.3 | 18.4 |
| $^{13}$C′–$^{15}$N | 1.14 | |

cific mutagenesis. Angular information describing the orientation of the individual peptide planes with respect to the magnetic field vector can be extracted from the measurements of heteronuclear dipole–dipole couplings. Generally, two different dipolar couplings are sufficient to characterize the orientation of the plane to within a few symmetry-related possibilities. Table II lists the solid-state NMR measurements that have been made for the Leu-41 and Phe-42 peptide planes of fd coat protein in an oriented solution of virus particles.[10] Each peptide plane includes the amide nitrogen of the amino acid residue.

The measurements are first analyzed in terms of the restrictions they place on the angles $\theta^{\text{H-N}}$ and $\Theta^{\text{C'-N}}$. Then the combined restrictions on each plane are analyzed in terms of the general polar angles $\alpha$ and $\beta$. The $\alpha$, $\beta$ plot in Fig. 18A graphically displays the angular restrictions placed on the orientation of the Leu-41 plane by the two measured dipolar couplings. Only four discrete allowed regions are found centered at $\beta = 90°$. An $\alpha$, $\beta$ plot for the Phe-42 plane is shown in Fig. 18B. In this case only one dipolar coupling was measured; however, it is a near-maximal splitting and thus it is sufficient to restrict the possible plane orientations to just two small circular regions on the plot. Note that the bottom of the $\alpha$,

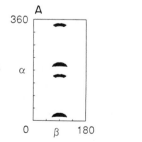
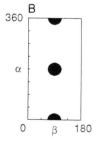

FIG. 18. Combined angular restrictions shown as $\alpha$, $\beta$ plots for the Leu-41 (A) and the Phe-42 (B) peptide planes in the fd coat protein.

$\beta$ plot at $\alpha = 0°$ is continuous with the top at $\alpha = 360°$. Both experimental error and tensor uncertainty are included in the plots in Fig. 18.

Eight possible plane orientations for the Leu-41 plane are chosen from the $\alpha$, $\beta$ plot in Fig. 18A: two centered around $\beta = 90°$ from each allowed region. Sixteen orientations for the Phe-42 plane are chosen from the $\alpha$, $\beta$ plot in Fig. 18B: eight from each circular allowed region. Coordinate sets for each of the possible plane orientations are generated and form the input to the program TOTLINK. Only four of the eight possible orientations for the Leu-41 plane, those with a $\theta^{N-H}$ angle smaller than 90°, are included in the analysis to ensure that duplicate structures with the direction of the magnetic field vector reversed are not created. The 4 orientations for the Leu-41 plane and the 16 orientations for the Phe-42 plane used in the analysis are shown in Fig. 19.

TOTLINK must test each orientation for the Leu-41 plane shown in Fig. 19A with each orientation for the Phe-42 plane shown in Fig. 19B. This gives 64 possible combinations of peptide plane orientations to evaluate and up to 128 possible structures. Thirty-eight structures with low-energy $\Phi$, $\Psi$ pairs were selected by TOTLINK as reasonable structures. All of these fell within the $\alpha$-helix and $\beta$-sheet regions of the Ramachandran plot. Since many of the plane orientations, especially for Phe-42, are similar, their $\Phi$, $\Psi$ pairs are also similar. The low-energy structures all fell within five small areas of the $\Phi$, $\Psi$ plot, one in the $\alpha$-helical region and four in the $\beta$-sheet region. The similar $\Phi$, $\Psi$ pairs were averaged together and the resulting five structures are plotted on a Ramachandran diagram in Fig. 20.

Other spectroscopic data, in particular from X-ray diffraction, circular dichroism (CD), Raman, and IR, have indicated that the coat protein of the class I bacteriophages like fd is nearly all $\alpha$-helical.[35] For this reason the $\alpha$-helical average NMR structure for the Leu-41 and Phe-42 peptide planes was selected. This structure is shown in Fig. 21 with standard peptide plane geometries. Fifteen $\alpha$-helical structures were averaged together to obtain the torsion angle values of $\Phi = -67$ and $\Psi = -44$. These values are close to the average $\alpha$-helical torsion angles found in proteins, $\Phi = -65$ and $\Psi = -41$.

When data from several adjacent peptide planes are included in the analysis, TOTLINK searches for stretches of the backbone forming secondary structural elements. For example, if three or more possible $\alpha$-helical linkages are found in a row, this is considered to be more likely than linkages not forming secondary structural elements. Thus including

[35] L. Makowski, in "Biological Macromolecules and Assemblies" (A. McPherson, ed.), Vol. 1, p. 203. Wiley, New York.

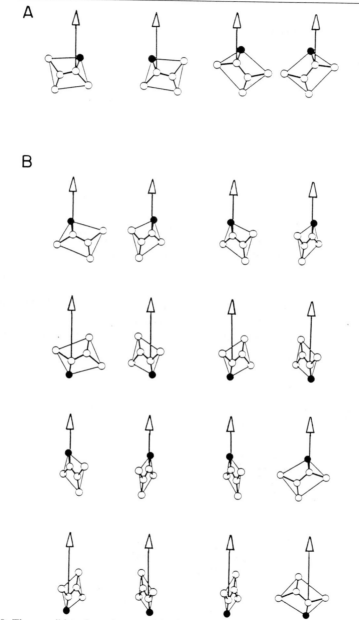

FIG. 19. The possible orientations used in the structural analysis for the Leu-41 (A) and the Phe-42 (B) peptide planes in the fd coat protein. The dark circle represents the hydrogen attached to the nitrogen. The arrow represents the magnetic field vector.

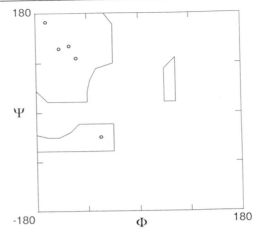

FIG. 20. The five possible average low-energy NMR backbone structures for the Leu-41 and Phe-42 peptide planes of the fd coat protein shown on a Ramachandran conformation plot.

information about other peptide planes can aid in the selection of $\Phi$, $\Psi$ pairs at any one linkage. This enabled the determination of an initial model of the coat protein structure based on solid-state NMR experimental measurements.[4]

## Protein Structure Determination

The goal of this solid-state NMR approach to protein structure determination is to find the relative orientations of adjacent peptide planes, i.e., the $\Phi$, $\Psi$ angles. The experimental measurements directly yield angular information describing the orientations of individual peptide planes with respect to the applied magnetic field. Further computer analysis is

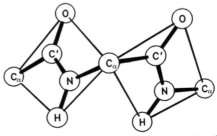

FIG. 21. The $\alpha$-helical average NMR structure for the Leu-41 and Phe-42 peptide planes of the fd coat protein.

needed to deal with the cylindrical symmetry of the information and to extract $\Phi$ and $\Psi$. In order to describe the orientation of a peptide plane, a general molecular axis system, $X$, $Y$, $Z$, is defined relative to the plane. The orientation of this axis system with respect to the magnetic field vector, $B_0$, can be described by the two polar angles, $\alpha$ and $\beta$, which are defined in Fig. 4. Since the goal is to determine relative orientations of peptide planes, the absolute orientation of the molecule within the magnetic field is irrelevant. This is demonstrated in Fig. 22 with the first two peptide planes in crambin in two different orientations with respect to the

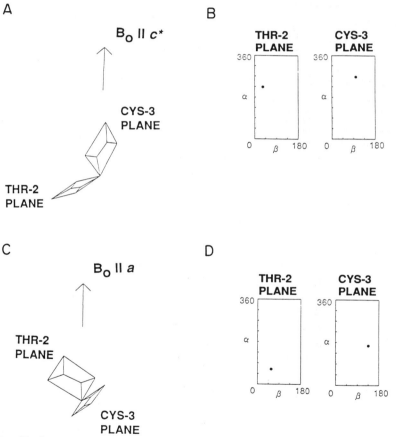

FIG. 22. (A) A representation of the first two peptide planes in crambin aligned with $B_0$ along the $c^*$ crystal axis.[17] (B) $\alpha$, $\beta$ plots for the two planes in (A). (C) A representation of the first two peptide planes in crambin as in (A) except aligned with $B_0$ along the $a$ crystal axis. (D) $\alpha$, $\beta$ plots for the two planes in (C). In both (A) and (C) $\Phi = -108$ and $\Psi = 144$.

magnetic field vector. In both orientations $\Phi$ and $\Psi$ are the same, while the $\alpha$ and $\beta$ values describing the individual plane orientations are different for the two molecular orientations. Figure 22A shows the two peptide planes aligned with $B_0$ along the $c^*$ crystal axis, and the corresponding $\alpha$, $\beta$ plots are shown in Fig. 22B. The second molecular orientation, with $B_0$ along the $a$ crystal axis, is shown in Fig. 22C with the $\alpha$, $\beta$ plots in Fig. 22D. Either molecular orientation could be used to determine $\Phi$ and $\Psi$ even though the $\alpha$, $\beta$ angles are different.

The solid-state NMR structural analysis method is demonstrated with a model calculation on the protein crambin. The magnetic field vector, $B_0$, is chosen to be aligned along the $c^*$ crystal axis. For this crystal orientation the two molecules in the unit cell of crambin are magnetically equivalent, thus they would give identical spectral parameters. Therefore, only one molecule is considered in the analysis. Every peptide plane of the protein has a unique orientation with respect to $B_0$ which is shown with respect to individual magnetic field vectors in Fig. 23. This information is represented equivalently in Fig. 24 in 45 $\alpha$, $\beta$ plots, 1 for each plane in crambin. Significantly, the information in Figs. 23 and 24 is not as complete as that in Fig. 1B because the cylindrical information is lost in going from the linked to separated peptide planes. There are two possible ways to link together each pair of sequential planes, resulting in $2^{44}$ possible ways to reconstruct the protein backbone from the unique plane orientations.

When experimental solid-state NMR data are used in structure determination there are always ambiguities as well as errors in plane orientation. Even with many solid-state NMR measurements, four possible plane orientations are generally obtained unless there are degeneracies. These four plane orientations are symmetry related and represent two mirror-image pairs with opposite $B_0$ directions. Four such symmetry-related orientations for each peptide plane in crambin are represented in $\alpha$, $\beta$ plots in Fig. 25. From this information over $10^{40}$ possible backbone structures can be generated, most of which are not chemically reasonable.

The program TOTLINK selected from the possible backbone structures on the basis of maximum amount of secondary structure, low-energy $\Phi$, $\Psi$ pairs, and agreement with disulfide bond information. Crambin has three disulfide bonds between residues 3 and 40, 4 and 32, 16 and 26.[17] The minimum and maximum acceptable distances between the $C_\alpha$ atoms of these residues for the calculated backbone structures were assigned as 3 and 14 Å, respectively, in the TOTLINK input file. These distances range from 4 to 6 Å in the X-ray crystal structure,[17] however loose restraints were used for TOTLINK in order to account for error propagation. Two additional options in TOTLINK were used. First, similar $\Phi$, $\Psi$

**THR 2**

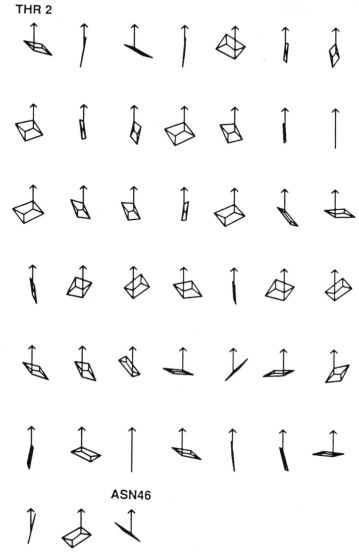

**ASN46**

FIG. 23. The 45 individual peptide planes of crambin with respect to magnetic field vectors. Orientations were calculated from the X-ray crystal structure[17] with $B_0$ along the $c^*$ crystal axis. Only the first and last peptide planes are labeled according to the residue name and number of the nitrogen in the plane.

THR 2

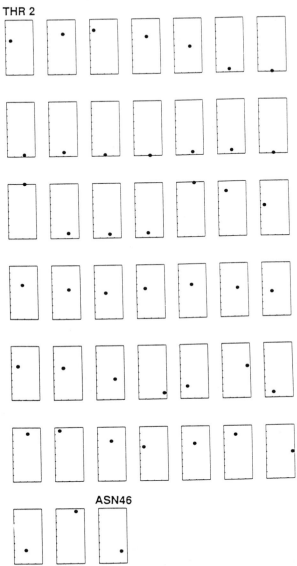

ASN46

FIG. 24. $\alpha$, $\beta$ plots showing the unique orientations of the 45 peptide planes of crambin calculated from the X-ray crystal structure[17] with $B_0$ along the $c^*$ crystal axis. Only the plots for the first and last peptide planes are labeled according to the residue name and number of the nitrogen in the plane. The dots are enlarged to make them visible.

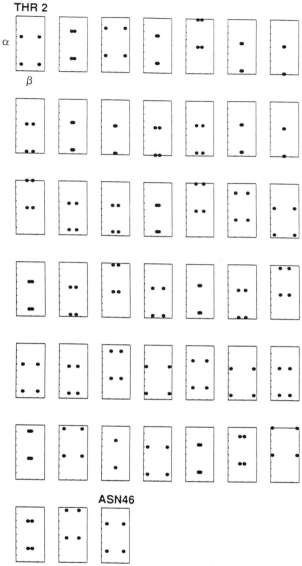

FIG. 25. α, β plots showing 4 symmetry-related orientations for each of the 45 peptide planes of crambin. The symmetry-related orientations are generated from the unique orientations represented in Fig. 24. Only the plots for the first and last peptide planes are labeled according to the residue name and number of the nitrogen in the plane. The dots are enlarged to make them visible.

pairs were averaged together to reduce the number of possible backbone structures. Second, all glycine linkages were automatically assigned Ramachandran energy values of one. This is done so that slightly higher energy glycine linkages involved in tight turns will not be discriminated against and will be considered as equally possible as low-energy glycine conformations.

Five hundred possible backbone structures, ranked in order of maximum secondary structure and lowest Ramachandran energy sum, were saved by TOTLINK. A second computer program, EVAL, was used to select further, based on overall shape and volume of the structure. The backbone was assumed to be globular with a maximum variation among principal inertial elements of 2.5 and an approximate volume calculated from the principal inertial elements of 5600–6400 Å. The X-ray backbone structure has a globular factor of 2.2 and an approximate volume of 6000 Å$^3$. EVAL also gave tighter restraints for the disulfide $C_\alpha$–$C_\alpha$ distances of 4 and 12 Å. The best calculated structure selected by TOTLINK and EVAL is shown in Fig. 26 superimposed on the X-ray structure. As before, only the $C_\alpha$ atoms are shown for simplicity. The overall folding of the calculated backbone is very similar to the X-ray structure and all of the major secondary elements are present in the calculated structure. The root-mean-square (rms) deviation between the $C_\alpha$ atoms of the two structures is 3.2 Å with the maximal positional disagreement at the two termini. After refinement using restrained molecular dynamics, the calculated structure agrees more closely with the X-ray crystal structure.[31]

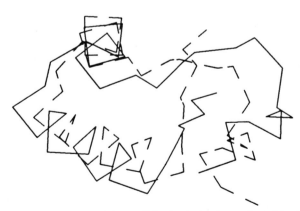

FIG. 26. The peptide plane backbone of the best structure selected by TOTLINK and EVAL (dashed line) is shown superimposed on the X-ray structure (solid line).[17] Only the $C_\alpha$ atoms are shown for simplicity.

## Prospects for Protein Structure Determination by Solid-State NMR Spectroscopy

The approach for determining protein structure by solid-state NMR spectroscopy is quite general in terms of the nuclear spin interactions that can provide useful angular restrictions, including those present in natural isotopic abundance. It should be possible to describe the structures of many proteins which cannot be studied by conventional X-ray diffraction and solution NMR spectroscopy methods with this approach.

The main hurdles to implementing this approach are in the practical aspects of sample selection and preparation and the performance of the NMR spectrometers. Even though uniaxially oriented samples, instead of single crystals, can be used, it is essential to obtain well-oriented, relatively large samples. However, the use of high-field spectrometers with their improved sensitivity should make it feasible to study more readily available samples. The additional sensitivity gain from low temperatures offers the promise of determining the structures of proteins in oriented preparations and single crystals of the same size used in X-ray diffraction.

### Acknowledgments

This research has been supported by grants from the NIH (GM-24266, GM-29754, and GM-34343). P.L.S. was supported by a National Science Foundation Graduate Fellowship.

# Section III

# Enzyme Dynamics

### A. Rate Constants
*(Chapters 14 through 17)*

### B. Molecular Motions
*(Chapters 18 through 22)*

# [14] Nuclear Magnetic Resonance Line-Shape Analysis and Determination of Exchange Rates

*By* B. D. NAGESWARA RAO

The information content of a high-field high-resolution NMR spectrum of a molecule in liquid state (for which the experimental conditions are such that all spin interactions depending on spatial orientation are motionally averaged to zero) is specified in terms of a total of six different types of parameters. Three of these are spectral parameters: chemical shifts ($\delta$), spin–spin coupling constants ($J$), and areas enclosed by the resonances. They determine the peak positions and intensities of the resonances in the spectrum. The other three are dynamic parameters: spin-lattice relaxation times ($T_1$), nuclear Overhauser effects (NOE), and line shapes (given by $T_2$ for simple resonances). They are governed by time-dependent interactions involving the spins. When chemical exchange is present in the system, the molecule (or a particular moiety in it) resides itinerantly in two or more states (or conformations) such that some or all of the above six types of parameters might change from one state to another. In such a case, the NMR parameters of the observed spectrum are determined by those in the individual states suitably "averaged" by the residence times and populations in these states. The averaging process often results in measurable changes in all the parameters, but the effects are most dramatic in the line shapes of resonances. A quantitative analysis of these line shapes is capable of yielding the exchange rates between different states, as well as the values of the NMR parameters in the individual states in the absence of the exchange.

Chemical exchange processes occur profusely in NMR experiments on enzyme systems due to a number of causes, e.g., the association and dissociation of substrates with the enzymes in enzyme–substrate complexes, slow conformational changes such complexes may undergo, or the interconversion of reactants and products catalyzed by the enzyme. The analysis of NMR spectra obtained in the presence of one or more of the exchange-causing processes enables the determination of the relevant rates. Such rates may sometimes be of considerable biochemical significance; for example, it is possible to determine the rate of interconversion of reactants and products on the surface of the enzyme rather simply, by the NMR method, and assess the role of the interconversion in governing the overall rate of the enzyme reaction.[1]

[1] B. D. Nageswara Rao, D. Buttlaire, and M. Cohn, *J. Biol. Chem.* **251**, 6981 (1976).

METHODS IN ENZYMOLOGY, VOL. 176

NMR line shapes arising from the canonical problem of two-site chemical exchange between simple lines [i.e., the spectrum of the spin(s) in either site is a single, nonoverlapping Lorentzian line without multiplicity] are well known, and are found in standard textbooks on NMR.[2] Analytical expressions may be obtained for the line shape (intensity as a function of frequency) for this case by a simple modification of Bloch equations. The results are described by considering three conditions: (1) *slow exchange* (chemical shift between the resonances in the absence of exchange is much larger than the exchange rates) leads to Lorentzian shapes for each resonance with the position unchanged but line width increased by an amount equal to the exchange rate out of the corresponding state (reciprocal lifetime), (2) *fast exchange* (chemical shift between resonance in the absence of exchange is much smaller than the exchange rates) leads to a Lorentzian resonance with a position and line width given by the population-weighted averages of the positions and widths in the absence of exchange, and (3) *intermediate exchange* (chemical shift between the resonance in the absence of exchange comparable to the exchange rates) leads to non-Lorentzian line shapes which depend on the values of the exchange rates, and the chemical shift differences. If the slow exchange condition is valid, the exchange rates are obtained simply in terms of the broadening of each resonance; for intermediate exchange a detailed comparison of the observed lineshapes with those predicted by the analytical expressions will be necessary to determine the exchange rate. However, if fast exchange prevails the spectrum does not contain measurable information about the exchange rates other than the implicit fact that these rates must be much larger than the chemical shift differences in the absence of exchange.

The simplicity of the description above is lost if the spins undergoing the exchange are spin coupled to other spins in the system which leads to a multiplicity of resonances in each state. It is generally known that the presence of spin–spin coupling is exceedingly common in NMR spectra. Some of the qualitative features of the slow, fast, and intermediate exchange conditions described above may be identified if all the spins are weakly coupled (spin–spin coupling constant for any spin pair is small compared to the corresponding chemical shift difference), but the quantitative dependence of the line shape on exchange rates is altered. Additional complications arise if the spins are strongly coupled or if there is overlap with other resonances or multiplets. Line-shape theories for coupled spins in the presence of chemical exchange, based on the density

[2] J. A. Pople, W. G. Schneider, and H. J. Bernstein, "High Resolution Nuclear Magnetic Resonance," p. 218. McGraw-Hill, New York, 1959.

matrix method, have been in vogue for some time.[3-6] These are tractable theories and are amenable to numerical computations of line shapes for a chosen set of conditions.[7] A comparison of such simulated line shapes with the experimentally observed spectra is the only reliable procedure available for extracting information on the exchange rates involving coupled spins from steady state NMR spectra.

The aim of this chapter is to describe, in some detail, and illustrate a somewhat simplified procedure for the computation of NMR line shapes of coupled spin-$\frac{1}{2}$ systems undergoing chemical exchange. The emphasis will be on setting up the matrices appropriate for the exchanging system of interest without the necessity of familiarity with density-matrix theories. These matrices can then be used with the help of standard computer programs to yield the required line shapes. A brief discussion will be given of the various aspects to be noted for obtaining experimental line shapes well enough to warrant the computational effort. The methodology will be illustrated by considering several examples for which useful information was obtained regarding spin systems of biochemical interest, in particular regarding enzyme–substrate complexes of ATP-utilizing enzymes.

Experimental Considerations

It is important to determine the experimental line shapes with the best possible accuracy, in order that the computational method of analysis described below provides meaningful information. The NMR spectra are usually obtained in the Fourier transform mode by present-day spectrometers, while the analysis simulates a frequency-swept spectrum, and a comparison is made by invoking the equivalence of the two spectra. The spin system must be fully relaxed by allowing sufficient delay before the application of each observing pulse so that partial relaxation will not distort the line shape. The data should be acquired with the maximum number of points per hertz to obtain a faithful representation of the line shape. Furthermore, the data should be obtained with the best signal-to-noise ratio, which implies accumulation of as large a number of free-induction decays as possible. All these three factors, namely, long relaxation delays, large data sizes, and large numbers of scans increase the time

[3] J. I. Kaplan, *J. Chem. Phys.* **28**, 278 (1958); **29**, 462 (1958).
[4] G. Binsch, *J. Am. Chem. Soc.* **91**, 1304 (1969).
[5] J. I. Kaplan and G. Fraenkel, "NMR in Chemically Exchanging Systems." Academic Press, New York, 1980.
[6] D. S. Stephenson and G. Binsch, *J. Magn. Reson.* **32**, 145 (1978).
[7] K. V. Vasavada, J. I. Kaplan, and B. D. Nageswara Rao, *J. Magn. Reson.* **41**, 467 (1980).

required for each spectrum, a fact which must be recognized in optimizing the sample conditions from the biochemical point of view. Finally, since exchange rates vary with temperature, pH, and in some cases the chosen buffer, the experiments must be performed under conditions in which these factors are controlled.

Every exchanging system of interest has its own special biochemical constraints which determine the sample conditions. In enzymatic systems, solubility (and sometimes availability) often limits the enzyme concentrations used, and stability of the enzymes or the possibility of degradation of the substrates limits the time over which an experiment can be performed. Other factors, such as the tendency of nucleotides to stack at high concentrations, should be kept in mind in choosing the concentrations in the sample. If exchange rates pertinent to enzyme-bound substrate complexes are of interest, the substrate concentrations are required to be sufficiently smaller than those of the enzyme such that over 90% of the substrate molecules are in the bound complexes. This requirement, taken along with the limitations on the highest manageable enzyme concentrations, and the smallest substrate concentrations detectable with the needed accuracy, will be a decisive factor in the success of such experiments. If the enzyme or substrate molecules bind cations such as Mg(II) or Ca(II), it is essential to scrupulously avoid the presence of paramagnetic impurity cations such as Mn(II), and Fe(II) even at minute concentrations. Paramagnetic cations produce dramatic changes in the NMR line widths of nuclei in their vicinity even if their concentrations are about $10^{-4}$–$10^{-5}$ times those of the substrates.

If the line shape computed for a designated exchange between spin systems should be meaningfully compared with experiment, sample conditions must be deliberately designed such that all other exchanges do not interfere with the measurements. This may sometimes make the experiment more difficult, but the results are likely to be readily interpretable. If the sample conditions are not optimized, and extraneous exchanges do occur, various assumptions will be required to account for them. Such assumptions are often difficult to validate and may lead to erroneous quantitative results.

## Computation of Line Shapes under the Influence of Exchange

### Basic Equations

The theoretical framework required for computing the line shapes of spin multiplets undergoing specific exchanges will now be described in terms of a set of rules to write down the required matrices without any familiarity with density matrix theory of NMR. We consider exchanges involving two rates (forward and reverse). Incorporation of multiple exchanges is straightforward (one such example will be considered later). The description will be restricted to weakly coupled systems of spin-$\frac{1}{2}$ nuclei, which may not be a limitation as most spin systems possess weakly coupled NMR spectra when recorded using modern high-field spectrometers. The theory is, nevertheless, fully capable of treating strong-coupling effects in an exact manner, if necessary.[4-7] We shall also ignore spin systems containing symmetric groups leading to chemical and magnetic equivalence to keep the formulation simple. Extension to these systems is straightforward in the framework of the original method. None of the examples discussed here contains symmetrical spin systems.

Consider a group of exchanging species each designated by a spin system $S_i$ ($i = 1, 2, 3, \ldots$) of concentration $[S_i]$, possessing $q_i$ spins, and $n_i = q_i 2^{q_i-1}$ *allowed* transitions in its NMR spectrum in the absence of exchange. The signal intensity at an observed frequency $\omega$ is given by

$$\mathcal{I}(\omega) = \sum_i [S_i] \left\{ \sum_{S_i} \text{Im}[\chi(\omega)]_{kk'} \right\} \tag{1}$$

where $\chi(\omega)$ is an $N$-dimensional column vector ($N = \sum_i n_i$) of complex numbers which are functions of $\omega$, $\text{Im}[\chi(\omega)]_{kk'}$ is the imaginary part of the $kk'$ element of $\chi$ representing the $k \rightarrow k'$ transition of the spin system $S_i$ for a given $\omega$, and $\sum_{S_i}$ designates a sum over all the $n_i$ transitions of $S_i$. The contributions of all the exchanging spin systems are superposed in Eq. (1) by summing over $i$. Equation (1) simply states that if the concentrations $[S_i]$ and the elements of $\text{Im}[\chi(\omega)]$ are known, $\mathcal{I}(\omega)$ can be calculated. $\chi(\omega)$ is given by the matrix equation

$$(i\omega\mathbf{1} + \mathbf{E})\chi = i\mathbf{W} \tag{2}$$

where $\mathbf{1}$ is a unit matrix, $\mathbf{E}$ is a complex matrix independent of $\omega$ representing the exchange, and $\mathbf{W}$ is a vector describing the weight factors (relative number of states) of the spin systems. If $\mathbf{E}$ can be diagonalized by a similarity transformation with a matrix $\mathbf{Z}$, i.e.,

$$\mathbf{D} = \mathbf{Z}\mathbf{E}\mathbf{Z}^{-1} \tag{3}$$

a formal solution of Eq. (2) may be written as

$$\chi = \mathbf{Z}^{-1}(i\omega\mathbf{1} + \mathbf{D})^{-1}\mathbf{Z}(i\mathbf{W}) \tag{4}$$

Note that since $\mathbf{E}$ is dependent of $\omega$, $\mathbf{Z}$ also is, so that the diagonalization of $\mathbf{E}$ can be performed only once, and need not be repeated as $\mathcal{I}(\omega)$ is calculated for different values of $\omega$. Thus for a given exchange if the exchange matrix **E** and the vector **W** are correctly composed, the line shape may be calculated using Eqs. (1) and (4) once **E** is diagonalized.

*Structure of* **E**, χ, *and* **W**

Given below is a step-by-step procedure for the specification of the elements of the matrix **E** and the vectors χ and **W**. The steps are briefly illustrated by considering the example of a three-spin system bifurcating into a two-spin system and a one-spin system:

$$\text{ABC} \underset{\tau_2^{-1}}{\overset{\tau_1^{-1}}{\rightleftharpoons}} \text{A}'\text{B}' + \text{C}' \tag{5}$$

This example is appropriate for the $^{31}\text{P}$ nuclei in a phosphoryl transfer reaction such as $\text{ATP} + X \rightleftharpoons \text{ADP} + XP$ where the second substrate on the left does not contain a $^{31}\text{P}$ nucleus.[7]

1. Identify the spin systems on both sides of the exchange process in a convenient order (e.g., $S_1$, $S_2$, $S_3$, ...). Numerically designate the spin product states of these systems in the same order. For example, in the exchange shown in Eq. (5) numbers 1 to 8 designate the states of ABC, 9–12 designate those of A'B', and 13 and 14 those of C' (see the list on p. 288).

2. Determine the allowed transitions of each spin system and list them in the same order as the spin systems. The dimension of **E**, χ, and **W** is $N = \sum_{S_i} n_i$. For the above example, there are 12 allowed transitions for ABC, 4 for A'B', and 1 for C'. Thus for this exchange, $N = 17$.

3. The element $\chi_{kk'}$ of the column vector χ corresponds to the transition $(k \rightarrow k')$.

4. The elements of **W** corresponding to the transitions of the spin system $S_i$ are all set equal to $(\frac{1}{2})^{q_i}$. Once again for the exchange in Eq. (5) with the first 12 elements of **W** are equal to $\frac{1}{8}$, the next 4 equal to $\frac{1}{4}$, and the last 1 equal to $\frac{1}{2}$.

5. The diagonal elements of **E** have the form

$$-i\omega_{kk'} - \frac{1}{2}(\Delta\omega)_{kk'} - \tau^{-1} \tag{6}$$

where $\omega_{kk'}$ is the transition frequency and $(\Delta\omega)_{kk'}$ [$= (2/T_2)_{kk'}$] is the line width (full width at half height) of the $k \rightarrow k'$ transition (in the absence of exchange) expressed in units of radians per second, and $\tau^{-1}$ is the reciprocal lifetime of the spin system. Note that each row of the **E** is designated in terms of a specific transition such as $k \rightarrow k'$, as are the elements of χ and **W**. The transition frequency, $\omega_{kk'}$, may be readily expressed in terms of the chemical shifts and spin–spin coupling constants of the spin system.

6. The off-diagonal elements of **E** represent exchange, and are proportional to the corresponding exchange rates. The coefficient of proportionality must be determined carefully. First, a correspondence must be made between the transitions of spin systems on one side of the exchange with those of the other. In exchanges involving large spin systems breaking up into smaller ones, the multiplicity (multiplicity is defined as the number of *allowed* transitions regardless of any coincidences) of the transitions of a given spin is reduced. In the reverse case the multiplicity is enhanced. In cases where the multiplicity of resonances associated with a spin in a spin system is reduced because of the break-up (regardless of whether the spin picks up additional multiplicity due to coupling with spins on the other side) the exchange rate in the direction of the break-up must be multiplied by the reduction factor and introduced as an off-diagonal element in the **E** matrix between the corresponding transitions on the two sides. The multiplicity is enhanced in the reverse direction due to interactions with other spins [i.e., when smaller spin systems group together to form a larger one, such as in the reverse reaction of Eq. (5)]. The exchange rate in this direction appears as an off-diagonal element between all corresponding transitions with a coefficient of unity.

In the exchange of Eq. (5), the 12 transitions of ABC coalesce into 5 transitions of A'B' and C' (4 for A'B' and 1 for C'). Since there are four transitions associated with each of the spins in ABC, for either of the two spins that will end up as A'B', four transitions are reduced to two. For the third spin ending up as C', the four transitions collapse into a single transition. Thus, for the first two spins the off-diagonal elements of **E** for the dissociation of $S_1$ is given by $(\tau_1^{-1}/2)$, whereas for the third spin the off-diagonal elements are given by $(\tau_1^{-1}/4)$. In the reverse direction the off-diagonal elements are all equal to $\tau_2^{-1}$ for each transition of A'B' and C' with every transition of ABC to which they are connected.

We shall now consider examples of exchanging spin systems of varying degrees of complexity, and set up **E**, $\chi$, and **W**. These matrices and vectors might serve to illustrate the various steps discussed above, so that they may be written down for any other cases of exchange not discussed here.

*Two-Site Exchange of Simple Lines (No Spin–Spin Coupling).* This is the simplest example of exchange (A ⇌ A'). There are four spin states, $1,\alpha$, $2,\beta$ for A, and $3,\alpha$ and $4,\beta$ for A' ($\alpha$ for spin up, and $\beta$ for spin down), and two transitions $1 \rightarrow 2$ and $3 \rightarrow 4$. It may be readily seen using Eqs. (2) and (6)

$$\chi = \begin{bmatrix} \chi_{12} \\ \chi_{34} \end{bmatrix}, \qquad \mathbf{W} = \tfrac{1}{2} \begin{bmatrix} 1 \\ 1 \end{bmatrix} \qquad (7)$$

and

$$\mathbf{E} = \begin{bmatrix} -i\omega_{12} - \frac{1}{2}(\Delta\omega)_{12} - \tau^{-1} & \tau_1^{-1} \\ \tau_2^{-1} & -i\omega_{34} - \frac{1}{2}(\Delta\omega)_{34} - \tau_2^{-1} \end{bmatrix} \qquad (8)$$

$\tau_1$ is the lifetime of A and $\tau_2$ is that of A'. At equilibrium these rates satisfy detailed balance, i.e.,

$$[A]\tau_1^{-1} = [A']\tau_2^{-1} \qquad (9)$$

*Two-Site Exchange of Spin-Multiplets (Coupling Preserved).* If the resonances at the two sites are spin multiplets arising from the same coupling pattern, e.g., AB $\rightleftharpoons$ A'B' where the spins are coupled at both sites, the line-shape calculation follows a procedure identical to that in the previous example for each transition in the multiplet. Thus if the multiplet has four lines at each site, a set of four 2 × 2 matrix equations similar to Eq. (2) will be obtained. The transition frequencies in **E** will now contain chemical shifts as well as spin–spin coupling constants at the respective sites.

*Exchanges Involving Spin Multiplets with Loss of Spin Coupling.* Consider the following example, in which a coupled two-spin system exchanges with a state in which the spins are not coupled, i.e., AB $\rightleftharpoons$ A' + B'. There are eight spin states: $1,\alpha\alpha$; $2,\alpha\beta$; $3,\beta\alpha$; and $4,\beta\beta$ for AB; $5,\alpha$ and $6,\beta$ for A; and $7,\alpha$ and $8,\beta$ for B'. There are four transitions for AB, and one each for A' and B'. The two A transitions exchange with the A' transition, and the two B transition with the B' transition. Equation (2) will have a dimension of six, which factors into two equations of dimension three, one for A $\rightleftharpoons$ A' and the other for B $\rightleftharpoons$ B'. $\chi$, **W**, and **E** for A $\rightarrow$ A' exchange are given by

$$\chi = \begin{bmatrix} \chi_{13} \\ \chi_{24} \\ \chi_{56} \end{bmatrix}, \qquad \mathbf{W} = \frac{1}{4}\begin{bmatrix} 1 \\ 1 \\ 2 \end{bmatrix} \qquad (10)$$

and

$$\mathbf{E} = \begin{bmatrix} -i(\omega_A + \frac{1}{2}J) - \frac{1}{2}(\Delta\omega)_{13} - \tau_1^{-1} & 0 & \frac{1}{2}\tau_1^{-1} \\ 0 & -i(\omega_A - \frac{1}{2}J) - \frac{1}{2}(\Delta\omega)_{24} - \tau_1^{-1} & \frac{1}{2}\tau_1^{-1} \\ \tau_2^{-1} & \tau_2^{-1} & -i\omega_{A'} - \frac{1}{2}(\Delta\omega)_{56} - \tau_2^{-1} \end{bmatrix}$$

$$(11)$$

where $\omega_A$ and $\omega_{A'}$ are the chemical shifts of A and A' and $J$ is the spin coupling constant between A and B (all in units of rad sec$^{-1}$). The **E** matrix for the B $\rightleftharpoons$ B' exchange is obtained by replacing A and A' with B and B', respectively in Eq. (11). The corresponding $\chi$ will have elements $\chi_{12}$, $\chi_{34}$, and $\chi_{78}$, and **W** is identical to the one above. The off-diagonal elements in **E** in Eq. (11) illustrate the occurrence of numerical coefficients associated with exchange rates which arise from a reduction in the multiplicity of the spin multiplets when the spin system breaks up.

*Exchanges Involving Multiplets in Which the Coupling Pattern Changes.* Consider the exchange AB + C $\rightleftharpoons$ A' + B'C' in which AB and B'C' are coupled spin systems. There are 12 spin states: $1,\alpha\alpha$, $2,\alpha\beta$, $3,\beta\alpha$, and $4,\beta\beta$ for AB; $5,\alpha$ and $6,\beta$ for C; $7,\alpha$ and $8,\beta$ for A'; and $9,\alpha\alpha$, $10,\alpha\beta$, $11,\beta\alpha$, and $12,\beta\beta$ for B'C'. There are a total of 10 transitions, 5 on each side. The exchange pattern for C $\leftrightarrow$ C' is the reverse of that for A $\leftrightarrow$ A', i.e., while the two A transitions of AB exchange with the single A' transition, the single C transition exchanges with the two C' transitions. On the other hand, B loses the multiplicity arising from coupling with A, and picks up the same multiplicity as B' by coupling with C'. For the three pairs of exchanges $\chi$, **W**, and **E** may then be written as follows:

A $\leftrightarrow$ A':

$$\chi = \begin{bmatrix} \chi_{13} \\ \chi_{24} \\ \chi_{78} \end{bmatrix}, \qquad \mathbf{W} = \tfrac{1}{4}\begin{bmatrix} 1 \\ 1 \\ 2 \end{bmatrix} \tag{12}$$

$$\mathbf{E} = \begin{bmatrix} -i(\omega_A + \tfrac{1}{2}J_{AB}) - \tfrac{1}{2}(\Delta\omega)_{13} - \tau_I^{-1} & 0 & \tfrac{1}{2}\tau_I^{-1} \\ 0 & -i(\omega_A - \tfrac{1}{2}J_{AB}) - \tfrac{1}{2}(\Delta\omega)_{24} - \tau_I^{-1} & \tfrac{1}{2}\tau_I^{-1} \\ \tau_2^{-1} & \tau_2^{-1} & -i\omega_A - \tfrac{1}{2}(\Delta\omega)_{78} - \tau_2^{-1} \end{bmatrix} \tag{13}$$

B $\leftrightarrow$ B':

$$\chi = \begin{bmatrix} \chi_{12} \\ \chi_{34} \\ \chi_{9,11} \\ \chi_{10,12} \end{bmatrix}, \qquad \mathbf{W} = \tfrac{1}{4}\begin{bmatrix} 1 \\ 1 \\ 1 \\ 1 \end{bmatrix} \tag{14}$$

$$
\mathbf{E} = \begin{bmatrix}
-i(\omega_B + \tfrac{1}{2}J_{AB}) & 0 & \tfrac{1}{2}\tau_1^{-1} & \tfrac{1}{2}\tau_1^{-1} \\
-\tfrac{1}{2}(\Delta\omega)_{12} - \tau_1^{-1} & & & \\
0 & -i(\omega_B - \tfrac{1}{2}J_{AB}) & \tfrac{1}{2}\tau_1^{-1} & \tfrac{1}{2}\tau_1^{-1} \\
& -\tfrac{1}{2}(\Delta\omega)_{34} - \tau_1^{-1} & & \\
\tfrac{1}{2}\tau_2^{-1} & \tfrac{1}{2}\tau_2^{-1} & -i(\omega_{B'} + \tfrac{1}{2}J_{B'C'}) & 0 \\
& & -\tfrac{1}{2}(\Delta\omega)_{9,11} - \tau_2^{-1} & \\
\tfrac{1}{2}\tau_2^{-1} & \tfrac{1}{2}\tau_2^{-1} & 0 & -i(\omega_{B'} - \tfrac{1}{2}J_{B'C'}) \\
& & & -\tfrac{1}{2}(\Delta\omega)_{10,12} - \tau_2^{-1}
\end{bmatrix} \quad (15)
$$

and $C \leftrightarrow C'$:

$$
\boldsymbol{\chi} = \begin{bmatrix} \chi_{56} \\ \chi_{9,10} \\ \chi_{11,12} \end{bmatrix}, \qquad \mathbf{W} = \tfrac{1}{4}\begin{bmatrix} 2 \\ 1 \\ 1 \end{bmatrix} \qquad (16)
$$

$$
\mathbf{E} = \begin{bmatrix}
-i\omega_C & \tau_1^{-1} & \tau_1^{-1} \\
-\tfrac{1}{2}(\Delta\omega)_{56} - \tau_1^{-1} & & \\
\tfrac{1}{2}\tau_2^{-1} & -i(\omega_{C'} + \tfrac{1}{2}J_{B'C'}) & 0 \\
& -\tfrac{1}{2}(\Delta\omega)_{9,10} - \tau_1^{-1} & \\
\tfrac{1}{2}\tau_2^{-1} & 0 & -i(\omega_{C'} - \tfrac{1}{2}J_{B'C'}) \\
& & -\tfrac{1}{2}(\Delta\omega)_{11,12} - \tau_2^{-1}
\end{bmatrix} \quad (17)
$$

*ABC* ⇌ *A'B'* + *C' Exchange.* This exchange was used earlier [see Eq. (5)] as an example to discuss the step-by-step procedure for the construction of **E**, **χ**, and **W**. In this exchange the spin coupling between A and B is preserved whereas spin couplings involving C vanish on the right-hand side. The matrices and vectors characterizing this exchange in full are given below.[7] The spin-product states for the spin systems are given by

| ABC | A'B' | C' |
|-----|------|-----|
| 1. $\alpha\alpha\alpha$ | 9. $\alpha\alpha$ | 13. $\alpha$ |
| 2. $\alpha\alpha\beta$ | 10. $\alpha\beta$ | 14. $\beta$ |
| 3. $\alpha\beta\alpha$ | 11. $\beta\alpha$ | |
| 4. $\beta\alpha\alpha$ | 12. $\beta\beta$ | |
| 5. $\alpha\beta\beta$ | | |
| 6. $\beta\alpha\beta$ | | |
| 7. $\beta\beta\alpha$ | | |
| 8. $\beta\beta\beta$ | | |

There are 4 transitions each for A, B, and C comprising a total of 12 for ABC, 2 transitions each for A′ and B′ comprising a total of 4 for A′B′, and 1 transition for C′. Thus Eq. (2) has a dimension of 17. However, it factors into several equations of smaller dimension. The four transitions of A exchanging with the two of A′ may be seen to be separable into two groups of two A transitions each for a given spin state of B (and split by coupling with C) exchanging with the A′ transition with the same spin state for B′. One group is $1 \to 4$ and $2 \to 6$ exchanging with $9 \to 11$, and the other $3 \to 7$ and $5 \to 8$ with $10 \to 12$. A parallel description applies for the exchanges involving B and B′ transitions. For C and C′, however, four C transitions exchange with a single C′ transition and the dimension of the corresponding Eq. (2) required becomes 5. Thus the dimension of 17 for Eq. (2) required to represent the entire exchange, factors into 4 equations each of dimension 3 and one with dimension 5. The various $\mathbf{E}$ matrices and $\chi$ and $\mathbf{W}$ vectors may be summarized as follows:

$A \leftrightarrow A'$:

$$\chi = \begin{bmatrix} \chi_{1,4} \\ \chi_{2,6} \\ \chi_{9,11} \end{bmatrix}, \qquad \mathbf{W} = \tfrac{1}{8} \begin{bmatrix} 1 \\ 1 \\ 2 \end{bmatrix} \tag{18}$$

and

$$\mathbf{E} = \begin{bmatrix} -i(\omega_A + \tfrac{1}{2}J_{AB} + \tfrac{1}{2}J_{AC}) \\ -T_{2A}^{-1} - \tau_1^{-1} & 0 & \tfrac{1}{2}\tau_1^{-1} \\ 0 & -i(\omega_A + \tfrac{1}{2}J_{AB} - \tfrac{1}{2}J_{AC}) \\ & -T_{2A}^{-1} - \tau_1^{-1} & \tfrac{1}{2}\tau_1^{-1} \\ \tau_2^{-1} & \tau_2^{-1} & -i(\omega_{A'} + \tfrac{1}{2}J_{A'B'}) \\ & & -T_{2A'}^{-1} - \tau_2^{-1} \end{bmatrix} \tag{19}$$

The other $3 \times 3$ equation for $A \rightleftharpoons A'$ exchange (for $\chi_{3,7}$ and $\chi_{10,12}$) is obtained by changing the signs of $J_{AB}$ and $J_{A'B'}$ in Eq. (19).

$B \rightleftharpoons B'$:

$$\chi = \begin{bmatrix} \chi_{1,3} \\ \chi_{2,5} \\ \chi_{9,10} \end{bmatrix}, \qquad \mathbf{W} = \tfrac{1}{8} \begin{bmatrix} 1 \\ 1 \\ 2 \end{bmatrix} \tag{20}$$

and

$$E = \begin{bmatrix} -i(\omega_B + \frac{1}{2}J_{AB} + \frac{1}{2}J_{BC}) \\ -T_{2B}^{-1} - \tau_1^{-1} & 0 & \frac{1}{2}\tau_1^{-1} \\ & -i(\omega_B + \frac{1}{2}J_{AB} - \frac{1}{2}J_{BC}) \\ 0 & -T_{2B}^{-1} - \tau & \frac{1}{2}\tau_1^{-1} \\ & & -i(\omega_{B'} + \frac{1}{2}J_{A'B'}) \\ \tau_2^{-1} & \tau_2^{-1} & -T_{2B'}^{-1} - \tau_2^{-1} \end{bmatrix} \quad (21)$$

The other $3 \times 3$ equation (for $B \rightleftharpoons B'$ exchange for $\chi_{4,7}$, $\chi_{6,8}$, and $\chi_{11,12}$) is obtained by changing the signs $J_{AB}$ and $J_{A'B'}$ in Eq. (21).

$C \rightleftharpoons C'$:

$$\chi = \begin{bmatrix} \chi_{1,2} \\ \chi_{3,5} \\ \chi_{4,6} \\ \chi_{7,8} \\ \chi_{13,14} \end{bmatrix}, \qquad W = \frac{1}{8}\begin{bmatrix} 1 \\ 1 \\ 1 \\ 1 \\ 4 \end{bmatrix} \quad (22)$$

and

$$E = \begin{bmatrix} -i(\omega_C + \frac{1}{2}J_{BC} + \frac{1}{2}J_{AC}) \\ -T_{2C}^{-1} - \tau_1^{-1} & 0 & 0 & 0 & \frac{1}{4}\tau_1^{-1} \\ & -i(\omega_C + \frac{1}{2}J_{BC} - \frac{1}{2}J_{AC}) \\ 0 & -T_{2C}^{-1} - \tau_1^{-1} & 0 & 0 & \frac{1}{4}\tau_1^{-1} \\ & & -i(\omega_C - \frac{1}{2}J_{BC} + \frac{1}{2}J_{AC}) \\ 0 & 0 & -T_{2C}^{-1} - \tau_1^{-1} & 0 & \frac{1}{4}\tau_1^{-1} \\ & & & -i(\omega_C - \frac{1}{2}J_{BC} - \frac{1}{2}J_{AC}) \\ 0 & 0 & 0 & -T_{2C}^{-1} - \tau_1^{-1} & \frac{1}{4}\tau_1^{-1} \\ \tau_2^{-1} & \tau_2^{-1} & \tau_2^{-1} & \tau_2^{-1} & -i\omega_{C'} - T_{2C'}^{-1} - \tau_2^{-1} \end{bmatrix} \quad (23)$$

## Input Parameters for Numerical Computation

From the structure of the $E$ matrices, it may be seen that the input parameters include all the individual chemical shifts (expressed as Larmor frequencies $\omega_A$, $\omega_B$, etc.), spin–spin coupling constants, line widths of all the transitions of the spin systems in the absence of the exchange, and the exchange rates in either direction. The exchange rates are usually related by an equilibrium constant [see Eq. (9), for example], which may be measurable from the spectrum itself (see B. D. Nageswara Rao, Vol. 177, [18], this series). Thus only one of the two rates is an independent parameter. The line shape is usually calculated for different values of the exchange rate and compared with experiment.

The spectral parameters of the spin systems, namely, the chemical shifts and spin–spin coupling constants, should be characteristic of the spectrum in the limit of vanishing exchange rate. In the case of enzyme-bound reactants and products, these parameters belong to the reaction complex in the absence of exchange. These may be difficult to determine because the reaction cannot be stopped when all components of the reaction complex are present on the enzyme, and interconversion is in progress. Some of the line shapes of resonances may obscure or modify the resonance. In such cases, the spectral parameters may also be adjustable, and may be determined through the fitting procedure by the criterion of agreement with experiment.

It may be noted that the various Larmor frequencies in $\mathbf{E}$ may all be entered with a chosen reference frequency, i.e., they can be entered as a chemical shift (in ppm) multiplied by $\omega_0$, the NMR operating frequency. Finally, it should be noted that all the frequencies, coupling constants, and line widths must be entered in units of radians per second. These data, usually given in hertz in NMR literature, must be multiplied by $2\pi$, and included in the matrix elements of $\mathbf{E}$.

It is evident that except in the special circumstance of an equilibrium constant exactly equal to unity, $\tau_1 \neq \tau_2$, and therefore, the $\mathbf{E}$ matrix is, in general, unsymmetrical in addition to containing complex numbers in its diagonal elements. A critical step in the computation is the diagonalization of $\mathbf{E}$ [see Eqs. (3) and (4)]. Computer programs for the diagonalization of complex unsymmetrical matrices, although not commonplace, are available with computation centers as sealed library programs. Such a program should be acquired (and tested) for the line-shape calculations considered in this chapter.

Simulated Line Shapes, Comparison with Experiment

*Exchanging Spin Multiplets with Coupling Preserved; [31]P Line Shapes of [β-P]ATP in the Presence of Mg(II) and Ca(II)*

[31]P NMR spectra of MgATP and ATP at pH 8.0 are shown in Fig. 1A and C, respectively.[8] Figure 1B shows the spectrum with [ATP]:[Mg] ≈ 2:1. Corresponding spectra obtained using Ca(II) instead of Mg(II) are shown in Fig. 2. The spectra of MgATP and CaATP are similar; chelation

[8] K. V. Vasavada, B. D. Ray, and B. D. Nageswara Rao, *J. Inorg. Biochem.* **21**, 323 (1984); correction: *ibid.* **25**, 293 (1985).

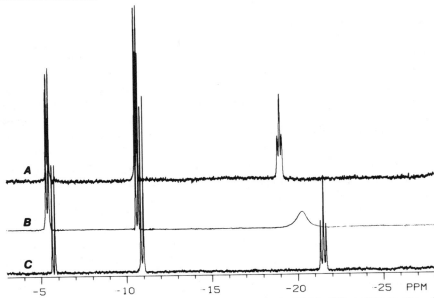

Fig. 1. $^{31}$P NMR spectra (at 121.5 MHz) of (A) MgATP, (B) MgATP + ATP (1 : 1), and (C) ATP, all in K$^+$–HEPES buffer at 20° and pH 8.0. (Reproduced with permission from Vasavada *et al.*[8])

with the cation causes a downfield shift (~2.5 ppm) in [$\beta$-P]ATP. The spin-coupling constants $J_{\alpha\beta}$ and $J_{\beta\gamma}$ are reduced by ~3–4 Hz upon chelation. Within experimental error $J_{\alpha\beta} = J_{\beta\gamma}$ in ATP, MgATP, and CaATP, leading to a 1 : 2 : 1 triplet for the respective $\beta$-P resonances. Note the significant contrast in the line shapes of $\beta$-P at intermediate cation concentrations (Figs. 1B and 2B).

Line shapes simulated using spectral parameters typical for [$\beta$-P]ATP (see legend for Fig. 3) as a function of the exchange rate are shown in Fig. 3. The exchange rate is expressed as a fraction of the chemical shift difference between the two sites so that the dependence on the spectrometer operating frequency may be readily noted. A similar simulation for a two-site exchange of simple lines at the same positions is shown in Fig. 4. It may be seen from Fig. 3 that for low rates of exchange the individual lines in both triplets broaden such that, for exchange rates larger than the spin–spin coupling constants ($\tau^{-1}/\Delta\omega = 0.1$), the multiplicity of the resonance is not discernible. A comparison of Figs. 3 and 4 shows that for $0.1 < \tau^{-1}/\Delta\omega < 1$ the resultant line shape is not significantly influenced by spin–spin coupling, leading to a possible misconception that the coupling

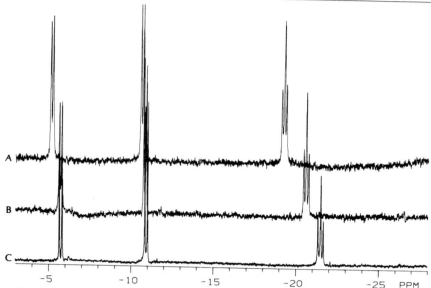

FIG. 2. $^{31}$P NMR spectra $J$ (at 121.5 MHz) of (A) CaATP, (B) Ca(II) + ATP (1 : 2), and (C) ATP, all in K$^+$–HEPES buffer at 20° and pH 8.0. (Reproduced with permission from Vasavada *et al.*[8])

is "washed out" by the exchange. However, for larger exchange rates ($\tau^{-1}/\Delta\omega > 10$) the inherent multiplet character of the resonances becomes evident, and for $\tau^{-1} \gg \Delta\omega$ a sharp 1 : 2 : 1 triplet with a chemical shift, spin–spin constant, and line width given by weighted averages of the respective quantities in the two species (in the absence of exchange) is obtained. It may be noted both from Figs. 3 and 4 that, while a single resonance centered at the average position is obtained for $\tau^{-1} \approx \Delta\omega$, it requires a much larger rate ($\tau^{-1}/\Delta\omega \approx 100$) to eliminate all broadening due to exchange and yield a resonance of average width. A comparison of the contrasting line shapes of [$\beta$-P]ATP in Figs. 1B and 2B with those in Fig. 3 readily suggests that the dissociation rate for Ca(II) from ATP is much larger than that for Mg(II). These rates (reciprocal lifetimes of the cation complexes) were estimated by simulation, and were found to be 2100 sec$^{-1}$ for MgATP, and $\geq 3 \times 10^5$ sec$^{-1}$ for CaATP.[9] The dissociation rate for Ca(II) is thus larger by at least a factor of 150 compared to that for Mg(II). The rates obtained by the NMR method are in general agreement

---

[9] The value of $3 \times 10^5$ sec$^{-1}$ should be considered a lower limit for the dissociation rate of CaATP because the calculated line shape does not change significantly for larger values.[8]

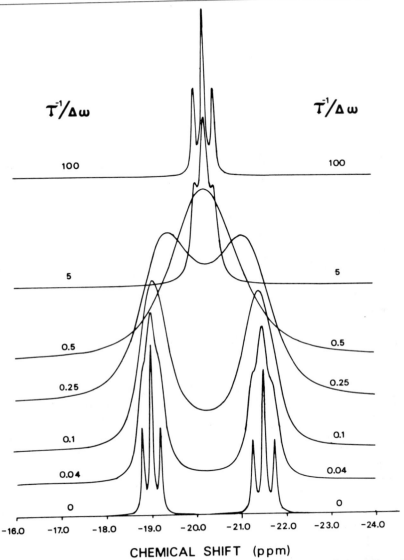

**CHEMICAL SHIFT (ppm)**

FIG. 3. Line shapes for a two-site exchange (B ⇌ B') with equal populations where the line shapes in the absence of exchange are given by 1 : 2 : 1 triplets as in the case of [β-P]ATP and [β-P]MATP, where M = Mg(II) or Ca(II). The exchange rate $\tau^{-1}$ is expressed as a multiple of the chemical shift $\Delta\omega$ (in rad sec$^{-1}$) at 121 MHz. The parameters are $\omega_B = 21.5$ ppm, $\omega_{B'} = 19.0$ ppm. $J_{\alpha\beta} = J_{\beta\gamma} = 19.5$ Hz, $J_{\alpha'\beta'} = J_{\beta'\gamma'} = 16.0$ Hz, and line widths of 6 Hz at both sites (in the absence of exchange). (Reproduced with permission from Vasavada *et al.*[8])

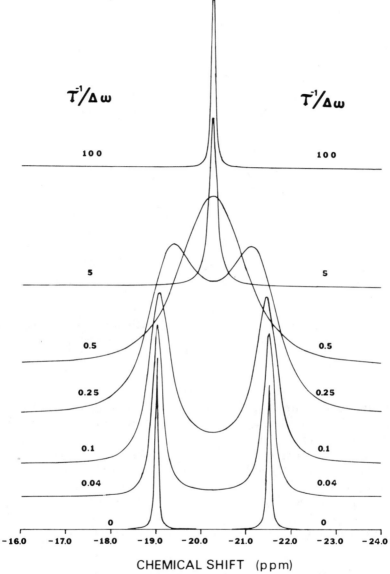

FIG. 4. Line shapes of two-site exchange (B $\rightleftharpoons$ B') in the absence of spin–spin coupling. All other parameters are identical to those in Fig. 3. (Reproduced with permission from Vasavada *et al.*[8])

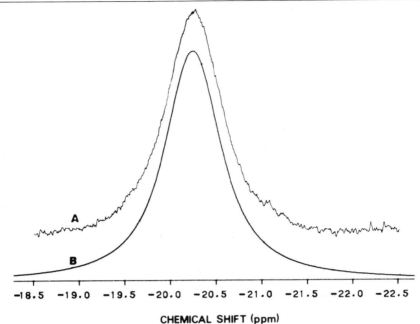

**CHEMICAL SHIFT (ppm)**

FIG. 5. Experimental (A) and simulated (B) $^{31}$P NMR line shapes of [$\beta$-P]ATP in the presence of Mg(II) such that [ATP] = [MgATP]. Experimental conditions for (A) are identical to those for Fig. 1B. Line shapes were calculated with $\tau_2^{-1} = 2100$ sec$^{-1}$. (Reproduced with permission from Vasavada *et al.*[8])

with those measured by the T-Jump methods.[10] A comparison of the simulated and experimental line shapes is shown in Fig. 5 for MgATP and Fig. 6 for CaATP.

It is useful to note one qualitative feature of line shapes of binary complexes such as MgATP. Line shapes of [$\beta$-P]ATP simulated for a set of seven different values of [MgATP] : [ATP] (including 0 and 100, which represent ATP and MgATP, respectively) are shown in Fig. 7. In all these cases $\tau_2^{-1}$ is set equal to 2100 sec$^{-1}$, the reciprocal lifetime of MgATP determined above. Note, however, that $\tau_1^{-1}$ changes according to the relation

$$\tau_1^{-1}/\tau_2^{-1} = [\text{MgATP}]/[\text{ATP}] \qquad (24)$$

It is evident from Fig. 7 that similar departures (in terms of concentration) from the two extreme cases of ATP and MgATP do not yield similar line

---

[10] M. Eigen and R. G. Wilkens, *Adv. Chem.* **49**, 55 (1965).

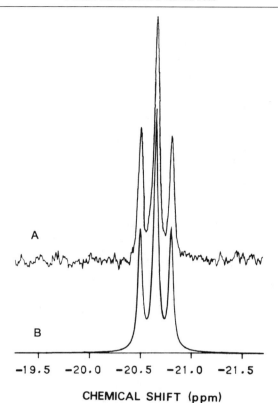

CHEMICAL SHIFT (ppm)

FIG. 6. Experimental (A) and simulated (B) $^{31}$P NMR line shapes [$\beta$-P]ATP in the presence of Ca(II) such that [ATP]:[CaATP] = 1.5:1. Experimental conditions for (A) are identical to those for Fig. 2B. Line shapes were calculated with $\tau_2^{-1} = 3 \times 10^5$ sec$^{-1}$. (Reproduced with permission from Vasavada *et al.*[8])

shapes. This asymmetry arises from the fact that as the Mg(II) concentration is varied $\tau_1^{-1}$ increases linearly with unchelated [Mg(II)] whereas $\tau_2^{-1}$ remains constant.

*Interconversion Rates of Arginine Kinase and Creatine Kinase Reactions.* Phosphoryl transfer enzymes catalyze the reversible transfer of the terminal phosphoryl group of ATP to a second substrate (*X*) to yield ADP, and a phosphorylated second substrate (*X*P):

$$\text{ATP} + X \underset{\longleftarrow}{\overset{\text{Mg(II)}}{\rightleftharpoons}} \text{ADP} + X\text{P} \tag{25}$$

e.g., for the arginine kinase *X* is arginine and *X*P is phosphoarginine. All

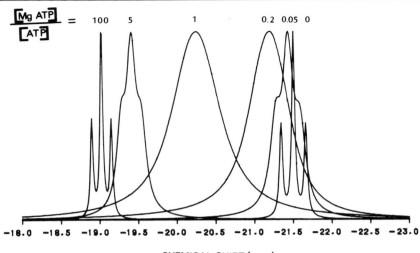

FIG. 7. $^{31}$P NMR line shapes of [$\beta$-P]ATP in the presence of Mg(II) at various values of the ratio [MgATP] : [ATP]. The line shapes were calculated using spectral parameters in Table I, $\tau_2^{-1} = 2100$ sec$^{-1}$ and $\tau_1^{-1} = \tau_2^{-1}$[MgATP]/[ATP]. (Reproduced with permission from Vasavada et al.[8])

kinases require a divalent cation, usually Mg(II), as an obligatory component of the reaction.

If the enzyme concentrations are chosen in sufficient excess such that all the four substrates in Eq. (25) are bound to the enzyme, the reaction will occur on the surface of the enzyme, which may be represented as[11–13]

$$E \cdot MgATP \cdot X \underset{\tau_2^{-1}}{\overset{\tau_1^{-1}}{\rightleftharpoons}} E \cdot MgADP \cdot XP \qquad (26)$$

If the second substrate, $X$, does not contain a $^{31}$P nucleus, the $^{31}$P NMR spectrum of this interconverting reaction complex is thus characterized by the ABC $\rightleftharpoons$ A'B' + C' exchange. The structure of **E**, $\chi$, and **W** for this exchange was described in the previous section.

Computer-simulated $^{31}$P NMR spectra for the enzyme-bound equilibrium mixture are shown along with the experimental spectra for arginine kinase and creatine kinase reactions in Figs. 8 and 9, respectively.[1,7,14]

[11] M. Cohn and B. D. Nageswara Rao, Bull. Magn. Reson. 1, 38 (1979).

[12] B. D. Nageswara Rao, in "Biological Magnetic Resonance" (L. Berliner and J. Reuben, eds.), Vol. 5, p. 75. Plenum, New York, 1983.

[13] B. D. Nageswara Rao, in "Phosphorus-31 NMR: Principles and Applications" (D. G. Gorenstein, ed.), p. 57. Academic Press, New York, 1984.

[14] B. D. Nageswara Rao and M. Cohn, J. Biol. Chem. 256, 1716 (1981).

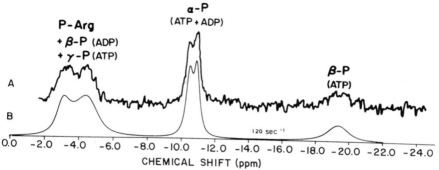

FIG. 8. Experimental (A) and simulated (B) $^{31}$P NMR spectra (at 40.3 MHz) for the equilibrium mixture of enzyme-bound substrates and products of arginine kinase. The simulated spectrum was obtained by using spectral parameters listed in Table I and $\tau_1^{-1} = 120$ sec$^{-1}$. (Reproduced with permission from Vasavada et al.[7])

The parameters chosen for the best fit between experimental and simulated spectra are listed in Table I. It may be noted that the chemical shifts chosen for the best fit differ from those of the enzyme-bound substrates in the absence of the reaction. The fitting procedure thus indirectly yields the chemical shifts in the reaction complex. It should be recognized that as the set of adjustable parameters gets enlarged, the confidence in the best fit might diminish. However, spectra such as those in Figs. 8 and 9

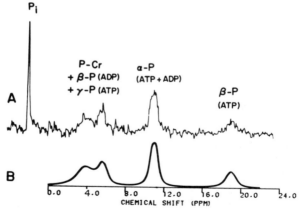

FIG. 9. Experimental (A) and simulated (B) $^{31}$P NMR spectra (at 40.3 MHz) for the equilibrium mixture of enzyme-bound substrates and products of creatine kinase. The resonance-labeled $P_i$ arises from a contaminant ATPase activity. The simulated spectrum was obtained by using the spectral parameters listed in Table I and $\tau_1^{-1} = 90$ sec$^{-1}$. Shifts upfield are positive. (Reproduced with permission from Vasavada et al.[7])

### TABLE I

SPECTRAL PARAMETERS USED FOR COMPUTER-SIMULATED $^{31}$P NMR SPECTRA OF ENZYME-BOUND EQUILIBRIUM MIXTURES OF ARGININE KINASE,[a] CREATINE KINASE,[a] AND ADENYLATE KINASE[b]; EXPERIMENTAL VALUES FOR ENZYME-BOUND SUBSTRATE COMPLEXES

| Complex | Chemical shift (in ppm) from 85% phosphoric acid | | | | | | Spin–spin coupling constant (in Hz) | | |
| | ATP | | | ADP | | AMP or P-arginine | ATP | | ADP |
| | $\alpha$-P | $\beta$-P | $\gamma$-P | $\alpha$-P | $\beta$-P or P-creatine | | $J_{\alpha\beta}$ | $J_{\beta\gamma}$ | $J_{\alpha\beta}$ |
|---|---|---|---|---|---|---|---|---|---|
| **Arginine kinase (pH = 7.25, $K'_{eq}$ = 0.8) (Fig. 8)** | | | | | | | | | |
| Enzyme-bound substrates + Mg(II) | 11.0 | 19.4 | 5.6 | 10.7 | 4.4 | 3.6 | 15.0 | 15.0 | 18.0 |
| Equilibrium mixture[c] | 11.0 | 19.4 | 4.8 | 10.4 | 4.4 | 2.9 | 15.0 | 15.0 | 18.0 |
| | (8) | (10) | (10) | (10) | (30) | (10) | | | |
| **Creatine kinase (pH = 7.8, $K'_{eq}$ = 1.0) (Fig. 9)** | | | | | | | | | |
| Enzyme-bound substrates + Mg(II) | 10.9 | 19.0 | 5.4 | 11.0 | 3.8 | 3.1 | 15.0 | 15.0 | 18.0 |
| Equilibrium mixture[c] | 11.2 | 19.0 | 5.8 | 10.7 | 3.8 | 3.5 | 15.0 | 15.0 | 18.0 |
| | (20) | (10) | (10) | (20) | (55) | (55) | | | |
| **Adenylate kinase[d] (pH 7.0, $K'_{eq}$ = 1.6) (Figs. 12–16)** | | | | | | | | | |
| A. No added Mg(II) | | | | | | | | | |
| Enzyme-bound substrates | 11.1 | 21.6 | 7.0 | 10.2 | 5.5 | −3.8 | 17.3 | 21.0 | 22.0 |
| Equilibrium mixture[c] | 11.1 | 21.6 | 7.0 | 11.0 | 7.5 | −3.9 | 17.3 | 21.0 | |
| | (10) | (15) | (10) | (10) | (10) | (12) | | | |
| | | | | 9.4 | 3.5 | | | | |
| | | | | (10) | (10) | | | | |
| B. With optimal Mg(II) | | | | | | | | | |
| Enzyme-bound substrates + Mg(II) | 10.9 | 171.7 | 6.1 | 9.8 | 3.5, 6.7 | −3.9 | 15.1 | 14.7 | |
| Equilibrium mixture[c] | 11.1 | 17.7 | 6.4 | 11.0 | 7.0 | −3.9 | 15.1 | 4.7 | 22.0 |
| | (10) | (15) | (10) | (10) | (10) | (12) | | | |
| | | | | 9.6 | 3.5 | | | | |
| | | | | (10) | (12) | | | | |

[a] Vasavada et al.[7]   [b] Vasavada et al.[16]   [c] Line widths in hertz given in parentheses.
[d] The two chemical shift values given for $\alpha$-P and $\beta$-P of ADP belong to ADP (donor) and MgADP, respectively. The coupling constant is not measurable for ADP on the enzyme because of exchange. The value given was measured in free solution.

consist of line shapes in three different regions and the sensitivity of these to various parameters is different. For example, the exchange rate $\tau_1^{-1}$ is fixed primarily by the line shape of the [$\beta$-P]ATP resonance at ~19 ppm in the spectrum. By using this value of $\tau_1^{-1}$ the shifts and line widths of $\alpha$-P of ATP and ADP may be fixed by comparing with the experimental line shape in the region of 10–12 ppm. Finally the shifts and line widths of [$\gamma$-P]ATP, [$\beta$-P]ATP, and phosphoarginine may be chosen exclusively on the basis of the complex line shape in the region of 2–6 ppm.

The chemical shift difference characteristic of the three [31]P exchanges (at 40.3 MHz) in the reaction complexes of arginine kinase and creatine kinase fall into three qualitatively distinct categories with respect to an exchange rate of about 100 sec$^{-1}$. The two $\alpha$-P resonances of ATP and ADP have a small chemical shift difference so that they are in fast exchange. The two $\beta$-P resonances, on the other hand, have a large chemical shift difference so that they are in slow exchange (for the same exchange rate). Finally, the [$\gamma$-P]ATP and $X$P (phosphoarginine or phosphocreatine) have an intermediate value of the chemical shift difference so that these resonances are in intermediate exchange. Thus, the three classes of exchange commonly used to describe the qualitative features of the effect of chemical exchange in NMR spectra all occur in the same spectrum. A comparison of the three line shapes in Fig. 8 with those in Fig. 4 should at once reveal this fact. While the qualitative features appear well described by the two-site exchange in Fig. 4 which does not include spin–spin coupling, it must be recognized that a quantitative analysis of the observed line shapes on such a basis leads to an incorrect estimate of the exchange rates. For example, when the enzyme-bound equilibrium mixture of arginine kinase was first published,[1] an exchange rate ($\tau_1^{-1}$) of 190 sec$^{-1}$ was estimated on the basis of the excess broadening of the [$\beta$-P]ATP resonance implicitly ignoring spin–spin coupling effects. However, computer simulation yields a value of 120 sec$^{-1}$ for $\tau_1^{-1}$ of the same spectrum (Fig. 8).[7]

The rates of the forward reaction (ADP formation) on the enzyme of 120 and 80 sec$^{-1}$ obtained, respectively, for arginine kinase and creatine kinase, and similar rates for the reverse reaction (the equilibrium constants are approximately equal to unity in both cases), are almost an order of magnitude larger than the overall rates of the reaction under the same conditions of pH, buffer concentration, and temperature. Thus the interconversion rate was found not to be rate limiting for either of the reactions under catalytic conditions. The ability to isolate and determine the kinetic parameters (in equilibrium) of the interconversion step on the surface of the enzyme is a straightforward consequence of choosing experimental conditions such that the substrates are in their enzyme-bound form. Analysis of the observed line shapes by the procedure of computer simulations using the density matrix theory was essential to reliably extract the kinetic information.

The complexity of spin–spin coupling effects on the spectrum of the exchanging system may be illustrated by the simulations of arginine kinase equilibrium mixture for values of $\tau_1^{-1}$ ranging from 10 to 600 sec$^{-1}$ shown in Fig. 10. The line shapes of [$\beta$-P]ATP resonance at ~19 ppm in the simulated spectra show evidence of spin–spin splitting up to $\tau_1^{-1} \sim 40$ to 60 sec$^{-1}$. For values of $\tau_1^{-1} \geq 80$ sec$^{-1}$ the spin-coupling effects are smoothed out and the resonance resembles a Lorentzian. However, the line-width change (with respect to $\tau_1^{-1} = 0$) is implicitly contributed in part by the spin–spin coupling effects, and exchange rates estimated merely

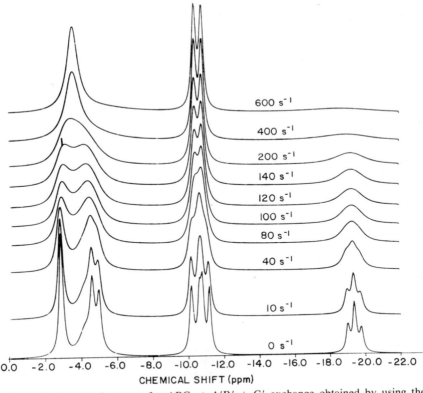

FIG. 10. Simulated spectra for ABC $\rightleftharpoons$ A′B′ + C′ exchange obtained by using the spectral parameters for arginine kinase and varying $\tau_1^{-1} = 10$ to 600 sec$^{-1}$. (Reproduced with permission from Vasavada *et al.*[7])

by considering line-width changes are likely to be larger than the true values $(\tau_1^{-1})$. It is also interesting to note that the coalescence of the $[\alpha\text{-P}]$ATP and $[\alpha\text{-P}]$ADP resonances for large values of $\tau_1^{-1}$ does not destroy the doublet structure of this resonance, demonstrating that the exchanges in the kinase reaction do not randomize the spin–spin coupling between $\alpha$-P and $\beta$-P (except for the small difference, $\sim2.5$ Hz, in $J_{\alpha\beta}$ between ATP and ADP). Finally, it may be noted that for values of $\tau_1^{-1}$ approaching 400 sec$^{-1}$ or higher, the $\beta$-P resonance is broadened to such an extent that it may not be detectable. The spectrum is somewhat deceptive in that it appears that there is no ATP in the sample.

## $^{31}P$ Exchanges in Adenylate Kinase Reaction: An Example of Complex Exchanges

Adenylate kinase catalyzes the reversible phosphoryl transfer from ATP to AMP leading to two ADP molecules, i.e., in Eq. (25), X = AMP.[15] In the analysis of the line shapes arising from exchanges in the $^{31}P$ spectrum of enzyme-bound reactants products there is a new and unique feature to be recognized. Of the two ADP molecules one is a phosphoryl donor and the other a phosphoryl acceptor, and since these molecules are indistinguishable off the enzyme, they may interchange their binding sites on the enzyme (through the medium of the bulk solution, i.e., by dissociation and reassociation). Such an interchange might affect the phosphoryl transfer reaction in the reverse direction. $^{31}P$ NMR studies[15] have established that the two nucleotide-binding sites are unequivocally differentiated on the enzyme, and that the cation Mg(II) plays an unmistakable role in providing the distinction; while both Mg-bound and free nucleotides bind at the acceptor ADP (or ATP) site, only free nucleotides bind at the donor ADP (or AMP) site. The $^{31}P$ exchanges in the adenylate kinase reaction are depicted in Fig. 11. These exchanges may be formally represented as[16]

$$\text{AB}\overset{\centerdot}{\text{C}} + \text{D} \underset{\tau_2^{-1}}{\overset{\tau_1^{-1}}{\rightleftharpoons}} \overset{\tau_3^{-1}}{\underset{\tau_4^{-1}}{\overgroup{\text{A}'\text{B}' + \text{A}''\text{B}''}}} \tag{27}$$

where $\tau_1^{-1}$ and $\tau_2^{-1}$ are the usual interconversion rates, and $\tau_3^{-1}$ and $\tau_4^{-1}$ represent the rates of interchange of donor and acceptor ADP molecules and are referred to as the interchange rates to avoid confusion. There are a total of 21 transitions for the spin systems in Eq. (27): 12 for ABC, 1 for

[15] B. D. Nageswara Rao, M. Cohn, and L. Noda, *J. Biol. Chem.* **254**, 1149 (1978).
[16] K. V. Vasavada, J. I. Kaplan, and B. D. Nageswara Rao, *Biochemistry* **23**, 961 (1984).

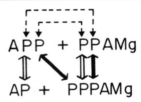

FIG. 11. $^{31}$P exchanges in the adenylate kinase reaction. The solid and open arrows represent fast and slow exchanges, respectively. The dashed lines represent the interchange of the ADP molecules between the acceptor and donor sites. (Reproduced with permission from Nageswara Rao *et al.*[15])

D, and 4 each for A′B′ and A″B″. Therefore, the dimension of Eq. (2) is 21. The **W** vector has $(\frac{1}{8})$ for the first 12 elements, $(\frac{1}{2})$ for the thirteenth element, and $(\frac{1}{4})$ each for the remaining 8. The structure of the **E** matrix is shown in Eq. (28) as a block diagram in which the occurrence of exchange rates in the off-diagonal positions is indicated.[16] The diagonal elements for the transitions of the different spins designated by

$$\begin{bmatrix} \begin{array}{c} \text{A Trans} \\ 4 \times 4 \end{array} & 0 & 0 & 0 & \tau_1^{-1}/2 & 0 & 0 & 0 \\[2mm] 0 & \begin{array}{c} \text{B Trans} \\ 4 \times 4 \end{array} & 0 & 0 & 0 & \tau_1^{-1}/2 & 0 & 0 \\[2mm] 0 & 0 & \begin{array}{c} \text{C Trans} \\ 4 \times 4 \end{array} & 0 & 0 & 0 & 0 & \tau_1^{-1}/4 \\[2mm] 0 & 0 & 0 & \begin{array}{c} \text{D Trans} \\ 1 \times 1 \end{array} & 0 & 0 & \tau_1^{-1} & 0 \\[2mm] \tau_2^{-1} & 0 & 0 & 0 & \begin{array}{c} \text{A' Trans} \\ 2 \times 2 \end{array} & 0 & \tau_3^{-1} & 0 \\[2mm] 0 & \tau_2^{-1} & 0 & 0 & 0 & \begin{array}{c} \text{B' Trans} \\ 2 \times 2 \end{array} & 0 & \tau_3^{-1} \\[2mm] 0 & 0 & 0 & \tau_2^{-1}/2 & \tau_4^{-1} & 0 & \begin{array}{c} \text{A'' Trans} \\ 2 \times 2 \end{array} & 0 \\[2mm] 0 & 0 & \tau_2^{-1}/2 & 0 & 0 & \tau_4^{-1} & 0 & \begin{array}{c} \text{B'' Trans} \\ 2 \times 2 \end{array} \end{bmatrix} \quad (28)$$

the letters in Eq. (27) may be written in the standard form [see Eq. (6)] in the blocks indicated in Eq. (28). The presence of an exchange rate with a coefficient in a block in Eq. (28) indicates that all the nonvanishing off-diagonal elements in that block have that value. The actual position of that off-diagonal elements depends on the labeling of the states and transitions of the different spin systems, and the resulting correspondence between them. It may be seen from Eq. (28) that the transitions of A and A′,

D and A″, B and B′, and C and B″ are connected in pairs by the interconversion as they would be in any phosphoryl transfer reaction, but the interchange further connects the first two pairs by connecting A′ and A″, and the last two pairs by connecting B′ and B″.

The spectral parameters used in the simulation are listed in Table I. The interchange rates $\tau_3^{-1}$ and $\tau_4^{-1}$ are set equal to each other in the absence of any other information regarding them. This appears reasonable because the interchange process is made up of steps of dissociation and reassociation for both the substrates, and the rate-determining step is likely to be the slowest of these.

[31]P NMR spectra of equilibrium mixtures of enzyme-bound substrates and products of adenylate kinase from two sources (porcine and carp) are shown in Fig. 12 along with a spectrum of the equilibrium mixture ob-

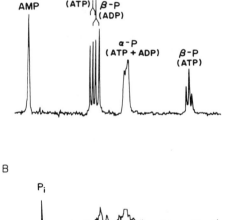

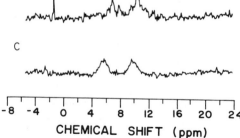

FIG. 12. [31]P NMR spectra (at 40.3 MHz) of equilibrium mixture of porcine and carp adenylate kinase at pH 7 = 0 and $T = 4°$. (A) Equilibrium mixture with catalytic enzyme concentration. (B) Equilibrium mixture of enzyme-bound substrates for porcine enzyme. (C) Same as (B) for carp enzyme. Shifts upfield are positive. (Reproduced with permission from Nageswara Rao et al.[15])

tained at catalytic enzyme concentrations. The experiment with the porcine enzyme (Fig. 12B) had to be terminated at a lower signal-to-noise ratio than that with the carp enzyme because of a contaminant ATPase activity leading to the accumulation of $P_i$ in the spectrum.[17] Nevertheless, neither spectrum showed any signals in the AMP and [$\beta$-P]ATP regions. Computer simulations of the line shapes should first explain these qualitative features of the spectrum of the equilibrium mixture before ascertaining whether the interchange rate influences the observed spectrum. Simulated spectra calculated by ignoring the interchange [i.e., by setting $\tau_3^{-1} = \tau_4^{-1} = 0$ in Eq. (28)] and varying $\tau_1^{-1}$ from 0 to 700 sec$^{-1}$ are shown in Fig. 13. The spectra for $\tau_1^{-1} > 400$ sec$^{-1}$ clearly indicate that the absence of observable resonances outside the two regions around 7 and 10 ppm requires interconversion rates in this range. For these rates the two exchanges, AMP $\rightleftharpoons$ [$\alpha$-P]ADP and [$\beta$-P]MgADP $\rightleftharpoons$ [$\beta$-P]MgATP, are still in slow exchange and the resonances become too broad to be observed. On the other hand, the other two exchanges [$\alpha$-P]MgATP $\rightleftharpoons$ [$\alpha$-P]MgADP and [$\beta$-P]ADP $\rightleftharpoons$ [$\gamma$-P]MgATP are in intermediate or fast exchange leading to the line shapes observed in the neighborhood of 7 and 10 ppm. While this qualitative feature of the spectra in Fig. 12B and C are explained, none of the simulated spectra in Fig. 13 agreed with the detailed line shapes in Fig. 12B and C, suggesting the need to inquire into the role of the interchange rate on the observed line shapes.

Figure 14 presents simulated spectra in which $\tau_1^{-1}$ was kept fixed at 600 sec$^{-1}$ and $\tau_3^{-1}$ varied from 0 to 300 sec$^{-1}$. The introduction of $\tau_3^{-1}$ does indeed alter the line shapes, and good agreement with the experiment is obtained for $\tau_3^{-1} = 100$ sec$^{-1}$. This result was the first clear evidence that the interchanges of the donor and acceptor molecules does occur under the conditions of the experiment.

Is the interchange rate affected by the Mg(II) concentration? The two spectra shown in Fig. 15 provide experimental evidence pertinent to this question. The spectrum in Fig. 15A is obtained in the absence of externally added Mg(II). Because of adventitious activating cations, a feeble interconversion at the rate of ~5 sec$^{-1}$ lead to an equilibrium mixture. The ADP resonance in Fig. 15A acquires a line shape arising from an interchange of the two ADP molecules. Spectral line shapes calculated by setting $\tau_1^{-1} = 0$ and varying $\tau_3^{-1}$ are shown in Fig. 16. A comparison of ADP line shapes with Fig. 15A leads to the conclusion that the inter-

---

[17] Although this contaminant ATPase activity created a difficulty in recording a good spectrum for the enzyme-bound equilibrium mixture, it is precisely this contaminant that serendipitously led to the identification of separate $^{31}$P signals for donor and acceptor ADP molecules bound to adenylate kinase.[15]

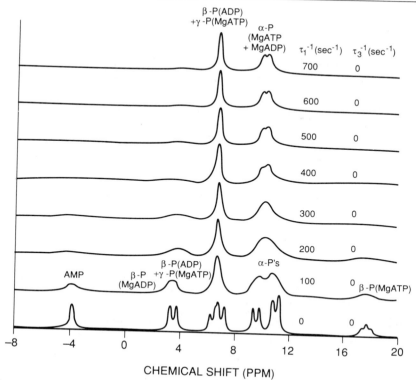

FIG. 13. Simulated $^{31}$P NMR spectra for the enzyme-bound equilibrium mixture of adenylate kinase for different values of $\tau_1^{-1}$ and $\tau_3^{-1} = 0$. The other parameters are listed in Table I. Shifts upfield are positive. (Reproduced with permission from Vasavada *et al.*[16])

change rate in the absence of Mg(II) is 1500 ± 100 sec$^{-1}$, an order of magnitude faster than that in the presence of optimal Mg(II).

The optimum concentration of Mg(II) in an adenylate kinase equilibrium mixture is [ATP] + ½[ATP]. The spectrum in Fig. 15B, obtained with Mg(II) concentration in excess of this optimum, shows that the line shapes in the region of ~7 and 10 ppm are narrower than those in Fig. 12B and C. At these high concentrations of Mg(II), there will be reduced availability of Mg(II)-free ADP for binding at the donor sites, and a greater likelihood for the formation of exchange-inert complexes such as E · MgADP · AMP. These factors effectively diminish the interchange rate and the simulated spectra in Fig. 14 show the line shapes in the regions of 7 and 10 ppm are narrower at smaller values of the interchange rate, thus

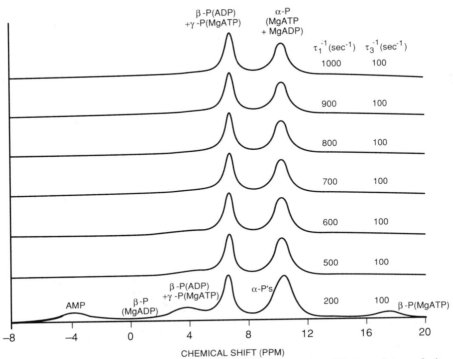

FIG. 14. Simulated $^{31}P$ NMR spectra for the enzyme-bound equilibrium mixture of adenylate kinase for $\tau_1^{-1} = 600$ sec$^{-1}$ and for different values of $\tau_3^{-1}$. The other parameters are listed in Table I. Shifts upfield are positive. (Reproduced with permission from Vasavada et al.[16])

providing a rationalization for the observed spectrum in Fig. 15B consistent with other findings regarding the interchange.

The striking reduction in the interchange rate from 1500 sec$^{-1}$ in the absence of Mg(II) to 100 sec$^{-1}$ at optimal Mg(II), points to a new dimension in the role of the cation in affecting differentiation of the two nucleotide-binding sites on this enzyme in addition to the normal catalytic role it has for all other ATP-utilizing enzymes. Considering that the interconversion rates are 600 sec$^{-1}$ (for ADP formation) and 375 sec$^{-1}$ (for ATP formation), it is apparent that the interchange rate is reduced from a value much larger (in the absence of Mg(II)) to a value sufficiently smaller [in the presence of optimal Mg(II)] than the eventual interconversion rate such that the catalytic rate is not severely hampered by the interchange.

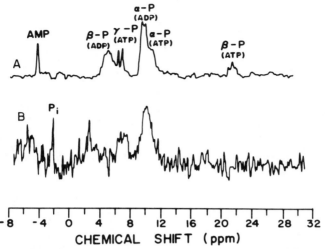

FIG. 15. $^{31}$P NMR spectra (at 40.3 MHz) of enzyme-bound equilibrium mixture of porcine adenylate kinase at $T = 4°$, obtained at different Mg(II) concentrations. (A) pH 7.0, no added Mg(II). (B) pH 8.0, [Mg]:[ADP] = 2.19. Shifts upfield are positive. (Reproduced with permission from Vasavada *et al.*[16])

## Summary

The fact that chemical exchange processes occur at rates that cover a broad range and produce readily detectable effects on the spectrum is one of the attractive features of high-resolution NMR. The description of these line shapes in the presence of spin–spin coupling requires the density matrix theory which is rather complex. Analysis of the line shapes usually needs computer simulations and is capable of providing reliable information on the exchange rates as well as spectral parameters in the absence of exchange. Simplified procedures, ignoring spin–spin coupling, often result in deviations in these exchange and spectral parameters determined.

A step-by-step procedure is detailed in this chapter for setting up the matrices required for computing the line shapes of exchanges involving weakly coupled spin systems on the basis of the density matrix theory without the need for a detailed understanding of the theory. A knowledge of the energy level structure and allowed transitions in the NMR spectra of the individual weakly coupled spin systems is all that is required. The procedure is amenable to numerical computation.

The group of illustrative examples chosen to demonstrate the develop-

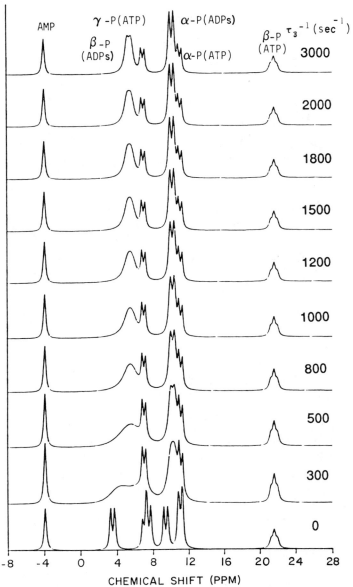

FIG. 16. Simulated $^{31}$P NMR spectra of enzyme-bound reactants and products of adenylate kinase in the absence of Mg(II), with $\tau_1^{-1} = 0$ and for different values of $\tau_3^{-1}$. Other parameters are listed in Table I. Shifts upfield are positive. (Reproduced with permission from Vasavada *et al.*[16])

ment of the computational tools cover some of the commonly encountered cases of exchange from simple systems to rather complex ones. Such exchanges occur frequently in biological molecules, especially those involving enzyme–substrate complexes.

In cases where the experimental line shapes are obtained with respectable precision, and the relevant exchange processes are unambiguously identifiable, the computer simulation method of line-shape analysis is capable of providing useful and incisive information. The example of the $^{31}P$ exchanges in the adenylate kinase is illustrative of this point. Not only has the line-shape analysis clearly indicated the role of the interchange process, but it has also produced evidence that the rates of interchange of the ADP molecules bound to the enzyme become relevant to the kinetics and mechanism of catalysis by this enzyme. It is probably difficult to obtain this information by any other experimental method or by any other method of analysis.

### Acknowledgments

Thanks are due to Drs. K. V. Vasavada, Bruce D. Ray, and Jerome I. Kaplan for some of the previous collaborations in the work abstracted in this chapter. The work was supported over the years in part by grants from the National Science Foundation (PCM 80 18725, PCM 80 22075, DMB 8309120 and 8608185) (1981 onward), the Research Corporation (1979–1982), the American Heart Association (1983–1985) and IUPUI. I wish to thank Ms. Margo Page for typing this manuscript.

# [15] Applicability of Magnetization Transfer Nuclear Magnetic Resonance to Study Chemical Exchange Reactions

By JENS J. LED, HENRIK GESMAR, and FRITS ABILDGAARD

The determination of rate constants and pathways of enzyme-catalyzed reactions is essential for the elucidation of the mechanisms and functions of biological systems at a molecular level. Historically, reaction rates have been studied by stopped flow, temperature jump, and radioactive tracer analyses.[1] Both stopped flow and temperature jump require that the system initially be placed in a nonequilibrium state, allowing the rate constant to be determined by subsequent monitoring of the system as it returns to equilibrium. Alternatively, radioactive tracer analysis allows enzymatic reactions to be studied at equilibrium but requires radioactive labeling and a subsequent assay procedure.

[1] A. Fersht, "Enzyme Structure and Mechanism," 2nd ed. Freeman, New York, 1985.

NMR spectroscopy offers the important possibility of measuring the rate constants by monitoring the system at equilibrium directly, without the need for special assay procedures. Furthermore, its ability to monitor nuclei in specific positions of the substrate and product molecules, as well as the transfer of nuclei between these positions, allows unprecedented, detailed analyses of reaction pathways. In addition, the available NMR techniques cover a large range of rate constants. Thus, the classical line-shape analysis[2] (see [14] by Nageswara Rao, in this volume) is applicable to relatively fast reactions, i.e., rate constants from 10 to $10^4$ sec$^{-1}$, while rate constants from $10^{-2}$ to $10^2$ sec$^{-1}$ can be measured by the magnetization transfer technique.[3,4]

In this chapter, we shall be concerned with the magnetization transfer technique in its general form. Rather than giving a detailed review, we shall focus on the present state of the technique, emphasizing its scope and limitations in the general, nonideal cases. This includes cases where the relaxation rates of all exchanging signals are different, unintentional signal perturbations occur, and the exponents of the involved multiexponential functions are almost identical. Also cases with phase- and base-line-distorted spectra, low signal-to-noise ratios, and partly overlapping signals are considered. In particular, we shall concentrate on the experimental procedure necessary to obtain spectra that are sufficiently informative, and the data analysis that must be used to retrieve the maximum information from these spectra.

## General Two-Site Magnetization Transfer Experiment

Consider a system in which a reversible molecular process transfers a nucleus $X$ back and forth between two different chemical environments, A and B, with the first order rate constants, $k_A$ and $k_B$. If $k_A$ and $k_B \ll \Delta\nu_{AB}$, where $\Delta\nu_{AB}$ is the chemical shift difference between the two sites, individual NMR signals are observed for $X$ in the A and B environments. In its general form the magnetization transfer experiment consists in an initial selective perturbation of one of the two signals, e.g., A, generated by an external radio frequency (rf) field; that is, the longitudinal magnetization, $M_A$, corresponding to the signal is selectively perturbed. Due to the exchange process, the perturbation will be transferred to the other

[2] J. I. Kaplan and G. Fraenkel, "NMR of Chemically Exchanging Systems," Chapter VI. Academic Press, New York, 1980.
[3] J. R. Alger and R. G. Shulman, Q. Rev. Biophys. **17**, 83 (1984).
[4] J. J. Led and H. Gesmar, J. Magn. Reson. **49**, 444 (1982).

signal. Assuming that the longitudinal relaxations of $X$ in the two sites are not considerably faster than the exchange rate, the transfer can be monitored by observing the change in intensity of one or both of the two signals at various times after the initial perturbation.

The return of $M_A$ and $M_B$ toward equilibrium after the perturbation is described by the McConnell equations[5] for the longitudinal magnetization. In the absence of the perturbing rf field, these equations take the form

$$dM_A/dt = -M_A k_{1A} + M_B k_B + M_A^\infty R_{1A} \qquad (1)$$
$$dM_B/dt = M_A k_A - M_B k_{1B} + M_B^\infty R_{1B} \qquad (2)$$

where $M_A^\infty$ and $M_B^\infty$ are the equilibrium longitudinal magnetizations of $X$ in the A and B site, respectively, and

$$k_{1A} = k_A + R_{1A} \qquad (3)$$
$$k_{1B} = k_B + R_{1B} \qquad (4)$$

$R_{1A}$ and $R_{1B}$ being the longitudinal relaxation rates. It should be noted that transfer of magnetization mediated by cross-relaxation (the nuclear Overhauser effect) between the A and B sites is described by a set of equations formally identical to Eqs. (1)–(4). Therefore, in order to obtain correct exchange rates from the experimental data, cross-relaxation must be either negligible or determined independently. An experimental procedure that allows a distinction between chemical exchange and cross-relaxation in special cases has recently been proposed.[6]

In the general case, Eqs. (1) and (2) have the following solution:

$$M_A(t) = C_1 e^{\lambda_1 t} + C_2 e^{\lambda_2 t} + M_A^\infty \qquad (5)$$
$$M_B(t) = C_3 e^{\lambda_1 t} + C_4 e^{\lambda_2 t} + M_B^\infty \qquad (6)$$
$$\lambda_1 = \tfrac{1}{2}\{-(k_{1A} + k_{1B}) + [(k_{1A} - k_{1B})^2 + 4k_A k_B]^{1/2}\} \qquad (7)$$
$$\lambda_2 = \tfrac{1}{2}\{-(k_{1A} + k_{1B}) - [(k_{1A} - k_{1B})^2 + 4k_A k_B]^{1/2}\} \qquad (8)$$
$$C_1 = [(\lambda_2 + k_{1A})(M_A^\infty - M_A^0) - k_B(M_B^\infty - M_B^0)]/(\lambda_1 - \lambda_2) \qquad (9)$$
$$C_2 = [-(\lambda_1 + k_{1A})(M_A^\infty - M_A^0) + k_B(M_B^\infty - M_B^0)]/(\lambda_1 - \lambda_2) \qquad (10)$$
$$C_3 = [-k_A(M_A^\infty - M_A^0) - (\lambda_1 + k_{1A})(M_B^\infty - M_B^0)]/(\lambda_1 - \lambda_2) \qquad (11)$$
$$C_4 = [k_A(M_A^\infty - M_A^0) + (\lambda_2 + k_{1A})(M_B^\infty - M_B^0)]/(\lambda_1 - \lambda_2) \qquad (12)$$

Here $M_A(t)$ and $M_B(t)$ are proportional to the observed time-dependent signal intensities, while $M_A^\infty$ and $M_B^\infty$ are proportional to the equilibrium intensities, and $M_A^0$ and $M_B^0$ are proportional to the initial intensities immediately after the perturbation.

[5] H. M. McConnell, *J. Chem. Phys.* **28**, 430 (1958).
[6] G. Wagner, G. Bodenhausen, N. Müller, M. Rance, O. W. Sørensen, R. R. Ernst, and K. Wüthrich, *J. Am. Chem. Soc.* **107**, 6440 (1985).

## Complementary Inversion Transfer Experiments

### Two-Site Case

Equations (5)–(12) describe the relaxation of a pair of exchanging signals after any perturbation of one or both of the signals; e.g., partial or complete saturation or complete inversion of one or both of the two signals. However, the maximum information about the rate constants is obtained only for the maximum selective perturbation of one of the two signals; i.e., a completely selective inversion. This is the essence of the inversion transfer experiment first described by Dahlquist et al.[7] In practice the experiment is performed using the pulse sequence 180° (selective)–$\tau$–90° (nonselective) sampling, where the selective 180° pulse has the frequency of the signal to be inverted, and $\tau$ is a variable delay between the pulse and the 90° nonselective sampling pulse. The pulse sequence is repeated for a number of different $\tau$ values.

The unknown parameters in Eqs. (5)–(12) are the rate constants $k_A$, $k_B$, $R_{1A}$, and $R_{1B}$, and the limiting values, $M_A^\infty$, $M_A^0$, $M_B^\infty$, and $M_B^0$. Since the system is at chemical equilibrium, the following relation holds:

$$k_A M_A^\infty = k_B M_B^\infty \tag{13}$$

This reduces the number of unknowns by one, leaving a total of seven parameters to be determined in the general case. However, the important point is as follows: Whereas the determination of the limiting values is straightforward, the rate constants are difficult to extract due to the double-exponential nature of the time dependencies [Eqs. (5) and (6)]. Thus the two exponents, $\lambda_1$ and $\lambda_2$, can be determined independently only if they are significantly different (in practice, $\lambda_1$ and $\lambda_2$ should differ by at least a factor of four[8]). For $\lambda_1 \simeq \lambda_2$ it has been shown[4] that only a combination of the two exponents can be obtained. The latter is clearly illustrated by the double exponentials in Fig. 1a, corresponding to the selectively inverted signal, e.g., the A signal. Despite the two exponentials differing both with respect to the $\lambda$ and the $C$ values, the curves they describe are almost identical. Consequently, the $\lambda$ and the $C$ values cannot be extracted from a set of data points describing either one of the curves. Aside from the limiting values, $M_A^0$ and $M_A^\infty$, the only information that can be obtained from such a data set is the combination $C_1\lambda_1 + C_2\lambda_2$. The intensity of the B signal (Fig. 1b), that varies due to the transfer of magnetization from the A signal, suffers from a similar ambiguity, yielding only the combination $C_3\lambda_1 + C_4\lambda_2$. Hence, in general, the two time

[7] F. W. Dahlquist, K. J. Longmuir, and R. B. Du Vernet, *J. Magn. Reson.* **17**, 406 (1975).
[8] H. Strehlow and J. Jen, *Chem. Instrum.* **3**, 47 (1971).

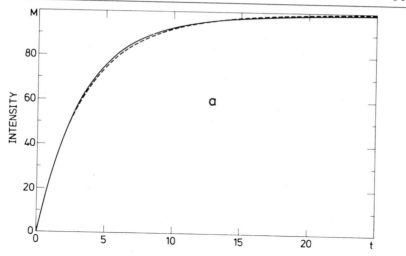

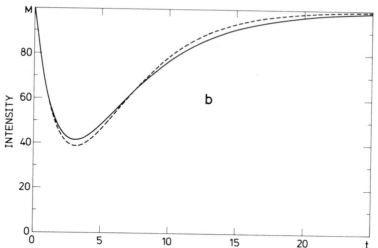

FIG. 1. (a) Time variation of the A signal after a selective inversion; $M_A(t) = -50e^{-0.2t} -50e^{-0.4t} + 100$ (---); $M_A(t) = -10e^{-0.1t} - 90e^{-0.322t} + 100$ (—); $\lambda_1 C_1 + \lambda_2 C_2 = -30$ in both cases. (b) Time variation of the B signal after a selective inversion of the A signal; $M_B(t) = -1080e^{-0.3t} + 1080e^{-0.35t} + 100$ (---); $M_B(t) = -180e^{-0.2t} + 180e^{-0.5t} + 100$ (—); $\lambda_1 C_3 + \lambda_2 C_4 = 54$ in both cases. For further explanation, see text.

dependencies of a two-site inversion transfer experiment provide only two independent combinations of the unknown rate constants. Although Eq. (13) reduces the number of these to three, the system is still underdetermined.

This deficiency can be remedied by a *complementary* experiment. As shown,[4] a second experiment in which the other signal (here the B signal) is being selectively inverted, provides another pair of independent combinations, resulting in a total of four combinations of the three unknown rate constants. Consequently, the system is now overdetermined. Therefore, inclusion of the complementary experiment results in a significant increase of the applicability of the magnetization transfer experiment, in general, and the inversion transfer experiment in particular. As demonstrated,[4] exchange rates covering about three orders of magnitude can be determined using this approach, as exchange rates that are more than seven times slower than the fastest relaxation rate involved can be evaluated.

### Multisite Case

The complementary inversion transfer experiment, described in the previous section, can be extended to the general $n$-site case, as described in detail recently.[9] Analogous to the two-site case, $n$ complementary experiments, corresponding to a selective inversion of each one of the $n$ signals, must be performed and the $n \times n$ recovery curves analyzed, simultaneously, in order to determine the $n(n - 1)/2$ possible exchange rates and the $n$ different relaxation rates. As in the two-site case, the recoveries of the $n$ signals are described by the McConnell equations, which in this case are conveniently written in matrix notation[10]

$$d\bar{M}/dt = \bar{\bar{K}}\bar{M} + \bar{M}^e \qquad (14)$$

where $\bar{M}$ and $\bar{M}^e$ are $n$-dimensional vectors whose $i$th elements, $M_i$, and $M_i^\infty R_{1i}$, refer to the $i$th site and are, respectively, the longitudinal magnetization and the equilibrium value of the longitudinal magnetization multiplied by the longitudinal relaxation of the site. Further, $\bar{\bar{K}}$ is an $n \times n$ matrix with the off-diagonal element, $K_{ij}$, denoting the pseudo-first-order exchange rate, $k_{ij}$, from the $j$th to the $i$th site. The diagonal elements of $\bar{\bar{K}}$ are given by

$$K_{ii} = - \left( R_{1i} + \sum_{j \neq i} k_{ji} \right) \qquad (15)$$

[9] H. Gesmar and J. J. Led, *J. Magn. Reson.* **68**, 95 (1986).
[10] J. Schotland and J. S. Leigh, *J. Magn. Reson.* **51**, 48 (1983).

When the value of $n$ exceeds two, Eq. (14) does not have an analytical solution. However, as shown[9] Eq. (14) can always be solved numerically.

The general complementary inversion transfer experiments described here have several advantages. First, they are experimentally simple, the perturbation consisting of only one selective pulse at a time, independent of the number of sites. Second, the absence of an rf field during the recovery period allows the time variations of *all* the involved signals to be monitored. Third, each experiment covers the maximum possible temporal and dynamical ranges since $M(t)$ can vary from $-M^\infty$ to $+M^\infty$. Consequently, the inversion transfer approach provides a maximum of experimental information about the involved rate constants, allowing a more accurate determination of kinetic parameters covering a larger range of rates than, e.g., the saturation transfer method (*vide infra*). Fourth, the combined analysis of complementary experiments allows all significant rate constants to be determined, independently, even if the exponents (the $\lambda$ values) are almost identical. Fifth, errors due to unintentional perturbations of noninverted signals are avoided, provided a correct data analysis, using Eqs. (3)–(12), is performed.

The drawback of the inversion transfer approach is the multiexponential time dependences that exclude simple data analyses based on semilogarithmic plots and initial slopes, in particular in multisite exchange cases. Only a computerized nonlinear least-squares analysis of the entire recovery period of all signals, using the complete theoretical model, can provide a correct estimation of the rate constants.

### Other Approaches

Besides the inversion transfer experiment, described above, other variants of the magnetization transfer experiment have been used, notably the saturation transfer and the two-dimensional NOESY experiments. In the following these two approaches shall be discussed briefly.

### Saturation Transfer

The basic idea behind the saturation transfer approach is to perform the experiment under conditions that simplify the data analysis. Proceeding with the two-site case of above, the experiment consists in a selective saturation of one of the two signals, e.g., the A signal, while the intensity of the other signal (B) is being monitored either as a function of time, as saturation is being transferred by the exchange process,[11,12] or more sim-

[11] S. Forsén and R. A. Hoffman, *J. Chem. Phys.* **30**, 2892 (1963).
[12] I. D. Campbell, C. M. Dobson, and R. G. Ratcliffe, *J. Magn. Reson.* **27**, 455 (1977).

ply, after steady state has been established.[13] Because of the saturation, the solution of Eq. (1) is reduced to a single exponential equation:

$$M_B(t) = (M_B^\infty - M_B^s)e^{-(k_B + R_{1B})t} + M_B^s \qquad (16)$$

$t$ being the time and $M_B^s$ the steady state value of the longitudinal magnetization given by

$$M_B^s = [R_{1B}/(R_{1B} + k_B)]M_B^\infty \qquad (17)$$

Hence, the change in intensity of the B signal in the steady state version is

$$\Delta M_B = M_B^\infty - M_B^s = [k_B/(R_{1B} + k_B)]M_B^\infty \qquad (18)$$

As it appears from Eq. (16) the exponents obtained in the temporal experiment contain both rate constants, $k_B$ and $R_{1B}$. Therefore, $k_B$ can be determined from the time dependence of the signal intensity only if $R_{1B}$ is known or can be determined independently. According to Eq. (18) the same applies to the steady state version of the experiment, since $\Delta M_B$ depends on both $k_B$ and $R_{1B}$. However, a combination of Eqs. (16) and (18), and thus a combination of the temporal and steady state experiments, provides sufficient information for the determination of both rate constants.[12] Alternatively, the steady state experiment can be combined with a conventional inversion–recovery experiment, measuring the relaxation rate of the B signal while saturating the A signal.[14] Under these conditions the effective relaxation rate is $k_{1B} = k_B + R_{1B}$. In any case, a temporal experiment must be included which may take the same amount of time as the corresponding inversion transfer experiment.

The saturation transfer approach suffers from a series of shortcomings and restrictions. First, unlike the inversion transfer experiment, only one recovery curve can be observed at a time. Furthermore, it covers smaller dynamical and temporal ranges than in the inversion transfer experiment. Second, the saturation transfer experiment is restricted by the conditions that the saturation be completely selective and instantaneous. Third, if more than two sites are involved, the saturation transfer experiment becomes considerably more complicated. For a total number of $n$ exchanging sites, $n - 1$ sites must be irradiated, simultaneously,[14,15] in order to retain the simple, single-exponential time dependence of the observed magnetization given by Eq. (16). While this, by itself, may impose an experimental problem,[3] it further increases the possibility of an unintentional irradiation of the observed signal, resulting in erroneous rate con-

[13] K. Ugurbil, R. G. Shulman, and T. R. Brown, in "Biological Applications of Magnetic Resonance" (R. G. Shulman, ed.), p. 537. Academic Press, New York, 1979.
[14] B. E. Mann, J. Magn. Reson. 25, 91 (1977).
[15] S. Forsén and R. A. Hoffman, J. Chem. Phys. 40, 1189 (1964).

stants. As discussed above, none of these restrictions applies to the inversion transfer experiment.

## Two-Dimensional NOESY Experiment

A powerful alternative to the magnetization transfer technique is the two-dimensional NOESY experiment.[16] This experiment has been applied to the study of enzyme-catalyzed reactions in a series of cases.[17-19] Boyd et al.[20] and Turner[21] compared the saturation transfer technique to the two-dimensional technique and found that the need for selectivity and "the requirement that the rate constant be greater than the relaxation rate"[20] were general limitations to the saturation transfer experiment. However, as explained above the inversion transfer experiment does not suffer from these inherent drawbacks and, thus, compares favorably with the NOESY experiment. Since $t$ in Eqs. (5) and (6) corresponds to the mixing time of the NOESY experiment, the number of complementary inversion transfer experiments, i.e., the number of exchanging sites, should be compared with the number of $t_1$ values needed to obtain sufficient resolution in the $\omega_1$ dimension of the NOESY experiment. Therefore, as long as the number of exchanging sites is limited, the inversion transfer experiment should be the most suitable.

## Applications

The first two-site inversion transfer experiment was performed by Dahlquist et al.[7] in 1975. In order to ensure single-exponential time dependence, the conditions $R_{1A} = R_{1B}$ and $k_A = k_B$ were required in the data analysis. In 1977 Alger and Prestegard[22] performed a similar experiment, applying a more complete data analysis. In particular, $M_A^0$, $M_B^0$, $R_{1A}$, $R_{1B}$, and $k_A$ were obtained as least-squares estimates while $M_A^\infty$, $M_B^\infty$, and $k_A/k_B$ were determined directly from the equilibrium spectrum. Brown and Ogawa used the inversion transfer[23] to study an enzyme-catalyzed reaction, applying a semiquantitative data analysis. Campbell et al.[24] sug-

[16] J. Jeener, B. H. Meier, P. Bachmann, and R. R. Ernst, J. Chem. Phys. **71**, 4546 (1979).
[17] R. S. Balaban and J. A. Ferretti, Proc. Natl. Acad. Sci. U.S.A. **84**, 1241 (1983).
[18] H. L. Kantor, J. A. Ferretti, and R. S. Balaban, Biochim. Biophys. Acta **789**, 128 (1984).
[19] G. L. Mendz, G. Robinson, and P. W. Kuchel, J. Am. Chem. Soc. **108**, 169 (1986).
[20] J. Boyd, K. M. Brindle, I. D. Campbell, and G. K. Radda, J. Magn. Reson. **60**, 149 (1984).
[21] D. L. Turner, J. Magn. Reson. **61**, 28 (1985).
[22] J. R. Alger and J. H. Prestegard, J. Magn. Reson. **27**, 137 (1977).
[23] T. R. Brown and S. Ogawa, Proc. Natl. Acad. Sci. U.S.A. **74**, 3627 (1977).
[24] I. D. Campbell, C. M. Dobson, R. G. Ratcliffe, and R. J. P. Williams, J. Magn. Reson. **29**, 397 (1978).

gested a combination of saturation transfer, inversion transfer, and nonselective experiments from which they derived not only the exchange and relaxation rates but also the cross-relaxation rate "without the need for difficult curve fitting to several variables." Since the analysis used in that study was based on techniques such as slope determination of initial gradients and semilogarithmic plots, the results should be regarded carefully. In particular, there is concern since the exchange rate and cross-relaxation rate, as mentioned previously, are formally identical and, therefore, inseparable, except for special cases. Furthermore, the procedure is unable to treat cases where $k_A \lesssim R_{1A} \simeq R_{1B}$, i.e., $\lambda_1 \simeq \lambda_2$.

The inversion transfer technique, using complementary sets of experiments, has, so far, been used only in a few studies,[25-28] despite its virtues. This is possibly a result of the computation and the computer algorithm necessary for the analysis. In the following the application of the complementary inversion transfer experiments will be illustrated by three examples, while the applied computer algorithm is described briefly in the Appendix to this volume.

### Mn(II)-Human Carbonate Dehydratase I

In the first example[25] the rate of the reversible $CO_2/HCO_3^-$ hydration/dehydration reaction

$$CO_2 + H_2O \rightleftharpoons HCO_3^- + H^+ \tag{19}$$

catalyzed by Mn(II)-human carbonate dehydratase I (Mn(II)-HCD I) was studied by monitoring the exchange of $^{13}C$ nuclei between $^{13}CO_2$ and $H^{13}CO_3^-$. The set of two complementary inversion transfer experiments necessary to solve this *two-site* exchange problem is shown in Fig. 2. The results of a simultaneous nonlinear least-squares analysis of all the experimental data, based on Eqs. (3)–(12) using the algorithm in the Appendix at the end of the volume, are shown in Fig. 3. Similar experiments were performed for the apoenzyme and the native Zn(II)-enzyme. The parameters obtained in the three experiments are given in Table I. When the differences between the pH of the samples are taken into account the activity of the Mn(II)-enzyme at the applied substrate concentration ($[HCO_3^-] = 0.100\ M$) was found to be about 5% of the activity of the native Zn(II)-enzyme.[29]

[25] J. J. Led, E. Neesgaard, and J. T. Johansen, *FEBS Lett.* **147,** 74 (1982).

[26] Aa. Hvidt, H. Gesmar, and J. J. Led, *Acta Chem. Scand., Ser. B* **B37,** 227 (1983).

[27] J. J. Led and H. Gesmar, *J. Phys. Chem.* **89,** 583 (1984).

[28] W. T. Ford, M. Periyasamy, H. O. Spivey, and J. P. Chandler, *J. Magn. Reson.* **63,** 298 (1985).

[29] J. J. Led and E. Neesgaard, *Biochemistry* **26,** 183 (1987).

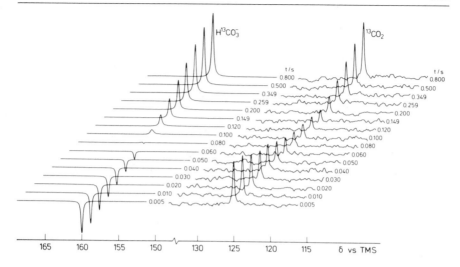

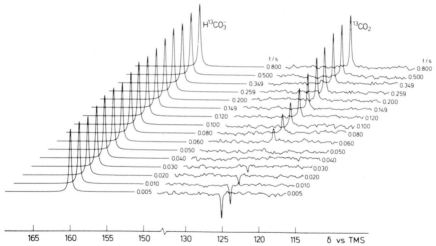

FIG. 2. Sets of complementary inversion transfer $^{13}$C NMR spectra at 67.89 MHz of 90% $^{13}$C-enriched H$^{13}$CO$_3^-$ (100 m$M$) and $^{13}$CO$_2$ (0.38 m$M$) dissolved in 50 m$M$ Tris-sulfate buffer at pH 8.5 and in the presence of Mn(II)-human carbonate dehydratase I (8.8 $\mu M$). The spectra were obtained at 19° and consist of 8500 scans, each accumulated with a delay time of 0.6 sec between scans. The chemical shift values are relative to the $^{13}$C signal of tetra-methyl silane (TMS). The selective inversion was accomplished using a DANTE pulse sequence [G. Bodenhausen, R. Freeman, and G. A. Morris, *J. Magn. Reson.* **23,** 171 (1976)], consisting of 16 consecutive pulses, each 26 $\mu$sec long and separated by 0.9 msec while placing the carrier frequency on the signal to be inverted. The 90° nonselective, analyzing pulse was 16.5 $\mu$sec. No proton decoupling was performed. (From Led *et al.*[25])

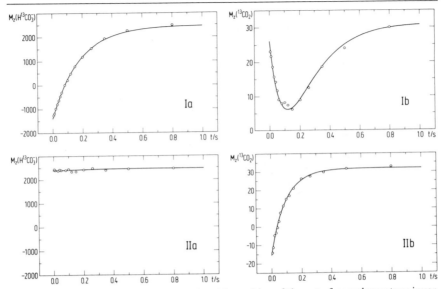

FIG. 3. Plot of the time-dependent signal intensities of the set of complementary inversion transfer spectra in Fig. 2. The curves represent the best fit obtained by the least-squares analysis. (From Led et al.[25])

As shown in Fig. 3 the agreement between the experimental data and the curves, calculated from the parameters obtained in the analysis, is excellent. Due to the large difference between the concentrations of $CO_2$ and $HCO_3^-$ ($[CO_2]/[HCO_3^-] \cong 1:300$) at the applied pH (pH = 8.5; the Mn(II)-enzyme is active only at relatively high pH), the $^{13}C$ signal of $HCO_3^-$ is unaffected by the inversion of the $^{13}C$ signal of $CO_2$. Consequently, the data in Fig. 3,IIa depend only on $M_{HCO_3^-}^{\infty}$ leaving three time-dependent signal intensities for the determination of the three rate constants $k^{CO_2}$, $R_1^{CO_2}$, and $R_1^{HCO_3^-}$. Also, since $\lambda_2/\lambda_1 \simeq 1.9$ only one combination of the three rate constants can be obtained from each of the time dependences. Thus, only the minimum information necessary for an independent determination of the rate constants is available, emphasizing the need for a complementary set of inversion transfer experiments in this case. Finally, it should be noted that the relaxation rates obtained provide detailed information about the enzyme–substrate complexes, as published previously.[25,29]

## Self-Assembled Structures of Na₂(5′-GMP)

The second example is a $^{31}P$ inversion transfer study of the interchanges between the self-assembled structures of disodium guanosine

TABLE I

EXCHANGE AND RELAXATION RATES[a,b] OBTAINED FROM LEAST-SQUARES ANALYSES OF THE COMPLEMENTARY INVERSION TRANSFER DATA

| Enzyme | pH | $k^{CO_2}$ (sec$^{-1}$) | $R_1^{HCO_3^-}$ (sec$^{-1}$) | $R_1^{CO_2}$ (sec$^{-1}$) |
|---|---|---|---|---|
| Mn(II)-HCD I[c,d] | 8.5 | $10.2 \pm 0.5$ | $5.50 \pm 0.07$ | $0.2 \pm 0.3$ |
| apoHCD I[d] | 7.4[e] | $0.67 \pm 0.05$ | $0.088 \pm 0.003$ | $0.04 \pm 0.03$ |
| Zn(II)-HCD I[f] | 7.5[e] | $56.0 \pm 9$ | $0.175 \pm 0.004$ | 0.2 |

[a] From Led et al.[25]
[b] In 0.05 $M$ Tris-sulfate buffer at 19° and [HCO$_3^-$] = 0.100 $M$.
[c] [Mn$^{2+}$] = 10 $\mu M$.
[d] [apoHCD I] = 100 $\mu M$. HCD, Human carbonate dehydratase.
[e] This pH value was chosen since the slow $^{13}$C relaxation rates in the apo- and Zn(II)-enzyme samples make it impractical to measure $k^{CO_2}$ at the very low [CO$_2$] at pH 8.5.
[f] [Zn(II)-HCD I] = 10 $\mu M$.

5'-monophosphate.[27] Although these interchanges are not enzyme catalyzed, the study serves as an example of a multisite exchange case.

The self-assembled structures, suggested by Bouhoutsos-Brown *et al.*,[30] are shown in Fig. 4, and some of the $^{31}$P spectra of the complementary inversion transfer experiments[27] are given in Fig. 5. The two outer signals in the $^{31}$P spectrum (2.7 and 4.0 ppm) are assigned to the two C$_4$ structures in Fig. 4 while the signals at 3.2 and 3.5 ppm are assigned to the D$_4$ structure (**III + VI** and **IV + V**, respectively), the former being a superposition of a D$_4$ signal and the signal from unstructured Na$_2$(5'-GMP). Exchanges occur if one of the two C$_4$ structures is transformed into the other by twisting the two tetramers 60° relative to one another, or in the case of the D$_4$ structures, if the transformation **III ↔ IV** (or **VI ↔ V**) takes place through a relative twist of the tetramers of only 30°.

These and other possible transformations between the self-assembled structures in Fig. 4 were explored by the four-site inversion transfer experiment indicated in Fig. 5. Due to the partly overlapping signals, perturbation of the noninverted signals could not be avoided. Nevertheless, a complete analysis of the 4 × 4 recovery curves shown in Fig. 6 reveals that the transformation **III ↔ IV** (or **VI ↔ V**) is active with a rate constant (at 1°) of $k = 5.7 \pm 0.5$ sec$^{-1}$, whereas exchange between any other pair of $^{31}$P sites is considerably smaller or negligible ($<0.20 \pm 0.08$ sec$^{-1}$). Although unnecessary for the determination of the exchange rate constants,

[30] E. Bouhoutsos-Brown, C. L. Marshall, and T. J. Pinnavaia, *J. Am. Chem. Soc.* **104,** 6576 (1982).

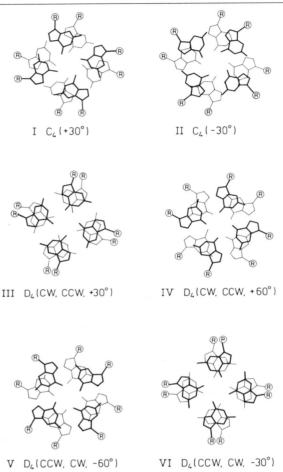

I    $C_4$ $(+30°)$     II    $C_4$ $(-30°)$

III    $D_4$ (CW, CCW, +30°)     IV    $D_4$ (CW, CCW, +60°)

V    $D_4$ (CCW, CW, −60°)     VI    $D_4$ (CCW, CW, −30°)

FIG. 4. The six possible diastereomeric octamers of $Na_2$(5′-GMP) in water. In the $C_4$ isomers the two tetramers are stacked in a normal head-to-tail arrangement with the same H-bonding directionality for the two tetramers while in the $D_4$ isomers the stacking is inverted (tail to tail or head to head); that is, the two tetramers have opposite H-bonding directionality. The heavier lines represent the upper tetramer unit, the lighter lines the lower tetramers. R represents the chiral ribophosphate group. The CW and CCW notations for the $D_4$ isomers designate the clockwise or counterclockwise sense of the hydrogen bonding for the upper and lower tetramer units. The twist angles are ±30° or ±60° as indicated. (From Bouhoutsos-Brown et al.[30])

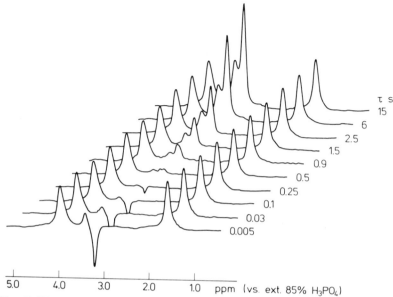

τ s
15
6
2.5
1.5
0.9
0.5
0.25
0.1
0.03
0.005

| | | | | |
5.0      4.0      3.0      2.0      1.0    ppm  (vs. ext. 85% H₃PO₄)

FIG. 5. ³¹P magnetization transfer NMR spectra at 109.4 MHz of a 0.49 *M* solution of Na₂(5'-GMP) in D₂O at pD 7.5 and 1°. Only one-third of the spectra monitoring the time dependence of the signal intensities after the selective inversion are shown. The selective inversion was accomplished using a DANTE pulse sequence [G. Bodenhausen, R. Freeman, and G. A. Morris, *J. Magn. Reson.* **23**, (1976)]. (From Led and Gesmar.[27])

the data obtained in a nonselective inversion-recovery experiment were included (Fig. 6) to further illustrate the reliability of the analysis.

*Nonideal Spectra*

In NMR studies of enzymes and other biological systems, overlapping signals often hamper the analysis. Due to the limited dynamic range of the spectrometer, further problems can arise when weak signals from biological molecules are present together with a strong solvent signal, in particular the proton signal from water. Although effective procedures for solvent suppression are available, the application of these procedures may often result in phase- and baseline-distorted spectra,[31,32] preventing a correct evaluation of the signal intensities from the spectra.

When these problems are present, the intensities necessary for the magnetization transfer experiments can be estimated together with the other spectral parameters by a least-squares analysis of the experimental

[31] P. Plateau, C. Dumas, and M. Guéron, *J. Magn. Reson.* **54**, 46 (1983).
[32] F. Abildgaard, H. Gesmar, and J. J. Led, *J. Magn. Reson.* **79**, 78 (1988).

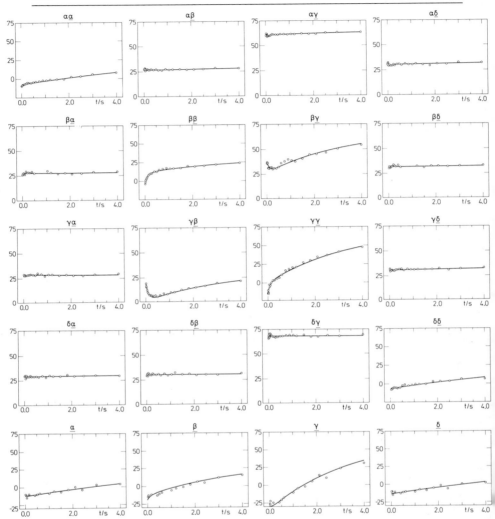

FIG. 6. Plots of the time-dependent signal intensities obtained in the set of four complementary magnetization transfer experiments. Signal intensities corresponding to 30 $\tau$ values in the range from 0.005 to 25 sec were measured. Only those between 0.005 and 4 sec are shown here. The order of the columns corresponds to the order of the signals in Fig. 4, $\alpha$, $\beta$, $\gamma$, and $\delta$ designating the signals from left to right. The first four rows show the variations of the intensities after a selective inversion of one of the signals. The inverted signal is given by the first Greek letter above each plot, and the observed signal is given by the second, underlined letter. The last row gives the time dependencies of the nonselective relaxation experiment. The curves represent the best fit obtained by a simultaneous nonlinear least-squares analysis of the total amount of data.

spectra, assuming the signals are Lorentzian. Recently it was demonstrated[32] that such a procedure is possible (and practical) if the correct theoretical expression for the discrete Fourier transform NMR spectrum is used in the analysis. By fitting this expression to both the real and imaginary part of a phase- and baseline-distorted Fourier transform spectrum, all four parameters characterizing each individual signal can be evaluated; i.e., the frequency $\nu_i$, the intensity $I_i$, the line width $\Delta\nu_{1/2,i}$, and the phase $\varphi_i$, including their estimated uncertainties. From these parameters the pure absorption mode spectrum can, of course, be constructed. However, the primary and most important information obtained from the analysis are the values of the four parameters with estimated uncertainties for each individual signal, notably the intensity in case of a magnetization transfer experiment.

Although the least-squares curve fitting mentioned here has not yet been applied in NMR studies of enzyme systems, it can, undoubtedly, greatly enhance the applicability of the magnetization transfer technique to such studies. In particular, studies of *in vivo* systems should benefit by such analyses since here the substrate signals are often broad and weak due to low mobility, low concentrations, and limited sampling time.

To indicate its potential applications, the least-squares curve fitting just described was used to analyze the spectra obtained in an inversion transfer study of the exchange rates of the amide protons of a 7.5 m$M$ solution of a heptapeptide in water. The algorithm applied in the curve fitting is described briefly in the Appendix at the end of the volume. Because of the intense water signal and the limited dynamic range of the NMR spectrometer, the weak amide proton signals and the intense water signal could not be observed simultaneously, but were recorded in different experiments. In a set of complementary inversion transfer experiments the spectra of the amide protons were obtained using a selective excitation pulse, namely a 1–$\bar{3}$–3–$\bar{1}$ pulse,[33] that leaves the water signal almost unperturbed and thus eliminates the intense water signal. A similar pulse was used for the selective inversion of the amide proton signals.

Spectra of the heptapeptide calculated from the parameters obtained by the least-squares curve fitting are shown in Fig. 7 together with the corresponding experimental spectra. In Fig. 8 are shown the results of a complementary set of inversion transfer experiments, based on signal intensities also obtained by the curve-fitting procedure. The rate determined in the experiment is that of the proton exchange between the water signal and the Phe amide signal at 8.28 ppm. Despite the relatively poor signal-to-noise ratio in the spectra (Fig. 7), the accuracy of the intensities

[33] P. J. Hore, *J. Magn. Reson.* **55,** 283 (1983).

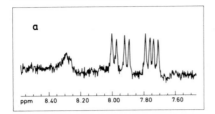

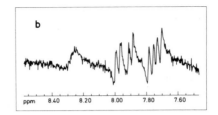

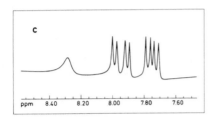

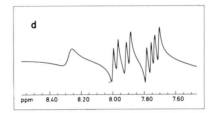

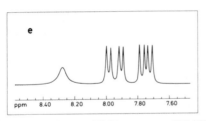

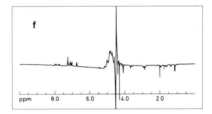

FIG. 7. Proton NMR spectra at 270 MHz of a 7.5 m$M$ solution of a heptapeptide (NH$_2$-Gly-Phe-Phe-Tyr-Thr-Pro-Lys-OH) in water at pH 2.98 and 27°. (a, b) The mutual 90° phase-shifted spectra of the amide region as obtained by Fourier transformation of the experimental time domain signal (FID); (c, d) as (a) and (b), respectively, but calculated from the frequencies, line widths, intensities, and phases obtained from a least-squares analysis of the experimental spectrum (see text); (e) pure absorption mode spectrum calculated as (c) and (d), but with the phases set to zero; (f) the complete experimental Fourier transform spectrum obtained using a 1–$\bar{3}$–3–$\bar{1}$ excitation pulse sequence for water suppression [P. J. Hore, *J. Magn. Reson.* **55**, 283 (1983)]. [From J. J. Led, H. Gesmar, and F. Abildgaard, *Alfred Benzon Symp.* **26** (1988).]

obtained is high (<5% uncertainty) as indicated by the agreement between the intensities and the calculated recovery curves (Fig. 8). Likewise, the uncertainties of the estimated rate constants vary from less than 20% ($k^{NH}$ = 0.22 ± 0.04 sec$^{-1}$) to less than 5% ($R_1^{NH}$ = 2.08 ± 0.08 sec$^{-1}$, $R_1^{H_2O}$ = 0.249 ± 0.005 sec$^{-1}$). Hence, accurate estimations of rate constants can be obtained even from phase- and baseline-distorted spectra, if

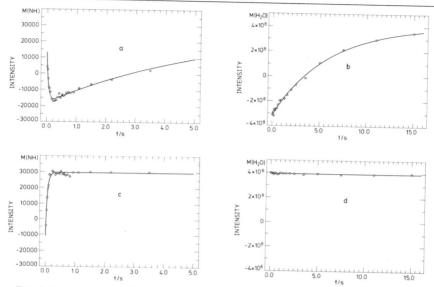

FIG. 8. Complementary inversion transfer experiments measuring the exchange between water and the amide proton of one of the Phe (8.28 ppm) of the heptapeptide in Fig. 7. The signal intensities indicated by ◯ were obtained from a nonlinear least-squares curve fitting to the experimental Fourier transform spectrum (see text). The full curves were calculated from the parameters derived in the magnetization transfer analysis based on Eqs. (3)–(12). [From J. J. Led, H. Gesmar, and F. Abildgaard, *Alfred Benzon Symp.* **26** (1988).]

the intensity estimations are based on a least-squares curve fitting to the experimental spectra, using the correct theoretical expression for a discrete Fourier transform spectrum. It should be emphasized that the computing time necessary for such curve fittings is significant and increases drastically with the number of signals. However, by confining the analysis to those parts of the spectrum that are of interest, e.g., the amide proton region in Fig. 8, the computations can be kept at a minimum.

### Acknowledgments

This work was supported by the Danish Technical Research Council, journal No. 16-3922.H and the Danish Natural Science Research Council, journal No. 11-3742, 11-5686, and 11-5916. F.A. acknowledges a students' scholarship and H.G. a postgraduate fellowship, both from the Carlsberg Foundation. Free access to the RC 8000 minicomputer at the H.C. Ørsted Institute is also acknowledged.

## [16] Two-Dimensional Nuclear Magnetic Resonance Studies of Enzyme Kinetics and Metabolites *in Vivo*

*By* BRUCE A. BERKOWITZ and ROBERT S. BALABAN

Two-dimensional nuclear magnetic resonance spectroscopy (2D NMR) has become a standard procedure for analyzing the structure and dynamics of isolated, purified compounds. However, the application of these techniques to more complicated systems has only recently begun. In this chapter the application of 2D NMR techniques to the study of enzyme kinetics and *in vivo* metabolite contents will be described.

### Enzyme Kinetics

Using the original procedures suggested by Jeener *et al.,*[1] the 2D NOE experiment has been used to follow the exchange rate of reactions catalyzed by enzymes *in vitro*[2,3] and in intact tissues *in vitro*[4] and *in vivo*.[5]

The simplest 2D NMR chemical exchange experiment consists of three radiofrequency (rf) pulses, as shown in Fig. 1. The basic experimental procedure is the same as that described in many other chapters in this volume to obtain 2D NOE spectra, leading ultimately to structural data. These pulses are 90° magnetization flip angles which provide optimal quantitative data acquisition. However, lower or spatially varying magnetization flip angles can be used for 2D studies *in vivo*.[6] The first pulse places the longitudinal magnetization in the $x-y$ plane, allowing phase and amplitude encoding of the magnetization with regard to the evolution period, $t_1$. This information is subsequently used to determine the frequency of the spins ($F_1$) before exchange has occurred. The evolution period is terminated with a second rf pulse which places magnetization on the $z$ axis. The subsequent fixed mixing period ($t_{mix}$) allows the exchange of the spins, previously labeled in the $t_1$ period, via the catalyzed reaction. The final pulse and acquisition of a free induction decay (FID) samples the magnetization after the exchange process and provides the frequency of the spins ($F_2$) with regard to $t_2$. Thus, the FID's collected after the third

[1] J. Jeener, P. Bachmann, and R. R. Ernst, *J. Chem. Phys.* **71,** 4556 (1979).
[2] R. S. Balaban and J. A. Ferretti, *Proc. Natl. Acad. Sci. U.S.A.* **80,** 1241 (1983).
[3] H. L. Kantor, J. A. Ferretti, and R. S. Balaban, *Biochim. Biophys. Acta* **789,** 128 (1984).
[4] P. Garlick and C. J. Turner, *J. Magn. Reson.* **51,** 536 (1983).
[5] R. S. Balaban, H. L. Kantor, and J. A. Ferretti, *J. Biol. Chem.* **258,** 12787 (1983).
[6] B. A. Berkowitz, S. D. Wolff, and R. S. Balaban, *J. Magn. Res.* **79,** 547 (1988).

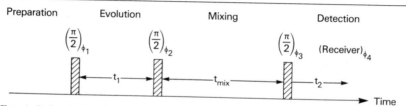

FIG. 1. Pulse-timing diagram for obtaining 2D NOE spectra. Pulses and receiver phase ($\phi_1$, $\phi_2$, $\phi_3$, $\phi_4$) are phase cycled according to Table I, yielding 64 steps/$t_1$ value. Experiment is conducted $N$ times incrementing $t_1$ for each $N$ to produce a 2D matrix. See text for more detail.

pulse are phase and amplitude encoded for the state of the spins just prior to exchange and frequency encoded for the spins after the exchange. After collecting a series of FID's with systematic increases in $t_1$, a 2D Fourier transform of this matrix with respect to $t_1$ and $t_2$ results in a 2D spectrum which is a function of the two frequencies $F_1$ and $F_2$. $F_1$ is the frequency of the spin before $t_{mix}$ and $F_2$ is the frequency of the spin after the exchange process. If a spin did not undergo exchange during $t_{mix}$ then the spin will have the same $F_1$ and $F_2$ frequency. However, if chemical exchange did occur during $t_{mix}$, then the spin will have the $F_1$ frequency from the molecule it started in at the beginning of $t_{mix}$ and the $F_2$ frequency of the molecule it became a part of after $t_{mix}$. An example of this is shown in Fig. 2, where the exchange of $^{13}C$ between $CO_2$ and $HCO_3^-$ in the presence of carbonate dehydratase is presented as a contour plot of $F_1$ versus $F_2$. Details of the experiment are presented in the figure legend. The "cross-peaks," where $F_1$ and $F_2$ frequencies are different, represent the spins which exchanged during $t_{mix}$. The "diagonal" peaks, where $F_1$ equals $F_2$, represent those spins which did not undergo exchange. Cross-peak A (i.e., $F_1$ of $CO_2$ and $F_2$ of $HCO_3^-$) represents the $^{13}C$ that went with $CO_2$ to $HCO_3^-$ in 780 msec. Cross-peak B (i.e., $F_1$ of $HCO_3^-$ and $F_2$ of $CO_2$) represents the $^{13}C$ that went from $HCO_3^-$ to $CO_2$ over $t_{mix}$.

To obtain the flux, or rate constant, the intensity of the cross-peaks or diagonal peaks must be measured as a function of mixing time. An example of this is shown in Fig. 3 where the cross-peak intensity build-up for the reaction of phosphocreatine (PCr) to ATP, catalyzed by creatine kinase, is plotted as a function of $t_{mix}$. Details of the experiment are presented in the legend for Fig. 3. The build-up of cross-peak intensity, $C_{ab}$ and $C_{ba}$, as a function of $t_{mix}$ for a simple $A \underset{k_2}{\overset{k_1}{\rightleftharpoons}} B$ reaction[1] is

$$C_{AB} = C_{BA} = X_a X_b \{1 - \exp[-(k_1 + k_2)t_{mix}]\}\exp(-t_{mix}/T_1) \qquad (1)$$

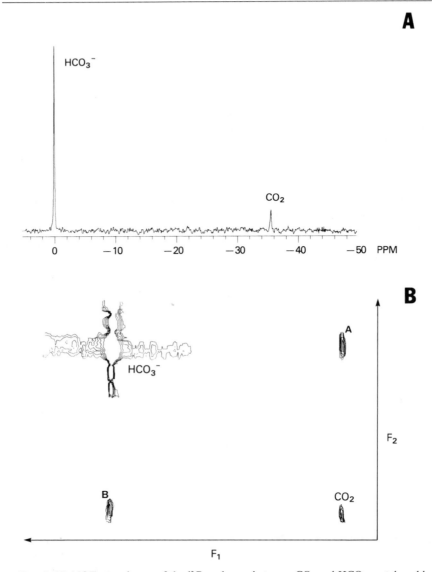

FIG. 2. 2D NOE experiment of the $^{13}C$ exchange between $CO_2$ and $HCO_3^-$ catalyzed by carbonate dehydratase. The reaction mixture was 297 mg of potassium hydrogen phthalate, 27 mg $NaHCO_3^-$, with 0.5 mg of carbonate dehydratase (Sigma lot #C-2522) in 3 ml of $H_2O$ and 2 ml of $^2H_2O$. This mixture, without enzyme, was then frozen in a 12-mm NMR tube. After freezing, the enzyme was added and the tube was sealed by fusing with heat. The sample was then warmed to 23°. (A) One-dimensional $^{13}C$ spectrum of the mixture, pH ~7.3. This spectrum is the average of 4 FID's acquired from 90° magnetization flip angles and a

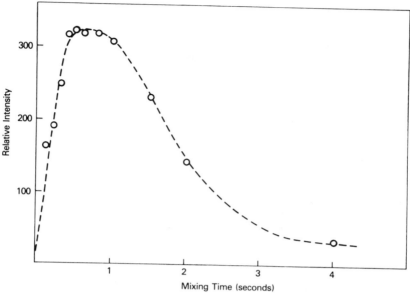

FIG. 3. A plot of mixing time versus phosphocreatine-to-ATP cross-peak intensity in a creatine kinase reaction mixture. The reaction mixture contained creatine, phosphocreatine, and MgATP in a ratio of 0.3 : 1 : 0.3, 0.5 m$M$ NaEDTA, and 330 m$M$ Na–HEPES buffer at pH 7.2. The solution also contained 20% $^2H_2O$ to enable $^2H$ locking. (Adapted from Kantor et al.[3])

where $X_a$ and $X_b$ are the mole fractions of A and B, respectively. The most efficient method of determining a rate constant is to rely on an initial rate condition. In this case Eq. (1) reduces to

$$(dC_{AB}/dt)_{t_{mix} \to 0} = X_a k_1 \qquad (2)$$

Thus, by determining the initial slope of the cross-peak intensity build-up, the rate constant can be determined assuming the initial rate conditions are met. Using this method only a few experiments with short $t_{mix}$ times, or for the bold at heart one experiment, can be conducted to determine the initial rate independent of spin–lattice relaxation times. For this reason the initial rate method is preferred for simple first order or pseudo-first-order reactions. However, at very short mixing times the cross-peak

---

delay between pulses of 250 sec ($T_1$ of $HCO_3^-$ is in excess of 30 sec). (B) 2D NOE spectrum of the reaction mixture. The mixing time was 780 msec. The delay between each NOE pulse sequence was 250 sec. Peaks are as assigned in the text. The spectrometer was frequency locked on the $^2H$ signal. This experiment was conducted with Drs. Howard Kantor and James Ferretti.

intensities are small, so the initial rate method requires an adequate signal-to-noise ratio to determine accurately the cross-peak intensity. In addition, the $T_1$ relaxation time must be sufficiently long relative to the reaction rate constant to permit the determination of an initial rate without any influence of $T_1$. Examples of initial rate measurements of enzyme-catalyzed reactions can be found for creatine kinase *in vitro*[3] and *in vivo*,[5] and for glucose-6-phosphate isomerase[2] and adenylate kinase *in vitro*.[3]

If the conditions are not appropriate for initial rate measurements then the entire curve of cross-peak intensities versus mixing times must be fit to a function similar to Eq. (1). This is generally the case for reactions which are not a simple one-step process since they often give rise to nonlinear initial portions of the cross-peak versus mixing time curve. In addition, when the $T_1$ values of the substrates are too short or the signal-to-noise ratio too low to accurately determine an initial rate, a complete fit is also required. In these cases, the intensity of the appropriate cross-peaks must be determined as a function of a wide range of mixing times. This is a very time-consuming process and requires the collection of several 2D matrices to characterize the entire curve. Examples of reactions analyzed in this fashion include glucose 6-phosphate isomerization[2] and some steps of the adenylate kinase reaction.[3]

There are several constraints on the 2D NOE experiment. Only exchange processes in the slow-exchange domain (i.e., exchanging species with discrete frequencies) can be analyzed with 2D NMR. In addition, the exchange rate must be $\gtrsim T_1$ to observe significant cross-peak build-up. Finally, each of the exchanging pools must be large enough to observe since the cross-peak intensities are dependent on the size of the exchanging pools [see Eq. (1)]. If one of the exchanging pools is too small to detect in a signal-averaged one-dimensional spectrum, then no cross-peak for this spin pool will be observed even if it is exchanging with a large pool.

In general, NOE experiments can be conducted in two manners. The first requires no phase cycling and destroys residual $x-y$ magnetization with a homospoil pulse after the second 90° pulse. This method also requires the carrier frequency to be placed on one side of the spectrum since quadrature detection in the $t_1$ domain requires phase cycling. This method is preferred when signal averaging is not required or cannot be performed due to an unstable sample (i.e., *in vivo* or sensitive enzyme preparations). In processing this type of data a magnitude calculation or specialized *post facto* phasing routine must be used. Using such techniques we have found an excellent correlation between the 2D NMR methods and more standard kinetic techniques.[2]

Under conditions where the sample is stable and signal averaging is required, phase cycling of the three 90° pulses and receiver to obtain pure

TABLE I
PHASE CYCLE[a] FOR 2D NOE EXPERIMENTS

| Phase step | $\theta_1$ | $\theta_2$ | $\theta_3$ | $\theta_4$ |
|:---:|:---:|:---:|:---:|:---:|
| 1 | $x$ | $x$ | $x$ | $x$ |
| 2 | $y$ | $x$ | $y$ | $y$ |
| 3 | $-x$ | $x$ | $-x$ | $x$ |
| 4 | $-y$ | $x$ | $-y$ | $y$ |
| 5 | $x$ | $x$ | $y$ | $y$ |
| 6 | $y$ | $x$ | $-x$ | $-x$ |
| 7 | $-x$ | $x$ | $-y$ | $y$ |
| 8 | $-y$ | $x$ | $x$ | $-x$ |
| 9 | $x$ | $x$ | $-x$ | $-x$ |
| 10 | $y$ | $x$ | $-y$ | $-y$ |
| 11 | $-x$ | $x$ | $x$ | $-x$ |
| 12 | $-y$ | $x$ | $y$ | $-y$ |
| 13 | $x$ | $x$ | $-y$ | $-y$ |
| 14 | $y$ | $x$ | $x$ | $x$ |
| 15 | $-x$ | $x$ | $y$ | $-y$ |
| 16 | $-y$ | $x$ | $-x$ | $x$ |

[a] This basic 16-step sequence is further cycled through the 4-step CYCLOPS sequence, producing a 64-step cycle per $t_1$ value. In addition, odd- and even-numbered steps were stored in different memory locations and analyzed according to States et al.[7] Data were collected in quadrature in the $t_2$ domain (on the GE CSI: QP on $y = 0$; AB off; AL = 2).

absorption mode 2D spectra,[7] without "phase twists," is the preferred method. This technique also prevents the need of a homospoil gradient pulse, reduces "$t_1$ ridge noise," and allows for quadrature detection in $t_1$. A phase-cycling scheme we have found effective for studying *in vivo* and *in vitro* samples on a 4.7-$T$ wide-bore (30 cm) General Electric system is listed in Table I.

There are several advantages of the 2D NMR technique over other methods which determine enzyme kinetics, such as saturation transfer and inversion transfer. First, the entire exchange map of the reaction is determined in one experiment. This is extremely important in complex reactions such as those involving adenylate kinase or glucose-6-phosphate isomerase. It should also be noted that since the 2D method is a "pulse label" experiment,[8] influences of small metabolite pools such as

[7] D. J. States, R. A. Haberkorn, and D. J. Ruben, *J. Magn. Reson.* **48**, 286 (1982).
[8] P. S. Hsieh and R. S. Balaban, *Magn. Reson. Med.* **7**, 56 (1988).

enzyme substrate complexes are minimized. The second advantage is that the technique helps resolve the exchanging spins in the 2D matrix by moving these spins out into the cross-peak regions away from other non-exchanging species. The third advantage is the lack of a calibrated second rf pulse which must either saturate or invert a particular spin. The limited frequency selectivity of these calibrated pulses make the use of these techniques difficult in poorly resolved spectra.

The disadvantages of the 2D NMR experiment center on the amount of time required to collect the entire 2D data set. Boyd et al.[9] suggest that for simple reactions either saturation transfer or selective inversion transfer is the preferred method by an order of magnitude in time. However, in this comparison the investigators compared a *complete* cross-peak versus $t_{mix}$ method (i.e., the accordion 2D NOE experiment[10]) versus a single-point equilibrium saturation-transfer method. Clearly, the comparison of an initial rate 2D NMR experiment which does not require the determination and error of a $T_1$ measurement with the $T_1$-dependent saturation-transfer method would be more reasonable. Or better yet, comparing the initial rate 2D method with an initial rate saturation-transfer method.[11] In our hands, to obtain a comparable signal-to-noise ratio for a rate determination, the 2D experiment is roughly only a factor of two longer than the equilibrium saturation-transfer method for a single exchange when the 2D initial rate can be measured. In comparison to initial rate inversion transfer[8] or saturation-transfer methods, the 2D method could take four to five times longer. However, for exchange networks and very complex spectra the time advantage of the one-dimensional methods quickly disappears.

Another disadvantage of the 2D method is that it does not observe exchanges with small metabolite pools or enzyme complexes which do not significantly contribute to the one-dimensional NMR spectrum[3,8] as you can with saturation-transfer techniques. This situation can sometimes be used to the experimenter's advantage, since the 2D method will provide the rate constants independent of the influence of small metabolite pools (i.e., measures bulk exchange properties only).

### Detection of Metabolites

The proton NMR spectrum *in vivo* is very difficult to analyze due to its limited spectral dispersion and may greatly benefit from the improved resolution provided by 2D methods. Principal among these techniques is

[9] J. Boyd, K. M. Brindle, I. D. Campbell, and G. K. Radda, *J. Magn. Reson.* **60**, 149 (1984).
[10] G. Bodenhausen and R. R. Ernst, *J. Am. Chem. Soc.* **104**, 1304 (1982).
[11] V. V. Kupriyanov, A. Ya. Steinschneider, E. K. Ruuge, V. I. Kapel'ko, M. Yu. Zueva, V. L. Lakomkin, V. N. Smirnov, and V. A. Saks, *Biochim. Biophys. Acta* **805**, 319 (1984).

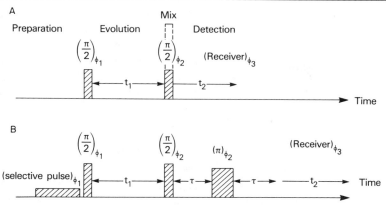

FIG. 4. Pulse-timing diagram for obtaining 2D COSY spectra. (A) Standard COSY sequence. (B) Water/fat suppression COSY sequence. Sequences (A) and (B) were cycled according to Table II, yielding 16 steps/$t_1$ value. See Fig. 5 legend and text for more detail.

homonuclear correlated spectroscopy (COSY), which provides a visual "road map" of $J$-coupled partners.[12] Analysis of this map can be used to monitor metabolites *in vivo* under certain conditions.

Recent 1D spectral editing methods have been used to detect one compound at a time in a complex spectrum.[13] However, this approach does not efficiently utilize all of the metabolite information contained in the FID. In addition, spectral editing methods require some knowledge about the compound to be studied, and this can limit the serendipity of the NMR experiment.

The basic COSY pulse sequence is present in Fig. 4A. To suppress the strong resonances from water and fat typically found *in vivo,* a water-selective, low-power pulse was applied during the preparation period and a spin-echo recovery was applied during the detection period (Fig. 4B). If additional water/fat suppression is necessary to reduce the annoying "$t_1$ ridges" then *post facto* data manipulations of the 2D matrix (e.g., symmetrization[14]) are effective but must be used with care since they tend to distort the intensity and connectivity information of the spectrum.

The first 90° pulse creates transverse magnetization from all allowed energy levels in a coupled spin system. A delay $t_1$, which is systematically varied as described previously, then allows the $x$–$y$ magnetization to

[12] A. Bax, "Two-Dimensional Nuclear Magnetic Resonance in Liquids." Reidel, London, 1982.

[13] K. L. Behar, J. A. denHollander, M. E. Stromski, T. Ogino, R. G. Shulman, and O. A. Prichard, *Proc. Natl. Acad. Sci. U.S.A.* **80,** 4945 (1983).

[14] R. Baumann, G. Wider, R. R. Ernst, and K. Wuthrich, *J. Magn. Reson.* **44,** 402 (1981).

evolve. The second 90° pulse abruptly ends the evolution period by returning some of the transverse magnetization back to the $z$ axis. Those magnetization components that are parallel to the second pulse are not affected by the second 90° pulse and left in the $x$–$y$ plane for detection. Thus, the second 90° pulse causes a redistribution, or mixing, of population among the energy levels by returning only some of the magnetization back to the $z$ axis. The time evolution of a cross-peak in a simple coupled system of two spins 1 and 2, neglecting relaxation, is given by[12]

$$S_{12}(t_1, t_2) = M_0 \sin(\pi J_{12} t_1)\sin(\pi J_{12} t_2)\exp(-iF_1 t_1)\exp(iF_2 t_2) \qquad (3)$$

where $J_{12}$ is the coupling constant, $M_0$ is the equilibrium magnetization, and $F_1$, $F_2$ are the Larmor frequencies of spins 1 and 2, respectively. After 2D Fourier transformation, the frequency position of the cross-peaks, given by $(F_1, F_2)$, maps out coupling partners. This map is usually unique for a compound which contains $J$-coupled spins. In addition, each cross-

TABLE II
PHASE CYCLE[a] FOR COSY EXPERIMENTS

| Phase step | $\theta_1$ | $\theta_2$ | $\theta_3$ |
|---|---|---|---|
| 1 | $x$ | $x$ | $x$ |
| 2 | $x$ | $y$ | $-y$ |
| 3 | $x$ | $-x$ | $x$ |
| 4 | $x$ | $-y$ | $-y$ |
| 5 | $y$ | $y$ | $y$ |
| 6 | $y$ | $-x$ | $x$ |
| 7 | $y$ | $-y$ | $y$ |
| 8 | $y$ | $x$ | $x$ |
| 9 | $-x$ | $-x$ | $-x$ |
| 10 | $-x$ | $-y$ | $y$ |
| 11 | $-x$ | $x$ | $-x$ |
| 12 | $-x$ | $y$ | $y$ |
| 13 | $-y$ | $-y$ | $-y$ |
| 14 | $-y$ | $x$ | $-x$ |
| 15 | $-y$ | $y$ | $-y$ |
| 16 | $-y$ | $-x$ | $-x$ |

[a] This basic 16-step sequence provides for quadrature detection in $t_1$, removes axial peaks, and corrects for receiver mismatches (CYCLOPS). Odd- and even-numbered steps were stored in different memory locations and analyzed according to States et al.[7] Data were collected in quadrature in the $t_2$ domain (on the GE CSI: QP on $y = 0$; AB off; AL = 2).

peak contains concentration information in the form of $M_0$. In general, in a crowded spectrum, several compounds (which may be unresolved in the 1D spectrum) may be detected simultaneously by this method.

As in the previous section, the 2D COSY pulse sequence may be phase cycled or not depending on the sample stability and experimental needs. It should be noted that the time restriction of the 2D NOE experiment is relaxed somewhat in the COSY experiment since the COSY sequence requires fewer phase cycle steps (compare Tables I and II), due to having one less pulse, and a very short ($\mu$sec) $t_{mix}$ period.

Phase-sensitive COSY data can be obtained using the method of States et al.[7] and phase cycling according to Table II. In this case it is impossible to obtain a pure absorption mode COSY spectrum due to a 90° phase shift between the cross- and diagonal peaks.[12] In high-resolution situations, a phase-sensitive COSY spectrum is presented with the diagonal peaks phased in dispersive mode so that the cross-peaks are phased absorptive. However, this presentation strategy may not work in vivo due to limited resolution (0.1–0.3 ppm). Such relatively broad resonances will result in reduced cross-peak intensity due to overlap and cancellation of the antiphase components of the cross-peaks. Presentation of the power spectrum avoids this problem although typically requires "reshaping" of the FID[12] to reduce the long tails present after the magnitude calculation.

Figure 5A is a water/fat suppressed COSY spectrum of a rabbit kidney in vivo obtained from a rabbit deprived of water for 3 days. Details of the experiment are presented in the figure legend. The strong resonances at (5.4, 2.0 ppm) and (5.4, 2.7 ppm) are not seen in the perchloric acid extract and probably represent methylene protons in lipids.[15] A blow-up of the region between 2.2 and 4.4 ppm is shown in Fig. 5B. Preliminary assignments of glycerolphosphocholine (GPC), inositol, phosphoethanolamine (PEA), and lactate, based on the cross-peak chemical shifts in the COSY spectrum, are made in Fig. 5. These chemical shifts agree with published values[13,16] as well as with shifts obtained in COSY phantom studies performed in this laboratory. Additional evidence for the assignment of inositol is presented in the Addendum at the end of this volume.

COSY offers several advantages for detection of metabolites in vivo. The observation of a compound's cross-peak in the 2D spectrum is suggestive that 1D spectral editing may also be usefully employed to study its metabolism. In addition, the 2D technique does not require a homoge-

[15] R. E. Block and B. C. Parekh, J. Magn. Reson. 75, 517 (1987).
[16] S. Bagnasco, R. Balaban, H. M. Fales, Y.-M. Yang, and M. Burg, J. Biol. Chem. 261, 5872 (1986).

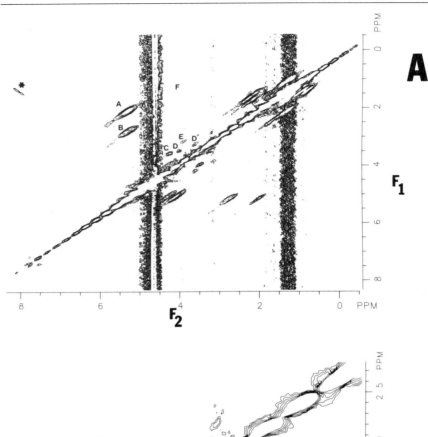

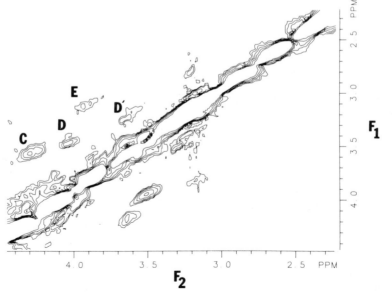

neous rf excitation profile and can be performed using a surface coil.[6] This can be useful in studying tissues which are difficult to investigate using a homogeneous $B_1$ field.

There are some limitations to the COSY experiment. Collection of the whole 2D data set is time consuming and only allows study of changes in steady state concentrations. However, several methods, e.g., 1D COSY[17] or "semiselective" COSY,[18] are available which can significantly reduce the experiment time. Another potential problem is that compounds in high concentration may have several cross-peaks which coincide with cross-peaks from other compounds, making careful identification necessary.

In summary, application of 2D methods (COSY and NOE) can provide important information concerning enzyme kinetics and metabolic processes *in vivo*. The major advantages of these techniques are the improvement in resolution and the ability to monitor numerous exchange processes and metabolites simultaneously.

[17] H. Kessler, H. Oschkinat, C. Griesinger, and W. Bermel, *J. Magn. Reson.* **70**, 106 (1986).
[18] J. Cavanagh, J. P. Waltho, and J. Keeler, *J. Magn. Reson.* **74**, 386 (1987).

---

FIG. 5. Water-suppressed proton 2D NMR spectrum of a rabbit kidney *in vivo*. (A) Complete 1K × 1K COSY spectrum obtained on a GE CSI system (4.7 *T*). An external deuterium oxide lock helped minimize static field drift. A modified Helmholtz (saddle) coil was used as previously described [S. D. Wolff and R. S. Balaban, *J. Magn. Reson.* **75**, 190 (1987)]. Using the pulse sequence of Fig. 4B, a 3-W water-selective pulse was gated on for 700 msec and then gated off for 1 msec prior to the first hard 90° pulse. A spin-echo recovery ($2\tau = 60$ msec) during the detection period helps to suppress broad spectral components, thus reducing baseplane roll. A composite 180° pulse ($90_x$–$180_y$–$90_x$), which compensates for $B_1$ inhomogeneities [M. H. Levitt and R. Freeman, *J. Magn. Reson.* **43**, 65 (1981)], was used to form the spin echo. The 90° pulse width was 28 $\mu$sec and the data acquisition time was 0.34 sec in $t_2$ and 0.17 sec in $t_1$. Data were acquired in 512 $t_1$ points and 1K $t_2$ points. Data processing was performed in the time domain as follows: Gaussian multiplication, apodization by the first half of a sine wave, zero fill, Fourier transformation, and presentation as the power spectrum. The chemical shift of water was referenced to 4.66 ppm and used as an internal chemical shift standard. The faint antidiagonal (marked by an *) is due to incomplete suppression of the antiecho signal. Six resonances are tentatively identified: A, methylene groups between two saturated hydrocarbon linkages (5.4, 2.0 ppm); B, methylene groups between two saturated hydrocarbon linkages (5.4, 2.7 ppm); C, glycerolphosphorylcholine (4.28, 3.61 ppm); D, inositol (4.04, 3.53 ppm); D', inositol (3.59, 3.24 ppm); E, phosphoethanolamine (3.91, 3.15 ppm); F, lactate (4.08, 1.29 ppm). Total data acquisition time was 3 hr. (B) Blow up of the region between 2.2 and 4.4 ppm.

## [17] Isotope Exchange

### By PAUL RÖSCH

#### Introduction

The study of enzyme kinetics by NMR observation of isotope-exchange processes is in many ways very similar to the study of enzyme kinetics with the magnetization and saturation transfer techniques described elsewhere in this volume [15]. In the application of magnetization transfer and saturation transfer techniques, a transient physical label of a chemical group is used as a marker which is transferred between different molecules, namely its high-frequency-induced nonequilibrium magnetization. In isotope-exchange studies an isotopic label is introduced chemically into the molecule or the molecular group under study and the transfer of this label to different molecules is observed. There are distinctive differences between the two methods, in particular as far as the range of kinetic parameters which may be determined is concerned. It should be noted in passing that all these methods may be used at least in principle with *in vitro* as well as *in vivo* systems.

Isotope exchange experiments are basically independent of NMR relaxation times as they depend on the concentration of permanently labeled chemical groups only. This is in contrast to the magnetization and saturation transfer studies as well as to the line-shape analysis method. The inverse of the pseudo-first-order exchange rates for isotope exchange reactions should be roughly in the order of magnitude of the time needed to take a decent NMR spectrum of the sample in order to allow the experimentalist to observe the exchange taking place in, roughly speaking, real time. The time needed to take an NMR spectrum of a biological sample is usually on the order of several minutes. It is also possible to take aliquots of a reaction mixture and quench the ongoing reaction in certain time intervals, thus extending the method to higher and lower exchange rates, respectively. The latter method also makes it possible to determine reaction rates under experimental conditions at which high-resolution NMR spectra may not be obtained satisfactorily, e.g., at temperatures below 273 K or in solutions with very high salt concentrations (several hundred millimolar).

The ways in which the exchange of isotopes may be observed by NMR are manifold. The simplest experimental protocol is the observation of a signal appearing or disappearing in the NMR spectrum of a reaction mixture by direct detection of the substitution of magnetically active nuclei with inactive isotopes or vice versa during the reaction. In practice, this almost always means observation of deuterium replacement with protons or the reverse reaction. There are also ways to probe exchanges of isotopes of nuclei which are not readily accessible to NMR detection. This is done by use of their influence on nuclei which are easily observed directly, usually protons or phosphorus. Examples for this type of experiment are the oxygen-16 for oxygen-18 exchange processes taking place in labeled inorganic phosphate ($P_i$) in the presence of ADP, metal ions, and catalytic concentrations of ATPases. These processes may easily be observed via the splitting of the $^{31}P$ resonance of the phosphate into five different resonances by the differential effect of $^{16}O$ and $^{18}O$ on the resonance frequency of $P_i$. Other indirect methods of isotope detection include the spin-echo technique allowing the observation of the exchange of spin-$\frac{1}{2}$ nuclei of low sensitivity such as carbon-13 or nitrogen-15 via attached protons. A review on isotope exchange studies *in vivo* appeared recently.[1] In the following, the experimental and data evaluation procedures are described with the help of one typical example per experimental category.

## Experimental

### *Proton-Exchange Experiments: Lactate Dehydrogenase (LDH)*[2-4]

*Direct Detection of Labels: Lactate($CH_3$)/Pyruvate($CH_3$)/Solvent Proton Exchange.*[2,3] As an example for the determination of enzyme-catalyzed isotope exchange by observation of the exchange of deuterium labels with protons, the study of the kinetics of the LDH is outlined.[2] For the description, it is irrelevant that the original experiment[2] was performed in erythrocytes and not *in vitro*. The reaction catalyzed by LDH is the interconversion between lactate and pyruvate:

$$
\begin{array}{ccc}
CH_3 & & CH_3 \\
\backslash & k_1 & | \\
HCOH + NAD^+ & \underset{k_{-1}}{\rightleftharpoons} & C\!-\!O + NADH + H^+ \\
/ & & | \\
CO_2^- & & CO_2^-
\end{array}
\tag{1}
$$

[1] K. M. Brindle and I. D. Campbell, *Q. Rev. Biophys.* **19**(3/4), 159 (1987).
[2] K. M. Brindle, F. F. Brown, I. D. Campbell, D. C. Foxall, and R. J. Simpson, *Biochem. Soc. Trans.* **8**, 646 (1980).
[3] R. J. Simpson, K. M. Brindle, F. F. Brown, I. D. Campbell, and D. L. Foxall, *Biochem. J.* **202**, 573 (1982).
[4] K. M. Brindle, I. D. Campbell, and R. J. Simpson, *Eur. J. Biochem.* **158**, 299 (1986).

In addition, protons from pyruvate are exchanged rapidly with solvent deuterons due to protein $\alpha$-$NH_2$ catalysis. In the presence of erythrocytes, this is expected to be of considerable magnitude. As the proton content of the solvent during the whole course of the reaction may be neglected in a first approximation, this solvent deuteron exchange reaction is essentially irreversible.

A simplified reaction scheme with respect to the methyl group, neglecting kinetic isotope effects, thus reads

$$\text{Lactate-}CH_3 \underset{k_{-1}}{\overset{k_1}{\rightleftharpoons}} \text{pyruvate-}CH_3 \overset{k_2}{\rightharpoonup} \text{pyruvate-}CD_3 \underset{k_1}{\overset{k_{-1}}{\rightleftharpoons}} \text{lactate-}CD_3 \qquad (2)$$

The parameters $k_1$, $k_{-1}$, and $k_2$ can be determined from the following experiment.

*Experimental Procedure*

1. Fill packed cells in deuterated buffer in a sample tube.
2. Equilibrate temperature of sample and stock solutions (deuterated solvent) of lactate and pyruvate.
3. Inject desired amount of lactate and pyruvate (e.g., 10 m$M$ lactate/ 10 m$M$ pyruvate final concentration as in Brindle *et al.*[5]) in sample tube and mix.
4. Put the sample into the spectrometer, tune probe, and take spectra at appropriate time points (dependent on necessary number of scans per spectrum).

The experiments were performed on a Nicolet NT 470 spectrometer with a proton resonance frequency of 470 MHz.

*Data evaluation:* The set of differential equations which describes the time dependence of the concentrations of the protonated species according to reaction (2) is

$$\begin{aligned} d/dt[L] &= -k_1[L] + k_{-1}[P] \\ d/dt[P] &= -(k_{-1} + k_2)[P] + k_1[L] \end{aligned} \qquad (3)$$

where L is lactate-$CH_3$ and P is pyruvate-$CH_3$. The analytical solution is straightforward[2]:

$$\begin{aligned} [L](t) &= [L](0)(ae^{-bt} - be^{-at})/(a - b) \\ [P](t) &= [P](0)[b(k_1 - a)e^{-at} - a(k_1 - b)e^{-bt}]/[k_{-1}(b - a)] \end{aligned} \qquad (4)$$

where $a = \frac{1}{2}\{k_1 + k_{-1} + k_2 + [(k_1 + k_{-1} + k_2)^2 - 4k_1k_2]\}^{1/2}$ and $b = k_1 + k_{-1} + k_2 - a$.

[5] K. M. Brindle, J. Boyd, I. D. Campbell, R. Porteous, and N. Soffe, *Biochem. Biophys. Res. Commun.* **109**, 864 (1982).

A least-squares fit of Eq. (4) to the experimental data by use of a standard minimization procedure (such as Nelder/Mead "Simplex") yields the rate constants. This can be performed with simple general fit programs fitting the experimental data to the solutions of the Eq. (3) differentials. The (inferior) alternative would be use of a program integrating these sets numerically and fitting the numerical solutions to the experimental data. Both types of program, running on a personal computer, are used in our laboratory.

### Homonuclear Spin-Echo Experiments: Lactate(2-CH) Exchange[4]

*Measurement principle.* Use of similar concentrations of reactants and products, as in the experiment described above, is often undesirable. For example, with LDH the product pyruvate is an inhibitor of the enzyme. In addition, pyruvate enables leakage of label by the exchange of methyl protons and solvent deuterons according to reaction (2). For both reasons it is desirable to keep the pyruvate concentration as low as possible. In cases like this, employment of a double-label experiment can be helpful.

In the study of LDH kinetics, the procedure for the double-label experiment is as follows: The experiment is performed in $H_2O$ rather than $D_2O$. A mixture of completely deuterated and completely protonated lactate results in a mixture of four differently labeled species in the presence of $NAD^+$ and LDH, namely:

$$
\begin{array}{cccc}
CD_3 & CD_3 & CH_3 & CH_3 \\
| & | & | & | \\
DCOH, & HCOH, & DCOH, \quad \text{and} \quad & HCOH \\
| & | & | & | \\
CO_2^- & CO_2^- & CO_2^- & CO_2^-
\end{array}
$$

This is the result of the following equilibria (again, kinetic isotope effects are neglected):

$$
\begin{aligned}
CD_3CDOHCO_2^- + NAD^+ &\underset{k_{-1}}{\overset{k_1}{\rightleftharpoons}} CD_3COCO_2^- + NADD \\
CD_3CHOHCO_2^- + NAD^+ &\underset{k_{-1}}{\overset{k_1}{\rightleftharpoons}} CD_3COCO_2^- + NADH \\
CH_3CDOHCO_2^- + NAD^+ &\underset{k_{-1}}{\overset{k_1}{\rightleftharpoons}} CH_3COCO_2^- + NADD \\
CH_3CHOHCO_2^- + NAD^+ &\underset{k_{-1}}{\overset{k_1}{\rightleftharpoons}} CH_3COCO_2^- + NADH
\end{aligned}
\tag{5}
$$

Obviously, in the methyl proton NMR spectrum, only the latter two species may be observed. A spin-echo experiment with the pulse sequence $(90°-\tau-180°-\tau)$ leads to inversion of the methyl resonances of the labeled lactate if $\tau$ is chosen so that $\tau = 1/(2J)$. $\tau = 68$ msec, corresponding to $J_{HH} = 7.2$ Hz, proved to be a good value for lactate.[4] As the methyl resonance of the 2-CD lactate is not inverted, the spin-echo sequence gives a signal intensity which is essentially the difference between both species. A spin-echo experiment with a sample containing equal amounts of labeled and unlabeled lactate at isotopic equilibrium, i.e., when $[CH_3—CDOH—CO_2^-] = [CH_3—CHOH—CO_2^-]$, would result in zero intensity for the methyl protein resonance.

*Sample preparation:* The preparation of completely deuterated lactate is described in ample detail in the original paper.[4] LDH was from Sigma, as were all other chemicals except $NAD^+$, which was obtained from Boehringer in preweighed vials.[4] In the original paper, the experiment was started with equal amounts of $CD_3CDOHCO_2^-$ and $CH_3CHOHCO_2^-$. A concentration of 10 m$M$ was used for both.

### NMR Procedure.

1. Put the sample without enzyme in the spectrometer.
2. Tune the probe by use of an HF-bridge or, preferably, by maximizing the free induction decay signal. In the latter procedure, care must be taken not to saturate the signal, i.e., very small pulse angles should be used. Maximizing the free induction decay proved to be the superior method of probe tuning in all cases in our laboratory.
3. Determine the 90° pulse angle according to one of the fashionable methods, for example by watching the signal phase inversion at a pulse angle of 180° and halving this value: (a) do a single-shot experiment, (b) Fourier transform, (c) phase correct to positive phase, (d) increase pulse width by a factor of two, (e) Fourier transform, (f) use automatic phase correction, (g) if the phase is positive, go to step (d), (h) decrease pulse width to determine the signal zero, (i) halve the value found in step (h).
4. Set up the spin-echo sequence.
5. Readjust the 90° pulse angle to give optimum spin-echo effect.
6. Inject LDH into the sample tube and take spectra at appropriate time points (2 min, 80 scans/time point in the original publication[4]).

The experiments were performed on a Nicolet NT 470 spectrometer with a proton resonance frequency of 470 MHz.

*Data evaluation:* The differential equations describing the Eq. (5) kinetics are

$$
\begin{aligned}
d/dt(\text{DH}) &= -k_1\text{DH} \cdot \text{N} + k_{-1}\text{HO} \cdot \text{ND} \\
d/dt(\text{HH}) &= -k_1\text{HH} \cdot \text{N} + k_{-1}\text{HO} \cdot \text{NH} \\
d/dt(\text{DD}) &= -k_1\text{DD} \cdot \text{N} + k_{-1}\text{DO} \cdot \text{ND} \\
d/dt(\text{HD}) &= -k_1\text{HD} \cdot \text{N} + k_{-1}\text{DO} \cdot \text{NH} \\
d/dt(\text{ND}) &= -k_{-1}\text{CO} \cdot \text{ND} + k_1(\text{DH} \cdot \text{N} + \text{DD} \cdot \text{N}) \\
d/dt(\text{NH}) &= -k_{-1}\text{CO} \cdot \text{NH} + k_1(\text{HH} \cdot \text{N} + \text{HD} \cdot \text{N})
\end{aligned}
\tag{6}
$$

with the abbreviations $[CD_3CHOHCO_2^-]$, HD; $[CH_3CHOHCO_2^-]$, HH; etc.; $[NAD^+]$, N; NADH, NH; NADD, ND; $CD_3COCO_2^-$, DO; $CH_3COCO_2^-$, HO.

It may be noted that under the given experimental conditions, HH + HD = constant, DH + DD = constant, and ND = NH. The time dependence of the difference DH − HH, which is observed with the spin-echo method, is then:

$$d/dt(DH - HH) = -k_1 N(DH - HH) \qquad (7)$$

After chemical equilibrium has been obtained, which is a very fast process compared to the establishment of the isotopic equilibrium, ND = NH = constant and OH = constant. Thus, the solution for Eq. (7) is a simple exponential:

$$(DH - HH)(t) = -HH(0)e^{-k_1 Nt} \qquad (8)$$

This solution can be fitted to the experimental data with $k_1$ and $k_{-1}$ as free parameters with one of the standard programs.

### Indirect Detection of Labels with Spin Echoes[5,7]

*Measurement Principle.* Indirect detection of nuclear labels with low NMR sensitivity can be performed by observation of spin echoes of attached protons. Although this technique shows its full advantages only when unwanted broad proton signals are to be suppressed, as in cells, it is described here since it proved to be worthwhile to apply this method also *in vitro*. In particular, in cases where the proton resonances cannot be resolved clearly the difference method described below proves to be very attractive. The principle of the measurement consists in the application of a (90°–τ–180°–τ–acquisition) pulse sequence at the proton frequency on the sample exactly as in a standard homonuclear spin-echo experiment. In addition, coincident with the proton 180° pulse, a 180° pulse is applied to the sample at the resonant frequency of the heteronucleus. The pulse sequence is shown in Fig. 1.

*Alanine Aminotransferase.[5]* Alanine aminotransferase catalyzes the reaction:

---

[6] R. Freeman, T. M. Mareci, and G. A. Morris, *J. Magn. Reson.* **42**, 341 (1981); R. Freeman and H. D. W. Hill, *J. Chem. Phys.* **54**, 301 (1971).

[7] K. M. Brindle, R. Porteous, and I. D. Campbell, *J. Magn. Reson.* **56**, 543 (1984).

$$\underset{\substack{\text{Alanine}}}{\overset{\substack{\text{CO}_2^-\\ |\\ \text{H}_3\text{N}^+\text{—C—H}\\ |\\ \text{CH}_3}}{\phantom{x}}} \;+\; \underset{\substack{\alpha\text{-Keto-}\\ \text{glutarate}}}{\overset{\substack{\text{CO}_2^-\\ |\\ \text{C}=\text{O}\\ |\\ (\text{CH}_2)_2\\ |\\ \text{CO}_2^-}}{\phantom{x}}} \underset{k_{-1}}{\overset{k_1}{\rightleftarrows}} \underset{\substack{\text{Pyruvate}}}{\overset{\substack{\text{CO}_2^-\\ |\\ \text{C}=\text{O}\\ |\\ \text{CH}_3}}{\phantom{x}}} \;+\; \underset{\substack{\text{Glutamate}}}{\overset{\substack{\text{CO}_2\\ |\\ \text{H}_2\text{N—C—H}\\ |\\ (\text{CH}_2)_2\\ |\\ \text{CO}_2^-}}{\phantom{x}}} \qquad (9)$$

Measurement of the reaction rate of alanine aminotransferase was performed with L-[3-$^{13}$C]alanine as labeled compound. The methyl spectrum of this compound is a doublet (due to the $^{13}$C splitting) of doublets (due to the $C_\alpha H$ splitting). A mixture of labeled and unlabeled alanine then gives a methyl spectrum of a central doublet from the unlabeled compound and two symmetrically located doublets from the labeled compound. A value $\tau = 1/(2 J_{HH})$ in the spin-echo sequence, where $J_{HH}$ is the proton–proton coupling constant (in the case of alanine, $J_{HH} = 7.3$ Hz), results in an inversion of all doublet signals in a mixture of labeled and unlabeled alanine. This is illustrated in Fig. 2a. Decoupling of $^{13}$C results in a collapse of the outer doublets into the central one giving, in essence, the sum of the signal from the labeled and unlabeled compounds (Fig. 2b). An echo delay of $\tau = n/(2 J_c)$ results in an inversion of the signals from protons attached to $^{13}$C relative to the signals from protons attached to $^{12}$C (Fig. 2c). $^{13}$C decoupling during acquisition again leads to the collapse of the $^{13}$C splitting and gives, in essence, a signal which corresponds to the difference between the signals from protons attached to the label and the corresponding protons from the unlabeled compound.

*Sample preparation:* The reaction mixture in the experiment described by Brindle et al.[5] contained 0.15 units/ml alanine aminotransferase as obtained from Sigma; 100 m$M$ glycylglycine–imidazole buffer, pH 7.4, 0.5 m$M$ EDTA, 0.15 m$M$ KCl, 10 m$M$ pyruvate, and 10 m$M$

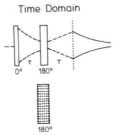

Time Domain

FIG. 1. Pulse sequence for heteronuclear spin-echo experiments. (After Brindle *et al.*[5])

Frequency Domain

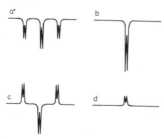

FIG. 2. (a) Effect of the proton spin-echo sequence with $\tau = 1/(2J_{HH})$ on the methyl protons of a mixture of 60% alanine $^{13}$C labeled in the methyl group (outer doublets) and 40% unlabeled alanine (central doublet). (b) Application of $^{13}$C decoupling during acquisition leads to a collapse of the outer doublets into the central doublet. (c) A 180° pulse $^{13}$C at the same time as the proton 180° pulse $[\tau = 1/(2J_0)]$ leads to inversion of the unlabeled resonance only. (d) $^{13}$C decoupling during acquisition in experiment (c) leads again to a collapse of the outer doublets into the central (inverted) doublet, thus resulting essentially in a difference spectrum. (After Brindle et al.[5])

L-[3-$^{13}$C]alanine. The measurements were performed in $H_2O$ as the alanine aminotranferase catalyzes the exchange of the alanine as well as the pyruvate methyl protons with the solvent.

*NMR procedure: Setup of spin-echo sequence:* Hardware requirements for the setup of a $^{13}$C–proton spin-echo sequence in most spectrometers currently available are as follow (the requirements are analogous for other $X$ nuclei): (1) a probe with a $^{13}$C detection coil and a proton decoupling coil; and (2) a computer-controlled, pulsed high-frequency power amplifier tuned to the $^{13}$C frequency.

Of course, the preferable equipment available in state-of-the-art spectrometers is a so-called $f_1/f_2$ switch. This allows interchange of decoupling and observation frequencies, thus making the external power amplifier superfluous. Also, in state-of-the-art instruments, an "inverse" probe with observation and decoupling coils interchanged for superior signal-to-noise ratio may be used.

Procedure with standard equipment for the setup of a proton-detected $^{13}$C measurement with the spin-echo method (again, the methods are analogous for other $X$ nuclei):

1. Connect the cable from the external computer-controlled $^{13}$C power amplifier to the observation coil plug of the probe.
2. Connect the proton observation channel to the decoupler coil of the probe.

3. Set the spectrometer to the $^{13}C$ observation mode.

4. Determine the $^{13}C$ 90° pulse width with the external power amplifier. Use a standard sample where the $^{13}C$ signal may be detected easily with a single-scan experiment. For the determination of the 90° pulse any of the currently used procedures may be used, for example the method outlined in the section, Homonuclear Spin-Echo Experiments.

5. Insert the sample without initializing the reaction, i.e., without enzyme. Set the spectrometer console to proton observation.

6. Set the spectrometer console to proton observation.

7. Tune the observation coil of the probe to $^{13}C$ using an HF-bridge with fast external power amplifier pulses. Tune the decoupling coil of the probe to protons as usual, i.e., using an HF-bridge or maximizing the FID with small excitation pulse angles.

8. Determine a value $n$ (odd) for which the equation $1/(2J_{HH}) = n/(2J_{CH})$ is approximately fulfilled. $J_{HH}$ is the proton–proton coupling constant, $J_{CH}$ the carbon–proton coupling constant of the signal under study. For the methyl signal of alanine, $J_{HH} = 7.3$ Hz and $J_{CH} = 130$ Hz; so $n = 15$ yields $t = 58$ msec as a good compromise.

9. Set up the pulse program.

10. Readjust the $^{13}C$ pulse angle to maximize the spin-echo effect.

11. Add enzyme to the sample.

12. Start the NMR measurement.

*Evaluation of data.* The differential equations governing the time development of the difference between $^{13}C$-labeled pyruvate (P*) and pyruvate (P) and between $^{13}C$-labeled alanine (A*) and alanine (A) can be derived from

$$
\begin{aligned}
d/dt[A^*] &= -k_1[A^*][\alpha] + k_{-1}[P^*][G] \\
d/dt[A] &= -k_1[A][\alpha] + k_{-1}[P][G] \\
d/dt[P^*] &= -k_{-1}[P^*][G] + k_1[A^*][\alpha] \\
d/dt[P] &= -k_{-1}[P][G] + k_1[A][\alpha]
\end{aligned}
\tag{10}
$$

where $\alpha$ is $\alpha$-ketoglutarate and G is glutamate. The time development for the difference between labeled and unlabeled compounds is (with the convention $[a] = [A^*] - [A]$; $[p] = [P^*] - [P]$:

$$
\begin{aligned}
d/dt[a] &= -k_1[\alpha][a] + k_{-1}[G][p] \\
d/dt[p] &= -k_{-1}[G][p] + k_1[\alpha][a]
\end{aligned}
\tag{11}
$$

At chemical equilibrium, i.e., $d/dt[G] = 0$ and $d/dt[\alpha] = 0$, an analytical solution of this set of differential equations can be obtained (omitting concentration brackets):

$$a = [a(0) + p(0)]k_{-1}(G/X) + \{a(0) - [a(0) + p(0)]k_{-1}G/X\}e^{-Xt}$$
$$p = [a(0) + p(0)]k_1(\alpha/X) + \{p(0) - [a(0) + p(0)]k_1\alpha/X\}e^{-Xt} \qquad (12)$$
$$X = k_{-1}G + k_1\alpha$$

These solutions may be fitted to the experimental data with the rate constants as free parameters to yield $k_1$ and $k_{-1}$. In more complicated cases, programs exist which integrate the set of differential equations numerically and fit the constants to the experimental equations in an iterative way.

### Creatine Kinase[8]

Creatine kinase catalyzes the reaction:

$$\text{Phosphocreatine} + \text{ADP} \underset{k_{-1}}{\overset{k_1}{\rightleftharpoons}} \text{creatine} + \text{ATP}$$

As an alternative to the numerous saturation transfer experiments performed on this system, an isotope-exchange experiment making use of an $^{15}$N-labeled creatine ([1-$^{15}$N]guanidino-1-methylethanoic acid) was used. The $^{15}$N label was observed via the $^{31}$P resonance.

*Sample preparation:* An excellent exposition of the synthesis of the labeled compound was given by Greenaway and Whatley.[9] The reaction mixture[8] contained 10 m$M$ phosphocreatine, 10 m$M$ [$^{15}$N]creatine, 5 m$M$ ATP, 200 m$M$ potassium HEPES buffer, pH 7.0, or triethanol-HCl, pH 8.0, 0.01 m$M$ EDTA, 2 m$M$ dithiothreitol, 120 $\mu M$ creatine kinase as obtained from Boehringer. MgCl$_2$ was added to result in 1 m$M$ free ion concentration as calculated from the known stability constants of the complex.

*NMR procedures:* The hardware requirements for a $^{31}$P-observed, heteronuclear $^{15}$N spin-echo experiment are a double-tuned probe ($^{31}$P and $^{15}$N) and an external power amplifier for the $^{15}$N frequency with the spectrometer console set for $^{31}$P observation (or vice versa). A home-built probe with the center coil tuned to the phosphorus frequency and the outer coil tuned to the nitrogen frequency was used.[8] The experiment was performed with a 10-mm tube on an essentially home-built spectrometer operating at a phosphorus frequency of 73.8 MHz. The instrumental setup for the experiment is analogous to the one described for the proton–$^{13}$C heteronuclear spin-echo experiment.

*Evaluation of data:* The differential equations describing the labeled (*) and unlabeled compounds phosphocreatine (PC) and creatine (C) are

[8] K. M. Brindle, R. Porteous, and G. K. Radda, *Biochim. Biophys. Acta* **786**, 18 (1984).
[9] W. Greenaway and F. R. Whatley, *J. Labelled Compd. Radiopharm.* **14**(4), 611 (1978).

$$d/dt[PC^*] = -k_1[PC^*][ADP] + k_{-1}[C^*][ATP] \tag{13}$$
$$d/dt[PC] = -k_1[PC][ADP] + k_{-1}[C][ATP], \text{ etc.}$$

With pc = $[PC^*]$ − $[PC]$ and c = $[C^*]$ − $[C]$, integration yields Eq. (12) with

$$\alpha = [ADP] \quad G = [ATP] \quad a = \text{pc} \quad p = \text{c} \tag{14}$$

*Detection of Oxygen Labels of Inorganic Phosphate:*
*Myosin $S_1$ ATPase*

*Measurement Principle.* The method of determination of kinetic parameters of nucleotide hydrolases by following the $^{18}O$ exchange catalyzed by these proteins rests on the fact that all known hydrolases cleave the $\beta$-bridge oxygen of nucleoside triphosphates at the terminal bond:

$$ADP-O-PO_3 + H_2O \underset{k_{-1}}{\overset{k_1}{\rightleftharpoons}} ADP + P_i \tag{15}$$

This means that with every ATP cleavage an oxygen atom from the surrounding water is incorporated into the $P_i$ molecule. The ATP cleavage and resynthesis may go back and forth on the hydrolase several times before product release. Thus, the average number of oxygens exchanged during the overall reaction may obtain any value. Labeling of the ATP terminal phosphoryl group with $^{18}O$ then yields a specific pattern of isotopic replacements in the product $P_i$. Of course, as the reaction is reversible, it may also be started with ADP and labeled $P_i$ and either the time course of incorporation of the labels into the terminal phosphoryl group of ATP or the change in the concentrations of the five different $P_i$ species, $P^{18}O_4$, $P^{18}O_3^{16}O_1$, . . . , $P^{16}O_4$, may be followed. The $^{31}P$ NMR spectrum shows different signals for the five different species, the signal spacing being about 0.025 ppm.[6,11] In the former case, product ATP must be removed by a secondary reaction to yield reasonable amounts of this compound, e.g., with hexokinase, which transfers the terminal phosphoryl group of ATP to glucose to yield glucose 6-phosphate essentially without disturbing the label content.[12] The concentration of the different $P_i$ species can also be detected by mass spectroscopy.[13] This latter method is somewhat more demanding as far as the chemistry involved is concerned but is much more sensitive than the NMR method, accordingly resulting in lower amounts of enzyme required. The major advantage of the use of NMR as the detection method is the simplicity of the experimental setup

[10] O. Lutz, A. Nolle, and D. Staschewski, *Z. Naturforsch.* **33A,** 380 (1978).
[11] M. Cohn and A. Hu, *Proc. Natl. Acad. Sci. U.S.A.* **75,** 200 (1978).
[12] J. J. Sines and D. D. Hackney, *Biochemistry* **25,** 6144 (1986).
[13] D. D. Hackney, K. E. Stempel, and P. D. Boyer, this series, Vol. 64, p. 60.

and, under certain conditions, the possibility of following the ongoing exchange reaction in virtually real time. One of the systems studied most intensively with the aid of $P_i$–$^{18}O$ isotope exchange (PIX) is the hydrolase part of myosin, myosin $S_1$. Here I describe the experimental setup for a simple experiment with observation of the time dependence of the five different $P_i$ species in the presence of myosin $S_1$ and ADP.

*Sample preparation:* The preparation of myosin $S_1$ is essentially standardized. The procedure was described earlier in this series.[14] Procedures for $^{18}O$ enrichment of $P_i$, based on isotope exchange, would require a large excess of isotopically enriched water and are thus not economical. An alternative synthesis of $^{18}O$-labeled $P_i$ was also described in this series.[13] In our laboratory we applied a somewhat different procedure based on hydrolysis of phosphorus pentachloride with $H_2^{18}O$ to phosphoric acid and hydrogen chloride and subsequent isolation of phosphoric acid as the crystalline monopotassium salt:

$H_2^{18}O$ (1 g, $^{18}O$ content approximately 98.8%, Monsanto, OH) is frozen with liquid nitrogen in a two-necked vial which contains a magnetic spin vane and is connected to an adjustable vacuum pump. Phosphorus pentachloride (2.9 g) is transferred quickly to the frozen enriched water. The equipment is protected against humidity while it is slowly warmed to room temperature. A reduced pressure of 66 kPa is applied and the mixture stirred and heated to 80° for a period of 90 min. After evolution of HCl is finished, the reaction mixture is allowed to cool to room temperature. Potassium hydroxide (2 M, in normalized water) is used to adjust the pH to 4.6 and the phosphate is precipitated by adding ethanol (70%, 15–20 ml), collected, and dried: 1.7 g $KH_2PO_4$, $^{18}O$ approximately 98%.

The chemical yield usually fluctuates between 75 and 90%. It may contain traces of potassium chloride. As suggested by Risley and van Etten,[15] a reprecipitation to get a virtually KCl-free product may be required, in particular if the product is used for purposes other than NMR observation of $^{18}O$ exchange. Synthesis of ATP ($^{18}O$) from this product was described earlier in this series.[13]

The reaction mixture in the experiment described by Rösch et al.[16] contained 0.18 m$M$ myosin $S_1$, 100 m$M$ labeled $P_i$, 0.2 m$M$ ADP, 0.02 $AP_5A$ (diadenosine pentaphosphate), 1 m$M$ $MgCl_2$, 50 m$M$ HEPES, pH 7.5, 0.1 m$M$ EDTA. Addition of $AP_5A$ to the reaction mixture was essential to suppress residual adenylate kinase activity. Traces of sodium azide were added to prevent bacterial growth. In order to obtain the

[14] A. G. Weeds and R. S. Taylor, *Nature (London)* **257**, 54 (1975). S. S. Margossian and S. Lowey, this series, Vol. 85, p. 55.
[15] J. M. Risley and R. L. van Etten, *J. Labelled Compd. Radiopharm.* **15**, 533 (1978).
[16] P. Rösch, R. S. Goody, and H. Zimmermann, *Arch. Biochem. Biophys.* **211**, 622 (1981).

necessary resolution, in general two things are essential: First, the pH of the solution must be well above (or below) the p$K$ region of inorganic phosphate, i.e., pH 7.5 or above should be used; second, the temperature must be equilibrated very well before data collection is started.

*NMR procedures.* The experiments were performed on a Bruker 360 HX NMR instrument with a $^{31}$P resonance frequency of 145.78 MHz. In our experiments, the spectral width was 60 Hz with a computer memory of 1K data, resulting in a scan repetition rate of 8.5 sec. The pulse angle was 90° and 200 accumulations/time point with a total of 100 time points required a measurement time of about 2 days. The relative concentration can be determined using the peak height in the spectrum. As the different $P_i$ species have virtually the same $T_1$ values, it is not necessary to take relaxation into account.

*Data evaluation:* Analysis becomes more laborious in exchange reactions with multiple isotope labels.[12,13,17] The minimum number of reaction steps for the $^{18}$O exchange catalyzed by ATPases is

$$E + ADP + P_i \underset{k_{-1}}{\overset{k_1}{\rightleftharpoons}} E \cdot ADP \cdot P_i \underset{k_{-2}}{\overset{k_2}{\rightleftharpoons}} E \cdot ATP \underset{k_{-3}}{\overset{k_3}{\rightleftharpoons}} E + ATP \qquad (16)$$

In general, experiments starting with labeled ATP and experiments starting with labeled $P_i$ may be described with the same methods. The description is also the same irrespective of which of the two compounds is the species whose time-dependent label incorporation or loss is followed by the experiment. With the notation $P^{18}O_n{}^{16}O_{4-n} = P_n$ and $ADP-O-P^{18}O_n{}^{16}O_{3-n} = ATP_n$, the kinetic scheme governing the oxygen isotope exchange as an extension of Eq. (16) (neglecting kinetic isotope effects) is

$$
\begin{aligned}
&E + ADP + P_4 \underset{k_{-1}}{\overset{k_1}{\rightleftharpoons}} E \cdot ADP \cdot P_4 \\
&\qquad\qquad\qquad\qquad {}^{v\,43}\!\!\diagdown\!\!\diagup{}^{w\,34} \\
&E + ADP + P_3 \underset{k_{-1}}{\overset{k_1}{\rightleftharpoons}} E \cdot ADP \cdot P_3 \underset{k_{-2}}{\overset{k_2}{\rightleftharpoons}} E \cdot ATP_3 \underset{k_{-3}}{\overset{k_3}{\rightleftharpoons}} E + ATP_3 \\
&\qquad\qquad\qquad\qquad {}^{v\,32}\!\!\diagdown\!\!\diagup{}^{w\,23} \\
&E + ADP + P_2 \underset{k_{-1}}{\overset{k_1}{\rightleftharpoons}} E \cdot ADP \cdot P_2 \underset{k_{-2}}{\overset{k_2}{\rightleftharpoons}} E \cdot ATP_2 \underset{k_{-3}}{\overset{k_3}{\rightleftharpoons}} E + ATP_2 \qquad (17) \\
&\qquad\qquad\qquad\qquad {}^{v\,21}\!\!\diagdown\!\!\diagup{}^{w\,12} \\
&E + ADP + P_1 \underset{k_{-1}}{\overset{k_1}{\rightleftharpoons}} E \cdot ADP \cdot P_1 \underset{k_{-2}}{\overset{k_2}{\rightleftharpoons}} E \cdot ATP_1 \underset{k_{-3}}{\overset{k_3}{\rightleftharpoons}} E + ATP_1 \\
&\qquad\qquad\qquad\qquad {}^{v\,10}\!\!\diagdown\!\!\diagup{}^{w\,01} \\
&E + ADP + P_0 \underset{k_{-1}}{\overset{k_1}{\rightleftharpoons}} E \cdot ADP \cdot P_0 \underset{k_{-2}}{\overset{k_2}{\rightleftharpoons}} E \cdot ATP_0 \underset{k_{-3}}{\overset{k_3}{\rightleftharpoons}} E + ATP_0
\end{aligned}
$$

[17] P. Rösch, *Prog. Nucl. Magn. Reson. Spectrosc.* **18**, 123 (1986).

The exchange of a label corresponds to the transition between different lines in the set of equations with reaction rates $k_2$ and transition probabilities between the differently labeled species $w^{mn}$ and $v^{mn}$. $w^{mn}$ is the probability of transition from $P_n$ to $ATP_m$, $v^{mn}$ the probability of transition from $ATP_n$ to $P_m$. Of course, $\sum_n v^{mn} = \sum_n w^{mn} = 1$.

The differential equations describing the above scheme are

$$d/dt[P_0] = -k_1[E][ADP][P_0] + k_{-1}[E \cdot ADP \cdot P_0]$$

$$\vdots$$

$$d/dt[P_4] = -k_1[E][ADP][P_4] + k_{-1}[E \cdot ADP \cdot P_4]$$
$$d/dt[ATP_0] = -k_{-3}[E][ATP_0] + k_3[E \cdot ATP_3]$$

$$\vdots$$

$$d/dt[ATP_3] = -k_{-3}[E][ATP_3] + k_3[E \cdot ATP_3]$$
$$d/dt[E \cdot ADP \cdot P_0] = -(k_{-1} + k_2)[E \cdot ADP \cdot P_0] + k_1[E][ADP][P_0])$$
$$\qquad + k_{-2}v^{00}[E \cdot ATP_0]$$
$$d/dt[E \cdot ADP \cdot P_1] = -(k_{-1} + k_2)[E \cdot ADP \cdot P_1] + k_1[E][ADP][P_1]$$
$$\qquad + k_{-2}(v^{11}[E \cdot ATP_1] + v^{10}[E \cdot ATP_0])$$

$$\vdots \tag{18}$$

$$d/dt[E \cdot ADP \cdot P_4] = -(k_{-1} + k_2)[E \cdot ADP \cdot P_4] + k_1[E][ADP][P_4]$$
$$\qquad + k_{-2}v^{43}[E \cdot ATP_3]$$
$$d/dt[E \cdot ATP_0] = -(k_3 + k_{-2})[E \cdot ATP_0]$$
$$\qquad + k_2(w^{01}[E \cdot ADP \cdot P_1] + w^{00}[E \cdot ADP \cdot P_0])$$
$$\qquad + k_{-3}[E][ATP_0]$$

$$\vdots$$

$$d/dt[E \cdot ATP_3] = -(k_3 + k_{-2})[E \cdot ATP_3]$$
$$\qquad + k_2(w^{34}[E \cdot ADP \cdot P_4] + w^{33}[E \cdot ADP \cdot P_3])$$
$$\qquad + k_{-3}[E][ATP_3]$$
$$d/dt[ADP] = \sum_n(d/dt\ P_n)$$

In practice, experimental conditions may be found under which one or the other of the rate constants may be neglected. For example, if the experiment is started from labeled $P_i$, usually $k_3$ and $k_{-3}$ may be zeroed in a first approximation. Thus, under decent assumptions for $w^{mn}$ and $v^{mn}$, the other parameters may be obtained directly by a least-squares fit of the numerical solutions of the Eq. (18) differentials to the experimental data. For a $P_i$ molecule with complete rotational freedom in the enzyme-bound state, the values for $v^{mn}$ and $w^{mn}$ are given by

$$
\begin{aligned}
v^{00} &= v^{11} = v^{22} = v^{33} = 1 - z \\
v^{10} &= v^{21} = v^{32} = v^{43} = z \\
w^{00} &= 1; \; w^{11} = 3/4; \; w^{22} = 1/2; \; w^{33} = 1/4 \\
w^{01} &= 1/4; \; w^{12} = 1/2; \; w^{23} = 3/4; \; w^{34} = 1
\end{aligned}
\tag{19}
$$

$z$ is the fractional $H_2^{18}O$ content of the solvent, i.e.,

$$
z = H_2^{18}O/(H_2^{16}O + H_2^{17}O + H_2^{18}O) \tag{20}
$$

All other $v^{mn}$ and $w^{mn}$ are zero. The fractional content of $H_2^{18}O$ may either be assumed to be invariant during the experiment (usually $z = 0.2$) or may be varied according to

$$
\begin{aligned}
z(t) &= (H_2^{18}O + K)/(H_2^{16}O + H_2^{17}O + H_2^{18}O) \\
K &= [\Sigma nP_n(0) + \Sigma nATP_n(0) - \Sigma nP_n(t) - \Sigma nATP_n(t)]
\end{aligned}
\tag{21}
$$

Of course, the values for $v^{mn}$ and $w^{mn}$ can be adapted to situations where the prerequisite of complete rotational freedom is not fulfilled. Extensive calculations along this line were done by Sines and Hackney.[12] The computer calculations may be performed with one of the mentioned programs being able to numerically integrate and least-squares fit the differential equations to the experimental kinetic parameters. In the simplest experiment, the one starting from $P_i$ and the $E \cdot ADP$ complex, the number of reversals of the ATP-synthesis step may be calculated as $N = k_2/k_{-1}$, and $k_1$ can be determined easily. This can be done to a very good approximation even with numerical procedures which are not based on the solution of the complete set of Eq. (18) differentials.[17]

An example of the results obtained with this method is given in Fig. 3. It shows in the upper part the calculated time dependence of the five different $P_i$ species in the presence of myosin $S_1$, ADP, and $MgCl_2$ under the conditions outlined above.[16] The lower part of Fig. 3 shows the final spectrum of the experiment as it was obtained after 23.5 hr (upper trace) and the computer simulation according to the fitted rate constants.

## Outlook

Observation of isotope exchange opened the possibility to measure kinetic parameters which could not be determined by any other method. NMR is a simple way to follow the time course of the isotopic exchange procedure. In most NMR experiments, not much chemistry is involved. Even in experiments where competitive methods such as mass spectrometry for the $^{18}O$ exchange experiments are available, NMR proves to be by far the simplest way to get the desired information. On the other hand,

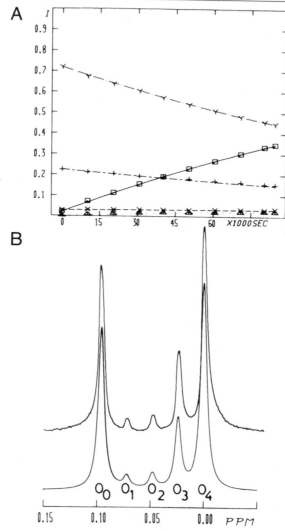

FIG. 3. (A) Time development of the concentration of the five different $P^{18}O_n{}^{16}O_{4-n}$ species under the experimental conditions described in the text. Y: $n = 4$; $+$: $n = 3$; X: $n = 2$; $\triangle$: $n = 1$; $\square$: $n = 0$. (B) Last measured spectrum from the reaction mixture (upper trace) and spectrum simulated according to the fitted parameters. (From Rösch $et\ al.$[16])

NMR requires sample amounts which worry many biochemists. In these cases, if alternative methods exist, researchers tend to employ these alternatives. A good example is again the $^{18}O$ exchange experiment. The groups most active in the field recently seem to employ the additional chemistry involved in the mass spectrometric approach in order to economize on the biological sample. Another point is that NMR studies at high field are still very expensive and it may be necessary to resort to methods which are not as simple as NMR, but are less costly.

### Acknowledgments

I would like to thank Dr. David D. Hackney (Pittsburgh, PA) for sending me manuscripts prior to publication. Mr. Herbert Zimmermann (Heidelberg, West Germany) worked out the preparation procedure for labeled inorganic phosphate.

## [18] Interpreting Protein Dynamics with Nuclear Magnetic Resonance Relaxation Measurements

*By* ROBERT E. LONDON

### Introduction

The extraction of a dynamic description of protein structure from NMR relaxation measurements constitutes a complex theoretical problem which has resisted reduction to the direct application of a set of simple methodological rules. This situation reflects the fact that the problem is not merely one of determining rates of motion within the framework of a particular dynamic model, but of formulating a description of the motion. Given the rather theoretical nature of this problem, it might seem inappropriate to treat this topic in a series devoted to methodology. However, there are two important reasons for the discussion of the effects of protein dynamics in this volume. First, one of the principle motivations for magnetic resonance studies of proteins has been the desire to characterize the "solution structure" of proteins, and to compare this structure with the more static picture provided by crystallographic analysis. Thus, the dynamic aspects of protein structure which strongly modulate the observed NMR spectra constitute one of the most unique contributions of the method. Second, since dynamic aspects of protein structure determine the NMR relaxation parameters, an appreciation of how these will affect the data becomes important at the design, execution, and interpretation stages of the NMR experiment. Hence, such considerations become of considerable relevance to the experimentalist.

One useful starting point in the analysis of protein dynamics is a consideration of the differences which exist between small, rigid molecules and larger, more complex molecules. In general, relaxation data obtained for flexible molecules such as peptides can be adequately described only if internal motion about the single bonds on a time scale shorter than that required for overall molecular diffusion is included in the description. Thus, attempts to describe the relaxation of peptides using the formalism for rigid molecules undergoing isotropic or anisotropic diffusion, as describes, for example, the motion of substituted benzenes, do not provide an adequate fit of the relaxation data. The folding of long peptide chains into globular structures with secondary and tertiary structure introduces additional dynamic constraints. First, much of the internal diffusion about the molecular bonds is characterized by low-amplitude motion consistent with the higher order structure, leading to the use of dynamic models which allow for restricted amplitude diffusion, as well as to models which allow discrete jumps between several significantly different, stable conformations. Second, the stabilization of particular conformations leads to chemical shift dispersion for the nuclei of each amino acid type located at different positions in the structure. Such shift dispersion is accompanied by conformational exchange effects which reflect the conformational flexibility of the macromolecule. These relatively slow processes are generally not observed in peptides, and are modulated by changes in temperature or by the addition of ligands which stabilize particular conformations upon binding to the protein. Such slow-exchange processes make the dominant contribution to the transverse relaxation rates of many, perhaps most, protein resonances.

## Relaxation Mechanisms

The interpretation of relaxation data obtained in NMR studies of molecules in solution depends first on either a determination of, or assumptions regarding, the relaxation mechanisms which contribute to the relaxation. Each of these mechanisms reflects the existence of a physical effect which can result in the production of a time-varying magnetic field in the vicinity of the nucleus under study. Such fluctuating fields can arise in a variety of ways.[1,2] One of the most generally significant mechanisms is the

[1] A. Abragam, "The Principles of Nuclear Magnetism." Oxford Univ. Press (Clarendon), London and New York, 1961.
[2] H. W. Spiess, *NMR: Basic Princ. Prog.* **15,** 1 (1978).

dipole–dipole interaction which can be viewed as resulting from the interaction of one nucleus (B) with the dipolar field produced by a second nucleus (A) (Fig. 1). Since the orientation of the dipolar field produced by nucleus A is quantized by the applied magnetic field $B_0$, molecular diffusion, such as the rotation illustrated by $R_1$, will lead to a change in the field observed at the location of nucleus B. The dipolar mechanism would not be present if the orientation of the fields also rotated with the molecule (rotation $R_2$). Additionally, as a consequence of the reduction in the strength of the dipolar field with the distance from the nucleus, the strength of the interaction drops off with a $1/r_{AB}^6$ dependence. Hence, the relaxation mechanism depends predominantly on interactions involving nearby spins. For the case of protonated carbon-13 nuclei, this means that only the directly bonded protons need to be considered.

In order to characterize the motion of large numbers of molecules in the sample, various statistical approximations are introduced, and the motion of the internuclear vector is described by a correlation function $G(t)$ which reflects how rapidly the vector connecting spins A and B changes orientation. The function $G(t)$ will approach 1.0 as $t$ approaches

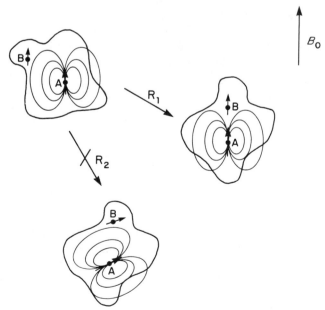

FIG. 1. Molecular diffusion modulates the dipolar interaction as illustrated by the rotation $R_1$, since the dipole fields are quantized in the laboratory frame of reference. If the fields rotated with the molecule, as illustrated by $R_2$, there would be no relaxation effect.

0, reflecting the fact that for sufficiently short times there is no change in orientation, and will approach 0 for sufficiently long values of $t$, reflecting the fact that after sufficient time has elapsed, there is no correlation between the orientation at that time and the initial orientation. A functional form of $G(t)$ can be calculated subject to various assumptions about the motion. For example, $G(t)$ can be determined subject to diffusion equations based on the assumptions of either isotropic or anisotropic diffusion. Alternatively, $G(t)$ can be calculated on the basis of a model in which internal motion is considered to be diffusive, but subject to boundary conditions which restrict the range of motion. $G(t)$ can also be determined using a model in which the orientation of the vector is considered to make instantaneous jumps between a limited number of possible orientations ("jump models"). These possibilities are discussed more fully in the following section.

As is shown in most texts on NMR and several specialized discussions, the observable NMR relaxation parameters can be expressed in terms of the Fourier transform of $G(t)$ with respect to frequency.[1,2] This Fourier transform is called the spectral density $J(\omega)$ and describes the frequency spectrum of the motion. For the specific example of a pair of interacting proton and carbon spins, in an experiment in which the proton resonances are decoupled, the $^{13}C$ spin-lattice relaxation is exponential. The spin-lattice relaxation time $(T_1)$, transverse or spin–spin relaxation time $(T_2)$, and proton–carbon nuclear Overhauser enhancement (NOE) have the functional dependence on $J(\omega)$ given below:

$$1/T_1 = (1/10)(\gamma_C^2 \gamma_H^2 \hbar^2 / r_{CH}^6)[J(\omega_C - \omega_H) + 3J(\omega_C) + 6J(\omega_C + \omega_H)] \tag{1}$$

$$\frac{1}{T_2} = \frac{1}{20} \frac{\gamma_C^2 \gamma_H^2 \hbar^2}{r_{CH}^6} [J(\omega_C - \omega_H) + 3J(\omega_C) + 6J(\omega_C + \omega_H) + 4J(0) + 6J(\omega_H)] \tag{2}$$

$$\text{NOE} = 1 + \frac{\gamma_H}{\gamma_C} \left[ \frac{6J(\omega_C + \omega_H) - J(\omega_C - \omega_H)}{J(\omega_C - \omega_H) + 3J(\omega_C) + 6J(\omega_C + \omega_H)} \right] \tag{3}$$

As discussed in detail elsewhere,[3,4] relaxation generally reflects more than a single interaction. This can include the dipolar interaction among several nuclear spins, as well as contributions arising from other relaxation mechanisms. These interactions do not vary independently, but change in a correlated manner. Hence, a complete description of the relaxation

[3] R. L. Vold and R. R. Vold, *Prog. NMR Spectrosc.* **12**, 79 (1979).
[4] L. G. Werbelow and D. M. Grant, *Adv. Magn. Reson.* **9**, 189 (1977).

process requires that the correlated nature of the time-dependent fields produced by the different interactions also be considered. In treating such effects, consideration can be given to both the autocorrelation functions describing the time-dependent variations for specific interactions, as well as cross-correlation functions, which describe the correlated nature of these changes. Such considerations invariably lead to predictions of non-exponential relaxation, as well as to differential relaxation among the lines of spin-coupled multiplets.[3,4] Unfortunately, the sensitivity attainable in most studies of biological macromolecules is seldom sufficient to characterize fully such detailed relaxation effects (see, however, Ref. 5), and hence these are not considered further in the present chapter. Instead, observed relaxation rates are generally viewed as a sum of the rates for each of the possible relaxation interactions.

In addition to the dipolar mechanism discussed above, several other relaxation mechanisms have been shown to play a significant role in protein systems. The chemical shift anisotropy (CSA) mechanism arises as a consequence of the dependence of the chemical shift of a particular nucleus on molecular orientation in the applied field. For an axially symmetric chemical shift tensor, the spin-lattice and spin–spin relaxation times have the form[6]

$$1/T_1 = (2\gamma^2/15)B_0^2(\Delta\sigma)^2 J(\omega) \tag{4}$$
$$1/T_2 = (1/45)\gamma^2 B_0^2(\Delta\sigma)^2[3 J(\omega) + 4J(0)] \tag{5}$$

where $\Delta\sigma = (\sigma_\parallel - \sigma_\perp)$.

This mechanism has been shown to be significant both for unprotonated carbon nuclei[6] as well as for phosphorus nuclei, which are observed for phosphorylated protein residues or for prosthetic groups,[7] and for $^{19}F$ nuclei in studies involving fluorinated proteins.[8] Since the magnitude of this relaxation mechanism is field dependent, the use of higher field magnets will continue to increase the significance of this relaxation mechanism in future studies of the NMR relaxation of proteins. Several other mechanisms, of less general significance in protein studies, are discussed in Refs. 1 and 2.

[5] A. J. M. S. Uiterkamp, I. M. Armitage, J. H. Prestegard, J. Slmoski, and J. E. Coleman, *Biochemistry* **17**, 3730 (1978).
[6] R. S. Norton, A. O. Clouse, R. Addleman, and A. Allerhand, *J. Am. Chem. Soc.* **99**, 79 (1977).
[7] H. J. Vogel, W. A. Bridger, and B. D. Sykes, *Biochemistry* **21**, 1126 (1982).
[8] W. E. Hull and B. D. Sykes, *J. Mol. Biol.* **98**, 121 (1975).

## Proton Relaxation in Macromolecules

Although proton relaxation is generally dominated by the dipolar interaction with other protons in the molecule, the interpretation of such relaxation data differs considerably from that of carbon-13. This difference reflects primarily two factors. First, although the dipolar interaction between a carbon-13 nucleus and the directly bonded proton is sufficiently dominant so that the effects of more remote protons need not be considered, no analogous simplification is generally possible for protons. Thus, dipolar interactions with remote spins must be considered even for the case of methylene and methyl groups. For such interactions, both the orientation and length of the internuclear vector become time dependent, so that the relaxation data are not simply interpreted in terms of local reorientational behavior.[9] Second, in contrast to the case of carbon-13 in which the interacting proton spins are typically decoupled, proton NMR studies carried out on a coupled spin system give rise to significant cross-relaxation effects.[10] In macromolecules undergoing slow molecular tumbling, i.e., systems for which $\omega\tau_0 > 1$, where $\tau_0$ is the rotational correlation time for overall motion, the dipolar interaction leads to spin diffusion. In such cases, the spin-lattice relaxation times for the strongly dipolar coupled spins are equalized, and the actual rate is determined by the group with dynamic behavior which most effectively couples the spin system to the lattice. Additionally, cross-relaxation processes with the solvent water protons may also become significant, with the protein acting as a relaxation "sink" for the solvent.[11] Consequently, proton relaxation studies have generally been less useful for the characterization of local dynamic behavior in proteins. Of course, other types of $^1$H NMR data, particularly amide proton–deuteron exchange (cf. [22], by Roder in this volume), have provided considerable insight into the dynamic aspects of protein structure.

## Modeling Spectral Density

As discussed in the previous section, the observed relaxation parameters can be related to the spectral density of the motion. However, as is apparent from Eqs. (1)–(3), the relationships actually involve only the values of the spectral density determined at several discrete frequencies.

[9] J. Tropp, *J. Chem. Phys.* **72**, 6035 (1980).
[10] A. Kalk and H. J. C. Berendsen, *J. Magn. Reson.* **24**, 343 (1976).
[11] S. H. Koenig, R. G. Bryant, K. Hallenga, and G. S. Jacob, *Biochemistry* **17**, 4348 (1978).

Hence, the experimental studies do not provide a complete description of the spectral density function unless they are carried out as a function of frequency for a broad range of frequencies. Although several groups have carried out such frequency-dependent "NMR dispersion" measurements,[12,13] such studies have generally been done only on water, and the use of low fields for measurement has precluded obtaining relaxation data as a function of chemical shift, which is necessary to interpret the dynamics of a complex molecule. A solution to this limitation involving high-field polarization, followed by low-field relaxation and then high-field observation, by rapidly moving the sample along the magnet bore, has recently been proposed.[14] One approach which has been used in the interpretation of relaxation data involves the use of dynamic models. The simplest model, which assumes isotropic motion, leads to an exponential correlation function and to a Lorentzian spectral density:

$$J(\omega) = \tau/(1 + \omega^2\tau^2) \qquad (6)$$

Carbon-13 relaxation studies make it clear, however, that a range of spin-lattice relaxation times and NOE values can be obtained which indicate that no protein studied to date can be considered to behave as a rigid, isotropically diffusing molecule. For example, it is clearly necessary to take into account the internal diffusion of methyl groups to analyze the corresponding relaxation data. The motion of such groups may be simply modeled as a superposition of overall motion and internal diffusion about the methyl axis.[15,16] The corresponding spectral density has the form

$$J(\omega) = \sum_{i=0}^{2} B_{i0}\{(6D_0 + i^2D_1)^{-1}/[1 + \omega^2(6D_0 + i^2D_1)^{-2}]\} \qquad (7)$$

where $D_0$ and $D_1$ are the diffusion coefficients for the overall, isotropic motion and the internal motion, and the coefficients $B_{00} = \frac{1}{4}(3\cos^2\beta - 1)^2$, $B_{10} = 3\sin^2\beta\cos^2\beta$, $B_{20} = \frac{3}{4}\sin^4\beta$, where $\beta$ is the angle between the methyl axis and the C–H bond (70.5°). Equation (7) can be plugged into Eqs. (1)–(3) for dipolar relaxation, but can be used only with Eqs. (4) and (5) for the case of an axially symmetric chemical shift tensor. The more general case is discussed in Refs. 2 and 8.

[12] S. H. Koenig and W. E. Schillinger, *J. Biol. Chem.* **244**, 3283 (1969).
[13] R. D. Brown III, C. F. Brewer, and S. H. Koenig, *Biochemistry* **16**, 3883 (1977).
[14] D. J. Kerwood and P. H. Bolton, *J. Magn. Reson.* **68**, 588 (1986).
[15] D. E. Woessner, *J. Chem. Phys.* **36**, 1 (1962).
[16] R. E. London, in "Magnetic Resonance in Biology" (J. S. Cohen, ed.), Vol. 1, p. 69. Wiley, New York, 1980.

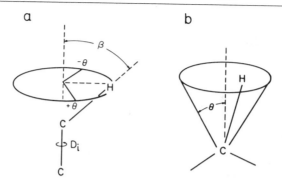

Fig. 2. Models for restricted internal motion. (a) Internal diffusion described using a diffusion equation subject to boundary conditions which limit the range of motion to $(+\theta, -\theta)$. (b) Diffusion within a cone with half-angle $\theta$. (Reprinted with permission from London,[16] p. 1.)

As may be directly inferred from crystallographic and other data, perhaps the most salient feature of the structure of globular proteins is the folding into preferred conformations. Hence, it may be anticipated that much of the internal motion, particularly for residues buried deep within the structure, will be characterized by a limited range of motion. This expectation is consistent with the observation that models for free diffusion, such as that used in treating the relaxation of methyl groups, do not generally provide reasonable predictions for the relaxation parameters of methine or methylene carbons. One approach to the problem of interpreting the relaxation data for proteins involves modeling the motion with a diffusion equation subject to boundary conditions which act to restrict the amplitude of internal motion. Two such types of calculations are illustrated in Fig. 2a, corresponding to motion about a carbon–carbon bond which is restricted to the range of azimuthal angles between $+\theta$ and $-\theta$, but with a fixed angle $\beta$, and Fig. 2b corresponding to "wobbling" motion restricted to fall within the boundaries of a cone of half-angle $\theta$.[16-19] Using such models, the corresponding spectral densities can be calculated and are given in Refs. 16–21. The results of a typical calculation corresponding to the model of Fig. 2a in which the spin-lattice relaxation time $T_1$ is plotted as a function of the half-angle of the range, $\theta$, for several different values of the overall (isotropic) rotational correlation time, are shown in

[17] R. J. Wittebort and A. Szabo, J. Chem. Phys. **69**, 1722 (1978).
[18] G. Lipari and A. Szabo, J. Chem. Phys. **75**, 2971 (1981).
[19] O. W. Howarth, J. Chem. Soc., Faraday Trans. 2 p. 863 (1979).
[20] R. E. London and J. Avitabile, J. Am. Chem. Soc. **100**, 7159 (1978).
[21] R. H. Newman, J. Magn. Reson. **47**, 138 (1982).

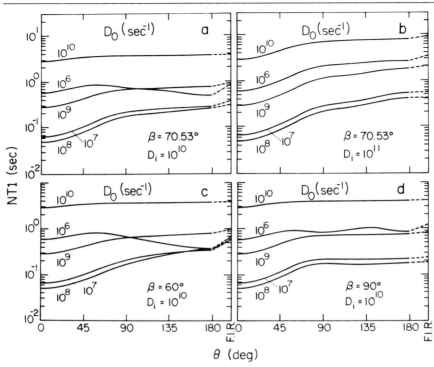

FIG. 3. Calculated $NT_1$ values ($N$ is the number of directly bonded protons) for carbon nuclei plotted as a function of $\theta$ for the model illustrated in Fig. 2a. Each curve corresponds to an overall (isotropic) diffusion coefficient $D_0$ indicated, and to the $D_i$ and $\beta$ values shown in the lower right-hand corners. The values corresponding to FIR correspond to the free internal diffusion model. (Reprinted with permission from London and Avitabile.[20])

Fig. 3. It is noted that increasing the value of $\theta$ to 180° does not correspond exactly to unrestricted internal diffusion, since motion between +180° and −180° is prohibited within the framework of the derivation.

An alternative formulation which has been of use for the characterization of the internal motion of protein systems involves discrete jumps between a small number of allowed orientations.[16,17,22–27] The validity of such an approach is supported by calculations such as those shown in Fig.

[22] R. E. London, *J. Am. Chem. Soc.* **100**, 2678 (1978).
[23] D. Wallach, *J. Chem. Phys.* **47**, 5258 (1967).
[24] H. Versmold, *J. Chem. Phys.* **58**, 5649 (1978).
[25] J. E. Anderson, *J. Magn. Reson.* **11**, 398 (1973).
[26] R. E. London and M. A. Phillippi, *J. Magn. Reson.* **45**, 476 (1981).
[27] R. E. London and J. Avitabile, *J. Am. Chem. Soc.* **99**, 7765 (1977).

3, which indicate that small, low-amplitude internal diffusion will have only a small effect on the relaxation, so that the comparatively rare but large-amplitude jump plays a more significant role. This situation is illustrated by the analysis of relaxation studies of the proline residue.[22] In this case, significant differences among the $NT_1$ values for the proline carbon nuclei have been interpreted to reflect the puckering of the five-membered pyrrolidine ring. Yet, the type of analysis shown in Fig. 3 would predict that such internal, low-amplitude motion might be insufficient to explain the observed $T_1$ data. This is indeed the case, and the relatively large differences in $NT_1$ can instead be interpreted on the basis of a bistable jump model in which only two "envelope" conformations (flap up and flap down) are allowed.

As is clear from the above discussion, the use of models implies that the interpretation of relaxation data in terms of molecular dynamics is not unique. Several groups have dealt with this problem by utilizing model-independent analyses. The simplest approach ignores geometrical considerations. A correlation time $\tau_0$ is calculated from the spin-lattice relaxation time determined for the most immobilized nuclei (typically the peptide $\alpha$ carbons) based on the use of an isotropic model formalism, and "effective" correlation times are similarly obtained for other carbons in the molecule using Eq. (6). Correlation times for the internal motion are then calculated from the relation[28]

$$1/\tau_i = (1/\tau_{\text{eff}}) - (1/\tau_0) \tag{8}$$

A somewhat more sophisticated approach developed by Jardetzky and co-workers utilizes a normalized sum of Lorentzian functions[29,30]:

$$J(\omega) = \sum_i A_i \tau_i/(1 + \omega^2 \tau_i^2): \quad \sum_i A_i = 1 \tag{9}$$

The model-free approach of Lipari and Szabo[31,32] is based on the introduction of a generalized order parameter $\langle S \rangle$ defined in Ref. 26, which reduces to the usual order parameter for axially symmetric motion:

$$\langle S^2 \rangle_{\text{axial symmetry}} = \langle P_2^2(\cos \theta) \rangle \tag{10}$$

with the spectral density having the form:

$$J(\omega) = [\langle S^2 \rangle \tau_0/(1 + \omega^2 \tau_0^2)] + \{[(1 - \langle S^2 \rangle)\tau]/(1 + \omega^2 \tau^2)\} \tag{11}$$

[28] D. A. Torchia and J. R. Lyerla, Jr., *Biopolymers* **13**, 97 (1974).
[29] R. King and O. Jardetzky, *Chem. Phys. Lett.* **55**, 15 (1978).
[30] R. King, R. Maas, M. Gassner, R. K. Nanda, W. W. Conover, and O. Jardetzky, *Biophys. J.* **6**, 103 (1978).
[31] G. Lipari and A. Szabo, *J. Am. Chem. Soc.* **104**, 4546 (1982).
[32] G. Lipari and A. Szabo, *J. Am. Chem. Soc.* **104**, 4559 (1982).

where $\tau_0$ is a correlation time corresponding to overall isotropic motion of the macromolecule, and $\tau^{-1} = \tau_0^{-1} + \tau_e^{-1}$, where $\tau_e$ is a correlation time corresponding to internal motion.

This is formally equivalent to utilizing two terms in the Jardetzky treatment although, as discussed by Lipari and Szabo, the interpretation of the analysis is somewhat different.[31,32] In the approach summarized by Eq. (9), a minimal number of terms is utilized to allow an adequate fit of the data, and the coefficients $A_i$ are interpreted to reflect the relative amplitude of dynamic processes characterized by correlation times $\tau$. However, a comparison of this formulation with the results of explicit models for restricted internal diffusion[16] indicates that the latter predict analogous expressions in which the $\tau_i$ are functions of both geometric and rate parameters. In the formulation summarized by Eqs. (10) and (11), only the parameters $S^2$, $\tau_0$, and $\tau$ are extracted from the data. Further interpretation of the values obtained for these parameters can then be achieved by directly calculating these parameters using particular dynamic models.

In general, the occasional misuse of models to describe situations in which the physical constraints make the analysis inapplicable, and the general tendency to overinterpret the results of relaxation studies, support this type of model-free analysis. Alternatively, the ultimate goal of these studies is the development of a physical characterization of the motion, rather than a mathematical fit of the data. It becomes tempting to mathematically fit the relaxation data and to subsequently invoke an attractive dynamic description without any mathematical connection to the data. Despite the number of parameters involved in some dynamic models, there are generally a sufficient number of constraints, particularly if $T_1$ and NOE data are obtained at several different frequencies, to rule out many possible models. For example, models allowing unrestricted diffusion about the bonds of constituent amino acids generally fail to adequately fit the data obtained for globular proteins. It is undoubtedly of value to carry out a series of analyses based on the use of a range of models to see whether these converge on a similar physical characterization of the motion. Although no fully comprehensive program for NMR relaxation times as a function of different dynamic models appears to be available, several are utilized in the program MOLDYN, available through the Quantum Chemistry Program Exchange, Indiana University, Bloomington, Indiana.

### Dependence of Chemical Shift on Molecular Conformation

The chemical shift of a particular nucleus in a flexible molecule is, in general, a function not only of the orientation of the molecule, but also of

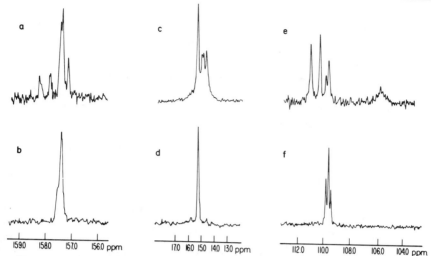

FIG. 4. Proton-decoupled $^{13}$C NMR spectra obtained at 25.2 MHz, for the native and 6 $M$ urea-denatured forms of *Streptococcus faecium* dihydrofolate reductase. (a) Guanido carbons of the eight arginine residues; (b) arginine residues after addition of urea; (c) methyl carbon resonances of the seven methionine residues of the enzyme; (d) methionine resonances after urea denaturation; (e) indole C-3 carbon resonances of the four tryptophan residues of the enzyme; (f) tryptophan resonances after urea denaturation. (Reprinted from London,[16] p. 1.)

the conformation. Thus, if one imagines a hypothetical experiment in which one rotates an amino acid side chain about a particular bond, the chemical shift for the nuclei of the amino acid may be expected to depend on a dihedral angle describing the rotation. However, as a consequence of the rapid diffusion which occurs for a molecule in solution near room temperature, only a single chemical shift is observed, which represents a weighted average of the shifts of all thermally accessible conformations. In globular proteins, the effects of conformation on chemical shift may be more readily observed.

Shown in Fig. 4[16,33–35] is a series of proton-decoupled $^{13}$C NMR spectra obtained for the enzyme dihydrofolate reductase ($M_r$ 20,000) derived from the bacterium *Streptococcus faecium,* which was grown on media containing either [*guanidino-*$^{13}$C]arginine (see Fig. 4a), [*methyl-*$^{13}$C]

[33] L. Cocco, R. L. Blakley, T. E. Walker, R. E. London, and N. A. Matwiyoff, *Biochemistry* **17,** 4285 (1978).

[34] R. L. Blakley, L. Cocco, R. E. London, T. E. Walker, and N. A. Matwiyoff, *Biochemistry* **17,** 2284 (1978).

[35] J. P. Groff, R. E. London, L. Cocco, and R. L. Blakley, *Biochemistry* **20,** 6169 (1981).

methionine (Fig. 4c), or [*indole*-3-$^{13}$C]tryptophan (Fig. 4e). The enzyme contains eight arginine residues, seven methionines, and four tryptophans. Although each of these amino acids would be expected to give rise to a single $^{13}$C resonance, incorporation into the enzyme leads to a range of shifts as a consequence of the differing local environments of the various residues. Of the eight arginine residues in the enzyme, five are unresolved and resonate at a position close to that of free arginine, while the remaining three are significantly shifted. For the four tryptophan residues in the enzyme, there are actually five resolvable resonances, as a consequence of a slow conformational exchange process which leads to a splitting of one of the resonances. This slow-exchange process has been proposed to correspond to cis–trans isomerism of one of the imide bonds involving a proline residue.[36] The resonances of the seven methionine residues in the enzyme are not well resolved at the field strength used in the study. Denaturation of the enzyme with 6 $M$ urea eliminates nearly all of the chemical shift heterogeneity observed for the arginine- and methionine-labeled enzyme. Some shift heterogeneity remains for the tryptophan-labeled enzyme, which may reflect the differences in the primary structure near each of the tryptophan residues, or, alternatively, the fact that the denatured enzyme is not truly a "random coil," but retains some residual structure. Hence, these data provide a clear illustration of the significance of higher order structure of the enzyme dihydrofolate reductase in determining the chemical shifts of the nuclei of individual residues. Furthermore, these contributions are not restricted to a particular residue type, but can occur for charged, aliphatic, or aromatic groups in the enzyme. In addition to the examples given above, chemical shift inequivalence for the two ortho or two meta protons of phenylalanine and tyrosine residues of globular proteins has frequently been observed.[16,37] Presumably, such shift effects arise from a combination of factors which include the ring currents produced by nearby aromatic side chains,[38] bond polarization due to nearby charges,[39] and changes in the solvation and electronic structure of the individual residues. Analogous observations on the dependence of the observed shift on conformation have been made for a number of other proteins.[16,37,40–43]

[36] R. E. London, J. P. Groff, and R. L. Blakley, *Biochem. Biophys. Res. Commun.* **86,** 779 (1979).
[37] I. D. Campbell, C. M. Dobson, and R. J. P. Williams, *Proc. R. Soc. London, Ser. B* **189,** 503 (1975).
[38] S. J. Perkins and R. A. Dwek, *Biochemistry* **19,** 245 (1980).
[39] J. G. Batchelor, *J. Am. Chem. Soc.* **97,** 3410 (1975).
[40] O. W. Howarth and D. M. J. Lilley, *Prog. NMR Spectrosc.* **12,** 1 (1978).
[41] A. Allerhand, *Acc. Chem. Res.* **11,** 469 (1978).

Conformational Exchange in Macromolecular Systems

As discussed in the previous section, the combination of the dependence of chemical shift on molecular conformation, which is presumably a general phenomenon, and the stabilization of particular folded conformation(s) of proteins, results in spectral characteristics which are unique to macromolecular systems. The branch of NMR which deals with effects due to the sampling by individual nuclei of sites with different chemical shifts and/or different coupling constants has generally come to be known as dynamic NMR (DNMR).[44] Although the corresponding physical processes are generally lumped under the term chemical exchange, it seems more appropriate to reserve such a description for situations, such as the exchange of amide protons with the solvent, in which spins physically migrate from one molecule to another, as stated, e.g., by Abragam.[1] In contrast, the sampling effect encountered in macromolecules could more appropriately be termed conformational exchange, where the term conformational emphasizes the physical nature of the effect, and the term exchange connotes processes treated by dynamic NMR and characterized by rate constants in the range $10^{-1}$ to $10^5$ sec$^{-1}$.

The analysis of the effects of conformational exchange in proteins differs from that which can be carried out on small molecules in two important respects: (1) the conformational equilibrium may involve a large number of states and (2) the range of physical parameters, particularly temperature, which can be studied is limited. Some aspects of the analysis which have been carried out are discussed below with illustrations from our studies of $^{13}$C-labeled dihydrofolate reductase.

The $^{13}$C NMR spectra of [guanidino-$^{13}$C]arginine-labeled bacterial *S. faecium* dihydrofolate reductase exhibit a marked temperature dependence[33,42] (Fig. 5). In particular, the arginine resonance furthest down field undergoes a chemical shift of 0.25 ppm over a 20° range. In contrast, the adjacent arginine resonance exhibits a negligible shift over this temperature range. Since the resonance is not observed to broaden significantly with temperature, it would appear that this temperature-dependent behavior reflects a fast-exchange situation in which the observed shift is a weighted average of the conformations involved. Since in this example and in a number of others as well, the high-temperature shift is toward the value for free arginine, the shift may reflect the increased probability at

[42] R. E. London, *in* "Topics in Carbon-13 NMR Spectroscopy" (G. C. Levy, ed.), p. 53. Wiley, New York, 1984.

[43] D. Hines and J. F. Chlebowski, *Biochem. Biophys. Res. Commun.* **144**, 375 (1981).

[44] J. Sandström, "Dynamic NMR Spectroscopy." Academic Press, New York, 1982.

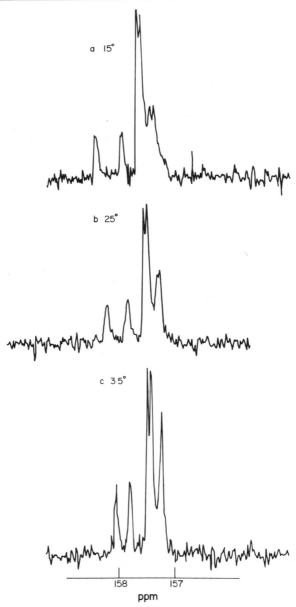

FIG. 5. Proton-decoupled $^{13}$C NMR spectra, obtained at 25.2 MHz, for the ternary complex of [*guanidino*-$^{13}$C]arginine-labeled dihydrofolate reductase–NADPH–methotrexate, obtained at the temperatures indicated. (Reprinted from London.[42])

high temperature of relaxed conformations in which the corresponding arginine residue has an environment more like that of a random coil. Such a transition could be entropically driven, so that at the higher temperature, the fractional population of such relaxed structures increases.

Temperature-dependent studies carried out on the [*indole*-3-$^{13}$C]tryptophan-labeled enzyme[36] (Fig. 6) indicate the existence of two different conformational exchange phenomena. Resonance 3 is split into two components, which presumably reflect a slow-exchange situation. Resonance 4 exhibits a significant increase in line width as well as a down-field shift as the temperature is increased. This exchange broadening is nearly eliminated upon the addition of an inhibitor such as 3′,5′-dichloromethotrexate which apparently stabilizes a particular conformation or set of conformations, altering the exchange behavior. The simplest model which can be used to analyze the temperature dependence of resonance 4 assumes that there are two classes of conformations, A and B, which are characterized by unique chemical shifts. A series of simulations can then be carried out for a series of values of the total chemical shift difference between the two classes of conformations, in order to relate the observed line width to the lifetime for one of the states (Fig. 7). A double-valued function is obtained with two values of $\tau_A$ corresponding to a given line width. These correspond to either the slow- or fast-exchange limit. Studies carried out at two frequencies can be used to resolve this uncertainty, and in the case of resonance 4, provide support for the interpretation of an intermediate-exchange situation yielding a frequency-dependent line width.[35]

Although in the above examples, particular resonances exhibit characteristics which seem to require a conformational exchange model for interpretation, such exchange effects are probably not restricted to these resonances. Thus, even tryptophan resonances 1 and 2 are found to exhibit line widths which are significantly broader than would be predicted by an isotropic diffusion model in which no internal motion is assumed. Conformational exchange processes may be responsible for this additional broadening. It is similarly clear, for example in the discussion given by Lipari and Szabo on the analysis of BPTI relaxation data,[32] that the attempt to include $T_2$ data in the analysis of molecular dynamics can lead to inconsistencies in some cases. Thus, the existence of conformational exchange processes in proteins and enzymes is of sufficient generality that equations such as (3) and (5) for $T_2$ will not generally provide a valid description of the relaxation. Some progress in sorting out such processes may result from $T_1\rho$ measurements[45] and from spin-echo studies using variable delay times.[3] As a consequence of the general significance of

[45] T. L. James, G. B. Matson, and I. D. Kuntz, *J. Am. Chem. Soc.* **100**, 3590 (1978).

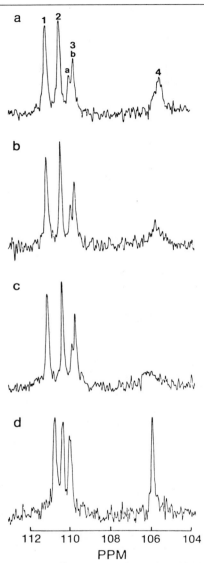

FIG. 6. Proton-decoupled $^{13}C$ NMR spectra, obtained at 25.2 MHz, of the indole C-3 resonances of [*indole*-3-$^{13}C$]tryptophan-labeled dihydrofolate reductase as a function of temperature: (a) 5°; (b) 15°; (c) 25°; (d) spectrum obtained at 15° after the addition of 3',5'-dichloromethotrexate. (Reprinted from London *et al.*[36])

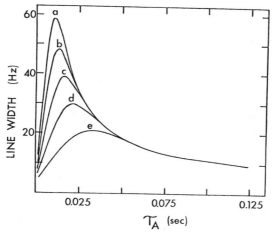

FIG. 7. Analysis of the line width of tryptophan 4 (Fig. 6) within the framework of a two-site exchange model. Line widths were measured from simulated spectra as $\tau_A$ was varied so that the exchange went from the fast to the slow limit. Line widths in the absence of exchange were set at 4 Hz, the fractional population of site A was set at 0.25, and the separation of the chemical shifts of states A and B was 150 Hz (a), 125 Hz (b), 100 Hz (c), 75 Hz (d), or 50 Hz (e). The variable shift differences used in the simulations can also be used to predict the frequency dependence of the line width. (Reprinted from Groff *et al.*[35])

conformational exchange processes in determining line widths of protein resonances, the use of higher field magnets will not always yield the desired improvement in resolution.

The distinction between conformational exchange processes considered in this section, and the internal jump or diffusion processes discussed earlier in the chapter, largely reflects differences in the rates characterizing these processes. Processes for which the rates of interconversion are similar to the chemical shift differences associated with the conformational change give rise to spectral effects described by dynamic NMR.[44] As the rate of interconversion increases, only a single resonance will be observed, with some additional broadening of the resonance reflecting the fact that the interconversion is not infinitely rapid. As the rate becomes similar to, or greater than the rate of overall molecular diffusion, the internal motion contributes to the relaxation parameters, as described earlier in the chapter. This unity is further emphasized by noting that at least one treatment views the bistable jump model as the limit of a rapid chemical exchange process.[46]

[46] H. Wennerström, *Mol. Phys.* **24**, 69 (1972).

## [19] Solid-State Deuterium Nuclear Magnetic Resonance Spectroscopy of Proteins

### By MAX A. KENIRY

Solid-state nuclear magnetic resonance cannot provide the elegant structural solutions that are now feasible for small water-soluble proteins via [1]H solution-state NMR, but it can contribute valuable information about restricted motion and, in favorable cases, orientation of the C–[2]H bond vector within membrane and structural proteins. Because solid-state NMR techniques take advantage of interactions that can be observed only when the macromolecule is undergoing restricted motion, naturally immobilized systems are studied in their native state, without resorting to artificial means such as detergent solubilization to obtain a high-resolution spectrum. Although the value of NMR for characterizing restricted motions in solids was recognized early,[1] the application to biological macromolecules had to await the advent of pulse Fourier transform techniques and parallel advances in instrumentation.

Deuterium offers several advantageous properties that make it particularly suitable as a solid-state NMR probe. The [2]H nucleus has spin $I = 1$, therefore the interaction between the nuclear quadrupole moment and the electric field gradient at the nucleus can be used to probe motional averaging and, in some cases, orientation of the C–[2]H bond vector. The field gradient tensor for most C–[2]H bonds is axially symmetric and the unique principal axis is along the C–[2]H bond direction.[2] Since the natural abundance of deuterium is only 0.016%, labeling is required, but when this is achieved the background is low and a certain degree of site selectivity is possible. Incorporation into a wide range of amino acids is feasible and, in general, labeling is easier and less expensive than with [13]C. The width of the [2]H quadrupolar splitting is sensitive to motions that have correlation times in the range $10^{-4}$–$10^{-7}$ sec and, finally, analysis of the relaxation is simple because it is dominated by the quadrupolar interaction.[3]

In this chapter we discuss the essential requirements to obtain successful high-resolution solid-state [2]H spectra. Next, we discuss the essential elements that are required to interpret [2]H spectra of oriented and

---

[1] G. E. Pake and H. S. Gutowsky, *J. Chem. Phys.* **16**, 1164 (1948).

[2] C. Brevard and J. P. Kintzinger, *in* "N.M.R. and the Periodic Table" (R. K. Harris and B. E. Mann, eds.), p. 119. Academic Press, New York, 1978.

[3] H. H. Mantsch, H. Saito, and I. C. P. Smith, *Prog. Nucl. Magn. Reson. Spectrosc.* **11**, 211 (1977).

unoriented systems. Then, we consider three different protypical systems that are accessible to analysis by solid-state ²H NMR. In this discussion we treat powder spectra which provide information about the nature, rate, and activation energies of restricted motion, oriented systems which provide spatial information relative to the laboratory frame and paramagnetic systems that have the added complication of large hyperfine shifts. Several reviews have appeared that treat in more depth many of the topics discussed below and the reader is referred to them for greater detail.[4–7]

Methodology

A source of specifically ²H-labeled protein, and instrumentation that is not normally available on commercial liquid-state spectrometers, are two essential requirements for the successful acquisition of broad-line ²H spectra of proteins. Labeled proteins are required because the low natural abundance of deuterium makes the collection of broad-line spectra of unlabeled macromolecules impractical. Special instrumentation and techniques are necessary because the spectra are broad, and consequently FID's decay rapidly, placing stringent requirements on the speed of data point collection and receiver electronics.

Facile interpretation of broad-line ²H spectra of proteins is only possible if labeled amino acids can be selectively incorporated into the proteins. This requirement severely restricts the number and types of system that may be studied. However, the systems that have been investigated are diverse and interesting. Oldfield et al.[8–10] and Griffin et al.[11] have studied the purple membrane, an easily isolated membrane fragment, which is produced in large quantities by *Halobacterium halobium*.[12] Tor-

[4] M. A. Keniry, R. L. Smith, H. S. Gutowsky, and E. Oldfield, *in* "Structure and Dynamics: Nucleic Acids and Proteins" (E. Clementi and R. Sarma, eds.), p. 435. Adenine Press, New York, 1983.

[5] R. L. Smith and E. Oldfield, *Science* **225**, 280 (1984).

[6] D. A. Torchia, *Annu. Rev. Biophys. Bioeng.* **13**, 125 (1984).

[7] M. Bloom, *in* "NMR in Biology and Medicine" (S. Chien and C. Ho, eds.), p. 19. Raven Press, New York, 1986.

[8] R. A. Kinsey, A. Kintanar, M.-D. Tsai, R. L. Smith, N. Janes, and E. Oldfield, *J. Biol. Chem.* **256**, 4146 (1981).

[9] M. A. Keniry, A. Kintanar, R. L. Smith, H. S. Gutowsky, and E. Oldfield, *Biochemistry* **23**, 288 (1984).

[10] M. A. Keniry, H. S. Gutowsky, and E. Oldfield, *Nature (London)* **307**, 383 (1984).

[11] B. A. Lewis, D. M. Rice, E. T. Olejniczak, S. K. Das Gupta, J. Herzfeld, and R. G. Griffin, *Biophys. J.* **45**, 213a (1984).

[12] B. M. Becher and J. Y. Cassim, *Prep. Biochem.* **5**, 161 (1975).

chia *et al.*[13–15] have incorporated labeled amino acids into cultured collagen and Opella[16] has performed similar experiments with bacteriophage. Seelig[17] has synthesized deuterated gramicidin and Bloom *et al.*[18] have studied a synthetic peptide in which the exchangeable protons have been replaced with deuterium. The most promising technique for the future, however, employs recombinant DNA technology. Ho *et al.*[19] have demonstrated the feasibility of this technology by incorporating $^{19}$F-labeled tryptophan into D-lactate dehydrogenase that was overproduced by *Escherichia coli.*

The availability of labeled amino acids is no longer a deterrent to this type of research. Many selectively labeled amino acids are commercially available and others are easily made from inexpensive precursors.[8,9] A word of caution: it should not be assumed that the label on a particular amino acid will remain in the same location or on the same amino acid after incorporation into the protein. Independent experiments should be performed to determine the metabolic fate of the label. For example, alanine and serine are metabolically labile amino acids in many organisms and labels on these amino acids have a high probability of being incorporated into phenylalanine and alanine, respectively.

All samples are thoroughly exchanged with deuterium-depleted water before measurement. Protein crystals are suspended in almost saturated ammonium sulfate buffer made up with deuterium-depleted water. Membrane samples are soft pellets formed by centrifugation from deuterium-depleted buffer. Sample volumes are typically 800 $\mu$l.

An excellent review of the instrumental and technical requirements for the successful performance of solid-state $^2$H NMR experiments with biological macromolecules has previously appeared in this series.[20] This chapter will treat only the more salient features of these requirements and the advances that have occurred since 1981.

The quadrupolar splitting of an immobile tryptophan in a membrane protein depends on orientation (*vide infra*) and is 139 kHz ($\theta = 90°$) or 278 kHz ($\theta = 0°$). To obtain a faithful representation of the spectrum, an

[13] L. W. Jelinski, C. E. Sullivan, L .S. Batchelder, and D. A. Torchia, *Biophys. J.* **10,** 515 (1980).
[14] L. W. Jelinski, C. E. Sullivan, and D. A. Torchia, *Nature (London)* **284,** 531 (1980).
[15] L. W. Jelinski and D. A. Torchia, *J. Mol. Biol.* **133,** 45 (1979).
[16] S. J. Opella, *Annu. Rev. Phys. Chem.* **33,** 533 (1982).
[17] J. Seelig, *Acc. Chem. Res.* **20,** 221 (1987).
[18] K. P. Pauls, A. L. MacKay, O. Soderman, M. Bloom, A. K. Taneja, and R. S. Hodges, *Eur. Biophys. J.* **12,** 1 (1985).
[19] C. Ho, G. S. Rule, and E. A. Pratt, *Biophys. J.* **49,** 113 (1986).
[20] R. G. Griffin, this series, Vol. 72, p. 108.

amplifier or a series of amplifiers with sufficient power output to excite uniformly over a 500-kHz spectral width and a data collection system capable of a point acquisition speed in excess of 1 $\mu$sec are required. Several amplifiers capable of such power output are commercially available, and some successful systems are described elsewhere.[13,21] A transient recorder is essential to digitize at sufficiently rapid speed; we and others have found the Nicolet Explorer II oscilloscope well suited for this purpose.[8,13] In the future, the adaption of versatile computer workstations to perform the operations now handled by dedicated data processors may remove the need for transient recorders.

It is universally accepted practice to acquire the quadrupolar echo rather than directly acquiring the FID.[22,23] A quadrupolar echo is induced by applying a second pulse that is phase shifted by $\pi/2$ a short time $\tau$ after the first pulse ($90_{\pm x}-\tau-90_y-\tau'$). The echo appears at time $2\tau$ after the first pulse. Sampling begins at some time $\tau' < \tau$. We typically employed 90° pulse widths in the range 2.5–3 $\mu$sec and $\tau$ values in the range 30–40 $\mu$sec. The FID is left shifted to the top of the echo, then apodized, zero filled (if desired), and Fourier transformed. Acquiring a quadrupolar echo alleviates the problems of finite receiver recovery time and probe ring down. This does not mean that every effort should not be made to remove these problems. Fukushima and Roeder[24] discuss in detail several practical ways to shorten receiver recovery and probe ring-down times. Pulse breakthrough and dc offset artifacts are also alleviated by proper phase cycling of the pulses.[20]

Spectral distortion will also occur because of effects caused by the quadrupolar interaction during the finite pulse length.[25] Until recently, the only solution to this problem was to make the pulses as short as possible by either increasing the transmitter power or by reducing the rotation angle of the pulse to less than 90°. Levitt et al.[26] introduced a composite pulse scheme that adequately excites a rigid deuterated lattice at relatively low radiofrequency (rf) field strengths but reduces the distortions in the line shape that arise from the effect of the quadrupolar interaction during a finite pulse time.

[21] T. M. Rothgeb and E. Oldfield, *J. Biol. Chem.* **256**, 1432 (1981).

[22] J. G. Powles and J. H. Strange, *Proc. Phys. Soc. London* **82**, 6 (1963).

[23] J. H. Davis, K. R. Jeffery, M. Bloom, M. I. Valic, and T. P. Higgs, *Chem. Phys. Lett.* **42**, 390 (1976).

[24] E. Fukushima and S. B. W. Roeder, "Experimental Pulse NMR: A Nuts and Bolts Approach." Addison-Wesley, Reading, Massachusetts, 1981.

[25] M. Bloom, J. H. Davis, and M. I. Valic, *Can. J. Phys.* **58**, 1510 (1980).

[26] M. H. Levitt, D. Suter, and R. R. Ernst, *J. Chem. Phys.* **80**, 3064 (1984).

Spectra of molecules undergoing anisotropic motion that have been acquired with any quadrupole echo sequence should be interpreted with care. When the rotational correlation time of the motion is comparable to the time $\tau$ between the pulses, line shapes obtained using the quadrupole echo sequence are severely distorted when compared to theoretical spectra that are calculated from a Fourier transform of the FID.[27] If a line shape is to be analyzed to obtain quantitative rate constants, the response to the solid echo should be simulated and compared to the experimental spectrum.

### Theoretical Background

In the case of $^2H$ ($I = 1$), where the quadrupolar interaction is dominant, there are three energy levels corresponding to $m = -1, 0, 1$; the energy of each level is given to first order by

$$E_m = \gamma \hbar B_0 m + (e^2 qQ/4)\{(3 \cos^2 \theta - 1)/[2(3m^2 - 1)]\} \tag{1}$$

where $eq$ is the field gradient, $eQ$ is the deuteron quadrupole moment, and $\theta$ is the angle between the principal axis of the electric field gradient tensor (to a good approximation the C–$^2H$ bond vector) and the external magnetic field $B_0$. The allowed transitions, given by the selection rule $\Delta m = \pm 1$, produce a frequency spacing of the absorption line, termed the quadrupolar splitting, $\Delta \nu_Q$, where

$$\Delta \nu_Q = (3/4)(e^2 qQ/h)(3 \cos^2 \theta - 1) \tag{2}$$

The above arguments apply only to a homogeneously oriented population of C–$^2H$ vectors (Fig. 1a). Experimentally, such populations are observed in samples of magnetically aligned proteins.[16,21]

In polycrystalline solids all orientations are possible and the sum of all the doublets results in a "powder pattern" (Fig. 1b) that has a singularity separation corresponding to $\theta = 90°$. Aliphatic compounds have a singularity separation of 127 kHz, and aromatic C–$^2H$ bonds have a separation of 138 kHz. Experimentally, such spectra are observed for deuterons on immobile amino acids in unoriented purple membrane.[8,28]

If there is molecular motion with a correlation time $\tau_C \leq \Delta \nu_Q$, only the average field gradient tensor is observed. The angular functions in Eqs. (1) and (2) are replaced by their averaged functions. Equation (2) becomes

$$\Delta \nu_Q = (3/4)(e^2 qQ/h)(3 \cos^2 \theta' - 1)\overline{(3 \cos^2 \beta - 1)} \tag{3}$$

[27] H. W. Speiss and H. Sillescu, J. Magn. Reson. 42, 381 (1981).
[28] R. A. Kinsey, A. Kintanar, and E. Oldfield, J. Biol. Chem. 256, 9028 (1981).

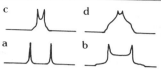

FIG. 1. Simulated spectra of (a) a homogeneously oriented population of C–²H bond vectors, (b) a polycrystalline sample with immobile C–²H bond vectors, (c) a polycrystalline sample with rotating methyl C–²H bond vectors, and (d) a polycrystalline sample with C–²H vectors undergoing rapid jumps between two equivalent sites.

where $\theta'$ is the angle between the axis of motional averaging and the magnetic field and $\beta$ is the angle between the instantaneous direction of the principal axis of the field gradient tensor and the axis of motional averaging (Fig. 2). Three-site methyl hops or rotation ($\beta = 70.5°$) cause a reduction in the quadrupolar splitting from 128 to $\approx 40$ kHz, but the spectra retain the appearance expected for an axially symmetric field gradient tensor (Fig. 1c). Any other motion or wobble that is superimposed on the methyl rotation will reduce the quadrupolar splitting even further and may induce a certain degree of asymmetry in the spectra.

Rapid two-site flips of aromatic rings are another, frequently observed reorientation of amino acid residues. In this case the axial symmetry of the spectrum is no longer retained; instead a spectrum similar to Fig. 1d is observed, in which the asymmetry parameter $\eta = 0.66$. Rotational diffusion can be distinguished from two-site jumps, since the former produces an axially symmetric spectrum which has a quadrupolar splitting of 17 kHz and the latter an axially asymmetric spectrum. If the correlation time $\tau \approx 1/\Delta\nu_Q$ (intermediate exchange) the spectral shape will deviate from that of Fig. 1d and the exact shape will be a function of the magnitude of $\tau$.[6,27]

Spectral shape cannot tell us all that we might want to know about C–²H bond reorientation. We must resort to relaxation parameters (specifically $T_1$) to extract the additional information. Fortunately, quadrupolar relaxation in restricted systems has been treated in detail for several models of molecular motion.[29] The capability of $T_1$ relaxation times to discriminate between motional models is exemplified by the orientation dependence of $T_1$ for jumping methyl deuterons as opposed to the orientation independence of $T_1$ for diffusing methyl deuterons.[29] A jump model cannot be distinguished from a diffusion model in the case of methyl deuterons simply by comparing the size of $\Delta\nu_Q$ or by the shape of the spectrum. Orientation-dependent $T_1$ values have been observed for the methyl deuterons of several amino acid residues of bacteriorhodopsin.[9]

[29] D. A. Torchia and A. Szabo, *J. Magn. Reson.* **49,** 107 (1982).

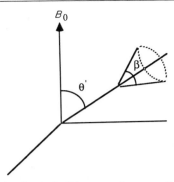

FIG. 2. Diagram illustrating the angle $\theta'$ between the magnetic field direction and the axis of motional averaging, and the angle $\beta$ between the instantaneous direction of the principal axis of the field gradient tensor and the axis of motional averaging.

### Protein Structure and Dynamics

Arguably, the most significant and complex processes in biochemistry occur in condensed phases. In the last decade, the first solid-state $^2$H NMR experiments were applied to proteins immobilized in these phases. The purpose of this chapter is not to treat in detail all this work but rather to discuss three very different prototypical systems, all of which have been studied extensively: oriented crystals of labeled myoglobin and hemoglobin, the purple membrane of *Halobacterium halobium,* and collagen.

### *Oriented Crystals of $^2$H-Labeled Myoglobin and Hemoglobin*

The application of $^2$H NMR to oriented crystals of proteins has not achieved the full potential of the technique. The obvious limitation is the small number of relevant systems that may be magnetically oriented. However, it is informative to consider the confirmatory work that has been published.

Small crystals of high-spin and low-spin ferric myoglobin orient readily in magnetic fields of relatively low magnitude.[21] In both cases the myoglobin was 50% labeled with deuterium at each of the methionine methyl groups, Met-55 and Met-131.[21] The $^2$H spectrum of either high-spin or low-spin myoglobin has two pairs of peaks with quadrupolar splittings that could be related to the orientation of the C–$^2$H bond vector with respect to the externally applied magnetic field. Addition of the stable organic free radical, TEMPAMINE, differentially broadened one pair of resonances which could then be assigned to the external methionine, Met-

55.[21] Since the orientation of the crystallographic axis with respect to the magnetic field could be visualized through a microscope, the orientation of the C–²H vectors with respect to the crystallographic axis could be determined and compared to the X-ray coordinates. In general, the agreement between the X-ray and NMR data is excellent. For example, a value of 17.5° for the angle between the $S^\delta$–$C^\varepsilon$ bond vector of Met-55 of aquoferrimyglobin and the crystallographic $c^*$ axis compares well with the X-ray value of 16.6°.[21] There are, however, two potential problems for an uncharacterized system. First, each quadrupolar splitting will in general yield two possible angles $\theta$, of which only one is correct, and second, librational motions and static disorder will introduce an element of uncertainty into the calculation of $\theta$.

Analysis of the ²H spectra of deuterons near a paramagnetic entity presents special problems. Oriented hemoglobin crystals in which the heme group was replaced with one of three different specifically deuterated hemes were studied by the magnetic ordering technique.[30] Spectra of deuterons close to the heme center were no longer symmetrically located about the Larmor frequency because of the large hyperfine shifts of these resonances.[30] Extrapolation to 0 K on a resonance position-vs-temperature plot does yield symmetric resonance locations at 0 K and aids in the assignment of each peak. Hyperfine shifts of the meso deuterons (4.5 Å from the $Fe^{2+}$ center) are ≈27.5 kHz, an order of magnitude larger than the solution-state hyperfine shifts of hemoglobin. In contrast, the 6,7-$b$ methylene protons (>9 Å from the $Fe^{2+}$ center) exhibit no observable hyperfine shift. In addition, the spectra unequivocally demonstrated the immobility of the methylene protons. This latter point illustrates the unique capacity of ²H NMR to simultaneously yield orientation and dynamic information. Despite uncertainties in the absolute orientation of the deuterons, agreement between the NMR data and the X-ray coordinates is excellent.[30]

## Purple Membrane of Halobacterium halobium

The plasma membrane of *Halobacterium halobium* contains extensive purple patches that may be isolated intact. The patches contain a single polypeptide, bacteriorhodopsin ($M_r \approx 26,000$) which is organized as seven approximately parallel helices. The first membrane protein to be characterized by electron microscopy,[31] it is also the most extensively studied by solid-state NMR spectroscopy.[8–11,28]

[30] R. W. K. Lee and E. Oldfield, *J. Biol. Chem.* **257**, 5023 (1982).
[31] R. Henderson and P. N. T. Unwin, *Nature (London)* **257**, 28 (1975).

The spectra of all the deuterated amino acids incorporated into bacteriorhodopsin contain at least two diverse components: a narrow component that arises from highly mobile residues in the nonhelical regions of the protein and a set of one or more broader components that are identified with residues that are immobile or undergoing restricted motion within the helical regions of the protein.[10] The first part of the discussion will focus on the latter components and the second will treat the narrow component.

The immobility of the helical regions of the backbone of bacteriorhodopsin is best exemplified by the almost "rigid" splitting of $\approx 126$ kHz for [$^2H_2$]glycine and [$\alpha$-$^2H_1$]valine incorporated into bacteriorhodopsin. Any fast librational motion of the backbone must be occurring over a small solid angle ($\pm 5$–$10°$). This means discrete side-chain motion in bacteriorhodopsin can be investigated independent of consideration of the motion of the backbone.

The mobility of the side chains will be dependent on many factors, including the degree of packing of the macromolecule, the extent of breathing motions of the protein and therefore its flexibility, the ability of the side-chain functional group to hydrogen bond to other amino acid residues, and the degrees of freedom available to the side-chain motion.

The environment around the side chains is one of compactness without causing total immobility. All methyl groups are free to rotate or at least jump between the three equivalent sites.[8,9] However, the valine, leucine, threonine, serine, tyrosine, tryptophan, and phenylalanine $C_\alpha$–$C_\beta$ bonds are not free to rotate.[8,9,28] The small aromatic rings of tyrosine and phenylalanine undergo two site flips about the $C_\beta$–$C_\gamma$ bond at the optimum growth temperature ($37°$) but the larger tryptophan ring is immobile under the same conditions.[8,28] This differential mobility of the aromatic rings is no doubt due to breathing motions of the protein and is a manifestation of the limited flexibility of the helical regions of the protein.

Hydroxyamino acids in the transmembrane pore of a protein such as bacteriorhodopsin are expected to have an important role in light-mediated proton transport. Almost all the internal threonine and serine side chains are immobilized within bacteriorhodopsin, which was in a resting state during these experiments. Breaking and reforming of these hydrogen bonds during light excitation may be an important pathway of proton transport.

Spectra of [$\gamma$-$^2H_6$]valine, [$\delta$-$^2H_3$]leucine, and [$\alpha,\beta,\gamma,\gamma',\delta$-$^2H_{10}$]isoleucine incorporated into bacteriorhodopsin cannot be simulated with spectra that assume $\Delta \nu_Q = 40$ kHz and $\eta = 0$. Instead, models that assume limited jump motions at bonds that separate the backbone from the methyl group produce better simulations. As the length of the amino acid side chain increases, the better it is able to accommodate the limited nonsymmetric motion that occurs by concerted movement about two or more intervening bonds.

The foregoing discussion gives the impression of a restricted environment for the amino acid side chains. However, this is not the case for the population of amino acid residues in the less structured regions of the protein. We found an excellent correlation between the intensity of the narrow central component of each spectrum and the percentage of that particular amino acid that resides at the surface of the protein and is thus exposed to a fluid aqueous environment.[10] The narrow component is diminished by chemical cross-linking or, alternatively, by enzymatically removing the C-terminal fragment (J. Bowers, M. A. Keniry, and E. Oldfield, unpublished results). Together, these results strongly suggest the surface amino acids are highly mobile on the $^2$H time scale ($>10^4$–$10^5$ sec$^{-1}$).

### Collagen

Collagen, a fibrous protein that accounts for one-third of all protein in humans, is the most important component of connective tissue. The basic subunit of collagen is a triple helix of polypeptide chains that is organized into a variety of macromolecular structures that have a broad range of biomechanical properties. Torchia *et al.* have studied extensively reconstituted chick calvaria collagen by $^2$H NMR.[6,13–15,32] These data can be divided into structural and dynamic aspects.

Unlike many proteins, collagen has a high degree of repetitive order which Jelinski *et al.*[13] utilized to determine an angle of 70° between the $C_\alpha$–$C_\beta$ bond of alanine and the long axis of the helix of [$^2$H$_3$]alanine-labeled collagen. They were also able to show that the fibers reorient about the long helical axis in a cone that has an average solid angle of approximately 15°.[13] The reorientation of the backbone of collagen is significantly larger than the backbone of bacteriorhodopsin, the latter being considered more prototypical of the majority of membrane proteins.

Collagen resembles bacteriorhodopsin very closely in the nature of the reorientation of amino acid side chains. Jump models simulate spectra better than the assumption of free rotation.[6] Spectra of [$^2$H$_{10}$]leucine-labeled collagen are best simulated by a two-site exchange model (assuming a rate ca. $8 \times 10^{-7}$).[6] Indeed, the movement of the backbone itself occurs by a jump process. Proline, an amino acid not studied in any other system, and an essential element of the collagen structural subunit, expe-

[32] S. K. Sarkar, Y. Hiyama, C. H. Niu, P. E. Young, J. T. Gerig, and D. A. Torchia, *Biochemistry* **26**, 6993 (1987).

riences motion through a root-mean-square (rms) angle of 25–30°.[6] Most recently, Torchia et al.[32] have shown there are two approximately equal populations of proline residues that differ only in their amplitude of motion. In this same study, deposition of mineral was shown to have a marked effect on the many-bond concerted motions of the backbone but very little effect on one or two bond motions of the amino acid side chains.[32]

### Other Studies

In a brief chapter such as this, the whole field cannot be adequately reviewed but some completed work and works in progress are worthy of mention. Opella[16] has selectively deuterated and magnetically aligned phage fd, although in this case the most extensive and detailed work has been performed with nuclei other than deuterium. Seelig[17] has chemically synthesized gramicidin selectively deuterated in the tryptophan rings and is beginning a systematic study of protein dynamics in lipid bilayers of well-defined composition. Pauls et al.[18] have demonstrated the feasibility of obtaining $^2H$ spectra of the exchangeable deuterons of small peptides in $^2H_2O$.

## [20] Protein Rotational Correlation Times by Carbon-13 Rotating-Frame Spin-Lattice Relaxation in Presence of Off-Resonance Radiofrequency Field

By Thomas Schleich, Courtney F. Morgan, and G. Herbert Caines

### Introduction

Insights into the structure and interaction of proteins in solution is afforded by the study of rotational diffusion. When combined with translational diffusion information the hydrodynamically equivalent shape of the diffusing particle may be determined. The rotational motion of a macromolecule is dependent on particle size, shape, and the location of the rotary diffusion axes, as well as solvent viscosity and temperature. An assortment of techniques is available for the evaluation of macromolecular rotational diffusion.[1] These include linear dichroism and birefringence,

[1] C. R. Cantor and P. R. Schimmel, "Biophysical Chemistry," Part II, p. 659. Freeman, San Francisco, California, 1980.

dielectric dispersion, fluorescence depolarization, depolarized quasi-elastic light scattering, electron spin resonance, and nuclear magnetic resonance relaxation. With the exception of NMR relaxation and depolarized light scattering, these techniques usually require either the covalent attachment of a spectroscopic "reporter" moiety to the macromolecule and/or the utilization of an external perturbation to achieve macromolecular orientation. For the evaluation of protein rotational motion in intact biological systems noninvasive techniques must be employed, thus eliminating the application of those methods which depend on the use of a covalently attached extrinsic "reporter" group. NMR spectroscopy, and in the case of a transparent tissue such as the lens of the eye, light scattering, are examples of two techniques appropriate for the noninvasive assessment of protein rotational dynamics in intact tissue.

Determination of the protein rotational correlation time ($\tau_0$) by NMR spectroscopy has been accomplished by the measurement of the field dependence of the water proton spin-lattice ($T_1$) relaxation time (NMR dispersion) of aqueous protein solutions,[2-4] the $T_1$ and, in limited instances, nuclear Overhauser enhancement (NOE) of polypeptide backbone $^{13}C$ $\alpha$-carbon amino acid residue resonances,[5] and by $^{13}C$ rotating-frame spin-lattice relaxation in the presence of an off-resonance radiofrequency (rf) field of the backbone peptide bond $^{13}C$ carbonyl resonances.[6,7]

The $^{13}C$ off-resonance rotating-frame spin-lattice relaxation technique introduced by James and colleagues[6-9] for the determination of protein rotational correlation times within the motional time window of $10^{-9}$ to $10^{-6}$ sec, i.e., longer than the inverse of the Larmor precessional frequency, offers a distinct advantage in that the occurrence of rotational diffusion can be expressed by the ratio of resonance intensities established in the presence and absence of an applied off-resonance rf field, thus obviating the time-consuming collection of magnetization intensities at timed intervals during the course of a relaxation experiment.

[2] S. H. Koenig and W. E. Schillinger, *J. Biol. Chem.* **244**, 3283 (1969).
[3] K. Hallenga and S. H. Koenig, *Biochemistry* **15**, 4255 (1976).
[4] S. H. Koenig, R. G. Bryant, K. Hallenga, and G. S. Jacob, *Biochemistry* **17**, 4348 (1978).
[5] D. J. Wilbur, R. S. Norton, A. O. Clouse, R. Addleman, and A. Allerhand, *J. Am. Chem. Soc.* **98**, 8250 (1976).
[6] T. L. James, G. B. Matson, and I. D. Kuntz, *J. Am. Chem. Soc.* **100**, 3590 (1978).
[7] T. L. James, R. Mathews, and G. B. Matson, *Biopolymers* **18**, 1763 (1979).
[8] T. L. James, G. B. Matson, I. D. Kuntz, R. W. Fisher, and D. H. Buttlaire, *J. Magn. Reson.* **28**, 417 (1977).
[9] T. L. James and G. B. Matson, *J. Magn. Reson.* **33**, 345 (1979).

Theoretical Background

The theoretical basis for the study of isotropic intermediate molecular motions by off-resonance rotating-frame spin-lattice relaxation has been described by James and co-workers.[6,8–10] The determination of $\tau_0$ for a tumbling protein is obtained by an assessment of the $^{13}C$ rotating-frame spin-lattice relaxation characteristics in the presence of an applied off-resonance rf field of the backbone $^{13}C$ carbonyl resonances.[6,7] The $^{13}C$ $\alpha$-carbon relaxation characteristics of the polypeptide chain backbone may also be utilized. The rationale for this selection is based on the assumption that polypeptide backbone motional dynamics are expected to mirror overall particle tumbling (rotational motion). This assumption has been confirmed experimentally for a globular protein,[11] muscle calcium-binding protein from carp. In addition, $\tau_0$ values determined by off-resonance rotating-frame spin-lattice relaxation are in excellent agreement with values determined by other techniques.[6] The theoretical basis for $^{13}C$ spin-lattice relaxation, in the presence and absence of an off-resonance rf field for both isotropic and anisotropic tumbling, arising from dipole–dipole (DD) coupling to protons, and chemical shift anisotropy (CSA) of the peptide bond carbonyl carbon, is described in this section. Inclusion of the latter relaxation mechanism is an extension of the original James formalism for proteins.[6] CSA relaxation was not included in the original formalism because the relaxation behavior was evaluated at relatively low magnetic field strength ($B_0 = 2.35$ T), as opposed to measurements made more recently at significantly higher values of $B_0$, where CSA relaxation is expected to be dominant for carbonyl carbons.[12] For the $^{13}C$ $\alpha$-carbon only contributions from dipolar relaxation need to be considered.

*Off-Resonance Rotating-Frame Experiment*

For a coordinate system rotating with angular velocity, $\omega$, the effective field, $\mathbf{B}_e$, in the rotating coordinate system is given by

$$\mathbf{B}_e = \omega_e/\gamma_C = [B_0 - \omega/\gamma_C]\mathbf{k} + B_1\mathbf{i} \tag{1}$$

where $\omega_e$ is the angular precessional frequency about the effective field for an ensemble of $^{13}C$ noninteracting spins with gyromagnetic ratio, $\gamma_C$. $B_0$ is the static magnetic field strength, $\mathbf{i}$ and $\mathbf{k}$ are unit vectors, and $B_1$ is an rf

[10] T. L. James, *in* "Phosphorus-31 NMR: Principles and Applications" (D. G. Gorenstein, ed.), p. 349. Academic Press, New York, 1984.

[11] D. R. Bauer, S. J. Opella, D. J. Nelson, and R. Pecora, *J. Am. Chem. Soc.* **97**, 2580 (1975).

[12] R. S. Norton, A. O. Clouse, R. Addleman, and A. Allerhand, *J. Am. Chem. Soc.* **99**, 79 (1977).

field off resonance by an amount $\Delta$:

$$\Delta = 2\pi\nu_{off} = -\gamma_C[B_0 - (\omega/\gamma_C)] \tag{2}$$

where $\nu_{off}$ is the difference in frequency of the applied off-resonance rf field from the resonance of interest. The effective field vector, $\mathbf{B}_e$, is inclined to the static magnetic field, $\mathbf{B}_0$, at an angle $\Theta$ defined by

$$\Theta = \tan^{-1}(\gamma_C B_1/\Delta) \tag{3}$$

The angular precessional frequency about the effective field vector is

$$\omega_e = \Delta/\cos\Theta \tag{4}$$

Assuming exponential relaxation of the magnetization aligned along $\mathbf{B}_e$, in the high-temperature limit $[kT \gg \gamma(h/2\pi)B_0]$, with $B_0 \gg \mathbf{B}_e$, the signal intensity ratio, $R = M_{eff}/M_0$, can be shown to be equal to $T_{1\rho}^{off}/T_1$.[13] $M_{eff}$ and $M_0$ are the respective steady state magnetizations along $\mathbf{B}_e$ and $\mathbf{B}_0$. $T_{1\rho}^{off}$ is the spin-lattice relaxation time for the decay (or growth) of magnetization along the effective field in the off-resonance experiment. $M_{eff}$ is obtained in the presence of, and $M_0$ is obtained in the absence of, the off-resonance rf irradiation field.

## Relaxation Expressions

*Random Isotropic Tumbling.* For the case of random isotropic molecular reorientation the following expressions for an individual $^{13}C$ spin in the absence of cross-relaxation may be written[6,10]:

$$\left(\frac{1}{T_1}\right)_{DD} = K \sum_1^N \left[J_0(\omega_H - \omega_C) + 3J_1(\omega_C) + 6J_2(\omega_H + \omega_C)\right] \tag{5}$$

$$\left(\frac{1}{T_{1\rho}^{off}}\right)_{DD} = \left(\frac{1}{T_1}\right)_{DD} + K\sin^2\Theta \sum_1^N \{2J_0(\omega_e)$$
$$+ \tfrac{3}{2}[J_1(\omega_H + \omega_e)] + \tfrac{3}{2}[J_1(\omega_H - \omega_e)]\} \tag{6}$$

where $K = (h/2\pi)^2\gamma_H^2\gamma_C^2/20r^6$. $J_n(\omega)$ is the spectral density function in which motional dynamics information is embedded, $\gamma_H$ and $\gamma_C$ are the gyromagnetic ratios for hydrogen and carbon, $\omega_H$ and $\omega_C$ are the respective angular Larmor precessional frequencies, $r$ is the distance between the carbon and the proton responsible for relaxation, and $N$ is the number of protons responsible for dipole–dipole relaxation. The summation over $N$ is necessary because of the possibility that more than one proton spin

[13] J. F. Jacquinot and M. Goldman, *Phys. Rev. B: Solid State* **8**, 1944 (1973).

may be responsible for relaxation, and that different spectral densities may be operant for each of the interactions. Analogous expressions may be written for nuclei other than the proton and carbon engaged in dipole–dipole relaxation (e.g., $^{13}C-^{14}N$).

CSA relaxation is dominant for the protein peptide carbonyl carbon at high magnetic field strength.[12] The relevant expressions are[10,14]

$$\left(\frac{1}{T_l}\right)_{CSA} = \tfrac{6}{40}(\gamma_C^2 B_0^2 \delta_Z^2) \sum_{j=0}^{2} J_j(\omega) \tag{7}$$

$$\left(\frac{1}{T_{1\rho}^{off}}\right)_{CSA} = \tfrac{1}{40}(\gamma_C^2 B_0^2 \delta_Z^2) \sum_{j=0}^{2} [6 J_j(\omega) + 4 \sin^2 \Theta \, J_j(\omega_e)] \tag{8}$$

where $\delta_Z$ is the principal component of the chemical shift tensor. The total relaxation rate for both relaxation mechanisms is

$$1/T_1 = (1/T_1)_{DD} + (1/T_1)_{CSA} \tag{9a}$$

and

$$\frac{1}{T_{1\rho}^{off}} = (1/T_{1\rho}^{off})_{DD} + (1/T_{1\rho}^{off})_{CSA} \tag{9b}$$

*Random Anisotropic Tumbling.* Anisotropic molecular reorientation is encountered with particles whose shape is represented by a hydrodynamic ellipsoid of revolution. The expressions appropriate for isotropic tumbling require modification to take into account differential motion about the major and minor axes of the ellipsoid. For an axially symmetric ellipsoid, with major semiaxis $a$, and minor semiaxis $b$, the following expressions for dipole–dipole relaxation are appropriate[5,10,15,16]:

$$(1/T_1)_{DD} = K[J_0'(\omega_H - \omega_C) + 3 J_1'(\omega_C) + 6 J_2'(\omega_H + \omega_C)] \tag{10}$$

$$(1/T_{1\rho}^{off})_{DD} = K\{\sin^2 \Theta[ 2J_0'(\omega_e) + \tfrac{3}{2} J_1'(\omega_H + \omega_e) \\ + \tfrac{3}{2} J_1'(\omega_H - \omega_e)]\} + (1/T_1)_{DD} \tag{11}$$

where $K$ is a constant identical to the one for dipolar relaxation of an isotropically tumbling particle, and $J_n'(\omega)$ is the spectral density function which differs from the isotropic case.

For CSA relaxation, the following expressions[10,14,15] hold:

$$(1/T_1)_{CSA} = \tfrac{3}{20}\gamma_C^2 B_0^2 \delta_Z^2[J_0(\omega_C) + J_1(\omega_C) + J_2(\omega_C)] \tag{12}$$

[14] P. H. Bolton, P. A. Mirau, R. H. Shafer, and T. L. James, *Biopolymers* **20**, 435 (1981).
[15] C. F. Morgan, T. Schleich, G. H. Caines, and D. Michael, *Biopolymers,* in press.
[16] D. E. Woessner, *J. Chem. Phys.* **37**, 647 (1962).

$$(1/T_{1\rho}^{\text{off}})_{\text{CSA}} = \tfrac{1}{10}\gamma_C^2 B_0^2 \delta_Z^2 \sin^2 \Theta[J_0(\omega_e) + J_1(\omega_e) + J_2(\omega_e)] + (1/T_1)_{\text{CSA}}$$

$$(13)$$

### Spectral Density Functions

*Random Isotropic Tumbling.* The rate of reorientational motion responsible for magnetic relaxation is characterized by the correlation time. For an isolated (i.e., at infinite dilution) uncharged spherical rigid rotor engaged in random isotropic motion the rotational correlation time is given by

$$\tau_0 = 4\pi a^3 \eta_0 / 3k_B T \tag{14}$$

where $a$ is the radius of the spherical particle, $\eta_0$ the solution viscosity in poise, $k$ the Boltzmann constant, and $T$ the absolute temperature. [The rotational (orientational) relaxation time, $\tau_r$, equals $3\tau_0$.] The rotational diffusion coefficient, $D_r$, is given by

$$D_r = 1/6\tau_0 \tag{15}$$

As is the case for translational diffusion, nonideal effects influence the rotational diffusion coefficient of macromolecules. While an exact treatment describing the magnitude of nonideal effects on rotational motion has not been devised, nonideal effects are not anticipated to be very severe, except for rod-shaped particles at high concentration. Considerations of excluded volume and hydrodynamic interaction are relevant. In the case of macroions, electrostatic interactions arising from particle charge become significant; thus solution pH and ionic strength are important variables. Contributions from Donnan effects may be present, but these are minimized at high ionic strength.

The spectral density function appropriate for dipolar relaxation is[6,9,10,14,17]

$$J_n(\omega) = 2\tau_0/(1 + \omega^2\tau_0^2) \tag{16}$$

where $\tau_0$ is the rotational correlation time for random isotropic tumbling. For CSA-induced relaxation the relevant expression for the spectral density function is[10,14,15]

$$J_j(\omega) = 2c_j\tau_j/(1 + \omega^2\tau_j^2) \tag{17}$$

where, for isotropic motion, $\tau_j = \tau_0$ for $j = 0, 1, 2$; $c_0 = 1$, $c_1 = 0$, and $c_2 = \eta^2/3$, where $\eta$ is the asymmetry parameter characterizing the CSA relaxation.

[17] W. E. Hull and B. D. Sykes, *J. Mol. Biol.* **98**, 121 (1975).

*Random Anisotropic Tumbling.* For an axially symmetric ellipsoid of revolution engaged in reorientational motion, $\tau_\perp$ is defined to be the relaxation time for the motion of semiaxis $a$ about semiaxis $b$, and $\tau_\parallel$ for the motion of semiaxis $b$ about semiaxis $a$. The following relations hold[5,10,16,18]:

$$\tau_A = \tau_\perp \tag{18a}$$
$$\tau_B = 6\tau_\perp\tau_\parallel(\tau_\perp + 5\tau_\parallel)^{-1} \tag{18b}$$
$$\tau_C = 3\tau_\perp\tau_\parallel(2\tau_\perp + \tau_\parallel)^{-1} \tag{18c}$$

The spectral density function for dipolar relaxation is[5,10,18]

$$J'_n(\omega) = A\,\frac{2\tau_A}{(1 + \omega^2\tau_A^2)} + B\,\frac{2\tau_B}{(1 + \omega^2\tau_B^2)} + C\,\frac{2\tau_C}{(1 + \omega^2\tau_C^2)} \tag{19}$$

where $A = (1/4)(3\cos^2\alpha - 1)^2$, $B = (3/4)(\sin^2 2\alpha)$, and $C = (3/4)\sin^4\alpha$. The angle $\alpha$ is defined by the orientation of the C–H internuclear vector relative to the unique symmetry axis of the diffusing particle. The coefficients $A$, $B$, and $C$ sum to unity at each value of $\alpha$.

For CSA relaxation $J_j(\omega)$ is identical to the isotropic case [see Eq. (17)], except that the correlation times $\tau_0 = \tau_A$, $\tau_1 = \tau_B$, $\tau_2 = \tau_C$ [see Eq. (18)] and the $c_j$ coefficients are different[10,14]:

$$c_0 = \tfrac{1}{4}[(3\cos^2\beta - 1) + \eta\sin^2\beta\cos 2\gamma]^2 \tag{20a}$$

$$c_1 = \tfrac{1}{3}\sin^2\beta[\cos^2\beta(3 - \eta\cos 2\gamma)^2 + \eta^2\sin^2 2\gamma] \tag{20b}$$

$$c_2 = [\sqrt{3/4}\,\sin^2\beta + [\eta/(2\sqrt{3})](1 + \cos^2\beta)\cos 2\gamma]^2$$
$$+ (\eta^2/3)\sin^2 2\gamma\cos^2\beta \tag{20c}$$

where $\beta$ and $\gamma$ are the Euler angles relating the rotation of the principal axis system for rotational diffusion into the principal axes of the peptide bond carbonyl chemical shift tensor (see below). The $c_j$ coefficients sum to $1 + \eta^2/3$ for each combination of the Euler angles $\beta$ and $\gamma$.

### Evaluation of Euler Angles $\beta$ and $\gamma$ and Internuclear Vector Angle $\alpha$

The analysis of the CSA relaxation contribution to the spin-lattice relaxation of a protein undergoing anisotropic tumbling is based upon the orientation of the carbonyl chemical shift tensor reported for the peptide bond of the model compound [1-$^{13}$C]glycyl[$^{15}$N]glycine · HCl · H$_2$O[19] relative to the unique diffusion axis. The required parameters are the Euler angles $\beta$ and $\gamma$ relating the CSA principal axis system of the $i$th residue

[18] F. Heatley, *Annu. Rep. NMR Spectrosc.* **17**, 179 (1986).
[19] R. E. Stark, L. W. Jelinski, D. J. Ruben, D. A. Torchia, and R. G. Griffin, *J. Magn. Reson.* **55**, 266 (1983).

peptide bond carbonyl carbon to the axially symmetric diffusion axis system, and the angle $\alpha$ defined by the major diffusion axis ($D_z$) and the C–H$_\alpha$ internuclear vector.[15,17,20] The details for the evaluation of these angles for each amino acid residue present in the protein structure (e.g., the Brookhaven Data Base) using protein X-ray structural information is summarized in Morgan et al.[15] and in Hull and Sykes.[17]

### Evaluation of $\delta_Z$ and Asymmetry Parameter $\eta$

The evaluation of $\delta_Z$ and $\eta$ from the principal values of the peptide bond carbonyl chemical shift tensor for [1-$^{13}$C]glycyl[$^{15}$N]glycine · HCl · H$_2$O[19] follows customary procedures.[17] The assumption is made that each peptide bond $i$ is characterized by the same value of the CSA tensor. Values of $-81.8$ ppm and $-0.82$ for $\delta_Z$ and $\eta$, respectively, are used in the computer simulations (see below).

### Simulation of the Off-Resonance Rotating-Frame Experiment

The off-resonance rotating-frame experiment is experimentally simple, entailing measurement of the spectral peak area in the absence ($\propto M_0$) and in the presence ($\propto M_{\mathrm{eff}}$) of the off-resonance rf field. The dependence of the spectral intensity ratio, $R$ ($= M_{\mathrm{eff}}/M_0 = T_{1\rho}^{\mathrm{off}}/T_1$) for a given resonance on $\tau_0$ (at a specified $\nu_{\mathrm{off}}$ and $B_1$ field strength) or $\omega_e$ (variable $\nu_{\mathrm{off}}$ and at fixed $B_1$ field strength) is simulated using programs written in Borland TURBO Pascal. (Copies of these programs are available upon request; see also Appendix at end of volume.) As described above the resonance intensity of protein backbone peptide bond carbonyl carbon atoms is assessed. Because of the relatively small differences in chemical shift of each of the peptide carbonyl carbons in the protein, an overall average resonance intensity ratio is experimentally determined. For the isotropic tumbling case of a sphere, each peptide bond carbonyl of the $i$th residue is equivalent to all other residues in terms of its contribution to the resonance intensity in an off-resonance rotating-frame experiment, assuming identical local peptide bond geometries and principal values of the peptide bond carbonyl carbon chemical shift tensor. By contrast, in the case of an anisotropically tumbling particle, each carbonyl carbon contributes to the resonance intensity in a way which is explicitly dependent on the orientation of the CSA principal axis system relative to the axially symmetric diffusion axis system (defined by the Euler angles $\beta$ and $\gamma$) as well as by the angle of the C–H internuclear vector with the unique axis of the anisotropic particle. The average value of the spectral intensity ratio is defined as follows:

---

[20] W. E. Hull and B. D. Sykes, J. Chem. Phys. **63**, 867 (1975).

$$\bar{R} = \left( \sum_i^N T^{\text{off}}_{1\rho,i}/T_{1,i} \right) \bigg/ N = \left[ \left( \sum_i^N M_{\text{eff},i}/M_{0,i} \right) \bigg/ N \right]$$

where $N$ is the number of amino acid residues in the polypeptide chain minus one.

The effects of polydispersity on the spectral intensity ratio is incorporated by considering the weight fraction of each protein species $j$.[15] Thus,

$$R = \sum_j f_j T^{\text{off}}_{1\rho,i}/T_{1,j}$$

where $f_j$ is the weight fraction of the $j$th protein species (equivalent to the mole fraction of the $j$th protein species with the assumption of a mean residue molecular weight).

The calculation of $R$ takes into account dipole–dipole interactions between the carbonyl carbon of the peptide bond and $H_\alpha$ (an average distance of 2.16 Å was assumed for $r_{\text{CH}}$), and a contribution from CSA relaxation of the peptide carbonyl carbon. Dipolar relaxation contributions from $^{13}C$–$^{14}N$ are negligible, and thus ignored.

For anisotropic tumbling, $R$ is calculated assuming values for the Euler angles, $\beta$ and $\gamma$, and the internuclear vector angle, $\alpha$, or by calculating these values for each amino acid residue, $i$, from X-ray crystallographic data as described above.[15] For a hydrodynamic ellipsoid of revolution, the dependence of $R$ on axial ratio requires values for $\tau_\perp$ and $\tau_\parallel$. The value of $\tau_\perp$ for a prolate ellipsoid of revolution with a given axial ratio ($a/b > 1$) relative to the rotational correlation time for a sphere of equivalent volume is calculated using corrected Perrin equations provided by Koenig.[21] Most literature citations (see Koenig[21]) of Perrin's equations are incorrect, including Woessner's equation [Eq. (58) in Ref. 16], which was employed by Wilbur et al.[5] for the evaluation of magnetic relaxation behavior of anisotropically tumbling particles from $^{13}C$ $\alpha$-carbon $T_1$ data.

## Experimental Procedures

### Pulse Schematic

The pulse diagram shown in Fig. 1 illustrates the basic features of the $T_{1\rho}$ off-resonance experiment. The low-power off-resonance rf ($B_1$ field strength $\approx 0.2$–0.6 G) is applied for a period of time of at least $3T_1$ to allow polarization of the nuclear spins along $\mathbf{B}_e$ at various frequencies relative to the resonance frequency of interest. Immediately following the applica-

[21] S. H. Koenig, *Biopolymers* **14**, 2421 (1975).

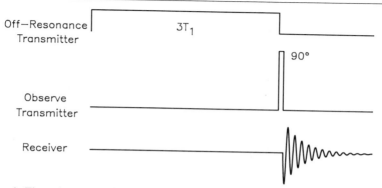

Fɪɢ. 1. The pulse sequence for the off-resonance rotating-frame spin-lattice relaxation experiment.

tion of the off-resonance field a hard 90° pulse is applied to sample the $z$ equilibrium magnetization in the rotating frame. (We have found it unnecessary to apply an intervening homospoil pulse.) A 90° pulse is used to achieve maximum signal intensity since there is a $3T_1$ equilibrium delay inherent to the experiment. Phase cycling may be used in sampling the magnetization to remove artifacts due to phase errors. For the observation of carbonyl carbon resonances proton decoupling is not usually necessary. However, studies of carbon resonances with attached protons or phosphorus coupled to protons require simultaneous proton (scalar) decoupling to avoid cross-relaxation effects which are not explicitly considered in the theoretical formalism described above.

### Block Diagram

Figure 2 is a block diagram of the instrumental implementation of the $T_{1\rho}$ off-resonance experiment utilizing an independent broad-band decoupler to supply the necessary low-power off-resonance irradiation. A watt meter may be used to monitor the forward and reflected power. Crossed diodes provide a filter to block background "hum" when the decoupler is gated off. A band-pass filter is in line to remove harmonics that may be present. The power coupler provides the means of protecting the decoupler amplifier from the high-power observe amplifier simultaneous with the routing of the off-resonance rf to the probe.

The diagram emphasizes the fact that the spectrometer observe frequency ($f_0$) and the low-power decoupler frequency ($f_1$) are obtained from independent sources and are not necessarily phase locked when $f_0 \cong f_1$. The lack of phase coherence between the two channels becomes important when the calibration of the off-resonance field is considered.

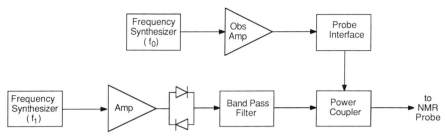

FIG. 2. Block diagram of the instrumentation required for the implementation of the off-resonance rotating-frame spin-lattice relaxation experiment. Frequency synthesizer $f_1$ provides the low-power off-resonance rf irradiation field. Crossed diodes and a band-pass filter block interfering background signals. The power coupler provides the means of protecting the off-resonance frequency amplifier from the high power rf pulses used for signal detection simultaneous with the routing of the off-resonance rf to the probe.

## Off-Resonance $B_1$ Calibration

To calibrate the off-resonance $B_1$ field strength a standard sample providing an adequate signal-to-noise ratio and of appropriate dielectric load is needed. A suitable standard is an aqueous solution of 2% sodium acetate-1-$^{13}$C (99 atom% $^{13}$C) containing ca. 35 $\mu M$ $MnCl_2$ (to reduce the carbonyl carbon $T_1 \cong 6.5$ sec) adjusted with NaCl to the appropriate ionic strength. Since the receiver is not phase locked to the decoupler, comparison of spectra taken with various decoupler pulse widths cannot be made using the same software phase correction. The comparisons are conveniently made, however, by recording the power spectrum at each pulse width. The calibration of the decoupler field strength for a given power setting is performed by determining the 180° pulse width at the acetate carbonyl resonance frequency. The 180° pulse width is used to estimate the nutational period of the carbonyl magnetization. The $B_1$ field strength is calculated by relating the frequency of nutation to the gyromagnetic ratio ($\nu_{\text{nutation}} = \gamma_C B_1$). The probe should be carefully tuned and matched, and any changes noted in going from the standard sample to the sample of interest, since changes in the quality factor ($Q$) of the coil will affect the $B_1$ strength.

## Experimental Procedure

It is recommended that studies of nuclei with low sensitivity and samples with low signal-to-noise ratio be preceded with computer simulations. Plots of intensity ratio ($R$) vs $\tau_0$ at various $B_1$ field strengths and $\nu_{\text{off}}$ values provide insight into the range of molecular motions that can be studied under the selected conditions. Intensity ratio vs $\omega_e$ plots at several

$B_1$ field strengths and $\tau_0$ values are used to determine optimum initial experimental parameters, i.e., at what $B_1$ field strengths and $\nu_{off}$ values is the intensity ratio most likely to be sensitive.

Simulations are also useful in estimating how large $\nu_{off}$ should be for the reference (control) spectrum. The reference spectrum is used to obtain the intensity ratio from spectra collected at the various values ($\nu_{off}$) of the off-resonance rf field. In practice the reference spectrum is collected in the presence of an off-resonance field of such large $\nu_{off}$ that there will be no off-resonance effect on the signal intensity. This is done to assure that the effects of rf heating, etc., are the same for each spectrum. Most protein samples require signal averaging. Therefore, to average changes that might occur during the course of the experiment, it is good practice to collect a small number of scans (e.g., 100 to 500) cycling through all values of $\nu_{off}$ until the total number of desired scans is obtained. Control studies using $^2H_2O$ in place of $H_2O$ indicate no appreciable intermolecular relaxation contribution to the intensity ratio.[15]

Quantitative analysis of the spectral intensity ratio ($R$) vs $\omega_e$ dispersion curve may be performed by nonlinear regression using a program such as NONLINWOOD.[22] With $R$ and $\omega_e$ the respective dependent and independent variables, the effective correlation time ($\tau_{0,eff}$) (see Simulation Studies) is extracted from a fit of experimental data using a theoretical model which assumes a single correlation time for random isotropic reorientation. More complex models may be employed such as those which, for example, recognize polydispersity arising from discrete mixtures or from an assumed model of association. Since $R$ is the ratio of the measured intensity in the presence of an off-resonance irradiation field to a reference intensity obtained very far off resonance, the reference intensity should also be used as a variable in fitted models to account for variations in this parameter.

Simulation Studies

The behavior of the off-resonance rotating-frame spin-lattice relaxation experiment for rigid hydrodynamic particles engaged in different motional tumbling modes is conveniently investigated by the use of simulation routines which incorporate the theory outlined in the section Theoretical Background. We consider in this section the off-resonance rotating-frame spin-lattice relaxation behavior for the peptide carbonyl carbon and the $\alpha$-carbon present in the backbone of rigid proteins involved in isotropic or anisotropic reorientation. The effect of polydispersity on the intensity ratio is also considered.

[22] C. Daniel and F. S. Wood, "Fitting Equations to Data," 2nd ed. Wiley, New York, 1980.

## Isotropic Reorientation

The spin-lattice relaxation time $(T_1)$ and the off-resonance spin-lattice relaxation time in the rotating frame $(T_{1\rho}^{\text{off}})$ both enter into the intensity ratio, $R$. A representative $^{13}$C NMR spectrum of the hemoglobin S peptide carbonyl carbons in the presence and absence of an off-resonance rf field is shown in Fig. 3. The theoretical dependence of the spin-lattice relaxation rate on the correlation time for isotropic reorientation of the peptide carbonyl carbon in a rigid tumbling protein is shown in Fig. 4. The characteristic maximum occurs when $\omega_C^2 \tau_0^2 = 1$, and is therefore dependent on $B_0$, moving to shorter $\tau_0$ values with increasing static magnetic field strength. In Fig. 5, $1/T_{1\rho}^{\text{off}}$ is plotted vs $\tau_0$ at three values of $\nu_{\text{off}}$ (at constant $B_1$ field strength); $1/T_{1\rho}^{\text{off}}$ is independent of $\nu_{\text{off}}$ at values of $\omega_C^2 \tau_0^2 \leq 1$. By contrast, for $\tau_0$ values such that $\omega_C^2 \tau_0^2 > 1$ the dependence of $1/T_{1\rho}^{\text{off}}$ on $\nu_{\text{off}}$ becomes appreciable. A minimum occurs at a value of $\tau_0$ which yields a value of 0.5 for $R$, i.e., the inflection point in a plot of $R$ vs $\tau_0$ (see below). Analogous behavior in the dependence of $1/T_{1\rho}^{\text{off}}$ on $\tau_0$ occurs at lower $B_0$ field strengths. Both simulations take into account dipole–dipole and CSA relaxation mechanism contributions. The simulations reveal that at a $B_0$ field strength of 2.35 T, the CSA relaxation mechanism contribution to the spin-lattice relaxation rate (or $1/T_{1\rho}^{\text{off}}$) is ca. 30%, whereas at 7.05 T, the contribution from this relaxation source rises to ca. 80%. The intensity ratio $(R)$, which incorporates both $T_1$ and $T_{1\rho}^{\text{off}}$, is plotted against $\tau_0$ (monodisperse) in Fig. 6 at various values of $\nu_{\text{off}}$ ($B_1$ is kept constant). The effect of increasing $\nu_{\text{off}}$ at constant $B_1$ is to shift the inflection point of the $R$ vs $\tau_0$

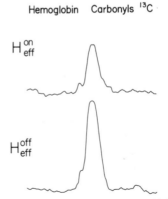

FIG. 3. $^{13}$C Fourier transform NMR spectrum (25 MHz) of the peptide carbonyl carbons of hemoglobin S in sickle red blood cells in the presence (top) and absence (bottom) of an rf field ($B_1 = 0.26$ G) applied 4.0 kHz off resonance (800 transients each). $T = 40°$. (Taken from James *et al.*[7])

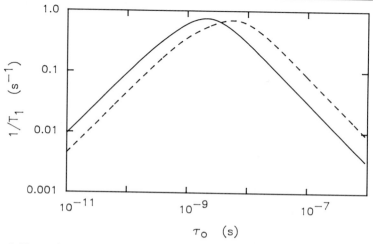

FIG. 4. Theoretical dependence of the $^{13}C$ spin-lattice relaxation rate ($1/T_1$) on the correlation time for random isotropic reorientation ($\tau_0$) at 2.35 T (---) and 7.05 T (—). The computed curves assume dipolar relaxation by a proton 2.16 Å from the $^{13}C$ and CSA relaxation contributions. Values of −81.8 ppm and −0.82 were used for $\delta_z$ and $\eta$, respectively (see the section, Evaluation of $\delta_z$).

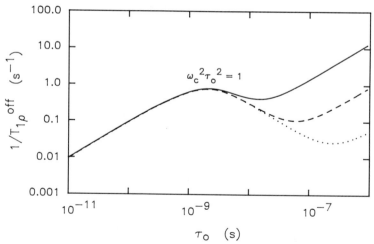

FIG. 5. Theoretical dependence of the $^{13}C$ off-resonance rotating-frame spin-lattice relaxation rate ($1/T_{1\rho}^{off}$) on $\tau_0$ at 7.05 T.[15] The computed curves assume dipolar and CSA relaxation contributions as described in Fig. 4. An off-resonance field of 0.5 G applied 3 (—), 12 (---), and 48 kHz (···) off resonance was assumed.

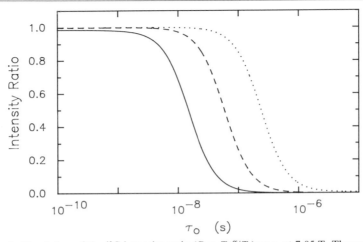

FIG. 6. Simulation of the $^{13}C$ intensity ratio ($R = T_{1\rho}^{off}/T_1$) vs $\tau_0$ at 7.05 T. The computed curves assume dipolar and CSA relaxation contributions using the parameters stated in Fig. 4. An off-resonance field of 0.5 G applied 3 (—), 12 (– – –), and 48 kHz ($\cdot\cdot\cdot$) off resonance was assumed.

curve to larger values of $\tau_0$. For the range of parameters adopted in this simulation, the motional window of sensitivity is ca. $10^{-9}$ to $10^{-6}$ sec. Analogous plots of $R$ vs $\tau_0$ under the constraint of fixed $\nu_{off}$ and variable $B_1$ are shown in Fig. 7. The effect of altering either $\nu_{off}$ or $B_1$ is to change both the angle that the effective field makes with the $z$ axis ($\Theta$) as well as the precessional frequency of the nuclear spins about $\mathbf{B_e}$. Decreasing $B_1$ results in shifting the inflection point of the $R$ vs $\tau_0$ curve to larger $\tau_0$ values. Alternatively, the intensity ratio can be plotted vs $\omega_e$ (at constant $B_1$), i.e., in the form of a dispersion curve. Representative plots for several different $\tau_0$ values (molecular weights) at constant $B_1$ are shown in Fig. 8. The advantage of the dispersion representation of an intensity ratio plot is that the effects of polydispersity may be readily incorporated by taking into account individual correlation time contributions to the overall value of $R$ from differently sized tumbling species. At 2.35 T, $R$ vs $\omega_e$ curves are virtually coincident in the presence and absence of CSA relaxation contributions, whereas at 7.05 T a slight difference, within the usually attained experimental error, is observed. This observation serves to illustrate the robust nature of the intensity ratio form of the experiment, in terms of its lack of appreciable sensitivity to relaxation mechanism contributions. Furthermore, the dispersion form of the experiment incorporates the results of a number of separate, individual off-resonance rotating-frame experiments, as opposed to the measurement under one set of experimental parameters.

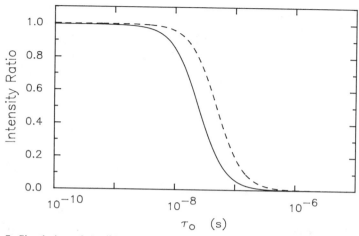

FIG. 7. Simulation of the [13]C intensity ratio ($R$) vs $\tau_0$ at 7.05 T. The computed curves assume dipolar and CSA relaxation using the parameters stated in Fig. 4. An off-resonance field of 0.5 (—) and 0.25 G (‑‑‑) was applied 5 kHz off resonance.

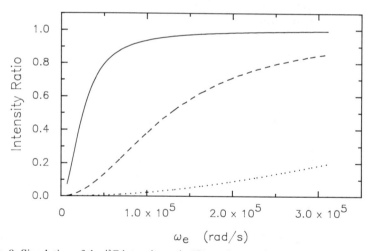

FIG. 8. Simulation of the [13]C intensity ratio ($R$) vs the angular precessional frequency of the effective field ($\omega_e$) produced by an rf field applied off resonance at three different values of $\tau_0$. The computed curves assume an off-resonance field strength of 0.5 G and a $B_0$ field strength of 7.05 T. Dipolar and CSA relaxation mechanism contributions were assumed using the parameters stated in Fig. 4. In aqueous solution at 20°, assuming a protein partial specific volume of 0.73 cm³/g, and an average hydration value of 0.51 g $H_2O$/g protein, $\tau_0$ values for isotropic reorientation of 20 (—), 100 (‑‑‑), and 500 (···) nsec represent molecular weights of 39,216, 196,078, and 980,392, respectively.

The evaluation of the rotational correlation time assuming monodispersity is achieved by the measurement of the intensity ratio performed at known values of $\nu_{off}$ and $B_1$, and evaluated from the theoretically generated plot of $R$ vs $\tau_0$. In the case where $\tau_0$ is much greater than $\omega_C^{-1}$, as shown by James et al.,[6] an estimate of $\tau_0$ accurate to within 10% can be obtained from

$$\tau_{0,eff} = \sqrt{[3(1 - R)]/2R} \, (\nu_{off}/\gamma_C B_1 \nu_c) \qquad (21)$$

This equation is also valid for CSA relaxation contributions. The concept of an effective correlation time, $\tau_{0,eff}$, is described below. Alternatively, the dispersion variation of the off-resonance rotating-frame experiment can be performed, and the data analyzed by nonlinear regression techniques to yield a value of $\tau_0$.

The assessment of $\tau_0$ described above assumes a monodisperse population of tumbling particles at infinite dilution. One indication of polydispersity is the determination of different $\tau_0$ values when $\nu_{off}$ or $B_1$ is altered. This has the effect of shifting the window of motional sensitivity for an $R$ vs $\tau_0$ plot along the $\tau_0$ axis (see Fig. 6). In the case of less extreme polydispersity, e.g., a polydispersity model which assumes variable amounts of dimer in a monomer–dimer mixture, the effect of size heterogeneity is not so readily apparent. Figure 9 depicts plots of $R$ vs $\omega_e$ as a function of the weight-average molecular weight ($M_r = \Sigma W_i M_i$, where $W_i$ and $M_i$ are the weight fraction and molecular weight, respectively, of the $i$th species). For this model, $R$ vs $\omega_e$ curves generated for a monodisperse molecular weight are virtually identical to curves obtained using an equivalent magnitude weight-average molecular weight. This observation stresses the point that the rotational diffusion information obtained in an off-resonance rotating-frame relaxation experiment is critically dependent on the model adopted for the analysis of the relaxation data.

The model adopted for the evaluation of $\tau_0$ from off-resonance rotating frame spin-lattice relaxation data assumes that all of the carbonyl carbons contributing to the resonance signal are part of the polypeptide backbone, and thus reflect the isotropic reorientation characteristics of the tumbling protein. Globular proteins generally contain ca. 15% amino acid residues which possess side-chain carboxyl or amide residues. The carbonyl carbons of these moieties contribute to the resonance signal envelope, and thus potential internal reorientational contributions need to be considered. The correlation time for side-chain motions are estimated to be ca. 0.1 nsec, whereas local motions producing changes in protein conformation are at least 10 to 100 times longer.[23]

[23] G. Gaveri, P. Fasella, and E. Gratton, Annu. Rev. Biophys. Bioeng. 8, 69 (1979).

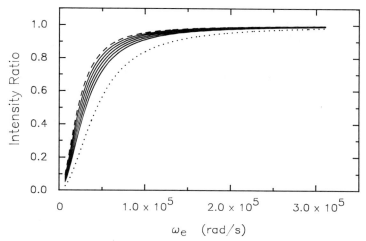

FIG. 9. Simulation of the $^{13}$C intensity ratio ($R$) vs $\omega_e$ produced by an rf field applied off resonance for a polydisperse mixture of isotropic, tumbling, unhydrated spheres.[15] The computed curves assume dipolar and CSA relaxation mechanism contributions as described in Fig. 4, a $B_0$ field strength of 7.05 T, and an off-resonance field strength of 0.5 G. A monomer ($M_r = 20,000$, ---)–dimer ($\cdots$) mixture of increasing weight fractions of dimer (0.1 to 0.5) (—) was assumed.

The effects of internal motion superimposed on overall isotropic reorientation is evaluated by the use of Eq. (19),[10,24,25] which defines the spectral density function. However, the definitions of $\alpha$, $\tau_A$, $\tau_B$, and $\tau_C$ must be modified. The angle $\alpha$ becomes the orientation of the C–H internuclear vector relative to the internal diffusion axis; $\tau_A$ is set equal to $\tau_0$, the correlation time for overall isotropic reorientation, and $\tau_B$ and $\tau_C$ are redefined as follows: $\tau_B = [1/\tau_0 + 1/(6\tau_i)]^{-1}$; $\tau_C = [1/\tau_0 + 2/(3\tau_i)]^{-1}$, where $\tau_i$ is the correlation time characterizing the internal motion.

Figure 10 depicts $R$ vs $\omega_e$ dispersion plots for various values of $\tau_i$ representing the internal motion contribution from 15% (by weight) of residues which contain carbonyl side chains superimposed on isotropic particle reorientation. The effect of internal motion is to increase the value of $R$ (at fixed $\omega_e$) relative to the totally rigid rotor case. As shown, the value of $R$ passes through a maximum with increasing $\tau_i$, as is expected from a consideration of the definitions for $\tau_B$ and $\tau_C$. The error in $\tau_0$ produced by an isotropic fit to a dispersion curve characteristic of a combination of isotropic reorientation and internal motion is assessed by

[24] D. E. Woessner, *J. Chem. Phys.* **36**, 1 (1962).
[25] T. L. James, *J. Magn. Reson.* **39**, 141 (1980).

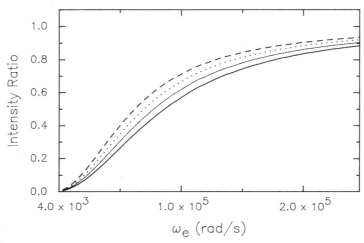

FIG. 10. Simulation of the $^{13}C$ intensity ratio $(R)$ vs $\omega_e$ applied off resonance for an isotropically reorienting sphere ($\tau_0 = 20$ nsec) containing 15% by weight of residues with side-chain carbonyl carbons. Internal reorientation times $(\tau_i)$ of 0 (—), 0.10 ($\cdots$), 1.0 (———), and 10 nsec (—·—) were assumed. The computed curves assume a $B_0$ field strength of 7.05 T, an off-resonance field strength of 0.5 G, and a C–H internuclear vector angle $\alpha$ of 109.5°. Dipolar and CSA relaxation contributions as described in Fig. 4 were assumed for the fraction of carbonyls engaged in isotropic reorientation, whereas for the carbonyl fraction involved in internal motion (superimposed on isotropic reorientation) dipolar relaxation by two protons located 2.16 Å was assumed (CSA relaxation was ignored). For $\tau_i$ values of 0.01 and 350 nsec, the theoretical dispersion curves are virtually coincident with the dispersion curve obtained for isotropic reorientation in the absence of internal motion.

a nonlinear regression analysis performed on a simulated dispersion curve (7.05 T) assuming only isotropic reorientation to obtain the effective correlation time for isotropic motion ($\tau_{0,\text{eff}}$). The result of internal motion in the case of an assumed isotropic fit is to decrease $\tau_{0,\text{eff}}/\tau_0$. For a particle with a $\tau_0$ of 20 nsec containing 15% residues with side-chain carbonyls the minimum value of $\tau_{0,\text{eff}}/\tau_0$ is ca. 0.90 ($\tau_i = 1$ nsec), whereas for a significantly larger particle with $\tau_0 = 70$ nsec the minimum value of $\tau_{0,\text{eff}}/\tau_0$ is ca. 0.75 ($\tau_i = 1$ nsec). Assuming a side-chain reorientation correlation time of 0.1 nsec,[23] $\tau_{0,\text{eff}}/\tau_0$ for $\tau_0 = 20$ nsec is 0.95, whereas for $\tau_0 = 70$ nsec, $\tau_{0,\text{eff}}/\tau_0 = 0.82$.

The $^{13}C$ $\alpha$-carbon resonance relaxation behavior of the protein polypeptide backbone is also sensitive to motion. Unlike the carbonyl carbon, the predominant relaxation mechanism at the aliphatic $\alpha$-carbon is dipolar. The nearly symmetric chemical shift tensor of aliphatic carbons makes CSA relaxation negligible. The dominant dipolar relaxation mechanism is quite efficient at the $\alpha$-carbon due to the directly bonded proton at a vibrationally averaged distance of 1.11 Å.

The NOE is small for the time scale of overall protein tumbling at high $B_0$ field strength values, and thus is primarily useful as a probe of internal motion.

*Anisotropic Reorientation*

Deviations from spherical shape of globular proteins are generally modeled for hydrodynamic purposes in terms of a rigid ellipsoid (prolate $(a/b > 1)$ or oblate $(a/b < 1)$) of revolution. Prolate ellipsoids rotate more easily about the long axis than a sphere of equivalent volume; however, rotation about the short axis encounters considerable friction. Oblate ellipsoids, by contrast, experience roughly the same friction regardless of rotation about either the short or long axis, but both modes of rotary motion involve more friction than a sphere of equivalent volume. These qualitative considerations form the basis for understanding the magnetic relaxation characteristics of anisotropically tumbling rigid particles.

The dependence of magnetic relaxation on the orientation of an individual peptide carbonyl bond within a protein relative to its diffusion axis becomes evident when rotational diffusion of an axially symmetric ellipsoid is considered. We consider the prolate case first since the greatest difference between $\tau_\perp$ and $\tau_\parallel$ occurs for this shape.

For the dipolar relaxation mechanism the angle $\alpha$ describes the orientation of the C–H internuclear vector $r_{CH}$ with respect to the principal diffusion axis $D_z$ $(= D_\parallel)$. The spectral density function is dependent on $\alpha$ through the coefficients A, B, and C as described by Eq. (19). Intuitively one would predict that orientations which place $r_{CH}$ parallel to $D_z$ $(\alpha = 0$ or 180°) correspond to minima in the relaxation rate curve; when $r_{CH}$ is perpendicular to $D_z$ $(\theta = 90°)$, however, we expect a maximum affect on relaxation rate. Computer simulations show these predictions to be correct.[15]

When CSA relaxation is considered the orientation of the CSA tensor for the peptide carbonyl with respect to the diffusion axis system becomes important. If the CSA tensor of all peptide carbonyls is assumed to be similar to that found in the model peptide bond, then the CSA tensor has the orientation relative to the atoms of the peptide bond as described in Fig. 11A. The two Euler angles $\beta$ and $\gamma$ completely describe the CSA orientation with respect to the diffusion axis for axially symmetric ellipsoids as depicted in Fig. 11B. The spectral density functions are modulated by $\beta$ and $\gamma$ through the coefficients $c_0$, $c_1$, and $c_2$ [see Eq. (20)]. Schematically the peptide carbonyl CSA tensor can be depicted as an

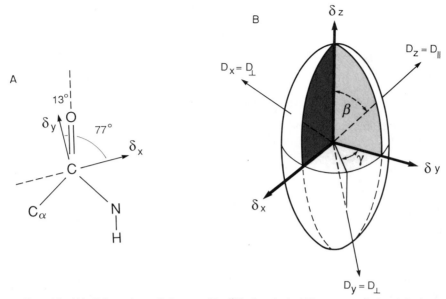

FIG. 11. (A) Orientation of the peptide $^{13}C$ chemical shift tensor of glycylglycine · HCl · $H_2O$. The $\alpha$-carbon, carbonyl carbon, and amide nitrogen are virtually coplanar; $\delta_x$ and $\delta_y$ are also in this plane, and form an angle of 77 and 13°, respectively, with the C–O bond. The direction of $\delta_z$ (not shown) is perpendicular to this plane. (Adapted from Ref. 18.) (B) The chemical shielding ellipsoid for the peptide bond carbonyl carbon indicating the relationship of the chemical shielding tensor principal axis system to the diffusion axis coordinate system. The Euler angles $\beta$ and $\gamma$ relate the CSA principal axis system to the axially symmetric diffusion axis system.

oblate spheroid with semiaxes $|\delta_z| \geq |\delta_x| \gg |\delta_y|$. The angle $\beta$ describes the orientation of the $\delta_z$ axis to $D_z$. Since $\delta_z$ and $\delta_x$ are components of the CSA tensor along which the greatest changes in chemical shift lie, orientations which place $D_z$ perpendicular to the $\delta_z\delta_x$ plane ($\beta = 90$ or 270° and $\gamma = 90$ or 270°) should show maxima in the relaxation rate curve. Additionally for $\beta = 90$ or 270° the relaxation rate should vary strongly with $\gamma$ and go through a minimum when $D_z$ is rotated into the $\delta_z\delta_x$ plane at $\gamma = 0$ or 180°. When $\beta = 0$ or 180°, $D_z$ is in the $\delta_z\delta_x$ plane. These values of $\beta$ should show minima in the relaxation rate curve and have weak dependence on $\gamma$. Computer simulations demonstrate these assumptions to be correct.[15]

Oblate ellipsoids have similar correlation times about both $D_\parallel$ and $D_\perp$ (i.e., $\tau_\parallel \approx \tau_\perp$). The similar values of the $\tau$'s should make the orientation effects less dramatic than in the case of prolate ellipsoids. Simulations where the axial ratio is varied over a range of oblate ellipsoids, through the spherical case, and into the range of prolate ellipsoids show just this behavior.[15]

The spin-lattice relaxation rate of an ellipsoid of revolution relative to that for an isotropically tumbling sphere of equivalent volume vs axial ratio ($a/b$) is plotted in Figs. 12 and 13. Deviations from spherical shape, either prolate or oblate, are manifested by substantial decreases in the relaxation rate from the spherical case value, in the nonextreme narrowing limit ($\omega_C^2 \tau_0^2 \gg 1$). Changes in the orientation of the internuclear C–H vector relative to the unique diffusion axis ($\alpha$) yield minor alterations in $1/T_1$ for the anisotropic tumbling particle at axial ratios greater than $\approx 2$ (prolate ellipsoid) (Fig. 12). With the Euler angle $\beta$ fixed at 0°, alterations in the Euler angle $\gamma$ produce no changes in $1/T_1$, as shown in Fig. 13. However, at values of $\beta$ greater than 0°, alterations in $\gamma$ result in significant changes in $1/T_1$ (Fig. 13), which is particularly evident for prolate ellipsoids. Figures 14 and 15 are similar plots of $1/T_{1\rho}^{\text{off}}$ (relative to that obtained for an equivalent volume sphere) vs axial ratio. As shown in Figs. 14 and 15 analogous behavior of $1/T_{1\rho}^{\text{off}}$ as a function of axial ratio, and the angles $\alpha$, $\beta$, and $\gamma$ is observed. As expected, the intensity ratio vs axial ratio plot (Fig. 16) displays similar behavior to the corresponding $1/T_1$ and $1/T_{1\rho}^{\text{off}}$ plots (Figs. 13 and 15).

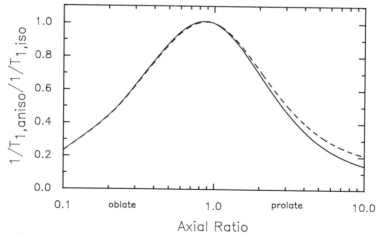

FIG. 12. Theoretical dependence of the [13]C spin-lattice relaxation rate for an anisotropically tumbling particle ($1/T_{1,\text{aniso}}$) relative to the spin-lattice relaxation rate of an equivalent volume sphere ($1/T_{1,\text{iso}}$) vs axial ratio for C–H internuclear vector angles $\alpha$ of 0° (—) and 90° (---).[15] The computed curves assume a $B_0$ field strength of 7.05 T, dipolar and CSA relaxation contributions as described in Fig. 4, $\beta$ and $\gamma$ Euler angles of 0°, and a $\tau_0$ value of 20 nsec for isotropic reorientation of the equivalent volume sphere.

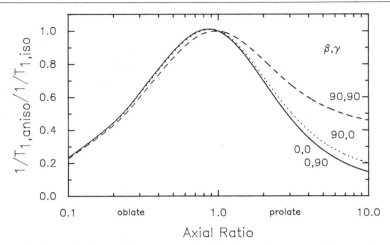

FIG. 13. Theoretical dependence of the $^{13}C$ spin-lattice relaxation rate for an anisotropically tumbling particle ($1/T_{1,aniso}$) relative to the spin-lattice relaxation rate of an equivalent volume sphere ($1/T_{1,iso}$) vs axial ratio for different $\beta$ and $\gamma$ Euler angles.[15] The computed curves assume a $B_0$ field strength of 7.05 T, dipolar and CSA relaxation contributions as described in Fig. 4, a C–H internuclear vector angle $\alpha$ of 0°, and a $\tau_0$ value of 20 nsec for isotropic reorientation of the equivalent volume sphere.

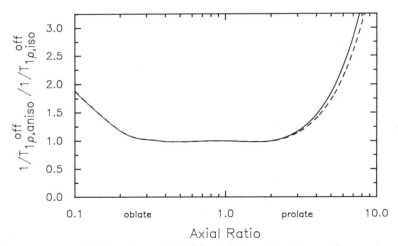

FIG. 14. Theoretical dependence of the $^{13}C$ spin-lattice relaxation rate for an anisotropically tumbling particle ($1/T_{1\rho,aniso}^{off}$) relative to the spin-lattice relaxation rate of an equivalent volume sphere ($1/T_{1\rho,iso}^{off}$) vs axial ratio for C–H internuclear vector angles $\alpha$ of 0° (—) and 90° (---).[15] The computed curves assume a $B_0$ field strength of 7.05 T, an off-resonance irradiation field of 0.5 G applied 5 kHz off resonance, dipolar and CSA relaxation contributions as described in Fig. 4, $\beta$ and $\gamma$ Euler angles of 0°, and a $\tau_0$ value of 20 nsec for isotropic reorientation of the equivalent volume sphere.

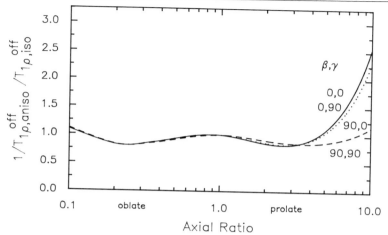

FIG. 15. Theoretical dependence of the $^{13}C$ off-resonance rotating-frame spin-lattice relaxation rate for an anisotropically tumbling particle ($1/T^{off}_{1\rho,aniso}$) relative to the off-resonance rotating-frame spin-lattice relaxation rate of an equivalent volume sphere ($1/T^{off}_{1\rho,iso}$) vs axial ratio for different $\beta$ and $\gamma$ Euler angles.[15] The computed curves assume a $B_0$ field strength of 7.05 T, an off-resonance irradiation field of 0.5 G applied 8 kHz off resonance, dipolar and CSA relaxation contributions as described in Fig. 4, a C–H internuclear vector angle $\alpha$ of 0°, and a $\tau_0$ value of 20 nsec for isotropic reorientation of the equivalent volume sphere.

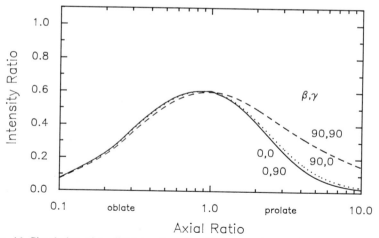

FIG. 16. Simulation of the $^{13}C$ intensity ratio ($R$) for an anisotropically tumbling particle ($T^{off}_{1\rho,aniso}/T_{1,aniso}$) vs axial ratio for different $\beta$ and $\gamma$ Euler angles.[15] The computed curves assume a $B_0$ field strength of 7.05 T, an off-resonance irradiation field of 0.5 G applied 5 kHz off resonance, dipolar and CSA relaxation contributions as described in Fig. 4, a C–H internuclear vector angle $\alpha$ of 0°, and a $\tau_0$ value of 20 nsec for isotropic reorientation of the equivalent volume sphere. The $\beta = 0°$, $\gamma = 0°$ and $\beta = 0°$, $\gamma = 90°$ curves are coincident.

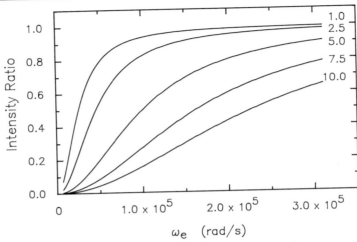

FIG. 17. Simulation of the $^{13}$C intensity ratio ($R$) for an anisotropic tumbling particle ($T^{off}_{1,aniso}/T_{1,aniso}$) vs $\omega_e$ at different axial ratios for prolate ellipsoids.[15] The computed curves assume a $B_0$ field strength of 7.05 T, an off-resonance irradiation field of 0.5 G, dipolar and CSA relaxation contributions as described in Fig. 4, a C–H internuclear vector angle $\alpha$ of 0°, $\beta$ and $\gamma$ Euler angles of 0°, and a $\tau_0$ value of 20 nsec for isotropic reorientation of the equivalent volume sphere.

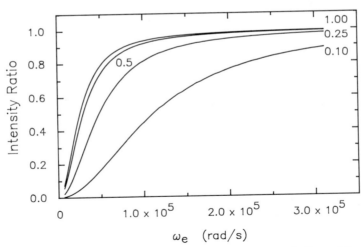

FIG. 18. Simulation of the $^{13}$C intensity ratio ($R$) for an anisotropically tumbling particle ($T^{off}_{1,aniso}/T_{1,aniso}$) vs $\omega_e$ at different axial ratios for oblate ellipsoids.[15] The computed curves assume a $B_0$ field strength of 7.05 T, an off-resonance irradiation field of 0.5 G, dipolar and CSA relaxation contributions as described in Fig. 4, a C–H internuclear vector angle $\alpha$ of 0°, $\beta$ and $\gamma$ Euler angles of 0°, and a $\tau_0$ value of 20 nsec for isotropic reorientation of the equivalent volume sphere.

Intensity ratio vs $\omega_e$ plots for prolate and oblate ellipsoids of different axial ratios (with constant volume) are shown in Figs. 17 and 18. Deviations from spherical shape (isotropic tumbling) for both prolate and oblate ellipsoids are characterized by reduced values of $R$ at fixed $\omega_e$ with increasing (prolate) or decreasing (oblate) axial ratio.[15] The effect is more substantial for the prolate ellipsoid as noted above.

Theoretical intensity ratio vs $\omega_e$ curves were calculated using the theory described above and the X-ray crystal structure coordinates for proteins of known structure. Three proteins were selected, as follows: myoglobin (high $\alpha$-helix content), concanavalin A (high $\beta$-sheet content), and lysozyme (mixture of different secondary structure types). In each case a primary diffusion axis was visually selected using molecular graphics display techniques of the protein X-ray structure.[15] Theoretical dispersion curves for these proteins using the hydrodynamically derived axial ratio are presented in Fig. 19. For comparison, the corresponding dispersion curve for assumed isotropic tumbling of each protein is given. In each case, the anisotropic dispersion curve is shifted to a smaller $R$ value

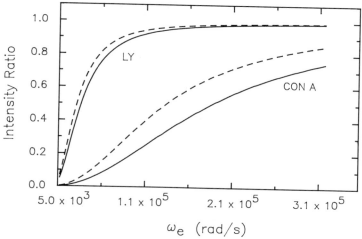

FIG. 19. Simulation of the $^{13}$C intensity ratio ($R$) for an anisotropically tumbling particle ($T_{1,aniso}^{off}/T_{1,aniso}$) (—) vs $\omega_e$ for the proteins lysozyme (LY) and concanavalin A (CON A).[15] Corresponding curves for assumed isotropic (axial ratio = 1.0) reorientation are also presented (---). The computed curves assume a $B_0$ field strength of 7.05 T, an off-resonance irradiation field of 0.5 G, dipolar and CSA relaxation contributions as described in Fig. 4; a C–H internuclear vector angle, $\alpha$, and internuclear vector, $\beta$ and $\gamma$ Euler angles calculated for each protein amino acid residue peptide bond present in the protein from the X-ray structure (see the section, Spectral Density Functions); $\tau_0$ for isotropic reorientation of the equivalent volume sphere and axial ratios of 20 nsec, 1.67; 100 nsec, 2.13, respectively, for lysozyme and concanavalin A.

at constant $\omega_e$. For the proteins myoglobin and lysozyme the anisotropic and assumed isotropic curves are fairly similar, whereas for concanavalin A the deviation between the isotropic and anisotropic cases is significantly more substantial. Analysis of $R$ (at fixed $\omega_e$) vs residue number plots for each of these proteins (calculated as described above from the X-ray structure) indicates, as expected, that the individual residue $R$ values are independent of the secondary structure type in which the particular amino acid residue is embedded, and reflects, as described above, the orientation of the internuclear dipole vector and CSA principal axis system relative to the particle diffusion axis.[15]

For an anisotropically reorienting protein the spectral density functions used for the calculation of the observed quantities ($T_1$, $T_{1\rho}^{off}$, $R$, etc.) require three correlation times ($\tau_A$, $\tau_B$, $\tau_C$) to describe the rotational motion. However, the spectral density function for an isotropic model requires only a single correlation time since there exists no unique axis of diffusion. If the observed quantities for an anisotropically tumbling protein are fitted using an assumed isotropic model, then only an "effective" correlation time, $\tau_{0,eff}$ is obtained representing some weighted value of $\tau_A$, $\tau_B$, and $\tau_C$.

This can be easily seen for an *individual* resonance by comparing the spectral density functions from the isotropic and anisotropic models for dipolar and CSA relaxation. For dipolar relaxation[5] in the extreme narrowing limit ($\omega_H \tau_{A,B,C} \ll 1$) the spectral density becomes magnetic field independent and Eqs. (16) and (19) yield $\tau_{0,eff} = A\tau_A + B\tau_B + C\tau_C$. In the slow motion limit ($\omega_C \tau_{A,B,C} \gg 1$) the spectral density remains magnetic field dependent and $\tau_{0,eff} = [A/\tau_A + B/\tau_B + C/\tau_C]^{-1}$. For CSA relaxation in the extreme narrowing limit Eqs. (17) and (20) yield $(1 + \eta^2/3)\tau_{0,eff} = C_0\tau_A + C_1\tau_B + C_2\tau_C$ and in the slow motion limit $\tau_{0,eff}/(1 + \eta^2/3) = [C_0/\tau_A + C_1/\tau_B + C_2/\tau_C]^{-1}$.

If the NMR signal is composed of an envelope of resonances, as is usually the case for the peptide bond carbonyl carbon and the $\alpha$-carbon of the polypeptide backbone, and because the coefficients are uniquely defined for each resonance ($A$, $B$, $C$, a function of $\alpha$ and $c_0$, $c_1$, $c_2$, a function of $\beta$ and $\gamma$), a summation over each resonance is required to calculate $\tau_{0,eff}$. The application of an isotropic model to a protein having anisotropic motion will thus result in the determination of an "effective" correlation time.

A more direct way of evaluating the error associated with the application of an isotropic reorientation model to an anisotropically tumbling particle is afforded by the following simulation.[15] For the proteins lysozyme, concanavalin A, and myoglobin $R$ vs $\omega_e$ dispersion curves were generated as a function of axial ratio at constant particle volume. Each

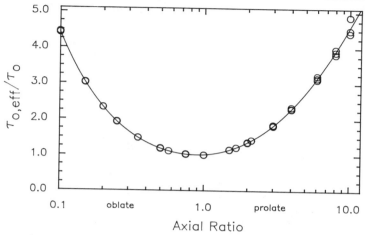

FIG. 20. The dependence of the ratio of the effective rotational correlation time ($\tau_{0,\text{eff}}$) to the assumed isotropic correlation time of an equivalent volume sphere ($\tau_0$) on axial ratio for the proteins lysozyme, concanavalin A, and myoglobin.[15] The solid line represents the fifth order polynomial fit to the data. See the section Anisotropic Reorientation for details of the $\tau_{0,\text{eff}}/\tau_0$ calculation and the numerical values of the coefficients of the polynomial.

plot in turn was analyzed using nonlinear regression techniques assuming isotropic reorientation to obtain $\tau_{0,\text{eff}}$. The dependence of $\tau_{0,\text{eff}}/\tau_0$ on axial ratio for each of the proteins is shown in Fig. 20. For a globular protein such as lysozyme, which has been characterized hydrodynamically as a prolate ellipsoid of axial ratio 1.67,[26] the application of an isotropic reorientation model for data analysis results in an overestimation of $\tau_0$ by 21%. The overestimation, as expected, increases significantly with departure from spherical shape. The error associated with an assumed isotropic analysis appears to be independent of protein secondary structure composition and shape in the axial ratio range of 0.1 to 4; at axial ratio values greater than four slight differences between the three test proteins are noted. The polynomial describing the dependence of $\tau_{0,\text{eff}}/\tau_0$ on axial ratio (ar) is[15]

$$\tau_{0,\text{eff}}/\tau_0 = 1.012 + 0.282x + 2.773x^2 + 0.244x^3 + 0.720x^4 - 0.446x^5$$

where $x = \log(\text{ar})$ ($0.1 \leq \text{ar} \leq 10.0$). Thus, the polynomial expression may be used to estimate the error introduced by the application of the isotropic analysis model to an anisotropically tumbling protein preparation.

[26] S. B. Dubin, N. A. Clark, and G. B. Benedek, *J. Chem. Phys.* **54,** 5158 (1971).

The carbonyl carbon, as discussed above, includes contributions from dipolar and CSA relaxation mechanisms. The $\alpha$-carbon is similar to the carbonyl carbon with the exception that CSA relaxation need not be considered. Therefore, the conclusions arrived at above in the treatment of the carbonyl carbon may be applied to the $\alpha$-carbon with the exclusion of CSA contributions.

### Applications

The off-resonance rotating frame spin-lattice relaxation technique for the determination of protein rotational correlation times has been applied to a variety of proteins in aqueous solution and *in vivo*. James *et al.*[6] determined the rotational correlation times of six monodisperse protein preparations, ranging in axial ratio from 1.2 to at least 3.5, and covering the molecular weight scale from 14,600 to 150,000. As summarized by James *et al.*,[6] these rotational correlation time values, obtained under the assumption of isotropic reorientation, are in excellent agreement with those determined by the use of established rotational diffusion based techniques.

Polydisperse protein samples may be analyzed if a model describing the molecular weight distribution is known or a reasonable one can be assumed. Figure 21 depicts the intensity ratio dispersion curves for monodisperse bovine serum albumin (BSA) and lysozyme preparations as well as a synthetic mixture of the two proteins (0.48 weight fraction BSA).[27] The use of nonlinear regression analysis yields a value of 46% (by weight) of BSA, demonstrating the utility of this approach. In another study, the pH-induced association of lysozyme was investigated. At pH values greater than 5.5 lysozyme undergoes significant association. Figure 22 shows the dispersion curves obtained under conditions favoring the monomer (pH 4.4) and the aggregated state (pH 8.0). Assuming a model of indefinite self-association (isodesmic) the apparent association constant obtained by nonlinear regression analysis ($K_{assoc} = 66 \ M^{-1}$) is in reasonable agreement with the value obtained by rigorous thermodynamic-based methods.[28] The isodesmic association model was applied to the lysozyme cold cataract protein model to obtain the conclusion of protein aggregation and an apparent association constant of $178 \ M^{-1}$ (see Fig. 23). These studies underscore the point that the derived rotational hydrodynamic information is dependent upon the model used to interpret the magnetic relaxation data.

[27] C. F. Morgan, T. Schleich, and G. H. Caines, *Biopolymers,* in press.
[28] P. R. Wills, L. W. Nichol, and R. J. Siezen, *Biophys. Chem.* **11,** 71 (1980).

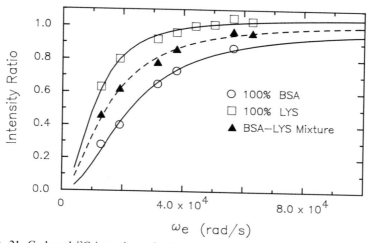

FIG. 21. Carbonyl [13]C intensity ratio ($R$) vs $\omega_e$ plots for 100% bovine serum albumin (BSA) ($\tau_{0,\text{eff}}$ = 22.1 nsec). 100% lysozyme (LYS) ($\tau_{0,\text{eff}}$ = 10.1 nsec), and a synthetic mixture of the two proteins (48% by weight BSA).[27] The solid lines for the two protein preparations represent the theoretical line calculated for isotropic tumbling with the indicated best fit $\tau_{0,\text{eff}}$ value, whereas for the mixture the dashed line represents the calculated curve for the best fit value of 45% (by weight) of BSA in the mixture, and the assumed $\tau_{0,\text{eff}}$ values for the constituent pure proteins. The calculations assume a $B_0$ field strength of 7.05 T and an off-resonance irradiation field of 0.4 G. The total protein concentration is 100 mg/ml.

The off-resonance rotating-frame spin-lattice relaxation technique can be employed for the elucidation of protein hydrodynamic behavior *in vivo* as illustrated by the assessment of $\tau_{0,\text{eff}}$ for hemoglobin in the red blood cell,[7] and for the crystallin proteins in the intact eye lens or its homogenates.

For hemoglobin in the normal adult oxygenated red blood cell the [13]C carbonyl region of the NMR spectrum is shown in Fig. 3 in the presence and absence of the off-resonance irradiation field. A mean value of 94 nsec was obtained for $\tau_{0,\text{eff}}$, whereas for hemoglobin S a mean $\tau_{0,\text{eff}}$ value of 130 nsec was observed. This increase in $\tau_{0,\text{eff}}$ is consistent with the well-known ability of hemoglobin S to associate.

The cold cataract hydrodynamic behavior of lens crystallin proteins, which constitute the major (90%) protein fraction in undiluted calf lens nuclear homogenates, was assessed by off-resonance rotating-frame spin-lattice relaxation.[29] Intensity ratio dispersion plots for calf lens nuclear

---

[29] C. F. Morgan, T. Schleich, G. H. Caines, and P. N. Farnsworth, *Biochemistry*, in press.

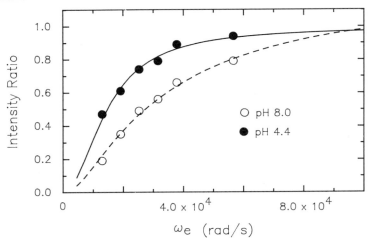

FIG. 22. Carbonyl $^{13}$C intensity ratio ($R$) vs $\omega_e$ plots for lysozyme at pH values of 4.4 and 8.0.[27] The solid line represents the best fit $\tau_{0,eff}$ value of 11.6 nsec at pH 4.4; the dashed line represents the results of a calculation assuming an isodesmic model of indefinite self-association at pH 8.8 with a best fit value of 66 $M^{-1}$ for $K_{assoc}$. The solution contains 150 m$M$ NaCl, protein concentration is 100 mg/ml, and the sample temperature is 15°; all other conditions are as described in Fig. 21.

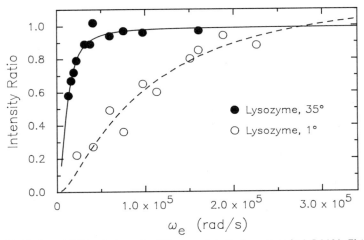

FIG. 23. Carbonyl $^{13}$C intensity ratio ($R$) vs $\omega_e$ plots for lysozyme in 0.5 $M$ NaCl (pH 5.5) at 35 and 1°.[27] The latter temperature induces a phase transition resulting in mimicking the "cold cataract" condition of juvenile mammalian eye lenses and lens protein solutions. The solid line represents the best fit assuming a model of isotropic reorientation with $\tau_{0,eff} = 7.7$ nsec; the dashed line is the best fit assuming a isodesmic model of indefinite self-association with $K_{assoc} = 178$ $M^{-1}$. The protein concentration is 100 mg/ml.

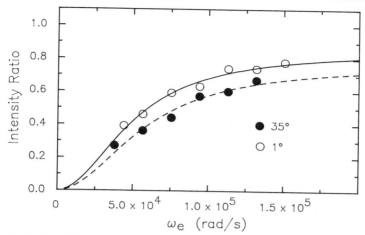

FIG. 24. Carbonyl $^{13}$C intensity ratio ($R$) vs $\omega_e$ plots for a calf lens nuclear homogenate at 35 and 1°.[29] The latter temperature induces a phase transition resulting in a "cold cataract," characterized by intense light scattering (opacification). The solid and dashed lines represent the best fit lines obtained by nonlinear regression assuming a model of monodisperse isotropic reorientation with $\tau_{0,\text{eff}}$ values of 43 and 57 nsec, respectively.

homogenates appear in Fig. 24. The effect of the low temperature is to induce a phase transition, manifested by pronounced opacification. Accompanying cold cataract induction is a decrease in $\tau_{0,\text{eff}}$ despite a decrease in temperature. Normally a nearly 3-fold increase would be expected on the basis of the change in solvent viscosity and temperature ($\eta/T$). This behavior is consistent with the differential removal of higher molecular weight species to form a solid-like and hence "invisible" fraction in this type of NMR experiment. Application to the intact calf lens gives $\tau_{0,\text{eff}}$ values which are intermediate between those obtained for nuclear and cortical homogenates.

$^{13}$C NMR rotating-frame spin-lattice relaxation has also been employed by James[25] for the study of internal motions in globular proteins. Using a model which assumed imposed internal motion on overall isotropic reorientation, James[25] was able to reconcile the experimental relaxation data for the threonine $C_\beta$ resonance of chymotrypsinogen A in terms of an overall isotropic reorientation time of $\tau_0 = 35$ nsec, and a correlation time, $\tau_i$, for the internal motion of 20 nsec. The value of $\tau_0$ determined in this study is in excellent agreement with the value determined using the backbone polypeptide chain carbonyl resonance relaxation behavior of the protein.[6] In another study, Goux et al.[30] investigated the molecular

[30] W. J. Goux, C. Perry, and T. L. James, *J. Biol. Chem.* **25**, 1829 (1982).

motional characteristics of $^{13}$C-enriched galactose covalently attached to the single carbohydrate chain of hen ovalbumin by monitoring the relaxational characteristics of the anomeric carbon as well as the value of $\tau_0$ for the protein, by measuring the relaxation of the $\alpha$-carbon in the backbone and the aromatic aryl carbon resonances. This study demonstrated the utility of the rotating-frame technique for the study of glycoprotein dynamics in circumstances where an extrinsically attached alien reporter group could potentially alter immunological behavior.

James and colleagues have also performed extensive studies on the relaxation behavior of nucleic acids. While beyond the scope of this chapter, an excellent review of the basic theory of $^{31}$P rotating-frame relaxation applied to nucleic acids, as well as a consideration of different internal dynamics models, is contained in Ref. 10. We note that $^{31}$P rotating-frame relaxation measurements may be also employed to study the characteristic rotational behavior of phosphorylated metabolites in intact tissue, thus affording insight into *in vivo* biochemistry.[31]

Acknowledgment

Supported, in part, by Grants EY 04033 and EY 05787, from the National Institutes of Health.

[31] G. H. Caines, T. Schleich, P. N. Farnsworth, and C. F. Morgan, *Biophys. J.* **55**, 447a (1989).

# [21] Measurement of Translational Motion by Pulse-Gradient Spin-Echo Nuclear Magnetic Resonance

*By* Ronald L. Haner and Thomas Schleich

## Introduction

Translational diffusion is a reflection of molecular shape and volume, intermolecular interactions, and is sensitive to the occurrence of phase transitions. Techniques which measure translational motion afford a powerful means for the study and biophysical characterization of complex biological systems.

Methods employed for the determination of translational diffusion coefficients include quasi-elastic light scattering (QELS),[1] radioactive tracer or optical measurement of concentration or concentration gradients[2] (absorption, schlieren, or interferometric) in free diffusion experiments, fluorescence bleaching photorecovery,[3] and magnetic field gradient NMR techniques.[4-6] The NMR method relieves constraints imposed by the scattering, optical, or radioisotope techniques, does not require lengthy measurement times, and is noninvasive, permitting investigation of both solute and solvent translational displacement dynamics. For example, optically dense samples are amenable to study which otherwise would be subject to thermal heating effects in laser light scattering experiments. Sample preparation is generally simpler for the NMR experiment than for other techniques. The field-gradient NMR method also offers the potential for evaluating restricted and anisotropic diffusion.

Field-gradient NMR has been employed for the investigation of solvent translational self-diffusion in polymer and cellular systems as a means of gaining insight into macromolecular structure, organization, and dynamics. However, few diffusion measurements of proteins have been made by NMR techniques, although the diffusion of nonbiological polymers has been extensively studied. Macromolecules such as globular proteins present difficult, though not insurmountable, problems for successful implementation of the field-gradient NMR technique. These problems include low sensitivity (in part due to limited solubility), slow diffusion, short transverse relaxation times ($T_2$), and interference from the solvent resonance.

Slow diffusion, combined with short $T_2$ values, require higher gradient strengths, shorter pulse separation times, and careful experimental design. Under such conditions, the potential for relatively large systematic errors exists unless certain precautions are taken. These difficulties and their circumvention have been addressed in detail. Despite the possibility of such problems, field-gradient NMR should be regarded as a potentially versatile technique that has been underutilized in the physical study of biological systems.

[1] C. S. Johnson, Jr., and D. A. Gabriel, in "Spectroscopy in Biochemistry" (E. J. Bell, ed.), Vol. 2, p. 178. CRC Press, Boca Raton, Florida, 1981.
[2] C. R. Cantor and P. R. Schimmel, "Biophysical Chemistry," Part II, p. 581. Freeman, San Francisco, California, 1980.
[3] B. R. Ware, in "Spectroscopy and the Dynamics of Molecular Biological Systems" (P. M. Bayley and R. E. Dale, eds.), p. 133. Academic Press, London, 1985.
[4] E. D. von Meerwall, Adv. Polym. Sci. **54**, 1 (1983).
[5] P. T. Callaghan, Aust. J. Phys. **37**, 359 (1984).
[6] P. Stilbs, Prog. NMR Spectrosc. **19**, 1 (1987).

The purpose of this chapter is to describe the theoretical considerations and experimental detail necessary for the implementation of the pulse-gradient spin-echo (PGSE) NMR experiment for the study of translational motion. The investigation of proteins is emphasized, although measurements of water and metabolite molecular diffusion in biological systems will also be considered. The application to proteins is still in its infancy. Several excellent review articles are available which describe the theoretical basis and application of magnetic field-gradient NMR techniques for the assessment of translational diffusion in a variety of systems. These include the recent articles by von Meerwall,[4] Callaghan,[5] and Stilbs.[6] The book by Fukushima and Roeder[7] provides an introductory account of both theoretical and practical aspects of pulse-gradient NMR techniques for the study of molecular diffusion.

Theoretical Background

*Translational Diffusion*

Fick's first law of diffusion states that particle flux $J_2$ is proportional to the solute concentration gradient:

$$J_2 = -D(\partial c_2/\partial x)_t \tag{1}$$

where the proportionality constant $D$ is the translational diffusion coefficient. For a single uncharged particle, i.e., at infinite dilution, or in very dilute solution where solute molecules diffuse independently of each other, the mutual diffusion coefficient is defined by Stokes–Einstein equation:

$$D^0 = k_B T/f \tag{2}$$

where $k_B$ is Boltzmann's constant, $T$ the absolute temperature, and $f$ is the hydrodynamic frictional coefficient, i.e., the frictional drag, of a single solute particle in pure solvent averaged over all orientations. For a sphere in a viscous medium, $f_0$ is given by the Stokes equation:

$$f_0 = 6\pi\eta a_h \tag{3}$$

where $\eta$ is the viscosity and $a_h$ is the hydrodynamic radius. Application to other particle shapes, such as an axially symmetric equivalent ellipsoid of revolution, is possible by rewriting the Stokes equation in the form:

$$f = 6\pi\eta a/G(b/a) \tag{4}$$

[7] E. Fukushima and S. B. W. Roeder, "Experimental Pulse NMR: A Nuts and Bolts Approach," p. 197. Addison-Wesley, Reading, Massachusetts, 1981.

where $G(b/a)$ is characteristic of a prolate or oblate ellipsoid with semiaxes $a$ and $b$. Expressions for $G(b/a)$ are given in Johnson and Gabriel.[1]

In more concentrated solutions solute–solute interactions contribute to the mutual diffusion coefficient, and a generalized Stokes–Einstein equation is required to describe the concentration dependence of the mutual diffusion coefficient[8]:

$$D_m = [(\partial\pi/\partial c)_{P,T}/f(c)](1 - \phi) \qquad (5)$$

where $(\partial\pi/\partial c)_{P,T}$ is the isothermal osmotic compressibility (at constant total pressure) of the solute molecules, $f(c)$ is the frictional coefficient in concentrated solution and is dependent on concentration, and $\phi$ is the volume fraction of solute in the solution. Equivalently, the mutual diffusion coefficient is given by:

$$D_m = [k_B T/f(c)][1 + (\partial \ln \gamma/\partial \ln c)](1 - \phi) \qquad (6)$$

The concentration dependence of $f(c)$ for real particles is exceedingly difficult to calculate accurately from hydrodynamic theory and instead is usually represented by a power series in $\phi$:

$$f(c) = f(1 + A_1\phi + A_2\phi^2 + \cdots) \qquad (7)$$

where $f$ is the frictional coefficient of the hydrodynamic particle at infinite dilution, and the coefficients $A_i$ are dependent on particle shape and interparticle interaction.

The mutual diffusion coefficient is a reflection of time-dependent degradation of a solute concentration gradient. By contrast, the self-diffusion coefficient (or tracer diffusion coefficient) reflects the random motions of a single, "labeled" solute in the absence of an established concentration gradient, i.e., the concentration of solute is uniform throughout the solution.

At infinite dilution the mutual and self-diffusion (tracer) coefficients equal one another, but at high concentration the self-diffusion coefficient is given by the Einstein equation which incorporates a concentration-dependent particle frictional coefficient:

$$D_s = k_B T/f(c) \qquad (8)$$

where $f(c)$ is again expressed by a power series. Thus, the concentration dependence of $D_s$ is seen to arise from the concentration dependence of the frictional coefficient. Assuming that the concentration dependence of $f(c)$ is the same for both self- and mutual diffusion[8] (although this may not

[8] G. D. Phillies, G. B. Benedek, and N. A. Mazer, J. Chem. Phys. 65, 1883 (1976).

always be a valid approximation[9]), the combination of Eqs. (5) and (8) leads to the following expression which relates $D_m$ to $D_s$:

$$D_m = (D_s/k_B T)(\partial \pi/\partial c)_{P,T}(1 - \phi) \tag{9}$$

or equivalently:

$$D_m = D_s[1 + (\partial \ln \gamma/\partial \ln c)](1 - \phi) \tag{10}$$

For charged macromolecules the effect of particle electrostatic charge needs to be included. Charge effects contribute significantly to the diffusion coefficient.[10]

Experimentally $D_m$ may be evaluated for macromolecular solutions by quasi-elastic light scattering.[1] In this situation the driving force for diffusion is the "microscopic" concentration gradient arising from thermally induced concentration fluctuations. By contrast, in the PGSE NMR experiment "labeling" of the solute nuclear spins is accomplished by the imposed field gradient, thereby imparting a spatially dependent characteristic Larmor precessional frequency. Thus, the PGSE NMR technique measures the self-diffusion (tracer) diffusion coefficient.

In practice, the diffusion coefficient of a macromolecule is corrected to standard conditions, usually pure water at 20° by use of the following expression:

$$D_{20,w} = D_{T,sol}(293.2/T)[\eta(T, sol)/\eta(20°, H_2O)] \tag{11}$$

where $D_{20,w}$ is the scaled diffusion coefficient, $D_{T,sol}$ is the diffusion coefficient at temperature $T$ and in the selected solvent, $\eta(T, sol)$ and $\eta(20°, H_2O)$ are the respective viscosities of the solvent at temperature $T$, and of water at 20°.

### Theory of the NMR Diffusion Measurement

A simple way to view the NMR diffusion measurement is to imagine an ideal Hahn spin-echo experiment, i.e., in the absence of significant static and $B_1$ magnetic field inhomogeneities, for the measurement of the transverse relaxation time ($T_2$). This experiment uses a 90°–τ–180°–τ– echo acquire sequence, where τ is the delay time between the two radiofrequency (rf) pulses. The 180° pulse effectively reverses the sign of the phase change of the dephasing isochromats, thus at the end of the second τ delay, the magnetization refocuses, and a spin echo is observed. Assuming no molecular diffusion and perfect rf pulses over the entire signal band width, the $T_2$ value is evaluated by repeating the pulse sequence for a

[9] J. A. Marqusee and J. M. Deutch, *J. Chem. Phys.* **73**, 5396 (1980).
[10] P. Doherty and G. B. Benedek, *J. Chem. Phys.* **61**, 5426 (1975).

range of $\tau$ values, and measuring the refocused echo amplitude for each $\tau$ value. Molecular diffusion is a source of error in the $T_2$ measurement. It is manifested by attenuation of the echo amplitude arising from imperfect magnetization refocusing, i.e., the loss of refocusable phase coherence, which occurs as a result of molecular displacement during the time $2\tau$[6,7,11] in the presence of significant magnetic field inhomogeneity. Thus, the value of $T_2$ measured in the presence of significant diffusion and magnetic field inhomogeneity is systematically low, somewhere between $T_2$ and $T_2^*$. Various techniques have been developed to minimize the effect of diffusion and rf inhomogeneity.

The adverse effect of diffusion on the Hahn spin-echo $T_2$ experiment is exploited to obtain translational self-diffusion coefficients and constitutes the basis for field-gradient NMR diffusion measurements. There are two basic types of field-gradient NMR diffusion experiments—the steady field gradient and the pulsed field-gradient methods. The steady gradient experiment involves measurement of echo amplitudes in the presence of a known, steady, linear magnetic field gradient. The expression for echo intensity which includes the effect of molecular self-diffusion is[12,13]

$$A(2\tau) = A(0)\exp(-2\tau/T_2 - 2\gamma^2 DG^2\tau^3/3) \tag{12}$$

where $A(2\tau)$ is the echo amplitude for a given $\tau$ delay in the presence of the field gradient, $A(0)$ is the amplitude of the free induction decay (FID) at time $t = 0$, $\tau$ is the delay between the two rf pulses, $T_2$ is the transverse relaxation time, $\gamma$ is the gyromagnetic ratio (rad/G sec), $D$ is the center of mass diffusion coefficient (cm²/sec), and $G$ the strength of the field gradient (G/cm).

The pulsed field-gradient spin-echo (PGSE) technique extends the range of applicability of the technique to smaller diffusion coefficients, multicomponent systems, and is also better suited for measurement of restricted or spatially dependent (anisotropic) diffusion.[14–16] The experiment most commonly used is depicted in Fig. 1. The Stejskal–Tanner relationship between echo intensity, gradient strength, and pulse timing parameters is given by[14]:

$$A(2\tau) = A(0)\exp[-2\tau/T_2 - (\gamma G\delta)^2 D(\Delta - \delta/3)] \tag{13}$$

[11] T. C. Farrar, "Pulse Nuclear Magnetic Resonance Spectroscopy: An Introduction to the Theory and Applications," p. 49. Farragut Press, Chicago, Illinois, 1987.
[12] E. L. Hahn, *Phys. Rev.* **80**, 580 (1950).
[13] H. Y. Carr and E. M. Purcell, *Phys. Rev.* **94**, 630 (1954).
[14] E. O. Stejskal and J. E. Tanner, *J. Chem. Phys.* **42**, 288 (1965).
[15] E. O. Stejskal, *J. Chem. Phys.* **43**, 3597 (1965).
[16] J. E. Tanner and E. O. Stejskal, *J. Chem. Phys.* **49**, 1768 (1968).

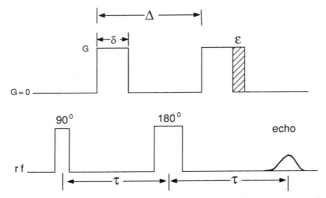

FIG. 1. The rf and gradient pulse sequence for the PGSE NMR experiment.

Equation (13) assumes that field gradients that exist when the gradient pulses are off are much less than $G$, and that the diffusion is unrestricted, isotropic, and Brownian.

If significant static $B_0$ gradients are present, Eq. (13) must be replaced by the following expression[14]:

$$A(2\tau) = A(0)\exp[-2\tau/T_2 - (\gamma G\delta)^2 D(\Delta - \delta/3)]\exp(-2\gamma^2 G_0^2 D\tau^3/3)$$
$$\times \exp\{-\gamma^2 \mathbf{G} \cdot \mathbf{G}_0 D\delta[t_1^2 + t_2^2 + \delta(t_1 + t_2) + 2\delta^2/3 - 2\tau^2]\} \quad (14)$$

where $t_1$ is the time between the first rf pulse and the leading edge of the first gradient pulse, $t_2$ is the time between the trailing edge of the second

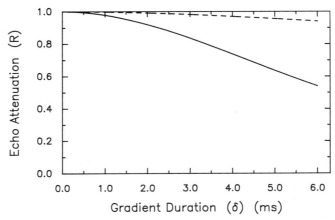

FIG. 2. Theoretical echo attenuation ($R$) plots for [1]H with self-diffusion coefficient values of $1 \times 10^{-5}$ cm$^2$/sec (—) and $1 \times 10^{-6}$ cm$^2$/sec (---). $G = 20$ G/cm and $\Delta = 8.0$ msec.

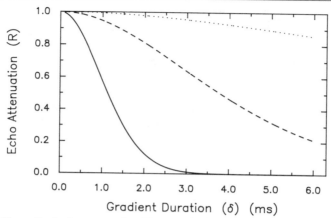

FIG. 3. Theoretical echo attenuation ($R$) plots for $^1H$ with self-diffusion coefficient values of $1 \times 10^{-5}$ cm$^2$/sec (—), $1 \times 10^{-6}$ cm$^2$/sec (–––), and $1 \times 10^{-7}$ cm$^2$/sec ($\cdots$). $G = 100$ G/cm and $\Delta = 8.0$ msec.

gradient pulse to the echo which occurs at time $2\tau$, and $G_0$ is the static field gradient (which usually is not linear). For a proper measurement $G$ should be greater than $G_0$ by at least an order of magnitude, and when $G_0\tau \ll G\delta$ the steady gradient is insignificant.

Simulations of the echo attenuation $[R = A(2\tau, G)/A(2\tau, G = 0)]$ illustrating the behavior of Eq. (13) for isotropic unrestricted Brownian motion are given in Figs. 2 and 3 for various combinations of $D$ and $G$. The simulations clearly show that relatively low gradient strengths are sufficient for the study of rapidly diffusing species, e.g., solvent, whereas, in contrast, larger hydrodynamic particles such as proteins require higher gradient strengths. At high gradient strengths the contribution of solvent is substantially reduced, as the simulations demonstrate, thereby permitting the investigation of macromolecular behavior in aqueous solution by proton NMR without the need for isotropic solvent substitution.

Various modifications to this sequence, including some which involve alternating or sinusoidal gradient pulses, have been proposed and tested.[17,18] Some of these variations may have applicability for the study of proteins.

Pulse sequences such as the stimulated echo sequence proposed by Hahn[12] may also be utilized in the measurement of self-diffusion. In this

[17] R. F. Karlieck and I. J. Lowe, *J. Chem. Phys.* **37**, 75 (1980).
[18] M. I. Hrovat, C. O. Britt, T. C. Moore, and C. G. Wade, *J. Magn. Reson.* **49**, 411 (1982).

case, the attenuation of the stimulated echo competes with the longitudinal relaxation as opposed to transverse relaxation. This is advantageous in those instances where $T_2$ is much less than $T_1$.

The stimulated pulse echo sequence consists of three $90^\circ_x$ rf pulses, where the second and third $90^\circ_x$ pulses are delivered at times $\tau_1$ and $\tau_2$, respectively. The first stimulated echo occurs at time $\tau_1 + \tau_2$. A pulse field gradient is applied between the first and second $90^\circ$ pulses and after the third $90^\circ$ pulse. The dependence of echo amplitude attenuation is given by[19]

$$A(\tau_1 + \tau_2) = 0.5A(0)\exp[-(\tau_2 - \tau_1)/T_1 - 2\tau_1/T_2 - (\gamma G\delta)^2 D(\Delta - \delta/3)] \quad (15)$$

The stimulated echo technique has also been employed as a sequence for magnetic resonance imaging.[20,21]

Multiple-quantum spin echoes may also be used for the measurement of self-diffusion coefficients[22,23] by employing a slightly modified PGSE NMR experiment. The advantage is that the *effective* field gradient experienced by the nucleus scales by the order of the multiple-quantum coherence, thus reducing problems associated with large pulse gradient field strengths, and thereby allowing study of slow diffusional processes.

### Diffusional Time Scale of PGSE Measurements

The PGSE NMR method for the measurement of translational diffusion is sensitive to displacements of 10 to 1000 $\mu$m or greater over time scales greater than 1 msec. For isotropic, unrestricted random-walk Brownian (Gaussian) motion, the mean square displacement in a given direction, e.g., parallel to the direction of the imposed pulse gradient, is given by

$$\langle (r - r_0)^2 \rangle = 2Dt \quad (16)$$

where $D$ is the translational diffusion coefficient, and $t$ the time during which diffusion occurs. For the PGSE experiment the effective observation time for Gaussian diffusive motion is[14,16,24]

$$\Delta_r = \Delta - \delta/3 \quad (17)$$

[19] J. E. Tanner, *J. Chem. Phys.* **52**, 2523 (1970).
[20] J. Frahm, K. D. Merboldt, W. Hänicke, and A. Haase, *J. Magn. Reson.* **64**, 81 (1985).
[21] A. Haase and J. Frahm, *J. Magn. Reson.* **64**, 94 (1985).
[22] J. F. Martin, L. S. Selwyn, R. R. Vold, and R. L. Vold, *J. Chem. Phys.* **76**, 2632 (1982).
[23] L. E. Kay and J. H. Prestegard, *J. Magn. Reson.* **67**, 103 (1986).
[24] R. L. Cooper, D. B. Chang, A. C. Young, C. J. Martin, and B. Ancker-Johnson, *Biophys. J.* **14**, 161 (1974).

where $\Delta > \delta$. Thus, the PGSE NMR method for the evaluation of translational diffusion extends over a distance defined by

$$\langle (r - r_0)^2 \rangle = 2D(\Delta - \delta/3) \tag{18}$$

For unrestricted diffusive motion, such as encountered in dilute macromolecular solutions, the dependence of ln $R$ (echo attenuation) on $G^2\delta^2(\Delta - \delta/3)$ is linear, with superposition of the echo attenuation curves occurring with variation in $\Delta$, i.e., echo attenuation is independent of $\Delta$. However, in the case where the distance of molecular movement occurring in the time $\Delta_r$ is comparable to the boundary spacing imposing restricted movement, the attenuation in the echo amplitude leading to the measurement of an apparent diffusion coefficient is now *dependent* on $\Delta$,[15,16] and constitutes a test for the presence of diffusion barriers. In practice, plots of ln $R$ vs $G^2\delta^2\Delta_r$ with $\Delta_r$ held constant and $G^2\delta^2$ varying, and vice versa, are compared, or the dependence of the echo amplitude attenuation on $\Delta_r$ is compared with the prediction afforded by theoretical models.[16] Figure 4 shows the theoretically calculated comparison of the echo amplitude attenuation curves vs $\Delta_r$ for restricted and unrestricted diffusion. Restricted diffusion results in a non-zero asymptotic value for the echo amplitude ratio at large values of $\Delta_r$. By contrast, for unrestricted diffusion the echo amplitude ratio approaches zero. Restricted

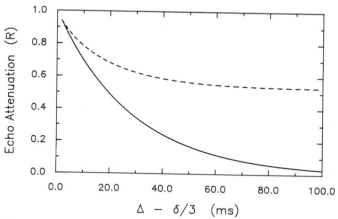

FIG. 4. Theoretical echo attenuation ($R$) vs ($\Delta - \delta/3$) plots for $^1$H undergoing unrestricted isotropic diffusion (—) and restricted anisotropic diffusion (---) created by a laminar system with parallel barriers. For both curves $D_s = 2 \times 10^{-5}$ cm$^2$/sec. The unrestricted curve was calculated using Eq. (13) in the text with $G = 50$ G/cm, whereas the restricted diffusion curve was calculated using Tanner and Stejskal's[16] Eq. (5) with $G_\parallel = 0$, $G_\perp = 50$ G/cm, and a barrier separation of $2.0 \times 10^{-3}$ cm. The gradient pulse width ($\delta$) is 1 msec.

diffusion for both macromolecules and small molecules (water and un-
bound metabolites) is likely to be encountered in the physical chemical
study of intact cellular or *in vivo* tissue systems.

## *Effect of Polydispersity*

A monodisperse macromolecular population engaged in unrestricted
Brownian motion is characterized by a single diffusion coefficient in the
PGSE NMR experiment. However, in those instances where size heterogeneity is present, i.e., polydispersity, an average diffusion coefficient is
obtained whose value is dependent on which portion of the ln $R$ (echo
attenuation) vs $\gamma^2 G^2 \delta^2 \Delta_r$ curve is analyzed. If the initial slope of this plot
is calculated, the value of $D_{max}$ is obtained, reflecting predominantly contributions from more rapidly diffusing species; conversely, the final slope
gives $D_{min}$, whereas a fit to the entire curve provides an average value of
the diffusion coefficient ($\overline{D}$).

Data from synthetic and biological polymer samples of known polydispersity[25-27] indicate the following relative magnitudes for the different
average diffusion coefficients potentially obtainable in a PGSE measurement:

$$D_{max} \approx \overline{D}_n > \overline{D} \approx \overline{D}_w > D_{min} \tag{19}$$

where $\overline{D}_n$ is the number-average diffusion coefficient, $\overline{D}_w$ is the weight-
average diffusion coefficient, and the other terms as described above.
Ignoring the effects of size heterogeneity usually introduces a bias in the
interpretation of PGSE-derived diffusion coefficients, particularly if an
insufficient range of echo attenuation values is used. This can be a problem even for paucidisperse preparations (small $\overline{D}_w/\overline{D}_n$). In those cases
where considerable differences in size exist between the smallest and
largest species, the differing $D$ values may be resolved by nonlinear regression, whereas for the case of a molecular weight distribution defined
by a continuum of discrete values, such as that produced by isodesmic
protein association, the individual $D$ values will not be resolved, and $\overline{D}_w$
will be determined. Additional complexity is introduced by the molecular
weight dependence of $T_2$ and by the long-range behavior of diffusion, the
result of coupled flow, in a multicomponent system.

[25] E. D. von Meerwall, *J. Magn. Reson.* **50**, 409 (1982).
[26] P. T. Callaghan and D. N. Pinder, *Macromolecules* **16**, 968 (1983).
[27] P. T. Callaghan and D. N. Pinder, *Macromolecules* **18**, 373 (1985).

Instrumentation

Few studies using field-gradient NMR as a tool for the measurement of protein translational diffusion[28–30] or for measurement of diffusion of other large biomolecules have been reported.[26,31] These, and virtually all other field-gradient NMR measurements, have been performed at low-field, i.e., static magnetic field strengths which are 2.35 T or less. Extension of the technique to higher static field strengths has been hampered because of potentially higher interactions between the gradient coil fringe fields and the superconducting and nonsuperconducting substructures of the magnet system, and because the required gradient coil and probe are of different geometry from conventional designs.

High-field NMR provides higher sensitivity and chemical shift dispersion, and therefore extends the applicability and usefulness of the PGSE technique. For example, higher $B_0$ fields facilitate the measurement of diffusion of species containing less sensitive nuclei, dilute or low signal-to-noise samples, or the extension to other variants of the basic technique such as spatially localized applications.

Instrumentation and procedures for low-field PGSE using low and high gradient strengths have been described in detail elsewhere.[6,29,32–34] We have developed high-field instrumentation for the translational diffusion study of proteins and protein systems,[35] and will describe the required instrumentation and experimental procedure.

This chapter describes the implementation of high-field pulsed gradient spin-echo (PGSE) NMR techniques on a General Electric (GE) GN-300 (7.05 T) wide-bore (89 mm clear bore access) spectrometer interfaced to a Nicolet 1280 computer. Use of other commercially available spectrometers will not significantly change the implementation procedure. The construction of specially designed probes and gradient control circuitry to provide the necessary linear magnetic field gradients (up to 100 G/cm) under conditions of temperature control and mechanical stability is described. Gradient pulse timing and current control is achieved by using one of the spare logic lines from the pulse programmer with a simple control circuit and amplifier.

[28] C. H. Everhart and C. S. Johnson, Jr., *Biopolymers* **21**, 2049 (1982).
[29] C. H. Everhart and C. S. Johnson, Jr., *J. Magn. Reson.* **48**, 466 (1982).
[30] C. H. Everhart, D. A. Gabriel, and C. S. Johnson, Jr., *Biophys. Chem.* **16**, 241 (1982).
[31] P. T. Callaghan and J. Lelievre, *Biopolymers* **24**, 441 (1985).
[32] M. I. Hrovat and C. G. Wade, *J. Magn. Reson.* **44**, 62 (1981).
[33] P. T. Callaghan, C. M. Trotter, and K. W. Jolley, *J. Magn. Reson.* **37**, 247 (1980).
[34] J. E. Tanner, Ph.D. Thesis, University of Wisconsin, Madison (1966).
[35] R. L. Haner and T. Schleich, unpublished experiments.

Software for controlling the experiment is incorporated into the GE GEM NMR software. The pulse programmer should be able to automatically increment and decrement pulse widths and intervals. The pulse programmer should also be capable of rapid rf phase switching so that composite rf pulses may be used.

### Gradient Probe

Low-gradient (2.3 ± 0.1 G/cm-A) and higher gradient (15.9 ± 0.6 G/cm-A) NMR field gradient probes were designed, constructed, and tested.[35] The probes are 10-mm, single-channel (tuned to 300 MHz), fitted with screened $Z^1$ gradient coils, and equipped for temperature control and regulation.

Field gradient probes are conveniently constructed from blank GN-300 probe bodies (71-mm o.d.). A cross-sectional drawing of the higher gradient probe is shown in Fig. 5. The upper 14 cm of the aluminum probe cover is removed and replaced with clear Plexiglas to reduce eddy current effects and to permit viewing of the probehead contents. Within the

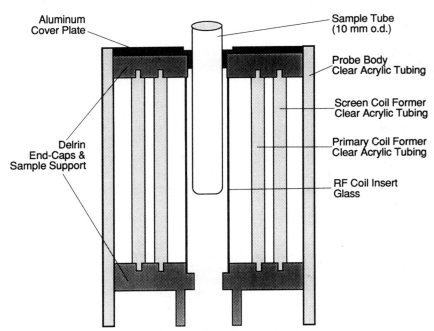

FIG. 5. Cross-section of gradient probe.[35]

probehead are two concentric clear Plexiglas tubes that support the gradient coils and a precision glass tube (15-mm o.d.) that supports the rf coil. The coil support tubes are rigidly held in place by two Delrin end caps. Samples are placed into flat-bottom NMR tubes (up to 10-mm diameter) and suspended from the upper end cap. The sample column height can be up to 1 cm, and is well within the region of sufficient gradient linearity.

Temperature control is achieved using the standard GE variable-temperature accessory, with the feedback thermocouple positioned 5 cm below the sample. To facilitate probe design and construction, the upper probe dewar is omitted, thus only the lower three-fourths of the probe is buffered against temperature gradients. Good stability ($\pm0.2°$ over 4 hr) is still achievable between 0 and 40°. The probe thermocouple is calibrated with a precision mercury thermometer positioned inside the rf coil.

The rf coil is designed for maximum $B_1$ rf field homogeneity, according to published design criteria.[36,37] Saddle coils are cut from 2-mil copper foil and secured to the outside of the glass insert with small strips of shrinkable Teflon tubing. Ninety degree pulse times for the gradient probes are typically 30 $\mu$sec using 100 W of transmitter power.

## Gradient Coil

The gradient coil design is based on the estimated required gradient strength (derived from computer simulations) and on the need for magnetic field gradient linearity over the desired sample volume. The selected design should minimize the coil fringe field, coil resistance, and inductance, and winding errors.

The maximum sample volume is determined by the anticipated experimental application, thus establishing the minimum diameter of the gradient coil. To provide adequate gradient linearity over the sample volume, the diameter of the gradient coils is at least three times greater than the diameter of the sample volume. Larger coil diameters permit greater achievable linearity, but drawbacks include lesser field strength per ampere, higher resistance higher inductance (which limits the gradient pulse rise and fall times), and potentially greater fringe fields.

The interaction of large, time-varying fringe fields with the various conductors in the magnet system potentially limits the accuracy of the PGSE experiment.[32,38] This problem is addressed by the use of an active

[36] D. M. Ginsberg and M. J. Melchner, *Rev. Sci. Instrum.* **41**, 122 (1970).
[37] E. J. Nijhof and H. J. M. ter Brake, *6th Annu. Meet. Soc. Magn. Reson. Med. Abstr.*, p. 37 (1986).
[38] M. I. Hrovat and C. G. Wade, *J. Magn. Reson.* **45**, 67 (1980).

screen, similar to the design introduced by Mansfield.[39,40] The basic *un-screened* $Z^1$ gradient coil is commonly referred to as an anti-Helmholtz coil and is composed of two circular wire loops whose centers are placed along the $z$ axis. The current direction of each loop is opposite, and the loops are separated equally about the $z = 0$ midplane, so that only odd-order gradients are generated at the coil (and sample) center. The number of turns $N$ in each wire loop may be increased to provide higher gradient strengths for a given current and coil radius. For wire loops with $N > 1$, the resulting wire bundles are referred to as "hoops." The separation of the two wire hoops is chosen so that the third order term of the axial field expansion is nulled. Assuming that the cross-sectional area of the wire in each hoop is small with respect to the radius of the coil, the separation between the midplanes of the wire hoops should be $1.732a$, where $a$ is the radius of the coil. As long as the diameter of the coil is at least three times the sample diameter and sample height, higher order odd gradients will not contribute significantly to gradient nonlinearity.

The active screen is realized by placing additional circular wire loops on a cylindrical former positioned over the primary coil at a radius that is at least 25% greater than the primary coil radius. The current direction in the screen is opposite to the current direction in the primary coil. A cross-sectional drawing of the upper half of the screened gradient coil[35] is shown in Fig. 6. Again, the wire loops are combined into bundles or "hoops," and the position and number of turns per hoop is chosen to minimize the fringe field, i.e., the field outside of the gradient coil assembly. As with the unscreened $Z^1$ coil, the lower half is placed symmetrically with respect to the upper half, about the $z = 0$ midplane, and is wound in the opposite direction to the upper half. Also, the upper half and lower half of the entire (primary and screen) assembly are separated about the coil midplane by a distance which minimizes the third order gradient. Because of the presence of the active screen, the separation between the mid-planes of the primary coil hoops will be less than $1.732a_p$, where $a_p$ is the radius of the primary coil. Due to the complexity of the expression of the third order term, the actual separation is solved numerically. All wire hoops (primary and screen) are connected in series.

The primary coil and active screen are wound into grooves cut into a pair of precision machined, concentric clear Plexiglas tubes using 26-gauge copper magnet wire with 0.5 kg winding tension. After securing the completed windings with epoxy, the ends of the Plexiglas coil formers are

[39] P. Mansfield and B. Chapman, *J. Magn. Reson.* **66**, 573 (1986).
[40] P. Mansfield and B. Chapman, *J. Phys. E* **19**, 540 (1986).

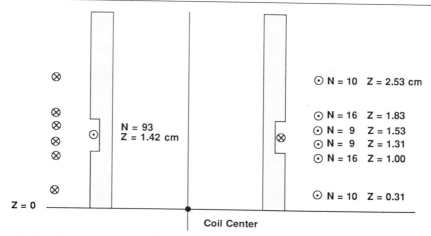

FIG. 6. Cross-section of the upper half of a screened $Z^1$ gradient coil. Primary coil radius = 1.84 cm; screen coil radius = 2.49 cm. $N$ is the number of turns per winding hoop, and $z$ is the vertical position of each winding hoop with respect to the coil center midplane. Lower half of gradient coil is of identical geometry except the winding direction is reversed. Grooves are cut into the Plexiglas coil support tubes to accommodate the number of turns and layers in each winding hoop.[35]

then pressed into circular grooves cut into a matched pair of Delrin end caps. NMR field maps are performed to determine and verify the approximate coil power and region of gradient linearity. This is done by measuring the resonant frequency of a small sample of water as a function of axial position ranging ±2 cm about the coil center, with a known amount of dc current applied to the gradient coil. Before moving the sample to the next position, the resonant frequency of the sample is measured with the dc current set to zero. The background field can then be subtracted from the field map measurements to obtain a map of the gradient coil field itself. The water sample is contained within the sealed tip of a 100-$\mu$l capillary tube, and is suspended axially within a 5-mm NMR tube using a vortex plug. The 5-mm tube is then repositioned after each measurement. The resulting map data is fit to a polynomial (sixth order or greater) to determine coil power (G/cm-A) and degree of linearity. Gradient strength calibration is ultimately achieved using reference samples of known diffusion coefficients.

## Gradient Coil Current Supply

Pulsed gradient NMR experiments are best performed with gradient coil current drivers containing very stable power supplies and simple switches. Here we demonstrate that diffusion measurements can also be

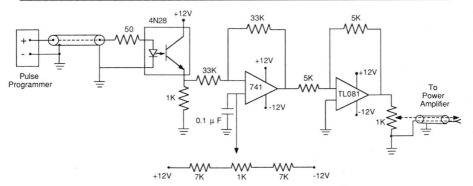

FIG. 7. Schematic of pulse gradient circuitry.[35]

performed using a high-quality dc-audio amplifier. Although more expensive, the use of such a gradient driver facilitates other types of experiments that extend the applicability of field-gradient NMR. In addition, this instrumentation forms the basis of the initial stage of an NMR microscopy and spatial localization system.

Gradient pulses are controlled with one of the spare logic lines of the GE 293D pulse programmer. A simple control circuit provides clean pulses with adjustable on/off levels. The output of this circuit serves as the input to a dc-audio power amplifier. For lower magnetic field gradient experiments, a Threshold model S/300 300-W amplifier (modified for dc) may be used (Threshold Corporation, Sacramento, CA). However, for higher magnetic field gradient experiments, a Techron 7550 750-W amplifier is used (Techron Industrial Products, Elkhart, IN). Contrary to expectation, the constant voltage output mode of the amplifier provides

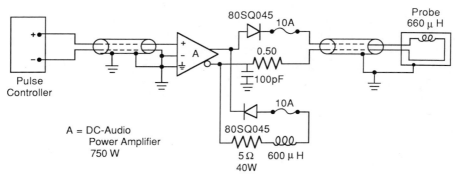

FIG. 8. Schematic of pulse gradient circuitry.[35]

performance superior to the constant current mode. Schematic drawings of the pulsed gradient circuitry[35] are given in Figs. 7 and 8. Diodes are used to establish and maintain zero current in the gradient coil when the amplifier output drops to a lower level or reverses sign. A separate output circuit, similar in impedance to the gradient coil circuit itself, serves as a load when the output of the power amplifier is of negative polarity. The circuit is assembled using care to properly ground, isolate, and shield leads to reduce high- and low-frequency noise to insignificant levels. Rise and fall times are limited by the large inductance of the screened gradient coil and were 150 $\mu$sec. The maximum coil current attainable with the Techron 750-W amplifier is 12 A with a 5-$\Omega$ load.

### Experimental Aspects

NMR pulse gradient self-diffusion determinations are achieved by measuring either the spin-echo intensity (amplitude or area) (PGSE), or the power spectrum integral (FT-PGSE) as a function of gradient pulse width (cf. Fig. 1). Initially, the gradient coil field strength is calibrated against a primary standard with physical properties similar to those of the sample of interest. The overall procedure involves first calibrating the 90 and 180° ($90_x^\circ$–$180_y^\circ$–$90_x^\circ$ composite) rf pulses for every sample. The calibration of the gradient pulse mismatch compensation time for each gradient pulse width is done using the reference sample, then FID's are collected over the appropriate range of gradient pulse widths for each sample, beginning with the reference sample.

Prior to any set of diffusion measurements, $T_1$ and $T_2$ relaxation time measurements serve as a guide for computer simulation so that the PGSE pulse timing parameters can be properly set (cf. Fig. 1) and so that the overall pulse sequence repetition rate can be determined. These measurements are made before designing the gradient coil, so that realistic simulations can be performed to determine the appropriate gradient strength requirements. For adequate signal to noise, the time $2\tau$ should not be much greater than $T_2$. The time between the trailing edge of the gradient pulse and the 180° rf pulse (or echo maximum) is generally 3 msec or greater, but shorter intervals are possible. The gradient strengths $G$ and gradient pulse durations $\delta$ are chosen to provide sufficient echo attenuation without yielding to noise. It is suggested that the experiment cover sufficient range in $\delta^2 G^2(\Delta - \delta/3)$ so that the echoes undergo an attenuation of 50–60% for the largest values of $\delta^2 G^2(\Delta - \delta/3)$. However, under favorable conditions, accurate diffusion measurements can be obtained using parameters that provide less attenuation.

The gradient coil is calibrated against pure water for low-gradient high-field PGSE measurements, and against $n$-decanol for higher gradient high-field measurements. The $n$-decanol (>98%, Sigma Chemical Co.) should be dried over a molecular sieve and then sealed under vacuum in a flat-bottom NMR tube. Once calibrated, the gradient strength can be obtained by measuring the gradient pulse voltage across the current sense resistor (cf. Fig. 8).

Radiofrequency pulse calibrations are performed with every sample. A composite 180° rf pulse is used to minimize the untoward effects of $B_1$ field inhomogeneity.

Gradient pulse mismatch compensation calibrations ["epsilon ($\varepsilon$) calibrations"] are performed whenever the overall experimental configuration is changed, such as the sample size, gradient strength, or PGSE timing parameters. For a given set of experimental conditions, the calibrations usually do not change significantly from sample to sample. The amount of mismatch is determined by measuring signal intensity (echo height or power spectrum integration) for a series of $\varepsilon$ values, and selecting the $\varepsilon$ value that results in the greatest signal intensity ($\varepsilon_{optimum}$). For the apparatus described above, the mismatch is approximately 0.3% of the gradient pulse width, and is resolvable to within 0.2 $\mu$sec. The amount of mismatch increases monotonically, but not linearly with increasing gradient pulse width; therefore, mismatch calibrations must be performed for every gradient pulse width in the series of FID measurements that comprise a diffusion measurement. A typical $\varepsilon$ calibration plot is shown in Fig. 9.

Once the mismatch is determined for a particular gradient pulse width, echo stability can be accurately assessed by measuring the standard deviation of the signal amplitudes for a series of repeated measurements with identical PGSE pulse parameters, using the optimum value of $\varepsilon$. For a given amount of inevitable current supply instability, echo instability increases dramatically as mismatch worsens, as was shown by Hrovat and Wade.[32]

Diffusion coefficients are determined by measuring signal attenuation as a function of gradient pulse width. Because the NMR signal shape is affected by ever-present residual gradients,[38] it was found that the best results are obtained if the trailing edge of the gradient pulse is kept as far away from the 180° pulse and subsequent echo as possible. To keep this time interval constant, gradient pulse widths are incremented from the leading edge of the gradient pulse.

It is useful to measure the intensity of the echo with the gradient pulses off. This measurement can flag systematic problems such as mismatch, instability, or drift. Certain experimental circumstances exist where it is necessary to measure $A_0$ after each increment in $\delta^2 G^2 (\Delta$ —

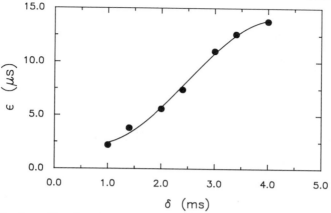

FIG. 9. Epsilon calibration plot. The sample is neat $n$-decanol, 10 mm o.d. at 20.0°. The gradient strength ($G$) is 95 G/cm, the gradient pulse separation ($\Delta$) is 8 msec, and the rf pulse separation ($\tau$) is 8 msec. The optimum epsilon value ($\varepsilon_{optimum}$) at a given gradient pulse duration ($\delta$) is determined by adjusting $\varepsilon$ for maximum echo intensity.[32,35] The data set is fit to a third-order polynomial.

$\delta/3$). Examples include reduction of the gradient current pulse duty cycle, or for situations of systematic drift.

## Signal Suppression

The measurement of protein diffusion by $^1$H PGSE NMR techniques in aqueous solution requires that unwanted signals be prevented from interfering with the measurement. The problem is compounded by the fact that the solvent $T_2$ is generally much longer than the $T_2$ of the protein protons, so an unsuppressed water ($H_2O$ or $HO^2H$) signal is further intensified relative to the protein signal, especially after the necessary $2\tau$ delay time between the 90° pulse and echo formation. Unwanted signals can be discriminated against by taking advantage of differences in chemical shifts, relaxation times, or diffusion coefficients. Other possible methods of handling this problem include alternative curve-fitting methods (if the difference in diffusion coefficients is a factor of three or greater), exchange or dilution with $^2H_2O$, and $^{13}C$ NMR spectroscopy.

Everhart and Johnson[29] have avoided $H_2O$ interference in hemoglobin–$H_2O$ solutions by using sufficiently high gradient strengths, effectively "washing out" the water proton resonance signal (cf. Fig. 3). We have resolved the solvent problem primarily by relying on the differences in $T_1$ between the solute and solvent protons. A 180°–$\tau$ prepulse nulls the unwanted signal.[41] To provide a more realistic calibration assessment, the $\tau$ delay is calibrated using a $180°-\tau-90°-\tau-180°-\tau'$-acquire sequence where $\tau'$ is equal to the delay time used in the actual spin-echo sequence. Because of the huge difference in signal intensity between the protein and solvent protons, we have also found it necessary to wait at least seven $T_1$ times between transients. We have found that this method eliminates most, but not all, of the residual solvent signal for lysozyme–$H_2O$ and lysozyme–$^2H_2O$ solutions. Therefore, a combination of higher gradient strengths and longer gradient pulse widths is required in addition to the $180°-\tau$ prepulse to totally eliminate the solvent signal. A combination of nonlinear (multiexponential) and linear (semilog) data analysis is used to obtain the diffusion information. For the case of unrestricted diffusion the signature of water interference in PGSE data is a rapid initial attenuation at small gradient pulse durations (or at smaller gradient strengths).

## Detection

In principle, under ideal conditions signal intensities can be determined by measuring echo heights or line shapes, or for FT-PGSE, by measuring peak heights or areas. As experimental requirements call for higher gradient strengths or shorter pulse spacings, the echo becomes increasingly prone to slight shifts in position, shape, and phase. Such instabilities can result in unreliable phase-sensitive signal averaging, thus phase-independent averaging methods are recommended.[32,38] Instabilities in echo shape (or line shape for FT-PGSE) are handled by integration and not echo (peak) amplitude measurement.

We have been able to obtain good precision and accuracy (<5%) by measuring echo heights for the determination of water diffusion in lens tissue. Because of the relatively fast solvent diffusion ($D > 4 \times 10^{-6}$ cm$^2$/sec), long $T_2$ values ($T_2 > 25$ msec), the spin echoes are reasonably stable under the required experimental conditions. The rf transmitter is placed on resonance and the spin echoes are collected off resonance at a high sampling rate, so that echoes are characterized adequately. However, if the echo line shape shows instability from scan to scan, the entire echo or spectral line shape should be measured and integrated.

For slower diffusion and more complicated systems such as proteins, we have found that best results are obtained by Fourier transforming the half echo. Here, the rf transmitter and receiver are placed at or near the center of the protein proton band. Sampling is performed with quadrature phase detection at a sampling rate of 20–40 kHz. Because of the higher required gradients strengths and shorter pulse intervals, the spin echoes

[41] B. D. Boss, E. O. Stejskal, and J. D. Ferry, *J. Phys. Chem.* **71**, 1501 (1967).

are somewhat more unstable in position; the rf receiver and digitizer are therefore triggered to begin sampling slightly prior (0.5–1.0 msec) to the anticipated echo maximum at $2\tau$. Phase difficulties are treated by computing the absolute value spectrum before taking the integral.

Presently, our experiments are run without an NMR lock, and with the resistive shim coils adjusted to maximize static field homogeneity. Because of the nonlinear behavior of the pulse mismatch calibration time as a function of gradient pulse width (cf. Fig. 9), the individual $\varepsilon$ values cannot be incremented automatically, so pulse parameter values are incremented by preparing and calling individual files. This also permits interleaving the set of gradient pulse widths that compose a diffusion measurement which helps to average out small time-dependent changes which may occur in biological samples. In practice, eight scans are individually measured and averaged.

### Sources of Error

The primary sources of error in PGSE experiments arise from inaccurate measurement of the gradient field strength and spin echo instability. Traditionally, the gradient field strength is calibrated against a reference of known diffusion coefficient. Other methods include FID line-shape analysis,[13,42] measurement of the sensitivity of the echo amplitude to gradient pulse mismatch,[32] and the use of NMR field maps.

The PGSE experiment requires a matched pair of gradient pulses. The FID amplitude, position, and phase is very sensitive to any mismatch. Small mismatches, even as low as 0.01%, have an effect, resulting in signal measurements which are systematically low. Also, echo intensity becomes more sensitive to power supply instabilities as the amount of mismatch increases.[31] However, the sensitivity of echo intensity and stability to gradient pulse mismatch decreases as the amount of background gradient ($G_0$) increases. The experimental approach to overcome this difficulty involves calibration and augmentation of the second gradient pulse, and the procedure for this has been outlined (*vide supra*).

Other sources of error include thermal and mechanical instability, instability of the NMR detector and digitizer, and a poor signal-to-noise ratio.

### Applications

Pulse field-gradient NMR spectroscopy is essentially an untouched technique for the assessment of protein self-diffusion, and many applica-

---

[42] D. M. Lamb, P. J. Grandinnetti, and J. Jonas, *J. Magn. Reson.* **72**, 532 (1987).

tions are foreseeable in the study of *in vitro,* intact cellular systems, and *in vivo* tissue preparations. While the literature contains many reports of diffusion coefficient measurements of synthetic polymers obtained by PGSE NMR methodology and proteins by various classical techniques, and more recently by quasi-elastic light scattering, few measurements of protein diffusion or other biologically relevant macromolecules by the PGSE NMR technique have appeared.

PGSE NMR has been employed to study the concentration and temperature dependence of dilute aqueous protein solutions by observation of the proton signal at a static field strength of 0.40 T.[28,29] These studies demonstrate that it is possible to virtually eliminate the unwanted $H_2O$ signal on the basis of the difference in the diffusion coefficients between the mobile water and the more slowly diffusing protein (cf. Fig. 3), thus achieving the selective monitoring of protein protons by the use of high gradient field strengths. Furthermore, in those instances where cellular components such as membrane and membrane fragments are present, as would be encountered in protein diffusion studies in intact cellular systems, little to no interference is expected because of their very short $T_2$ values arising from a higher degree of immobility.[30]

Measurements of the self-diffusion coefficient for hemoglobin as a function of concentration by PGSE NMR spectroscopy compare favorably with those obtained by use of other techniques. It is important to remember that the concentration dependence of the self-diffusion and mutual diffusion coefficients are not identical,[43] and that the QELS method yields the $z$-average mutual diffusion coefficient of the diffusing species. Temperature-dependent studies of hemoglobin self-diffusion indicate no significant deviations from predicted Stokes–Einstein behavior.[28]

The measurement of protein diffusion in intact cells by PGSE NMR is feasible as demonstrated by the determination of the self-diffusion coefficients for oxyhemoglobin A and oxyhemoglobin S in the red blood cell.[30] The considerations noted above for discrimination against both rapidly diffusing and immobile components apply to PGSE measurements performed on intact cells. In addition, the complexities introduced by restricted diffusion become relevant in the study of intact biological systems.[16,24]

For packed red blood cells in saline, plots of log $R$ vs $G^2$ reveal marked curvature, in contrast to linear behavior observed for hemoglobin free in solution.[30] The substitution of $^2H_2O$ in the red cell suspending medium reduces the interfering water to an acceptable level, such that the

---

[43] R. S. Hall and C. S. Johnson, Jr., *J. Chem. Phys.* **72,** 4251 (1980).

diffusion coefficient may be evaluated from the linear portion of the plot obtained at high $G$ values. The nonlinear echo attenuation plot obtained for hemoglobin is attributed to the reduction of the water apparent self-diffusion coefficient arising from restricted motion imposed by the red blood cell.[30] For these experiments, the spatial distance of the PGSE water diffusion measurement is ca. 7 $\mu$m; the reported diameter of the red cell is 7.6 $\mu$m, with a minimum thickness of 1.4 $\mu$m. Thus, the restricted motion imposed by the cell membrane results in the reduction of the water apparent diffusion coefficients to a value where it interferes with the hemoglobin diffusion measurements. By contrast, for hemoglobin the spatial distance of the measurement is ca. 0.7 $\mu$m. Because of the strong dependence of the hemoglobin self-diffusion coefficient on concentration, it is not possible to rigorously compare the values of $D$ obtained in the red blood cell with that for aqueous solution.[29,44] Qualitatively the value for hemoglobin diffusion in the red blood cell is somewhat lower than the value in aqueous solution assuming approximately equivalent protein concentrations in both cases.

The majority of PGSE NMR-derived self-diffusion coefficient measurements reported in the literature were obtained at low static magnetic field strengths. Only a very few measurements of this kind have been obtained on high-field (>2.35 T) instrumentation (see, e.g., Stilbs[6]). The few reported measurements of protein self-diffusion were obtained at 0.4 T, as described above. Extension of the technique to higher $B_0$ field strengths permit a new range of measurements to be made on biological systems.

High-field (7.05 T) echo attenuation vs $\gamma^2 G^2 \delta^2(\Delta - \delta/3)$ plots for dilute aqueous solutions of lysozyme in $^2H_2O$ and $H_2O$ are shown in Figs. 10 and 11, respectively.[45] In both cases the contribution of the interfering water resonance is markedly reduced by the application of a 180°–$\tau$ delay prior to the PGSE NMR experiment. The delay, $\tau$, is determined in a separate experiment such that the $H_2O$ contribution to the FID is minimized (*vide supra*). This strategy is only effective in those cases where the $T_1$ for the protein protons is significantly shorter than the $T_1$ for solvent protons. As shown in Figs. 10 and 11 the contribution of the residual solvent (HO$^2$H or $H_2O$) to the echo attenuation curve is evident at smaller values of $\gamma^2 G^2 \delta^2(\Delta - \delta/3)$, but becomes insignificant at larger values. This poses no significant problem for the data analysis provided that the appropriate multicomponent models are used in the data fitting, either by nonlinear methods or by the use of semilog plots. The results of the concentration

[44] A. P. Minton and P. D. Ross, *J. Phys. Chem.* **82**, 1934 (1978).
[45] R. L. Haner, C. F. Morgan, and T. Schleich, unpublished experiments.

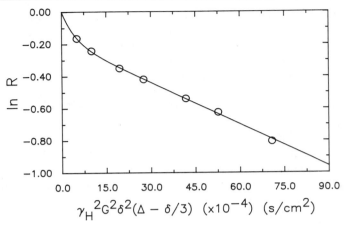

FIG. 10. Natural log $R$ (echo attenuation) vs $\gamma^2 G^2 \delta^2 (\Delta - \delta/3)$ for lysozyme[45] (5 mg/ml) in $^2H_2O$. The solution contains 0.1 $M$ NaCl adjusted to pH 5.5. Temperature = 20 ± 0.2°. The gradient strength ($G$) is 95 G/cm, the gradient pulse separation ($\Delta$) is 8 msec, and the rf pulse separation ($\tau$) is 8 msec. The solid line represents the best fit assuming exponential contributions from solvent (HO$^2$H) and protein diffusion. $D_{solvent}$ = 2.3 ± 1.2 × 10$^{-5}$ cm$^2$/sec and $D_{protein}$ = 8.65 ± 0.28 × 10$^{-7}$ cm$^2$/sec (uncorrected).

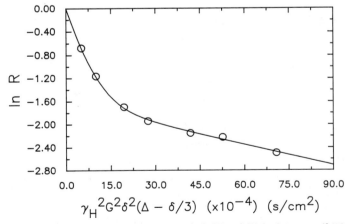

FIG. 11. Natural log $R$ (echo attenuation) vs $\gamma^2 G^2 \delta^2 (\Delta - \delta/3)$ for lysozyme[45] (10 mg/ml) in $H_2O$. Other conditions are stated in the caption to Fig. 10. The solid line represents the best fit assuming exponential contributions from solvent ($H_2O$) and protein diffusion. $D_{solvent}$ = 1.79 ± 0.12 × 10$^{-5}$ cm$^2$/sec and $D_{protein}$ = 1.20 ± 0.13 × 10$^{-6}$ cm$^2$/sec (uncorrected).

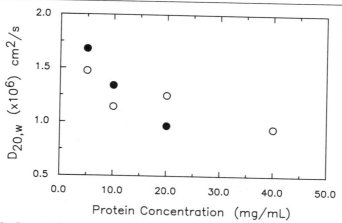

FIG. 12. Corrected self-diffusion coefficients ($D_{20,w}$) for lysozyme[45] as a function of protein concentration in $H_2O$ (●) and $^2H_2O$ (○). Other conditions are stated in the caption to Fig. 10.

dependence of lysozyme self-diffusion coefficient (Fig. 12), extrapolated to zero concentration,[45] are in reasonable agreement with the infinite dilution extrapolated value of the mutual diffusion coefficient obtained using QELS.[46] The standard deviation in $D_s$ for a series of repeated self-diffusion measurements at a particular protein concentration is between 5 and 10%. The major source of systematic error originates in the magnetic field gradient strength calibration, and is estimated to be less than 5%. Because of the high-field and wide-bore design, sample tubes up to 10-mm o.d. can be accommodated in the probe, and thus protein diffusion coefficients at concentrations as low as 0.5% (w/v) may be measured with 5–10% accuracy by averaging as few as eight FIDs. This is a factor of at least 10 better than the lower concentration limit attainable at 0.4 T,[29] suggesting that high-field PGSE NMR methodology is necessary for the study of proteins and macromolecules of lesser solubility.

Callaghan and Lelievre[31] employed the PGSE NMR method for the measurement of amylopectin and solvent self-diffusion in solutions of $^2H_2O$ and dimethyl sulfoxide. These studies demonstrate that the self-diffusion coefficient of a solvent in the presence of a macromolecular solute, when interpreted using the modified Wang model which accounts for solute-induced "obstruction" and "solvation" solvent effects,[47] pro-

[46] S. B. Dubin, N. A. Clark, and G. B. Benedek, *J. Chem. Phys.* **54**, 5158 (1971).
[47] M. E. Clark, E. E. Burnell, N. R. Chapman, and J. A. M. Hinke, *Biophys. J.* **39**, 289 (1982).

vides an effective probe of polymer size and shape. Extension of this approach to biological systems shows promise.

Oriented macromolecular systems which form an organized supramolecular structure are prime candidates for the investigation of anisotropic solvent diffusion. By use of a magnetically oriented chloroform solution of poly($\gamma$-benzyl-L-glutamate), which forms a nematic suprastructure, Moseley[48] demonstrated that solvent diffusion is 20–30% faster in a direction parallel to the aligned helices, i.e., parallel to the nematic director, than for the perpendicular direction. The assessment of anisotropic diffusion is facilitated by the ability to orient the sample in a defined direction relative to the pulse gradient in the PGSE experiment.

There have been numerous NMR studies of cellular and protein systems where water diffusion has been investigated. As noted above, the study of water diffusion is in general much easier because of generally longer $T_2$ values, faster diffusion, and strong signal strength. Wesbey et al.[49] have summarized many of these measurements.

The utility of solvent diffusion measurements in the physical study of complex macromolecular systems is illustrated by the application to bovine lens homogenates.[50] We have measured the diffusion of water in calf lens homogenates as a function of temperature (Fig. 13). In accordance with a previous study employing the steady gradient method,[51] the self-diffusion constant of water was observed to decrease with decreasing water content. However, the diffusion coefficients measured with the PGSE NMR method are approximately two times smaller than those obtained by the steady gradient technique. This difference may arise because of the effects of restricted diffusion.[16,24] The temperature dependence of $D_s$ for water in nuclear and cortical regions of the calf lens was also investigated (Fig. 13), and for the nuclear lens homogenate displayed anomalous behavior in the temperature region of the phase transition leading to cold cataract formation.[52,53] By contrast, the temperature dependence for water self-diffusion in the cortical lens homogenate, which does not form a cold cataract, showed a fairly monotonic decrease with decreasing temperature. The temperature-dependent self-diffusion data contained in the plot were obtained from fits using the Stejskal–Tanner

[48] M. E. Moseley, J. Phys. Chem. **87**, 18 (1983).

[49] G. E. Wesbey, M. E. Moseley, and R. L. Ehman, in "Biomedical Magnetic Resonance" (T. L. James and A. R. Margulis, eds.), p. 63. Radiology Research & Education Foundation, San Francisco, California, 1984.

[50] R. L. Haner, T. Schleich, C. F. Morgan, and J. M. Rydzewski, Exp. Eye Res., in press.

[51] M. C. Neville, C. A. Patterson, J. L. Rae, and D. E. Woessner, Science **184**, 1072 (1974).

[52] J. I. Clark and G. B. Benedek, Biochem. Biophys. Res. Commun. **95**, 482 (1980).

[53] M. Delaye, J. I. Clark, and G. B. Benedek, Biochem. Biophys. Res. Commun. **100**, 908 (1981).

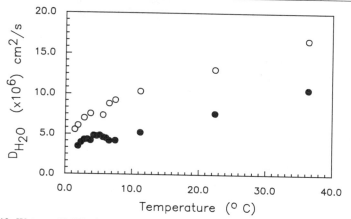

FIG. 13. Water self-diffusion coefficient in calf lens cortical (○) and nuclear (●) homogenates[50] as a function of temperature. The gradient strength ($G$) is 11.4 G/cm, the gradient pulse separation ($\Delta$) is 10 msec, and the rf pulse separation ($\tau$) is 10 msec.

relation,[14] assuming unrestricted isotropic diffusion. The difference in behavior of the nuclear and cortical lens homogenates is undoubtedly a reflection of the temperature-induced change in the protein macromolecular organization of the nuclear homogenate. These data from lower field pulse gradient experiments result in better experimental precision (3–5%) as compared to the higher gradient experiments on lysozyme, due to smaller required gradient strengths ($G$) and longer pulse intervals ($\Delta$). We note that the PGSE NMR technique has been employed for the assessment of restricted water diffusion, i.e., the characteristic diffusion length, in a variety of animal tissues.[24]

[31]P PGSE NMR measurements of the diffusion rates of the metabolites phosphocreatine and inorganic phosphate have been performed in intact frog muscle, thus demonstrating the applicability to metabolic investigations of *in vivo* systems.[54] The diffusivities of both metabolites was reduced by a factor of 2.5 relative to a control aqueous sample. The decreased diffusional rate was attributed to greater intracellular viscosity. Translational diffusion measurements of metabolites in intact biological systems, in conjunction with the concomitant assessment of rotational diffusion by spin-lattice off-resonance rotating-frame relaxation (this volume [20]) offers a powerful methodology for the noninvasive investigation of free vs bound metabolites in tissue, and for the assessment of tissue organization. Such investigations assume particular importance because

[54] K. Yoshizaki, Y. Seo, H. Nishikawa, and T. Morimoto, *Biophys. J.* **38**, 209 (1982).

of the proposal by Srivastava and Bernhard[55] that crucial metabolic intermediates of a coordinated pathway may not exist free in the intracellular aqueous phase, but rather in a bound state. Thus, the passage of metabolite from one enzyme to another in a coordinated pathway would be mediated by "protein conduction" rather than by Brownian motion of a metabolite free in solution.

### Acknowledgments

Supported in part by Grant EY 04033 from the National Institutes of Health. We thank Drs. Courtney F. Morgan and G. Herbert Caines for helpful discussion.

[55] D. K. Srivastava and S. A. Bernhard, *Annu. Rev. Biophys. Chem.* **16**, 175 (1987).

# [22] Structural Characterization of Protein Folding Intermediates by Proton Magnetic Resonance and Hydrogen Exchange

*By* HEINRICH RODER

### Introduction

In trying to understand the principles of protein folding we are faced with the seemingly intractable challenge of determining the order of conformational states encountered during refolding, and of describing each intermediate in structural, energetic, and kinetic terms. NMR has long been recognized as a promising tool for this complex problem. The ability to resolve resonances from a large number of individual nuclei, especially by two-dimensional NMR methods, permits a detailed characterization of the conformational changes associated with folding–unfolding transitions. The nuclear Overhauser effect (NOE) can in principle provide quantitative structural information on stable analogs of folding intermediates. The primary advantage of NMR for protein folding applications lies, however, in its temporal information content. The dynamics of structural fluctuations and transitions can be probed over a wide time scale by using relaxation methods, line-shape analysis, magnetization transfer, and hydrogen exchange.

There are numerous possible NMR applications for protein folding studies. The selection of NMR techniques covered in this chapter is limited to a number of one- and two-dimensional $^1$H NMR techniques: NMR observation of folding–unfolding transitions, determination of protein stability by hydrogen exchange, NMR methods to characterize unfolded proteins, magnetization transfer measurement of folding and unfolding rates at equilibrium, and hydrogen exchange labeling techniques for the characterization of folding pathways.

These methods rely on the ability to monitor changes in the structural environment of individual protons and require prior assignment of the proton resonances. Comprehensive assignments are becoming available for a growing number of proteins. Two of these are used as examples in this chapter: the basic pancreatic trypsin inhibitor (BPTI) and horse heart cytochrome c. BPTI was the first protein to be completely assigned by [1]H NMR.[1] The assignment work on horse heart cytochrome c was recently completed, both in the reduced form[2,3] and in the oxidized form.[4]

### Protein Folding–Unfolding Transitions at Equilibrium

Much of the early protein folding work centered around the search for partially folded intermediates populated under equilibrium conditions. Evidence for folding intermediates is found when distinct transition curves are observed by different physical techniques or probes (equilibrium ratio test). [1]H NMR is a particularly sensitive method for this test since each resolved resonance constitutes a probe for monitoring changes in the local structural environment.

The interconversion between folded and unfolded forms is usually slow on the NMR time scale, giving rise to separate lines for the native and denatured states at equilibrium. This is illustrated in Fig. 1 for the thermal unfolding transition of HPI, a trypsin inhibitor from the snail *Helix pomatia* with 50% sequence homology to BPTI. The [1]H NMR spectrum of the native protein at 43° displays a number of resolved and assigned resonances[5] that are characteristic of the folded conformation. In particular, some of the groups in the hydrophobic interior (Ala-9, Tyr-23, and Phe-45) exhibit significant ring-current shifts, and three $C_\alpha H$ located in the central $\beta$ sheet are shifted downfield. The spectrum recorded at 65°, where the inhibitor is about 90% unfolded, resembles the random coil spectrum expected on the basis of the chemical shifts measured in model

[1] G. Wagner and K. Wüthrich, *J. Mol. Biol.* **155**, 347 (1982).
[2] A. J. Wand and S. W. Englander, *Biochemistry* **25**, 1100 (1986).
[3] A. J. Wand, D. L. Di Stefano, Y. Feng, H. Roder, and S. W. Englander, *Biochemistry* **28**, 186 (1989).
[4] Y. Feng, H. Roder, and S. W. Englander, A. J. Wand, and D. L. Di Stefano, *Biochemistry* **28**, 195 (1989).
[5] G. Wagner, H. Tschesche, and K. Wüthrich, *Eur. J. Biochem.* **89**, 367 (1979).

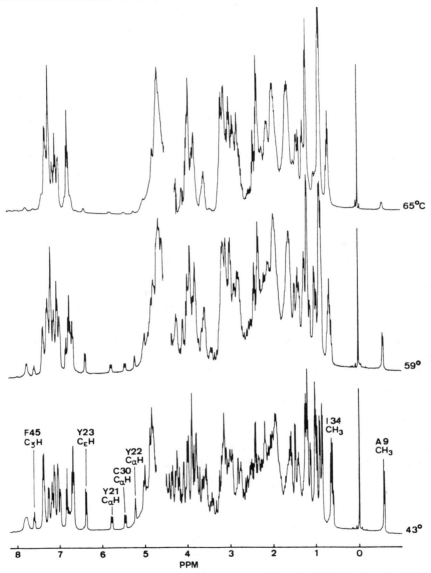

FIG. 1. Thermal denaturation of HPI, a 58-amino acid trypsin inhibitor related to BPTI, monitored by ¹H NMR at 360 MHz. The spectra were recorded in $D_2O$ (pD 2.5) after complete exchange of all labile protons. Some resolved resonances of the native protein at 43° are labeled with the assignments of Wagner et al.[5] At 59° the protein is 50% unfolded and at 65° it is about 90% unfolded. TSP was added as a chemical shift standard.

peptides.[6] The NMR properties of unfolded proteins will be discussed in the following section. A spectrum recorded after lowering the temperature back to 43° is indistinguishable from the original spectrum, which represents a stringent test for reversibility. The spectrum at the intermediate temperature, near the midpoint of the thermal unfolding transition, is essentially a superposition of the two extreme spectra. As we go through the transition the resolved lines of the folded conformation decrease without significant line broadening or shifts, indicating that the rate of interconversion is small compared to the frequency difference of corresponding resonances in the two states. This was confirmed by using the magnetization transfer method described below. At the midpoint of the transition, folding and unfolding rates of about 3 sec$^{-1}$ were measured.

The intensity of a resolved line in the native spectrum, normalized on a per proton basis, provides a direct measure of the fraction of molecules with the native conformation in the vicinity of a particular proton. Since the absolute spectral intensity is not necessarily constant at different temperatures, it is necessary to use an internal intensity standard. In our example the group of lines near 0.7 ppm represents a suitable reference, since the only protons resonating in this spectral region are the six methyl protons of the single Ile residue, both in the folded and in the unfolded state. In other cases, the total area under the aromatic region of the spectrum recorded in $D_2O$ (after complete exchange of labile protons), divided by the known number of aromatic protons, may be used as an intensity standard.[7]

All lines in spectra recorded in the unfolding transition of the trypsin inhibitor can be attributed to either the folded or the unfolded state. This indicates that there are no discrete partially folded states with sufficient lifetime and population to give rise to separate resonances. Any existing intermediates would have to be in rapid exchange with either the folded or the unfolded conformation. The ratio test for folding intermediates is also negative in the case of HPI. Figure 2A shows that the transition curves for three resolved resonances are identical within experimental error. Analogous results were obtained for BPTI, while deviations from a cooperative two-state transition were found with RCOM-BPTI, a destabilized derivative of BPTI in which the 14–38 disulfide bond is selectively reduced and carboxymethylated.[7]

The interpretation of transition curves in terms of a simple two-state model is illustrated in Fig. 2. The normalized intensity of a resolved native

[6] A. Bundi and K. Wüthrich, *Biopolymers* **18**, 285 (1979).
[7] K. Wüthrich, H. Roder, and G. Wagner, *in* "Protein Folding" (R. Jaenicke, ed.), p. 549. Elsevier, Amsterdam, 1980.

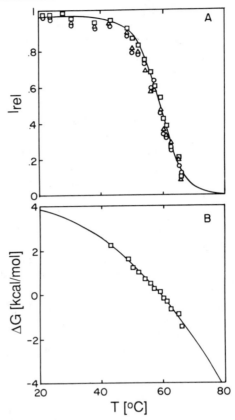

FIG. 2. Thermal folding–unfolding equilibrium of HPI in $D_2O$ at pD 2.5. (A) Temperature dependence of the normalized $^1H$ resonance intensities for Ala-9 $CH_3$ (□), Tyr-21 $C_\alpha H$ (△), and Tyr-23 $C_\varepsilon H$ (○) in the folded form (see Fig. 1). Intensities were determined by integration of resolved resonances, using the area under the Ile-34 methyl lines as an internal intensity standard. (B) Temperature dependence of the apparent unfolding free energy, calculated with Eq. (1) for the Ala-9 data in (A). The curve represents a fit of Eq. (2) with $T_m = 59°$, $\Delta H_m = 56$ kcal/mol, and $\Delta C_p = 1.2$ kcal/mol/K. The curve in (A) was calculated for the same parameters, using Eq. (2) and solving for $f_n$ in Eq. (1).

resonance directly yields the fraction of molecules in the native form, $f_n$, and the equilibrium constant for unfolding is given by

$$K_u = (1 - f_n)/f_n \qquad (1)$$

The unfolding free energy, $\Delta G_u = -RT \ln K_u$, is conveniently parameterized in terms of the midpoint temperature ($T_m$), the enthalpy change at $T_m$

($\Delta H_m$), and the heat capacity change ($\Delta C_p$), by the relationship given in Eq.[8]:

$$\Delta G_u = \Delta H_m(1 - T/T_m) - \Delta C_p[T_m - T + T \ln(T/T_m)] \tag{2}$$

One of the main advantages of using NMR to observe protein denaturation is that in the baseline zones below and above the transition the resonance intensities are independent of the variable used to induce unfolding (e.g., temperature, denaturant concentration, or pH). It is therefore not necessary to extrapolate changing baseline values into the transition zone, which is a major source of systematic errors in the analysis of transition curves observed by optical and other physical probes. One shortcoming that NMR shares with other methods is the limited range over which the folding–unfolding equilibrium can be monitored. A method for estimating unfolding equilibrium parameters under conditions far removed from the transition zone is discussed in the next section.

### Hydrogen Exchange and Protein Stability

*The Structural Unfolding Model.* Since the pioneering work of Linderstrøm-Lang,[9] hydrogen exchange has been recognized as a highly informative method to study internal motions in proteins.[10-14] While some of the details underlying the mechanism of hydrogen exchange are still being debated, there is general agreement that under destabilizing conditions major structural unfolding is responsible for the exchange of internal labile hydrogens with the solvent. The structural unfolding model[11,15,16]

$$F(H) \underset{k_f}{\overset{k_u}{\rightleftharpoons}} U(H) \underset{D_2O}{\overset{k_c}{\rightleftharpoons}} U(D) \underset{k_u}{\overset{k_f}{\rightleftharpoons}} F(D) \tag{3}$$

lends itself to a particularly straightforward interpretation in situations where structural unfolding occurs via a cooperative transition between an ensemble of folded states, F, and unfolded conformations, U. The opening transition can then be characterized by an unfolding rate, $k_u$, and a refolding rate, $k_f$, or an opening equilibrium constant, $K_{op} = k_u/k_f$. The

[8] P. L. Privalov, *Adv. Protein Chem.* **33**, 167 (1979).

[9] K. U. Linderstrøm-Lang, *Spec. Publ.—Chem. Soc.* **2**, 1 (1955).

[10] A. Hvidt and N. W. Nielsen, *Adv. Protein Chem.* **21**, 287 (1966).

[11] S. W. Englander, N. W. Downer, and H. Teitelbaum, *Annu. Rev. Biochem.* **41**, 903 (1972).

[12] C. K. Woodward, I. Simon, and E. Tüchsen, *Mol. Cell. Biochem.* **48**, 135 (1982).

[13] G. Wagner, *Q. Rev. Biophys.* **16**, 1 (1983).

[14] S. W. Englander and N. R. Kallenbach, *Q. Rev. Biophys.* **16**, 521 (1984).

[15] A. Berger and K. Linderstrøm-Lang, *Arch. Biochem. Biophys.* **69**, 106 (1957).

[16] G. Wagner and K. Wüthrich, *J. Mol. Biol.* **130**, 31 (1979).

well-founded assumption[14,17] that a hydrogen initially buried within the native protein structure can exchange only after it is brought into contact with water by a structural unfolding transition then leads to the scheme in Eq. (3). As a specific example, let us consider a buried amide proton involved in an internal hydrogen bond within the folded protein. Transient unfolding ruptures the H bond and the site becomes freely accessible to the solvent. When the protein is transferred into $D_2O$, hydrogen–deuterium exchange occurs from U at an intrinsic rate $k_c$.

The chemistry underlying the intrinsic exchange event is well understood.[18,19] Peptide NH exchange is acid- and base-catalyzed, and $k_c$ can be written as

$$k_c = k_H[H^+] + k_{OH}[OH^-] \tag{4}$$

General catalysis by $H_2O$ appears to be negligible.[20,21] Base catalysis is much more efficient than acid catalysis and the minimum in $k_c$ is shifted down to about pH 3. Representative rates are those derived by Englander et al.[22] for poly(DL-alanine) in $D_2O$ at 20°, $k_H = 0.5$ sec$^{-1}$ $M^{-1}$ and $k_{OH} = 5.6 \ 10^8$ sec$^{-1}$ $M^{-1}$, using $pK_w = 15.06$ for $D_2O$.[23] The apparent activation energy is 1.5 kcal/mol for $k_H$[20] and 17 kcal/mol for base catalysis (2.6 kcal/mol for $k_{OH}$ and 14.4 kcal/mol due to the temperature dependence of $K_w$[23]). Molday et al.[24] studied model peptides and calibrated the effect of different amino acid side chains and neighboring amino acids on $k_c$. Exchange rates measured for individual protons in unfolded BPTI[21] are in excellent agreement with the mechanism in Eq. (4) and are predicted quite well by the rules of Molday et al.[24]

*Kinetic Limits.* Under conditions that energetically favor the folded state ($k_f \gg k_u$), Eq. (3) leads to the following expression for the observed exchange rate, $k_{ex}$:

$$k_{ex} = k_u k_c/(k_f + k_c) \tag{5}$$

Two limiting situations can be distinguished, depending on the ratio between $k_f$ and $k_c$. In the limit of rapid structural fluctuations compared to the intrinsic exchange ($k_f \gg k_c$), Eq. (5) becomes

$$k_{ex} = K_{op} k_c \tag{6}$$

[17] S. W. Englander, *Ann. N.Y. Acad. Sci.* **244**, 10 (1975).
[18] A. Berger, A. Loewenstein, and S. Meiboom, *J. Am. Chem. Soc.* **81**, 62 (1959).
[19] M. Eigen, *Angew. Chem., Int. Ed. Engl.* **3**, 1 (1964).
[20] S. W. Englander and A. Poulsen, *Biopolymers* **7**, 329 (1969).
[21] H. Roder, G. Wagner, and K. Wüthrich, *Biochemistry* **24**, 7407 (1985).
[22] J. J. Englander, D. B. Calhoun, and S. W. Englander, *Anal. Biochem.* **92**, 517 (1979).
[23] A. J. Covington, R. A. Robinson, and R. G. Bates, *J. Phys. Chem.* **70**, 3820 (1966).
[24] R. S. Molday, S. W. Englander, and R. G. Kallen, *Biochemistry* **11**, 150 (1972).

The exchange rate is then determined by the relative concentration of unfolded protein, which is approximated by the unfolding equilibrium constant, $K_{op}$, and the rate-limiting intrinsic exchange rate. This exchange-limited situation, often also termed the $EX_2$ limit,[10] is typically encountered for proteins under conditions where the folded state is stable and intrinsic exchange is relatively slow. The quantity of interest is $K_{op}$ which contains information on the structural contribution to hydrogen exchange. With $\Delta G_{op} = -RT \ln K_{op}$ we can write Eq. (6) as

$$\Delta G_{op} = -RT \ln(k_{ex}/k_c) \tag{6a}$$

In the limit of slow structural fluctuations relative to the intrinsic exchange rate ($k_f \ll k_c$), the measured exchange rate is determined by the rate of unfolding, and Eq. (5) simply becomes

$$k_{ex} = k_u \tag{7}$$

In this opening limited case ($EX_1$ limit), exchange occurs with each structural opening and the measured rate directly gives the rate of unfolding. Conditions favoring opening limited exchange are basic pH, where $k_c$ is fast, and/or destabilizing conditions, where folding rates tend to be slow (see below).

The distinction of the two kinetic exchange mechanisms is crucial for the interpretation of hydrogen exchange data. Indications for opening limited exchange can be obtained from the pH dependence of exchange.[25–27] In the $EX_2$ limit the pH dependence is determined by the strong variation of the intrinsic exchange rate [Eq. (4)], modulated by any pH dependence of the structural equilibrium. On the other hand, in the $EX_1$ limit only the unfolding rate contributes to the pH dependence. A more rigorous distinction between the two exchange mechanisms is based on the fact that only in the $EX_1$ limit can exchange of different protons occur in a structurally correlated way; in the $EX_2$ limit the protein has to undergo many unfolding transitions before a particular proton is exchanged, giving rise to a random distribution of exchanged sites. The first clear evidence for opening limited exchange was found for BPTI at elevated temperatures,[27] using an NOE method developed by Wagner.[28]

The $EX_2$ mechanism is expected to prevail under most common circumstances. For example, the most slowly exchanging amide protons in the core of the BPTI $\beta$ sheet were shown to exchange via $EX_2$ processes over the pH range between pH 1 and 10 at 45°.[27] Opening limited ex-

[25] R. Richarz, P. Sehr, G. Wagner, and K. Wüthrich, *J. Mol. Biol.* **130**, 19 (1979).
[26] C. K. Woodward and B. D. Hilton, *Biophys. J.* **32**, 561 (1980).
[27] H. Roder, G. Wagner, and K. Wüthrich, *Biochemistry* **24**, 7396 (1985).
[28] G. Wagner, *Biochem. Biophys. Res. Commun.* **97**, 614 (1980).

change was only observed above pH 7 under more destabilizing conditions (at 68° or in the presence of 3 $M$ GuHCl at 55°). According to Eq. (6) exchange rates measured in the EX$_2$ regime, together with the known values for $k_c$ (estimated with the rules of Molday et al.,[24] or measured directly in the unfolded form as in Roder et al.[21]), provide equilibrium information on structural unfolding transitions. When individually assigned protons are observed by [1]H NMR, it becomes apparent that exchange can be caused by a wide spectrum of structural fluctuations, ranging from local distortions that break a small number of hydrogen bonds (for examples, see Refs. 14 and 29), to global transitions approaching general unfolding transitions. In the following we will concentrate on the global end of the spectrum and its relationship to protein stability and denaturation.

*Correlation between Hydrogen Exchange and Protein Stability.* In a study of homologous and chemically modified trypsin inhibitors with variable stability, Wagner and Wüthrich[16] found that the exchange rates for the slowly exchanging amide protons in the $\beta$ sheet were correlated with thermal stability, indicating that global fluctuations are responsible for the exchange of these protons. This correlation was further investigated by measuring the effect of various denaturants on exchange.[30] Exchange rates were measured at 55°, pH 6.5, for seven of the most slowly exchanging amide protons in BPTI, all of which are part of the hydrogen bond network in the central $\beta$ sheet. These rates are plotted in Fig. 3[30a] as a function of the midpoint temperature of the thermal denaturation transition, $T_m$, in different solvents (D$_2$O and different concentrations of ethanol, urea, and GuHCl). $T_m$ was determined from thermal transition curves measured by circular dichroism (CD) at 285 nm with BPTI solutions containing the same solvent additives. A clear correlation between $k_{ex}$ and protein stability is observed; $k_{ex}$ increases dramatically with decreasing $T_m$, by a factor of up to 6000 between the extremes. We further note that the difference in rate for the protons studied is small and decreases somewhat with increasing destabilization, indicating that this set of protons is exchanged via a common cooperative unfolding transition. Other protons at more peripheral locations in the structure exchange more rapidly (too rapid for quantitative rate measurements in the more denaturing solvents), suggesting exchange via less cooperative local fluctuations.

The exchange of the BPTI core protons considered here is governed by EX$_2$ processes under all solution conditions,[27] and Eq. (6) can be

---

[29] A. J. Wand, H. Roder, and S. W. Englander, *Biochemistry* **25**, 1107 (1986).

[30] H. Roder, Ph.D. Thesis No. 6932, ETH Zürich (1981).

[30a] E. Moses and H. J. Hinz, *J. Mol. Biol.* **170**, 765 (1983).

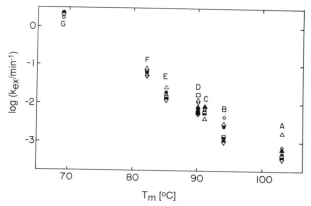

FIG. 3. Effect of destabilizing solvent additives on the exchange rates of BPTI $\beta$-sheet amide protons. Lyophilized BPTI was dissolved in $D_2O$, 0.1 $M$ $NaPO_4$, pD 6.5 (uncorrected pH meter reading) and incubated for H exchange at 55° in the presence of the following additives: none (A), 10% (v/v) ethanol (B), 4 $M$ urea (C), 20% ethanol (D), 8 $M$ urea (E), 3 $M$ GuHCl (F), and 6 $M$ GuHCl (G). H–D exchange rates measured by ¹H NMR for the NH of Arg-20 (●), Tyr-21 (■), Phe-22 (▽), Tyr-23 (□), Gln-31 (▲), Phe-33 (○), and Phe-45 (△) are plotted logarithmically as a function of the midpoint temperature of the thermal unfolding transition, $T_m$, of BPTI in the different solvents. $T_m$ was determined from equilibrium unfolding curves monitored by CD at 285 nm. The value for $T_m$ in the absence of denaturing agents (A) was taken from the calorimetric data of Moses and Hinz.[30a]

applied to extract opening equilibrium constants. Exchange data measured in thermally unfolded BPTI for the same set of protons[21] were used for $k_c$, after correction for the effect of the solvent additives on intrinsic exchange.[30,31] The free energy changes, $\Delta G_{op}$, calculated with Eq. (6a), are compared in Fig. 4 with the free energy of unfolding, $\Delta G_u$, determined from the CD melting data in the respective solvents. Equation (2) was used to obtain values for $\Delta G_u$ at 55° from the CD data measured near $T_m$ over a range of about 30° (cf. Fig. 2). The data in Fig. 4 fall close to a straight line through the origin with slope 1, indicating close agreement between the free energies determined by hydrogen exchange and those obtained from thermal unfolding curves. (The vertical displacement of the data, about 1 kcal above the diagonal, is most likely due to the uncertainty of the extrapolation of $k_c$ values.) This leads to the following conclusions: (1) exchange of the core protons in the BPTI $\beta$ sheet requires major unfolding, comparable to the conformational changes associated with denaturation; (2) the structural unfolding model [Eq. (3)] provides a firm basis for the quantitation of free energy changes and protein stability; (3)

---

[31] D. J. Loftus, G. O. Gbenle, P. S. Kim, and R. L. Baldwin, *Biochemistry* **25**, 1428 (1986).

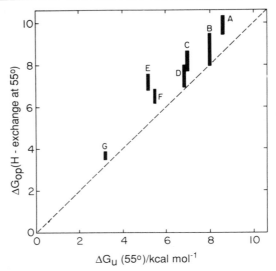

FIG. 4. Correlation between the opening free energy, $\Delta G_{op}$, calculated with Eq. (6a) from the BPTI $\beta$-sheet NH exchange data of Fig. 3, and the unfolding free energy, $\Delta G_u$, determined by extrapolation of thermal unfolding data to 55°, using Eq. (2). The letters refer to the different solvent additives indicated in Fig. 3. The spread of exchange rates among the seven amide protons shown in Fig. 3 is indicated by the length of the bars. For the calculation of $\Delta G_{op}$ with Eq. (6a), the $k_c$ values measured in unfolded BPTI for the same set of amide protons[21] were used after minor corrections for the effect of the solvent additives on free-peptide exchange rates.[30,31]

exchange rates measured in the fully unfolded protein[21] or derived from model peptide data[24] are good approximations of the actual chemical exchange rates in transiently unfolded states.

The application of hydrogen exchange to measure protein stability promises to solve some of the problems associated with conventional methods. Calorimetry and spectroscopic melting methods are sensitive only over a limited range of conditions around $T_m$, and the extrapolation required to obtain the desired stability under ambient conditions introduces significant uncertainty. This problem is particularly serious in denaturant studies where the result is strongly model dependent.[32] Another limitation is the requirement for complete reversibility of denaturation, which is difficult to achieve for many proteins due to limited solubility of the unfolded form. Both of these problems can be avoided by the hydrogen exchange technique. Exceedingly small unfolding equilibrium constants can be measured, especially at basic pH. For example, at pH 9, $k_c$ is

[32] C. N. Pace, this series, Vol. 131, p. 266.

about $10^3$ sec$^{-1}$ at room temperature, and the measurement of an equilibrium constant of $10^{-9}$ ($\Delta G_u$ = 12 kcal/mol) would require exchange times of about 10 days; 100 fold longer exchange times needed for a measurement of pH 7 are still quite feasible and longer times are only limited by the long-term stability of the protein solution and the patience of the investigator. Conformational stability can thus be measured directly under ambient conditions. By the same token, irreversible denaturation is not a problem, since one does not have to cross the unfolding transition. It is anticipated that these methods will be particularly useful to study the effect of mutations on protein stability.

Other than the problem of resonance assignment the most difficult aspect in applying these methods to a new protein is finding a suitable set of exchangeable protons that serve as indicators of global stability. Such groups are likely to be among the most slowly exchanging protons in the molecule, provided these are not involved in residual structure remaining after the major unfolding transition. Some of these issues are illustrated by the following example from our current work with cytochrome $c$.[29,33] Among about 50 amide protons in oxidized cytochrome $c$ observed by two-dimensional NMR,[29] there is a subset of about a dozen NH groups with very similar exchange rates. All exchange about $10^6$ times slower than $k_c$, corresponding to a $\Delta G_{op}$ of about 8 kcal/mol [Eq. (6a)]. This value is close to the free energy expected on the basis of equilibrium melting studies,[34,35] suggesting that the exchange of these protons reflects global unfolding. This correlation can be rationalized on structural grounds as many of these amide protons are deeply buried in the structure and are expected to become accessible only after drastic unfolding. A representative case is the indole NH of Trp-59, which forms a hydrogen bond to the heme propionate 7 in the center of the molecule. On reduction of cytochrome $c$, many of the protons belonging to the 8-kcal class in the oxidized form are stabilized by 2–3 kcal/mol. $\Delta G_{op}$ for the Trp-59 indole NH, for example, increases from 8.4 kcal in the oxidized form to about 11 kcal in the reduced form. This increase is consistent with estimates of the free energy change associated with the redox transition based on comparison of the redox potentials of cytochrome $c$ and model hemes.[36] The cytochrome $c$ results also illustrate the capability to study irreversible systems. There are no reliable stability data available for ferrocytochrome $c$ due to the difficulty of keeping the protein reduced in the unfolded form.

[33] H. Roder, A. J. Wand, and S. W. Englander, in preparation.
[34] J. A. Knapp and C. N. Pace, *Biochemistry* **13**, 1289 (1974).
[35] P. L. Privalov and N. N. Kechinashvili, *Biofizika* **19**, 10 (1974).
[36] H. A. Harbury, J. R. Cronin, M. W. Fanger, T. P. Hettinger, A. J. Murphy, Y. P. Myer, and S. N. Vinogradov, *Proc. Natl. Acad. Sci. U.S.A.* **54**, 1658 (1965).

Measuring exchange and $\Delta G_{op}$ in the reduced form near room temperature is, however, straightforward.

### Structural Characterization of Unfolded Proteins

The conformational properties of peptide fragments have attracted considerable interest in recent years due to their implications for the early phases of protein folding.[37,38] NMR is again among the most promising tools for the characterization of marginally stable peptide conformations. The problem of obtaining specific resonance assignments in such systems is often more difficult than in stable native proteins, due to structural flexibility and limited spectral dispersion. The assignment problem becomes somewhat easier for reversibly unfolded proteins, since resonance assignments obtained in the folded form can be carried over into the unfolded form by means of magnetization transfer methods.

*Resonance Assignment in Unfolded Proteins.* NMR methods for studying chemical exchange or structural isomerization by magnetization or saturation transfer are well established.[39–42] They are applicable to systems where exchange or structural interconversion is slow on the NMR time scale, but not too slow relative to spin-lattice relaxation times. The unfolding–refolding transition of many globular proteins falls into this category (cf. Fig. 1). Typical interconversion times near the midpoint of the transition are of the order of 1 sec[43,44] (see also following section). If the NMR signal of a nucleus is resolved in either the folded or the unfolded conformation, one-dimensional double-resonance methods, using selective presaturation or inversion,[30,40] can be applied to correlate the corresponding resonances. An interesting application of this technique to a system with more than two interconverting states is the work of Dobson and co-workers[45,46] with staphylococcal nuclease.

[37] K. R. Shoemaker, P. S. Kim, D. N. Brems, S. Marqusee, E. J. York, I. M. Chaiken, J. M. Stewart, and R. L. Baldwin, *Proc. Natl. Acad. Sci. U.S.A.* **82**, 2349 (1985).

[38] H. J. Dyson, K. J. Cross, R. A. Houghten, I. A. Wilson, P. E. Wright, and R. A. Lerner, *Nature (London)* **318**, 480 (1985).

[39] S. Forsen and R. A. Hoffman, *J. Chem. Phys.* **39**, 2892 (1963).

[40] I. D. Campbell, C. M. Dobson, R. G. Ratcliffe, and R. J. P. Williams, *J. Magn. Reson.* **29**, 397 (1978).

[41] J. Jeener, B. H. Meier, P. Bachman, and R. R. Ernst, *J. Chem. Phys.* **71**, 4546 (1979).

[42] J. J. Led, H. Gesmar, and F. Abildgaard, this volume [15].

[43] H. Roder, G. Wagner, R. Richarz, and K. Wüthrich, *Experientia* **35**, 942 (1979).

[44] C. M. Dobson and P. A. Evans, *Biochemistry* **23**, 4267 (1984).

[45] R. O. Fox, P. A. Evans, and C. M. Dobson, *Nature (London)* **320**, 192 (1986).

[46] P. A. Evans, C. M. Dobson, R. A. Kautz, G. Hatfull, and R. O. Fox, *Nature (London)* **329**, 266 (1987).

For assignment purposes it is more efficient to use two-dimensional NMR methods (see Basus, this series, Vol. 177, [7]). When applied to a molecule undergoing chemical exchange or structural isomerization, the 2D NOE (NOESY) experiment[41,47] produces cross-peaks between resonances that are connected via exchange-mediated magnetization transfer. With sufficiently rapid folding–unfolding rates ($>5$ sec$^{-1}$) magnetization transfer by site exchange can be more efficient than cross-relaxation, and exchange cross-peaks dominate over most NOE cross-peaks. On the other hand, exchange broadening occurs when conformational changes are too rapid.

Spectral resolution is a major concern in such experiments, even with relatively small proteins, since up to twice the number of resonances can be present in a two-state equilibrium. Using phase-sensitive 2D NOE spectra,[48] the cross-assignment of nuclei with resolved resonances in either the folded or the unfolded form is straightforward, but additional information is required if both positions are poorly resolved. It is necessary to combine magnetization transfer experiments with $J$-correlated spectroscopy (COSY) in the fully folded and the fully unfolded protein. Thus, assignments based on magnetization transfer can be extended to complete spin systems.

Recent 2D NMR experiments from our laboratory[49] have shown that the application of isotropic mixing or rotating-frame experiments (TOCSY or ROESY)[50–52] can be very useful in site-exchange situations. In addition to the usual $J$-coupling cross-peaks, 2D isotropic mixing spectra contain exchange cross-peaks and relayed connectivities due to the combination of $J$ mixing and exchange[53] (see also Brown and Farmer [11], this volume). All of these effects have the same sign and are in phase, while rotating-frame NOE (ROE) cross-peaks have the opposite sign. Relayed connectivities appear between the frequency of a proton in one state and the frequency of a proton that is $J$ coupled to the corresponding proton in the other state. These cross-peaks are generally removed from the crowded region near the diagonal and provide useful confirmatory evidence for exchange connectivities. Compared to the analogous relayed NOESY scheme,[54] the sensitivity of isotropic mixing experiments ap-

[47] A. Kumar, R. R. Ernst, and K. Wüthrich, *Biochem. Biophys. Res. Commun.* **95**, 1 (1980).

[48] D. J. States, R. A. Haberkorn, and D. J. Reuben, *J. Magn. Reson.* **48**, 286 (1982).

[49] Y. Feng and H. Roder, *J. Magn. Reson.* **78**, 597 (1988).

[50] L. Braunschweiler and R. R. Ernst, *J. Magn. Reson.* **53**, 521 (1983).

[51] A. A. Bothner-By, R. L. Stephens, J. Lee, C. D. Warren, and R. W. Jeanloz, *J. Am. Chem. Soc.* **106**, 811 (1984).

[52] A. Bax and D. G. Davis, *J. Magn. Reson.* **65**, 355 (1985).

[53] B. T. Farmer II, S. Macura, and L. R. Brown, *J. Magn. Reson.* **72**, 347 (1987).

[54] G. Wagner, *J. Magn. Reson.* **57**, 497 (1984).

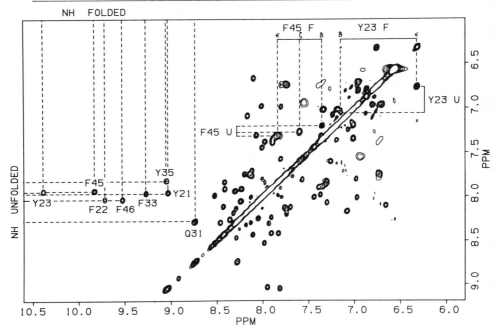

FIG. 5. Application of 2D magnetization transfer spectroscopy for the chemical shift correlation between protons in folded and unfolded forms of a protein. The contour plot shows an expanded region of an NOESY spectrum[47] recorded in the phase-sensitive mode[48] at 500 MHz (mixing time 100 msec) on a 15 mM RCAM-BPTI sample in $H_2O$, pH 2.5, 58°. Under these conditions the protein is about 50% unfolded. The majority of the cross-peaks arise from magnetization transfer between the folded and the unfolded form. Selected examples of NH and aromatic resonance assignments are shown. Dashed lines indicate the resonance positions in the folded form (vertical) and the corresponding shifts in the unfolded form (horizontal).

pears to be superior. At optimal mixing times, relay peaks can be as intense as exchange and diagonal peaks.

*Characterization of Unfolded BPTI.* The methods described above were applied to BPTI, resulting in a number of sequence-specific assignments in the $^1$H NMR spectrum of the unfolded form.[30] The chemically modified RCAM-BPTI, obtained by selective reduction and carboxyaminomethylation of the 14–38 disulfide bond, was chosen for these studies because of its lower thermal stability.[55] Detailed $^1$H NMR studies showed that the structure of RCAM-BPTI is very similar to the native inhibitor, with only minor perturbations in the vicinity of the cleaved

[55] J. P. Vincent, R. Chicherportiche, and M. Lazdunski, *Eur. J. Biochem.* **23,** 401 (1971).

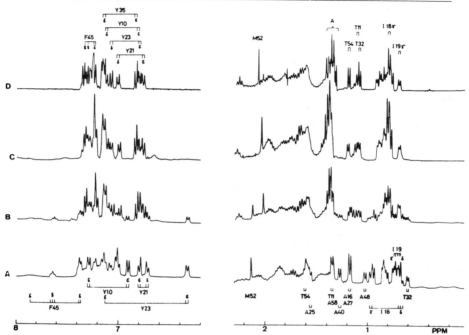

FIG. 6. Aromatic and methyl region of the 360-MHz $^1$H NMR spectrum of various forms of BPTI in $D_2O$ (pD 2.5, 41°): native RCAM-BPTI (A); partially unfolded RCAM-BPTI in 4 $M$ urea (B); unfolded RCAM-BPTI in 8 $M$ urea (C); BPTI unfolded by reduction of all disulfide bonds (R-BPTI) in $D_2O$ (D).

disulfide.[56] An example of a 2D magnetization transfer spectrum, recorded in $H_2O$ near the midpoint of the thermal unfolding transition of RCAM-BPTI, is given in Fig. 5. The spectral region containing the NH and aromatic resonances is displayed along with a selection of assignments in the folded and unfolded forms. Additional assignments are indicated in the 1D spectrum shown in Fig. 6, in which native RCAM-BPTI is compared with the urea-unfolded protein and a denatured derivative obtained by complete reduction of all three disulfides (R-BPTI).

Upon unfolding, the spectral features characteristic of the folded state disappear and the resonances generally move closer to the random coil position expected on the basis of model peptide data.[6] But a significant amount of spectral dispersion remains in all unfolded forms studied (the most extreme case was R-BPTI in 6 $M$ GuHCl at 90°). For example, the

[56] C. I. Stassinopoulou, G. Wagner, and K. Wüthrich, *Eur. J. Biochem.* **145,** 423 (1984).

ring protons of all four tyrosines and the methyl groups of all three threonines are resolved, and the $C_\gamma H_3$ of Ile-19 gives rise to a separate high-field shifted resonance. Very similar patterns were observed in unfolded BPTI with all three disulfide bonds intact, in unfolded RCAM-BPTI with two disulfides, and in R-BPTI with no disulfides, indicating that the spectral dispersion is not related to cross-linking of the unfolded chain. The side chains of Ile-19, Tyr-21, and Thr-32 display the largest deviations from the expected intrinsic chemical shifts.[6] It is interesting that all these groups occur close to aromatic residues in the amino acid sequence. Some of the resolved resonances in the spectrum of R-BPTI gradually move toward their random coil position when the temperature was increased from 10 to 90°. These observations suggest that unfolded BPTI assumes a disordered, but not a random, conformation. Residual structure appears to persist in some parts of the structure even under extreme conditions. The observation that the hydrogen exchange rates in thermally unfolded RCAM-BPTI are close to those expected for an unstructured peptide[21] suggests that no stable hydrogen-bonded structure is present. More likely, the residual structure is due to short-range interactions between hydrophobic side chains.

## Dynamics of Folding and Unfolding at Equilibrium

The magnetization transfer techniques used in the previous section to obtain assignments in unfolded proteins can be extended to provide quantitative kinetic information on the structural isomerization under equilibrium conditions (see Led *et al.* [15], this volume). In a time-resolved one-dimensional saturation transfer experiment a resolved line is selectively saturated for a period $\tau$, followed by a nonselective 90° pulse and data acquisition. The magnetization transferred across the conformational equilibrium is monitored as a function of $\tau$.[40] Difference spectroscopy with alternating on- and off-resonance irradiation is used to observe saturation transfer into crowded spectral regions. This approach was first applied to a protein-folding equilibrium some time ago.[30,43] A more recent application to lysozyme has been reported by Dobson and Evans.[44]

The *time-resolved saturation transfer* technique is illustrated in Fig. 7 with a series of experiments at different temperatures across the thermal unfolding transition of RCAM-BPTI ($T_m = 59°$). The $C_\varepsilon$ ring proton resonance of Tyr-23 in the unfolded state was irradiated and the normalized intensity of the corresponding resonance in the folded state is plotted as a function of the saturation time $\tau$. At each temperature the intensity decays exponentially from the equilibrium value, $M_0$, to a limiting value, $M_\infty$, indicating that the situation is adequately described by a two-state model.

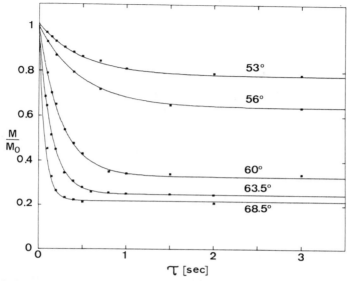

FIG. 7. Temperature dependence of the time-resolved saturation transfer between folded and unfolded forms of RCAM-BPTI ($D_2O$, pD 3.0), induced by selective irradiation at the resonance position of Tyr-23 $C_\varepsilon H$ in the unfolded form (see text).

The decay rate is then given by the sum of the unfolding rate, $k_u$, and the relaxation rate, $R_f$, in the folded form, and $M_\infty$ is given by[56a]

$$M_\infty/M_0 = R_f/(R_f + k_u) \tag{8}$$

From the decay rate $k_u + R_f$ and $M_\infty$, determined by an exponential fit to the data, both $k_f$ and $R_f$ can thus be independently determined. Conversely, the rate of folding, $k_f$, is measured by irradiating a resonance in the folded state and observing the unfolded state. Therefore, both the forward and the backward rate constant can be measured independently for a system at thermodynamic equilibrium. This capability is probably the main advantage of saturation transfer over conventional kinetic techniques like stopped flow or temperature jump that measure effective relaxation rates consisting of the sum of forward and backward rate constants.

The existence of multiple forms of the unfolded protein with a fraction of more slowly refolding molecules may not be directly manifested in the

[56a] It is assumed that the irradiated resonance is saturated instantaneously and completely. For a discussion of transient effects at short times see Refs. 40 and 57.

[57] G. Wagner and K. Wüthrich, *J. Magn. Reson.* **33**, 675 (1979).

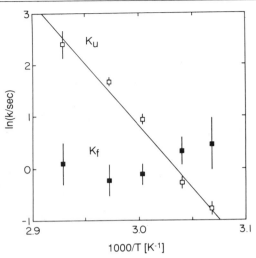

FIG. 8. Temperature dependence of the folding rate $k_f$ (■) and the unfolding rate $k_u$ (□), measured by saturation transfer in the folding–unfolding equilibrium of RCAM-BPTI at pD 3. $k_u$ was determined from exponential fits to the data in Fig. 7 (irradiation in the unfolded form), and $k_f$ from similar data (not shown) obtained by irradiation of the Tyr-23 $C_\varepsilon H$ resonance in the folded form. Linear regression of $\ln(k_u)$ vs inverse absolute temperature (solid line) yields a value of $47 \pm 5$ kcal/mol for the activation energy of unfolding.

time course of magnetization transfer. However, a constant background corresponding to the relative amplitude of the slow phase would be added to the steady state magnetization at long times and Eq. (8) would lead to erroneous results. The presence of slowly folding molecules is revealed by the following tests: (1) the apparent equilibrium constant $K = k_u/k_f$ based on Eq. (8) deviates from the equilibrium constant calculated with Eq. (1) from normalized resonance intensities; (2) $R_f$ and $R_u$ from the saturation transfer data differ from spin-lattice relaxation rates measured by conventional methods. In the case of RCAM-BPTI these criteria lead to an estimated upper limit of 15% for the amplitude of the slow folding phase. This value is consistent with the stopped flow data of Jullien and Baldwin.[58]

The temperature dependence of $k_f$ and $k_u$ measured in RCAM-BPTI is presented in Fig. 8 as an Arrhenius plot. The rate of unfolding is strongly temperature dependent, while the refolding rate remains essentially constant at about 1 $sec^{-1}$ across the unfolding transition. The apparent activation energy for unfolding is 50 kcal/mol. This value is close to the equilib-

[58] M. Jullien and R. L. Baldwin, J. Mol. Biol. 145, 265 (1981).

rium enthalpy change for unfolding of RCAM-BPTI.[30] Thermal unfolding is induced primarily by a steep increase in unfolding rates, and the rate of refolding contributes little to the temperature dependence. This type of behavior may be a general feature of protein folding. Very similar results have been obtained in analogous magnetization transfer experiments with lysozyme.[44] Refolding reactions with complex temperature dependencies have also been observed by stopped flow and temperature jump for a number of proteins.[59-62] The anomalous temperature dependence of the folding rate appears to be related to the large unfavorable entropy change associated with refolding and is an indication that folding is not a simple two-state reaction.[63]

### Characterization of Folding Pathways by Hydrogen Exchange Labeling

The NMR methods discussed so far are applicable to equilibrium situations. Their application to small globular proteins has generally shown that folding–unfolding transitions are highly cooperative, indicating that folding intermediates are short lived and only marginally populated at equilibrium. One has a better chance of gaining insight into the sequence of intermediate structures if refolding is carried out far below the transition zone where partially folded conformations are likely to be more stable. Since NMR is inherently slow and insensitive it cannot be directly applied to study transient protein conformations. A promising approach for increasing the time resolution without sacrifice of structural resolution is to combine rapid mixing methods with hydrogen exchange to apply proton or deuteron labels to transient conformations encountered during refolding. The labels are trapped within the refolded protein and their location can be determined at leisure by ¹H NMR.

Backbone and side-chain amide protons are ideal conformational probes: (1) they are distributed throughout the protein structure; (2) amide protons reflect important aspects of protein structures since their exchange rates are primarily determined by intramolecular hydrogen bonding[14] (3) hydrogen–deuterium exchange is associated with negligible structural perturbation; (4) individual protons can be monitored by one- and two-dimensional ¹H NMR.[25,29,64] When hydrogen exchange labeling is

[59] P. McPhie, *Biochemistry* **11**, 879 (1972).
[60] P. J. Hagerman and R. L. Baldwin, *Biochemistry* **15**, 1462 (1976).
[61] F. M. Pohl, *FEBS Lett.* **65**, 293 (1976).
[62] S. Kato, N. Shimamoto, and H. Utiyama, *Biochemistry* **21**, 38 (1982).
[63] H. Utiyama and R. L. Baldwin, this series, Vol. 131, p. 51.
[64] G. Wagner and K. Wüthrich, *J. Mol. Biol.* **160**, 343 (1982).

combined with rapid mixing methods and $^1$H NMR analysis one can achieve both structural and temporal resolution.[65]

The application of hydrogen exchange labeling to kinetic folding studies was introduced by Schmid and Baldwin.[66] They labeled the slow-folding species of unfolded RNase A with tritium and induced refolding by manual mixing with nonradioactive buffer to set up a competition between $^3$H–$^1$H exchange and refolding. More $^3$H was trapped in the refolded protein than expected on the basis of optically monitored folding rates, indicating that structure capable of retarding hydrogen exchange is formed early during refolding. These results were confirmed by using pulse labeling,[67] a complementary approach in which the protein is allowed to refold for variable time periods before exchange labeling is applied. A detailed review of hydrogen exchange labeling methods and applications appeared in this series.[68]

When amide proton assignments in BPTI became available[1,69] it became possible to apply $^1$H NMR to observe individual protons trapped during refolding.[30,65] Since BPTI refolding is, in contrast to RNase A, dominated by rapid folding on a 10-msec time scale,[58] it was necessary to extend the time resolution of the competition method by using rapid mixing techniques. Interpretation of the competition data in terms of a minimal kinetic model suggested that the protons in the center of the $\beta$ sheet were protected early during refolding, on the time scale of the fast folding phase observed by tyrosine absorbance.[58] Apparent folding rates for several $\beta$-sheet protons ranged from 25 to 60 sec$^{-1}$ while one NH on the C-terminal $\alpha$-helix was protected at a rate of 15 sec$^{-1}$.

*Competition Method.* The strategy of the competition method[65,66] is to balance the rate of NH exchange against the rate of refolding so that exchangeable protons are trapped in parts of the protein that refold early. The flow system of a rapid mixing apparatus used for such experiments is schematically shown in Fig. 9.[69a,b] The protein is initially unfolded in $H_2O$ in the presence of a suitable denaturing agent or at extreme pH (syringe S1). Refolding and H–D exchange are induced simultaneously in a first mixer (M1) by rapid dilution of the denaturant with a $D_2O$ buffer solution (S2), causing a jump to native conditions. After a reaction time $\tau$ which can be adjusted by changing the flow rate and/or the reaction volume, the

[65] H. Roder and K. Wüthrich, *Proteins* **1**, 34 (1986).

[66] F. X. Schmid and R. L. Baldwin, *J. Mol. Biol.* **135**, 199 (1979).

[67] P. S. Kim and R. L. Baldwin, *Biochemistry* **19**, 6124 (1980).

[68] P. S. Kim, this series, Vol. 131, p. 136.

[69] A. Dubs, G. Wagner, and K. Wüthrich, *Biochim. Biophys. Acta* **577**, 177 (1979).

[69a] Preparative quench system from Hi-Tech Scientific, Ltd., Salisbury, England.

[69b] R. L. Berger, B. Balko, and H. Chapman, *Rev. Sci. Instrum.* **39**, 493 (1968).

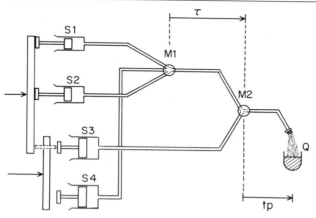

FIG. 9. Schematic diagram of the flow system of a commercial preparative rapid mixing apparatus[69a] adapted for competition and pulse-labeling experiments. The solutions from syringes S1 and S2 are mixed in the first mixing chamber, M1. After a variable aging time, $\tau$, the mixture is combined in M2 with a third reagent from syringe S3. Highly efficient ball mixers of the Berger type[69b] are used. Aging times from about 5 to 200 msec are achieved by continuous flow, for which syringes S1, S2, and S3 are advanced simultaneously by a motor-driven cam drive (a pin, indicated by dashed lines, is inserted at S3). For longer aging times interrupted flow is used. First the cam drive is activated to advance S1 and S2 (pin removed), filling the tubing between M1and M2 with reaction mixture; after an electronically timed interval, a pneumatic drive advances syringes S3 and S4 to push the solution through M2 where it is mixed with a third reagent (S3). For competition experiments H exchange is quenched in M2 by mixing with an acidic buffer solution (from S3). In pulse-labeling experiments hydrogen or deuterium labeling is induced in M2 and after a delay, $tp$, quenching is achieved by injecting the mixture through a nozzle into excess buffer solution in an external container (Q).

exchange process is quenched by rapid lowering of the pH in the second mixer (M2). This halts further hydrogen–deuterium exchange but allows refolding to continue to completion, thereby protecting the remaining amide protons against further exchange. The protein is then recovered and NMR samples are prepared under conditions chosen to minimize loss of NH label (low temperature, moderately acidic pH). The amount and location of trapped hydrogen is analyzed by measuring the areas of resolved NH resonances in 1D $^{1}$H NMR spectra or cross-peak volumes in 2D spectra. Relative proton occupancies, $P$, at each site are calculated by normalizing the measured signal intensities, $I_m$, as follows:

$$P = [(I_m/I_0) - f_h]/(1 - f_h) \tag{9}$$

where $I_0$ is the signal intensity of the fully protonated group, and $f_h$ is the residual fraction of $H_2O$ present in the reaction mixture.

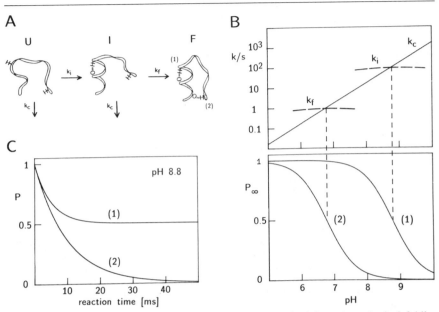

FIG. 10. Schematic illustration of the competition method for a hypothetical folding pathway (A). The principle of competition experiments at long reaction times and variable pH is shown in (B). The expected results of competition experiments at constant pH and variable reaction times are illustrated in (C). Details are described in the text.

The principle of the competition method is illustrated in Fig. 10 for a hypothetical folding pathway in which an exchangeable proton (1) is protected at a rate $k_i$ by rapid formation of an intermediate $I$ containing a helical segment; another proton (2) is protected at a slower rate $k_f$ as it forms a tertiary hydrogen bond at a later stage during refolding. In this example, pH is used to vary the free-peptide exchange rate $k_c$, and the limiting degree of protonation, $P_\infty$, is measured at long reaction times ($\tau \gg 1/k_c$). Within the limitations of the simple kinetic scheme,[70] $P_\infty$ measured at site 1 is given by

$$P_\infty = k_i/(k_i + k_c) \tag{10}$$

For site 2, $k_i$ is replaced by $k_f$ in Eq. (10). Assuming that the pH dependence of $P_\infty$ is dominated by $k_c$ according to Eq. (4), both protons are trapped at low pH where $k_c < k_f$. As $k_c$ is increased through $k_f$ the proton

---

[70] This analysis assumes that refolding proceeds essentially irreversibly. More general kinetic schemes involving marginally stable intermediate states are discussed in Refs. 65 and 66.

occupancy measured at site 2 drops while the more rapidly formed intermediate structure protects proton 2 ($k_c < k_i$). Finally, at more basic pH where exchange is fast, this proton also enters the competition ($k_c \sim k_i$). The effective rate of protection for each probe can be estimated from the value of $k_c$ at $P_\infty = 0.5$. $k_c$ should be determined independently, e.g., by exchange measurements in the fully unfolded protein,[21] or it can be estimated from model peptide data.[24]

In a variation of the competition method[65] illustrated in Fig. 10C, the reaction time is varied at a fixed pH value, e.g., at pH 8.8 where $P_\infty \sim 0.5$ for proton 1 in our example. The relative protonation decays exponentially at a rate $k_c + k_i$ from 1 at short $\tau$ to $P_\infty = k_i/(k_i + k_c)$. Exchange and folding rates can thus be measured independently from the decay rate and $P_\infty$ determined by an exponential fit.

Some practical aspects of the competition method are illustrated by an example from our work with horse cytochrome $c$ (cf. Ref. 71). Ferricytochrome $c$ was unfolded in $H_2O$ solutions with 4.5 $M$ GuHCl and refolded at 10° in 85% $D_2O$ containing 0.75 $M$ GuHCl at variable pH. Exchange and refolding were allowed to proceed for relatively long time periods (between 0.2 and 10 sec, depending on pH) so that limiting exchange levels were reached before H exchange was quenched (pD 5, 0°). Excess ascorbate was included in the quench buffer to reduce the refolded cytochrome $c$. This procedure was repeated at several different pH values in the range between pH 6 and 9.5 and a total correlation spectrum[50,52] was recorded for each sample. Expanded contour plots containing a selection of NH–$C_\alpha$H cross-peaks are shown in Fig. 11 for three representative refolding preparations. A reference spectrum of a fresh cytochrome $c$ sample in $D_2O$ is included for comparison. Some of the cross-peak in the sample refolded at pH 7 (Fig. 11B) have intensities comparable to the reference spectrum, indicating that these protons are protected on a time scale that is short compared to the intrinsic exchange time at this pH (0.1 to 0.5 sec). Other protons are already partially exchanged, even at more acidic pH (data not shown), which may be due to the presence of a small fraction of slow-folding cytochrome $c$ molecules.[72] With increasing pH the cross-peak intensities decrease as the exchange

[71] H. Roder, G. A. Elöve, and S. W. Englander, *Nature (London)* **335**, 700 (1988).

[72] Absorbance-detected stopped flow data recorded under similar folding conditions[71,73] display a major folding phase (80% of the amplitude) in the 100-msec time range and a minor process with a 10-sec relaxation time. Ridge *et al.*[74] showed that the two phases are due to fast folding and slow folding forms of the unfolded protein, probably as a result of proline isomerization.

[73] A. Ikai, W. W. Fisher, and C. Tanford, *J. Mol. Biol.* **73**, 165 (1973).

[74] J. A. Ridge, R. L. Baldwin, and A. Labhardt, *Biochemistry* **20**, 1622 (1981).

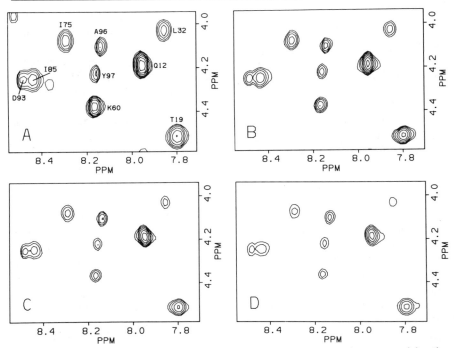

FIG. 11. Two-dimensional NMR analysis of cytochrome *c* samples prepared by the competition method at variable pH. Expanded contour plots of TOCSY spectra containing a selection of NH–C$_\alpha$H cross-peaks are shown for a reference sample of reduced cytochrome *c* freshly dissolved in D$_2$O (A), and three representative competition samples that were refolded at pH 7.0 (B), pH 7.6 (C), and pH 8.5 (D) as described in the text. Identical sample conditions were used for NMR analysis (D$_2$O solutions of 5 m$M$ ferrocytochrome *c*, 10 m$M$ ascorbate, 0.1 $M$ NaPO$_4$, pD 5.2, 20°). TOCSY spectra in pure absorption mode (TPPI method, 512 $t_1$ increments) were recorded at 500 MHz, using a 20-msec MLEV-17 mixing sequence.[52] All spectra are plotted at identical contour levels relative to a reference cross-peak (not shown) between nonexchanging protons.

rates progressively increase [Eq. (4)] and begin to compete with refolding. The sample refolded at pH 7.6 (Fig. 11C) displays larger intensity differences among different cross-peaks compared to the reference spectrum, which is an indication of differential folding behavior at the different sites in the protein. At basic pH, exchange dominates over refolding and most cross-peaks drop down to the background level determined by the residual amount of H$_2$O present in the mixture (15%); at pH 8.5 (Fig. 11D) exchange from the unfolded protein occurs in the range of 3 to 15 msec and most amide protons exchange before formation of H-bonded structure can protect them.

*Pulse Labeling.* The competition method is well suited for the investigation of early folding events, but the later stages of folding are more conveniently studied with the pulse-labeling method.[67,71,74a] The strategy is to allow the initially deuterated protein to refold in $D_2O$ for a variable time period before pulse labeling with excess $H_2O$ buffer. NH sites still exposed at this stage are selectively protonated. After the pulse, exchange is quenched, the protein refolds to the native form, and the labeled amide sites are assayed by NMR. The pulse should be chosen so that exposed sites become fully labeled and protected groups remain deuterated, which can be accomplished by briefly jumping to basic pH. Since three mixing stages are involved, pulse labeling is slower and requires higher dilution of the protein, but it offers greater flexibility compared to the competition scheme. Refolding can be done at fixed pH and the labeling pulse can be applied at any time during refolding. Furthermore, quantitative knowledge of intrinsic exchange rates is not required as long as the duration and pH of the pulse is chosen such that all exposed amide sites are completely labeled. The method is also advantageous in terms of background and dynamic range. Two recent studies on cytochrome $c$[71] and RNase A[74a] show that the pulse labeling method, together with 2D NMR analysis, can yield detailed structural information on transient folding intermediates. The cytochrome $c$ results provide clear evidence for partially folded forms.[71] For example, amide sites in the N- and C-terminal helices are protected within the first 20 msec of folding, while other sites remain accessible, indicating that formation of the two helices and their mutual stabilization are among the first events in folding.

*Experimental Considerations.* Much of the initial effort in such protein folding projects will have to be devoted to exploring conditions for unfolding and refolding, and testing of the reversibility and quench efficiency. Some of the methods discussed earlier in this chapter can be applied at this stage. It is important to choose unfolding conditions where the protein is as much as possible devoid of residual structure or aggregation. The presence of intra- or intermolecular H bonds would retard hydrogen exchange and lead to erroneous interpretation of the trapping results. The most pertinent test is to perform hydrogen exchange measurements under the conditions chosen for unfolding, and to compare the results with free-peptide rates. Under most circumstances rapid mixing is required for these measurements.

For refolding it is desirable to jump as far as possible away from the transition zone. When working with denaturants this requires large dilution factors. The dilution factor is also important for hydrogen exchange

---

[74a] J. B. Udgaonkar and R. L. Baldwin, *Nature (London)* **335,** 694 (1988).

since it determines the dynamic range of the observable change in NH resonance intensity.

Any loss of proton label occurring during the quench step and subsequent sample handling can be accounted for by recording the spectrum of a folded control sample that was never exposed to unfolding conditions, e.g., by leaving out the denaturant in the initial protein solution, but is otherwise processed in the same way.

Isotope effects on hydrogen exchange kinetics are small.[22] Except for possible solvent effects ($H_2O$ versus $D_2O$) on protein stability,[10] complementary results should be obtained in competition experiments where the protein is initially unfolded in $D_2O$ instead of $H_2O$. It is useful to compare both protocols, especially if 1D NMR is used for analysis; when exchange-out is used, the protons protected early in folding are well resolved at high pH, while exchange-in results in selective protonation of slowly protected amide groups. Similar considerations apply to pulse-labeling experiments.

The interpretation of H exchange labeling experiments may be complicated by the fact that unfolded proteins are generally kinetically heterogeneous and contain slow folding and fast folding species (for recent reviews, see Refs. 75 and 76). If in a competition experiment a fraction of the molecules would remain completely unfolded for minutes, a corresponding fraction of all proton labels would be lost. In the example of Fig. 10, the degree of protonation would level off at a value smaller than 1 at low pH. However, the presence of slow folding species can be detected more directly by the pulse-labeling method, which allows the slow folding population to be selectively labeled by an $H_2O$ pulse applied after completion of the fast folding phase in $D_2O$.

A number of 2D NMR experiments can be applied for the quantitation of NH intensities. The simplest one is the 2D $J$-correlated (COSY) experiment[77,78] in the absolute magnitude presentation. NH–$C_\alpha$H cross-peaks are convenient markers for quantitating amide proton population.[64] For instance, a COSY spectrum recorded in 6 hr on a fresh sample of reduced cytochrome c in $D_2O$ (10 m$M$ protein, pD 5.0, 20°) contained 55 distinct NH–$C_\alpha$H cross-peaks,[29] all of which have been assigned.[3] COSY spectra contain relatively few artifacts and a flat base plane (when processed with an unshifted sin or $sin^2$ filter). The spectral resolution can be improved by

[75] B. T. Nall, *Comments Mol. Cell. Biophys.* **3,** 123 (1985).
[76] P. S. Kim and R. L. Baldwin, *Annu. Rev. Biochem.* **51,** 459 (1982).
[77] W. P. Aue, E. Bartoldi, and R. R. Ernst, *J. Chem. Phys.* **64,** 2229 (1976).
[78] K. Nagayama, R. R. Ernst, and K. Wüthrich, *J. Magn. Reson.* **40,** 321 (1980).

using the double-quantum-filtered COSY in 2D absorption mode,[79] but this experiment requires longer acquisition times. Other possible 2D schemes are the NOESY[47] and total correlation (TOCSY) methods.[50,52] An attractive feature of these experiments is the in-phase nature of cross-peaks which, however, goes along with more serious artifacts ($t_1$ and $t_2$ ridges) and base plane distortions. Compared to COSY spectra the finger-print region in NOESY spectra is also more crowded and some NH–$C_\alpha H$ cross-peaks are very weak. A TOCSY experiment recorded at a short mixing time (about 20 msec) has the appearance of a COSY spectrum, but the cross-peak multiplet structure is in phase, which leads to significant gains in sensitivity. However, this advantage is offset to some extent by more severe artifacts and reduced reproducibility.

### Acknowledgments

I would like to thank G. A. Elöve, S. W. Englander, Y. Feng, B. T. Nall, G. Wagner, and K. Wüthrich for their contributions and discussions. Some of this research was performed as part of my Ph.D. work in the laboratory of Dr. K. Wüthrich at the ETH Zürich. The work performed at the University of Pennsylvania was supported by a grant from the National Institutes of Health (GM-35926).

[79] M. Rance, O. W. Sørensen, G. Bodenhausen, G. Wagner, R. R. Ernst, and K. Wüthrich, *Biochem. Biophys. Res. Commun.* **117**, 479 (1983).

Appendix

# Computer Programs Related to Nuclear Magnetic Resonance: Availability, Summaries, and Critiques

This Appendix provides a brief overview of selected computer programs discussed in Volumes 176 and 177 and information regarding their acquisition, operation, computer compatibility, maintenance, etc. By virtue of the dynamic nature of the field, any such listing will be incomplete; there are obviously many more computer programs devoted to all aspects of NMR spectroscopy than are listed here. Nonetheless, we hope that the information on the programs included will serve as a resource that will aid investigators in the design of experiments and in the selection of methods for data analysis.

## Determination of Solution Structure of Proteins

PROTEAN: Available from Dr. Oleg Jardetzky, Stanford University, Stanford Magnetic Resonance Laboratory, Stanford, California 94305-5055.

PROTEAN is a computer program designed to sample systematically the conformational space accessible to a protein and to determine the entire set of positions for each atom that are compatible with the given set of experimentally derived constraints. It solves a protein structure in a series of hierarchical steps starting with the calculation of a coarse topology of the folded structures and culminating in an iterative calculation of the original data to verify the correctness of the resulting family of structures.

The program is written primarily in LISP with a number of subroutines in C and FORTRAN. The coarse sampling can be run on microcomputers including VAX, $\mu$VAX, SUN, and even a Macintosh II (the latter for small proteins with a backbone containing $<50$ C$\alpha$'s). The full refinement requires a CRAY or a comparable minisupercomputer. See Chapter [11] by Altman and Jardetzky in Volume 177 of this series.

EMBED and VEMBED, Distance Geometry Calculations of Solution Structures: Available from Dr. I. D. Kuntz, Department of Pharmaceutical Chemistry, University of California San Francisco, San Francisco, CA 94143-0446.

EMBED and VEMBED are programs designed to produce one or more molecular structures that meet a set of constraints that can be derived from a wide variety of experimental or theoretical sources. It provides a means for a relatively high speed and wide ranging search of conformational space to supply starting structures as input for more sophisticated refinement methods.

EMBED is written in FORTRAN and designed for use in the UNIX or VMS operating system environment. VEMBED is the vectorized FORTRAN version designed to run on a Cray or Convex minisupercomputer. See Chapter [9] by Kuntz, Thomason, and Oshiro in Volume 177 of this series.

GROMOS, Molecular Dynamics Calculations: Available from Dr. Wilfred F. van Gunsteren, Laboratory of Physical Chemistry, University of Groningen, 9747 AG Groningen, The Netherlands.

The molecular dynamics approach, based on the molecular dynamics program GROMOS, determines solution structures of some initial set of coordinates, e.g., coordinates generated by a Distance Geometry analysis of the experimental data, by the sequential application of an energy minimization routine followed by a molecular dynamics calculation. The first operation is the Distance bounds Driven Dynamics (DDD) calculation which increases the variety of structures that satisfy the set of constraints, e.g., by removing "obvious" steric inconsistencies that would otherwise cause the molecular dynamics calculations to fail. The Restrained Molecular Dynamics calculations on the resulting structures provide high quality molecular structures consistent with the NMR data and other constraints.

GROMOS is written in FORTRAN 66 and currently runs under the VMS operating system (a UNIX-based version is being developed). The program contains special optimization routines for modern vector supercomputers. Practical considerations regarding computer time limit the number of atoms to <20,000. No graphics programs are provided, however, interfacing to graphics packages is straightforward. See Chapter [10] by R. Scheek, van Gunsteren, and Kaptein in Volume 177 of this series.

### Basic NMR Programs

These programs cover the basic operations needed for analyzing one- and two-dimensional spectra.

NMR1, NMR2: Available from New Methods Research, Inc., 719 East Genesee Street, Syracuse, NY 13210. It is written in FORTRAN 77 and designed for use with VAX, $\mu$VAX(VMS and ULTRIX), SUN(UNIX), Tektronics and similar color graphics work stations. This package is supported and has a software tool kit including programs for maximum entropy and linear prediction methods and spectral simulations.

FTCGI and FELIX: Available from Hare Research, Inc., 14810 216th Ave. N.E., Woodinville, WA 98072. These programs are written in FORTRAN (VMS or UNIX) and designed to run on microcomputers including VAX and $\mu$VAX as well as SUN and comparable workstations.

## Sequence Assignments

SEQASSIGN: Available from Dr. Martin Billeter, Institut für Molekularbiologie und Biophysik, Eidgenössiche Technische Hockschule-Hönggerberg, CH-8093 Zurich, Switzerland.

SEQASSIGN is a program designed for the automated assignment of 2D NMR spectra of proteins. It has two underlying concepts, first that all assignments consistent with currently available data are retained until unambiguously excluded. Second, the process of assignment is split into formal steps with "unambiguous" data, and the interpretation of "ambiguous" NMR data. The latter involves an interactive dialog with the user which can indicate what kind of data would be most important for further assignments. The program is written in Pascal and has been implemented on VAX (using VMS and Berkeley UNIX) and SUN (UNIX). See M. Billeter, V. J. Basus, and I. D. Kuntz, *J. Magn. Reson.* **76,** 400 (1988); Chapter [8] by Billeter in Volume 177 of this series.

## Spectra Simulations

SPHINX: Available from Dr. Kurt Wüthrich, Institut für Molekularbiologie und Biophysik, Eidgenössiche Technische Hockschule-Hönggerberg, CH-8093 Zurich, Switzerland.

SPHINX is a program capable of calculating homonuclear and heteronuclear 2D NMR experiments using a wide variety of pulse sequences and application of common phase-cycling and data-handling routines. It can also handle strongly coupled spin systems with up to six spins $I = 1/2$. An additional program is available called LINSHA that creates lineshapes and allows graphical presentation of the simulated 2D spectra.

The programs, as well as a stacked plot program, have been written in FORTRAN 77 and a contour plot program has been written in Pascal. The computations have been run on a DEC 10, working with the TOPS 10 system. See H. Widmer and K. Wüthrich, *J. Magn. Reson.* **70,** 270–279 (1986).

MARS: Available from Dr. Paul Rösch, Max-Planck-Institut für Medizinische Forschung, Abteilung Biophysik, Jahnstrasse 29, D 6900 Heidelberg, Federol Republic of Germany.

MARS provides for the simulation of 2D spectra of amino acids including glutamate with 8 coupled spins and leucine with 10 spins (computer memory may impose some limitations for spin systems of more than eight coupled spins). The program is written in Pascal and has been adopted to UNIX and MS-DOS operating systems. See I. Bock and P. Rösch, *J. Magn. Reson.* **74,** 177–183 (1987).

### Relaxation and Exchange Rates

Lineshape Analysis and Determination of Exchange Rates: Available from Dr. B. D. Nageswara Rao, Department of Physics, Indiana University–Purdue University at Indianapolis, 1125 East 38 Street, P.O. Box 647, Indianapolis, IN 46223.

This program will calculate lineshapes for NMR spin systems undergoing exchange between two magnetically distinct environments. Input parameters include all the individual chemical shifts (expressed as Larmor frequencies $\omega_A$, $\omega_{BS2A}$, etc.), spin–spin coupling constants, linewidths of all the transitions of the spin systems in the absence of exchange, and the exchange rates in either direction. The lineshapes are then calculated for different values of the exchange rate and compared with experiment. The program is written in FORTRAN and requires the availability of an external computer program for the diagonalization of complex unsymmetrical matrices (usually available at most computer centers, e.g., from IMSL). See Chapter [14] by Rao in Volume 176 of this series.

NMR50 and NMR55AUTO: Available from Dr. Peter P. Chuknyisky, Laboratory of Cellular and Molecular Biology, National Institutes of Health, Gerontology Research Center, Baltimore MD 21224. It is written in FORTRAN and designed for use with VAX 11/780 under VMS. These programs allow for the semiautomatic calculation of the parameters involved in the procedure for distance determination in NMR experiments. The programs are quite general and allow several different types of input. It is important that the number of frequencies used to measure $T_{1p}$ data be at least one more than the number of parameters chosen for computer searching for minimization. The programs can be used to compute the probe-nucleus distance for various combinations of probes and NMR nuclei (C, F, H, P) or to search for a preselected number of parameters ($\tau_m$, B, $\tau_v$, $\tau_r$, r) among those that are important in paramagnetic relaxation phenomena.

*CORMA* COmplete Relaxation Matrix Analysis: Available from Dr. Thomas L. James, Department of Pharmaceutical Chemistry, University of California San Francisco, San Francisco, CA 94143-0446.

CORMA is a FORTRAN program for calculating expected 2D NOE intensities (actually mixing coefficients) for any user-supplied molecular structure. CORMA requires as input: Cartesian coordinates supplied in Protein Data Bank (PDB) format, a correlation time, and (optionally) experimental intensities. A routine for generating proton coordinates from all other atomic coordinates is also provided. This information is supplied in two input files: model.PDB and model.INT.

The program calculates dipolar relaxation rates based on the proton–proton distances and the supplied correlation time. The correlation time can be a single effective isotropic correlation time for the whole molecule. Alternatively, correlation times are calculated for each proton–proton pair based on a local diffusion time assigned to each residue in the molecule, i.e., $\tau_{ij} = (\tau_i\tau_j)/(\tau_i + \tau_j)$. Correlation times involving methyl groups are further modified by a user-supplied multiplier ($\leq 1.0$) in order to approximate their shorter effective correlation times. The matrix of relaxation rates is then diagonalized and the exponential function $a = e^{-R\tau_m}$ is evaluated by left- and right-multiplying the exponentiated eigenvalue matrix by the unitary matrix of orthonormal eigenvectors: $a = \chi e^{-\lambda\tau_m}\chi$.

A summary of the calculated intensities (and RMS errors between calculated and experimental intensities if supplied), and a listing of cross-peaks (in the same format as the model.INT file) including the calculated intensity, the distance between protons, and the relaxation rate constants ($\sigma_{ij}$ or $\rho_i$), is generated. Additionally, PostScript files are created for plotting (on a laser printer such as the Apple LaserWriter) a representation of the intensity matrix in a schematic gray-scale plot and optionally a logarithmic number plot (with intensities represented by a numeral 0–9). Additional software for drawing contour plots of the intensities on true chemical shift axes for greater ease in making visual comparisons with experimental data will also shortly become available.

The program can do a 100 proton problem in about 15 sec on a VAX 8650 (UNIX f77 compiler), or about 100 sec on a VAX 11/750. Versions of CORMA are available for both UNIX and VMS compilers.

MTFIT for Calculation of Exchange Rates: Available from Dr. Jens J. Led, Department of Chemical Physics, University of Copenhagen, The H. C. Ørsted Institute, 5 Universitetsparken, DK-2100 Copenhagen Ø, Denmark.

MTFIT is a program that calculates the $n(n - 1)/2$ exchange rates and the $n$ longitudinal relaxation rates from the experimental data obtained in an $n$-site magnetization transfer experiment, consisting of $n$ complementary experiments.

The recovery curves are described by the modified Bloch equations for exchange between $n$ sites

$$\frac{d\overline{M}}{dt} = \overline{\overline{K}}\overline{M} + \overline{M}^e \tag{1}$$

where $M_i$ is the time-dependent spin magnetization along the magnetic field, $B_0$, corresponding to the $i$th site and $M_1^e = R_{1i}M_i^\infty$, $M_i^\infty$ being the equilibrium value of $M_i$, and $R_{1i}$ the longitudinal relaxation rate in the $i$th site. $\overline{\overline{K}}$ is an $n \times n$ matrix where the off-diagonal elements, $K_{ij}$, denote the pseudo first-order exchange rate $k_{ij}$ from the $j$th to the $i$th site while the diagonal elements are given by

$$K_{ii} = -\left(R_{1i} + \sum_{j \neq i} k_{ji}\right) \tag{2}$$

The solution of Eq. (1) is straightforward and can be written as

$$\overline{M} = \overline{\overline{Q}} \begin{bmatrix} b_1 e^{\lambda_1 t} \\ b_2 e^{\lambda_2 t} \\ \vdots \\ b_n e^{\lambda_n t} \end{bmatrix} + \overline{M}^\infty \tag{3}$$

where $\overline{\overline{Q}}$ diagonalizes $\overline{\overline{K}}$. The $b_i$'s are determined from the initial condition through the relation

$$\overline{b} = \overline{\overline{Q}}^{-1}(\overline{M}^0 - \overline{M}^\infty) \tag{4}$$

$M_i^0$ being the initial value of $M_1$.

An iterative least squares procedure that fits a number of functions of the form given in Eq. (2), only differing in $\overline{M}^0$ and $\overline{M}^\infty$, is applied to the $n^2$ recovery curves. The columns of $\overline{\overline{Q}}$ are calculated as the eigenvalues to the $\overline{\overline{K}}$ matrix, resulting from the latest iteration or from the initial guess. The first-order derivatives, with respect to the rate constants and relaxation rates, needed to calculate the next step in the iteration process, are evaluated numerically since $\overline{\overline{Q}}$ has no analytical form. The remaining derivatives, however, are evaluated analytically. Since only first-order derivatives are applied, the so-called Hartley's modification [H. O. Hartley, *Technometrics* **3**, 269 (1961)] was implemented to assure convergence of the iteration.

Input to the program is the experimental data defining the $n^2$ recovery curves, and an initial guess of the exchange rates and the longitudinal relaxation rates. Output from the calculation is the rate constants including an estimate of the uncertainties of the parameters.

The program, written in ALGOL and VAX PASCAL/FORTRAN, is available on request from the authors.

Relaxation Parameter Calculations: Available from Dr. Thomas Schleich, Department of Chemistry, University of California, Santa Cruz, Santa Cruz, CA 95064.

A family of computer programs was developed to examine the behavior of the theoretical spectral intensity ratio, $R$, and the spin-lattice relaxation times, $T_1$ and $T_{1\rho}^{off}$, under different experimental conditions. The mathematical expressions, furnished in Chapter [20] by Schleich *et al.* in Volume 176 of this series, were coded in Borland TURBO PASCAL, version 5.0, assuming the presence of a math coprocessor chip, and run on IBM Model 70, 80 or AT computers.

These programs allow the effects of isotropic and anisotropic tumbling of rigid hydrodynamic particles, including those of internal motion, on magnetic relaxation times to be studied as a function of axial ratio, rotational correlation time, or precessional frequency in the effective magnetic field. The variety of input parameters, in addition to NMR constants, include protein molecular weight, hydration, partial specific volume, protein x-ray coordinates, Euler angles, rotational correlation time, and off-resonance rf irradiation power and frequency. Depending on the simulation, a particular program requires only a subset of the parameters listed above that are specific to that case.

Program listings will be supplied on request, and can be made available on either 5.25 or 3.5″ diskettes (IBM format) or sent electronically via E-mail.

### Data Processing and Analysis

NMRFIT: Available from Dr. Jens J. Led, Department of Chemical Physics, University of Copenhagen, The H. C. Ørsted Institute, 5 Universitetsparken, DK-2100 Copenhagen Ø, Denmark.

NMRFIT is a program that extracts the parameters characterizing a Fourier transform NMR spectrum, i.e., the resonance frequency, the intensity, the linewidth, and the phase for a specified number of signals. The program takes into account the effect of nonideal sampling of the free induction decay (FID) and the consequences of processing the FID by a discrete Fourier transformation. Also, extensively overlapping spectral lines can be handled by the program. The output of the program includes an estimate of the uncertainties of the parameters.

The spectral parameters are obtained by a nonlinear least-squares fit of the expression for the discrete Fourier transform of a sum of exponentially damped sinusoids

$$S(\nu_j) = \sum_{l=1}^{L} A_{0,l} T e^{i\varphi_l} \frac{1 - e^{(i2\pi(\nu_{0,l}-\nu_j)-R_{2,l})NT}}{1 - e^{(i2\pi(\nu_{0,l}-\nu_j)-R_{2,l})T}}$$

where $\nu_{0,l}$, $A_{0,l}$, $R_{2,l}$, and $\varphi_l$ are the resonance frequency, the intensity, the transverse relaxation rate, and the phase, respectively, of the $l$th signal in the spectrum, consisting of $L$ lines. Further, $T$ is the time between two consecutive measurements, and $N$ is the number of points used in the Fourier transformation. The real and imaginary part of the frequency spectrum are fitted simultaneously using a Newton-type minimization procedure. In this procedure the first- and second-order derivatives are calculated analytically. In the input to the program the predicted number of lines must be specified together with an initial guess of the four parameters for each line.

The program is especially suited for complicated, nonideal spectra with a limited number of signals, or spectra that can be divided into minor parts containing isolated groups of lines.

The program written in ALGOL and VAX PASCAL/FORTRAN is available on request from the authors.

Processing, postprocessing, and storage of 2D NMR data: Available from Operations Assistant, National Magnetic Resonance Facility at Madison, Biochemistry Dept., 420 Henry Mall, Madison, WI 53706.

*Software for the Bruker Aspect 1000/3000*

The following programs have been written by Zsolt Zolnai to be run on the Aspect computer.

"MAKEUP"[1] is a program that removes noise in phase-sensitive or absolute-value 2D NMR spectra. The relationship between noise and signal is linear in phase sensitive spectra. This means that the average noise level (dc offset) can be subtracted directly from line intensities. The average noise level is calculated from a reference spectral strip selected to contain no spectral lines along the full length of the $f_2$ dimension. The standard deviation of the noise also is calculated. The calculations pro-

---

[1] Zs. Zolnai, S. Macura, and J. L. Markley, *Computer Enhanced Spectroscopy* **3,** 141 (1986).

duce two profiles along the $f_2$ dimension, along with the dc offset and the rms noise level. Each point of the 2D spectrum is corrected for the corresponding dc offset. The correction is constant within an $f_1$ column but varies from column to column. In the next step, each spectral point is compared with a number related to the rms noise level: points that fall in the range, $\pm$ (rms noise) $\times$ (weighting factor), are zeroed; points outside this range are left intact. The overall effect is selective elimination of background $t_1$ noise irrespective of the variability of its value from column to column. The horizontal ridges ($t_2$ noise) can be eliminated by carrying out the same treatment rowwise. In practice, all steps are performed simultaneously. Because a linear relationship between signal and noise is required, absolute value spectra must be converted first into the power mode. MAKEUP enables the simultaneous observation of all cross-peaks and diagonal peaks from a single plot, even if very strong noise bands are present in the original spectrum. Spectra processed in this way are amenable to automatic computer analysis and can be compressed into a smaller storage size without loss of information.

"STANLIO" and "OLIO"[2] are programs for the compression and decompression of 2D NMR spectra after MAKEUP. Since MAKEUP replaces noise-bearing points by zeros, compression can be achieved by storing nonzero points normally while representing strings of zeroes by two numbers: the first being the leading zero itself, and the second the number of zeros in the sequence. Depending on the spectra, compression ratios between 5 and 100 can be realized. Reduction in the size of 2D NMR spectra speeds up data storage, retrieval and transmission, and reduces the cost of long-term data storage. Data retrieval, i.e., expansion of compressed data into a format that can be recognized by DISNMR (the Bruker NMR program for the Aspect computer) is performed by OLIO.

"SPELOG"[2] is a program for logarithmic scaling of 2D NMR spectra. The relative error in the final, Fourier-transformed spectrum rarely is less than 1%. Therefore, a large number of bits is not necessary to represent the data, and compression can be achieved without appreciable loss of information by squeezing more than one data point into a single computer word. A convenient way of achieving this is to scale the data logarithmically with a suitably chosen base, $b$. For example, a 24-bit two's complement word can be compressed into 12 bits via logarithmic scaling by using $b = 1.007815$ (and log 0 = 0), with the maximum relative error due to compression of only 0.8%. Data retrieval, into a format that can be recognized by the DISNMR program, is performed in one or two steps, de-

---

[2] Zs. Zolnai, S. Macura, and J. L. Markley, *J. Magn. Reson.* **80**, 60 (1988).

pending on whether the data are to be viewed in a linear or logarithmic mode.

"SPEMAN"[3] is a routine for making a linear combination of two 2D NMR spectra. It multiplies the intensity of one 2D spectrum by a specified positive or negative factor and adds it to the intensity of another 2D spectrum.

"DIST"[4] is a program for calculating distances between protons from NOEs and ROEs. The program consists of several modules: (1) A module for entering coordinates for each cross-peak to be analyzed. The coordinates are stored in a file and can be used to identify the corresponding two cross-peaks and two diagonal peaks each time they are needed. (2) A module for finding intensities. This module uses the file with cross-peak coordinates, and creates a file containing the intensities of the two cross-peaks and corresponding two diagonal peaks from each spectrum recorded with different mixing times. All intensities are refined by searching for local maximum around actual cross or diagonal peak—local minimum for ROESY cross-peaks. (3) A module that calculates linewidth at half height for each selected cross-peak in both dimensions. (4) A module for distance calculations. This module calculates build-up rates and distances on the basis of intensities or volume integrals, intensity $\times$ width$_1$ $\times$ width$_2$, of cross and diagonal peaks. (5) A module for inspection and printout of files created in the other modules. Files are not text files because of space considerations, and some information is coded. (6) A module for visual checking of intensities or volume integrals vs. mixing times and the fit of experimental data to calculated build-up rates.

"CHOPPER"[3] is a program which allows one to "chop out" any part of a 2D NMR spectrum whose size is a power of two in both dimensions. Spectra may be in the absolute-value or pure-phase mode. The "chopped" part is in a format that is recognized by DISNMR and can be treated like any other 2D spectra. Often only some parts of 2D spectra carry relevant information (e.g., spectra obtained with the carrier at one end). CHOPPER retains the imaginary parts of phase-sensitive spectra; thus time and space can be saved by phasing only the important part of the spectrum.

"ROTATE"[3] is a program for transposition of various types of 2D NMR spectra: phase sensitive, absolute value, and square and nonsquare spectra. Since the transpose is done in place, no additional disk space is required. It also is useful for faster plotting. In DISNMR, plotting of horizontal strips is much faster than plotting of vertical strips (because of

[3] Zs. Zolnai, S. Macura, and J. L. Markley, to be published.
[4] Zs. Zolnai, S. Macura, and J. L. Markley, *J. Magn. Reson.* in press (1989).

the way the contour plot routine is written). Therefore, time can be saved if one transposes a vertical strip and plots it as a horizontal strip instead of directly as a vertical strip. After plotting, the data can be transposed back.

"STROLLER"[4] is a program for baseplane correction. Baseplane distortions arise from instrumental artifacts, aliasing of dispersive tails of large peaks (i.e., solvent), and/or $t_1$ noise. Baseplane correction [26] is an essential step before many kinds of postprocessing: intensity measurement for distance calculation, measurement of volume integrals, automatic linewidth measurement, and peak picking for assignment purposes. In 2D plots it is common to find that some peaks are "lost" because they lie on the negative tail of a large peak whose negative intensity is greater than the positive intensity of the "lost" cross-peak. For baseplane correction, one must first select a grid of control points that best determine the true baseplane. On the basis of this grid, the baseplane is calculated for each point in the spectrum by using the "Cardinal bicubic surfaces algorithm" and is subtracted from this point. The determination of each baseplane point, which requires 20 floating point multiplications and 12 floating point additions, is slow since floating point operations are not efficient on the Aspect computer. Because of this, and because it is often enough to correct the baseplane only in selected regions, the program allows the procedure to be performed on just a subset of the 2D spectrum.

"INTENZY"[3] is a program that gives as output an integer value (stored on the disk) of any point in 2D NMR spectra. This routine is necessary for SPEMAN since DISNMR reports only relative intensities. This program is useful as a subroutine in many other programs. It is embedded in STROLLER and DIST, for example.

### Software for Silicon Graphics Iris

Prashanth Darba has rewritten preliminary versions of the programs "PROC2D" and "ANALYS2D"[5] to run under the UNIX operating system on the SGI Iris 2400T and SGI 4D-Series workstations.[6] The main new software package is called "MADNMR".[7] The package supports the usual 2D NMR processing operations plus user-friendly graphics features for interactive analysis of 2D data that take advantage of the graphics engine on the Iris workstations.

Spectral processing can be carried out in either an interactive or batch mode. The package supports both the TPPI[8] and States–Haberkorn–

[5] P. Darba and L. R. Brown, unpublished.
[6] P. Darba and J. L. Markley, to be published.
[7] B. H. Oh, W. M. Westler, P. Darba, and J. L. Markley, *Science* **240,** 908 (1988).
[8] D. Marion and K. Wüthrich, *Biochem. Biophys. Res. Commun.* **113,** 967 (1983).

Ruben[9] method for generating phase-sensitive spectra and linear combination procedures for efficient generation of multiple quantum filtered 2D NMR spectra.[10] Symmetrization and filtering of noise can be carried out with respect to a user-defined $S/N$ threshold. Peaks above the threshold that do not have a counterpart (of the proper sign and above the specified threshold) in a region related by the symmetry specified are zeroed; all other signals are left as is. Interactive graphics features support the creation of spin-system connectivity databases; peak picking and inking of connectivities is carried out in multiple viewport modes designed to facilitate the analysis of different classes of 2D NMR data. Spin system connectivities chosen for one data set can be overlaid upon another data set. Inking of spin connectivities is updated automatically in all windows during peak picking procedures; all user-selected portions of previously defined spin system data bases can be inked in all windows in colors of the user's choice. Two 2D spectra can be overlaid and zoomed to resolve higher detail. Specified sections of a 2D spectrum can be plotted along with the spin-system connectivities identified.

Another program for the Silicon Graphics Iris, "GENINT," has been developed to assist in refinement and intensity measurement of the NOESY peaks picked interactively in MADNMR. The user picks peaks interactively by placing the crosswire cursor approximately at the center of a contour and pressing a button on the mouse; the coordinates of the crudely picked peaks are stored in a spin system database. GENINT uses these coordinates as the starting coordinates for searching the extremes of peaks. With a 2D matrix containing ~200 peaks, the routine takes only a few seconds to create a database of precise chemical shift coordinates and cross-peak intensities.

The Rowland NMR Toolkit: Available from Dr. Jeffrey C. Hoch, Rowland Institute for Science, 100 Cambridge Parkway, Cambridge, Massachusetts 02142.

The implementations of the Burg MEM and LPZ methods are used in the present work are part of the Rowland NMR Toolkit. The Toolkit consists of a suite of programs for processing one- and two-dimensional NMR data on Digital Equipment Corporation VAX computers running the VMS operating system. Although designed as a "workbench" for developing NMR data processing methods, the Toolkit may also be used for routine processing and plotting of NMR data. The Toolkit accommodates data from a variety of sources and supports a wide variety of

[9] D. J. States, R. A. Haberkorn, and D. J. Ruben, *J. Magn. Reson.* **48,** 286 (1982).
[10] R. Ramachandran, P. Darba, and L. R. Brown, *J. Magn. Reson.* **73,** 349 (1987).

graphic display devices. The Toolkit is available to academic institutions for a nominal handling fee. For commercial enterprises the fee is $4000.00 U.S. (version 2.0).

Linear Prediction Singular Value Decomposition and Hankel Singular Value Decomposition: Available from Dr. Ron de Beer, University of Technology Delft, Department of Applied Physics, P.O. Box 5046, 2600 GA Delft, The Netherlands.

The implementations of the LPSVD and HSVD methods were written by H. Barkhuijsen and R. de Beer.

Maximum Entropy Reconstruction: Available from Maximum Entropy Data Consultants, Ltd., 33 North End, Meldreth, Royston SG8 6NR, England.

The implementation of maximum entropy reconstruction used was written by John Skilling.

# Addendum

## Addendum to Article [16]

*By* BRUCE A. BERKOWITZ and ROBERT S. BALABAN

To further verify that peak D (Fig. 5) corresponds to inositol, a kidney perchloric acid extract was performed. An inositol concentration was obtained using both HPLC and 2D COSY methods. Theoretical justification that the 2D experiment can provide concentration information is found in Eq. (3) where the time evolution of a cross peak is seen to be linearly proportional to the equilibrium magnetization (i.e., concentration) under the same acquisition parameters. Figure 6 is a graph of the results from phantom studies which demonstrate this linear correlation between cross-peak volume and concentration. A lower limit of cross-

**Cross-Peak Evolution**
$$M_0 \sin(\pi J t_1) \sin(\pi J t_2)$$

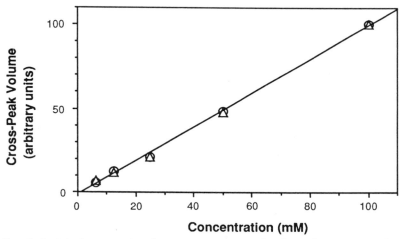

FIG. 6. A plot of cross-peak volume versus concentration for various concentrations of inositol (○) and phosphoethanolamine (triangle) in D20. In each case an internal standard of lactate (50 m$M$) was used for normalization. Cross-peak volumes were measured by taking the sum of the areas of individual slices through a cross peak in the F1 domain after baseline correction. The volumes thus obtained for the cross peaks on both sides of the diagonal were then added together to improve the measurements' precision. The basic COSY pulse sequence was used (Fig. 4A) with a sweep width of 380 Hz producing a 45-min total experimental time. Data processing was described in Fig. 5.

peak detectability, based on this work, is roughly 1–2 m$M$. Thus, given that an appropriate internal or external standard is used, and correcting for T1 differences, absolute concentrations may be obtained from the 2D COSY experiment.

To obtain a concentration from the kidney extract we added a known amount of sodium 3-trimethylsilylpropionate 2,23,3d4 (TSP) in D20 as an internal standard. By first measuring the diagonal-to-cross peak volume ratio in a separate inositol phantom study and then using this ratio to compare, in the kidney extract, the diagonal volumes of TSP to inositol obtained under the same acquisition parameters as the phantom study, we obtained an inositol concentration of 1.5 $\mu$mol/g wet weight. This is in good agreement with the inositol concentration measured on the same extract by HPLC of 1.3 $\mu$mol/g wet weight. These results further support assignment of peak D in Fig. 5 as inositol and clearly demonstrate the quantitative nature of the 2D COSY experiment.

# Author Index

Numbers in parentheses are footnote reference numbers and indicate that an author's work is referred to although the name is not cited in the text.

## A

Abildgaard, F., 325, 328, 329
Abragam, A., 184, 359, 361(1), 371(1)
Ackerman, J. C., 259
Addleman, R., 362, 387, 388, 390(5), 392(5), 394(5)
Affolter, M., 20
Akaike, H., 221
Akhrem, A. A., 19
Albrand, J. P., 73, 74
Alexander, S., 70, 73
Alger, J. R., 312, 319
Allerhand, A., 65, 185, 362, 370, 387, 388, 390(5), 392(5), 394(5)
Altona, C., 178
Ancker-Johnson, B., 426, 440(24), 444(24), 445(24)
Anderson, J. E., 366
Anderson, N., 220
Anderson, W. A., 61
Andrianov, A. M., 19
Anil Kumar, 18, 31, 32
Appleby, C. A., 20
Arata, Y., 77
Armitage, I. M., 149, 362
Armitage, I., 238
Arseniev, A. S., 18, 19, 20
Aue, W. P., 24, 31(28), 93, 118, 133(17), 472
Avitabile, J., 365, 366

## B

Bach, A. C., 18
Bacher, A., 45
Bachman, P., 458, 459(41)
Bachmann, P., 6, 24, 31(28), 111, 319, 330, 334(1)
Bachovchin, W. W., 70, 71(56), 73
Bagnasco, S., 339
Balaban, R. S., 319, 330, 333(3), 334(2, 3, 5), 335, 336 (3, 6), 339, 341
Baldwin, R. L., 455, 458, 464, 465, 466, 469, 471(67),472
Balko, B., 466
Bangerter, B., 107
Barker, P. B., 99, 140
Barker, P., 66
Barkhuijsen, H., 61
Barkuijsen, H., 224
Barna, J., 227, 239
Barni-Comparini, I., 199
Baron, M., 18
Bartholdi, E., 93, 118, 133(17)
Bartoldi, E., 472
Basas, V. J., 62
Basus, V., 17
Batchelder, L. S., 378, 379(13)
Batchelor, J. G., 370
Batta, G., 107, 197
Bauer, C., 66
Bauer, D. R., 388
Baumann, R. G., 61
Baumann, R., 19, 337
Bax, A., 20, 22, 31, 32, 33, 37(43), 39, 48, 49(68), 50(68), 52, 55, 57(43), 64, 73, 74, 75, 76, 99, 100, 102, 104, 105, 106, 107, 110, 111, 114, 116, 117(10), 118, 134, 135, 137, 139, 140, 142, 143, 144, 147, 149, 150, 151, 152, 157, 162, 163, 164, 166(29), 200, 208, 209(9), 212(4), 213, 337, 338(12), 339(12), 459, 469(52), 470(52), 473(52)

# Subject Index

## A

# N

NAVL. *See* N-Acetyl-L-valyl-L-leucine

NERO methods, 70–71

α-Neurotoxin, from *Dendroapsis polylepsis polylepsis*, sequence-specific proton NMR assignment, 16

Neutrophil peptide 5, from rabbit, sequence-specific proton NMR assignment, 16

Nitrogen-15, labeling of proteins, 46, 53

NMR. *See* Nuclear magnetic resonance

NMR55AUTO (program), 480

NMR1 (program), 478–479

NMR2 (program), 478–479

NMR50 (program), 480

NMRFIT (program), 483–484

Nochromix, 79

NOE. *See* Nuclear Overhauser effect

NOESY. *See* Nuclear Overhauser effect spectroscopy

NONLINWOOD (program), 397

Nuclear magnetic resonance spectroscopy
areas enclosed by resonances, 279
characterization of motions in solids, 376
characterization of protein structure and dynamics, 93
complementary to crystallography, 12
computer programs related to, 477–489
basic, 478–479
database management, 62
databases, 63
diffusion measurement, theory of, 422–426
digitization error, 7
four-dimensional, 59–61
$^1$H. *See* Proton magnetic resonance
higher-dimensional, 59–61
high-field, high-resolution, spectral parameters, 279
$^1$H solution-state, 376
instrumental noise, 7–9
for noninvasive assessment of protein rotational dynamics in intact tissue, 387
one-dimensional (1D) pulse-Fourier transform experiment, 23
peak intensity measurements, 6–11
systematic errors in, 6–7

for protein folding studies, 446–473
advantage of, 446
relaxation of spins in, theory of, 4–5
semiautomated analysis in, 62
solid-state. *See* Solid-state NMR spectroscopy
structural and dynamic information from, 63
for study of enzymatic reaction rates, 311–312
three-dimensional. *See* Three-dimensional NMR spectroscopy
truncation error, 7–8
tube
direct lyophilization of sample in, 83–86
drying of, 81
with vacuum applied, shielding against implosion, 86
two-dimensional. *See* Two-dimensional NMR spectroscopy
two dipolar-coupled spin-½ nuclei, energy levels of, 4–5
unknown phase of peak, 7

Nuclear Overhauser effect, 12, 48, 279, 446
definition of, 3–4
to determine internuclear distances, 11
estimate of, 4
estimate of $\eta$, coefficient of variation of, 11
and exchange rate determination, 313
factors, variability of, 11
measurements, 4–6
bias in, 7–10
degradation in precision, 7, 10
error in, 6–7

Nuclear Overhauser effect spectroscopy, 45–48, 77, 94, 111, 143, 153, 199–200. *See also* Rotating-frame nuclear Overhauser effect spectroscopy
combination with heteronuclear correlation, 150
cross-relaxation in, 199–201
heteronuclear experiment, 107–108
iterative structure refinement programs, 177–178
magnetization during mixing period of, time course of, 170–171